CBS PROBLEMS and SOLUTIONS SERIES

Problems and Solutions in STRENGTH OF MATERIALS

Problems and Solutions in
STRENGTH OF MATERIALS

- This book is suitable for examinations of all Indian universities.
- Solutions are given in most simplified manner.
- Each type of problems is included and the solutions given in detail.
- To explain the mechanism lucidly, computerised figures have been liberally added in the solutions. This volume contains more than 450 line diagrams.
- Problems from the last ten years Indian Engineering Service and Civil Service examinations are added till date in each chapter.

CBS PROBLEMS and SOLUTIONS SERIES

Problems and Solutions in STRENGTH OF MATERIALS

S C SINGHAL

Engineer

CBS

CBS Publishers & Distributors Pvt. Ltd.

New Delhi • Bengaluru • Chennai • Kochi • Kolkata • Mumbai

Hyderabad • Nagpur • Patna • Pune • Vijayawada

ISBN: 81-239-1090-8

First Edition: 2004
Reprint: 2007, 2009, 2011, 2013, 2016

Published by:
Satish Kumar Jain for CBS Publishers & Distributors Pvt. Ltd.,
4819/XI Prahlad Street, 24 Ansari Road, Daryaganj, New Delhi - 110002
delhi@cbspd.com, cbspubs@airtelmail.in • www.cbspd.com
Ph.: 23289259, 23266861, 23266867 • Fax: 011-23243014

Corporate Office: 204 FIE, Industrial Area, Patparganj, Delhi - 110 092
Ph: 49344934 • Fax: 011-49344935
E-mail: publishing@cbspd.com • publicity@cbspd.com

Branches:
- ***Bengaluru:*** 2975, 17th Cross, K.R. Road, Bansankari 2nd Stage, Bengaluru - 70 • Ph: +91-80-26771678/79 • Fax: +91-80-26771680 E-mail: cbsbng@gmail.com, bangalore@cbspd.com
- ***Chennai:*** No. 7, Subbaraya Street, Shenoy Nagar, Chennai - 600030 Ph: +91-44-26681266, 26680620 • Fax: +91-44-42032115 E-mail: chennai@cbspd.com
- ***Kochi:*** Ashana House, 39/1904, A.M. Thomas Road, Valanjambalam, Ernakulum, Kochi • Ph: +91-484-4059061-65 Fax: +91-484-4059065 • E-mail: cochin@cbspd.com
- ***Kolkata:*** 6-B, Ground Floor, Rameshwar Shaw Road, Kolkata - 700014 Ph: +91-33-22891126/7/8 • E-mail: kolkata@cbspd.com
- ***Mumbai:*** 83-C, Dr. E. Moses Road, Worli, Mumbai - 400018 Ph: +91-9833017933, 022-24902340/41 • E-mail: mumbai@cbspd.com

Representatives:
- Hyderabad: 0-9885175004
- Nagpur: 0 9021734563
- Patna: 0-9334159340
- Pune: 0-9623451994
- Vijayawada: 0-9000660880

Printed at:
J.S. Offset Printers, Delhi

Preface

Strength of Materials is one of the fundamental subjects for the students of civil engineering. The author feels great pleasure in presenting the subject of STRENGTH OF MATERIALS through typical problems and their solutions. The book covers every topic taught under strength of materials at the undergraduate level in an engineering college. All the problems are discussed at length, and line diagrams and figures have been given liberally to help the reader in understanding the logic. This book is suitable for all university examinations. Also questions of the past ten years from various competitive examinations like Indian Engineering Service and Civil Service examinations, have been added till date.

Although every care has been taken to ensure accuracy, yet some errors might have crept in. The author will be grateful if these are brought to his notice so that these could be rectified in the subsequent printings and editions of the book. Readers are requested to send their suggestions for further improvement and revision at the following e-mail address: cbspubs@del3.vsnl.net.in.

S C Singhal

Contents

Chapter 1

Simple Stress and Simple Strain

Problem 1.1: A hollow steel tube with an inside diameter of 100 mm must carry a tensile load of 400 kN. Determine the outside diameter of the tube of the stress is limited to 120 MN/m^2.

Solution:

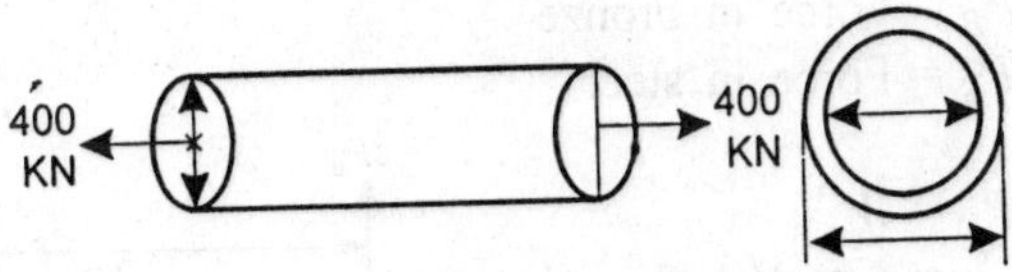

Fig. P. 1.1

$$\therefore \quad \text{Stress on tube } (p) = \frac{F}{A} = \frac{4F}{\pi\left[D^2 - d^2\right]}$$

$$\left(\text{where } A = \frac{\pi}{4}\left[D^2 - d^2\right]\right)$$

$$\therefore \quad [D^2 - d^2] = \frac{4F}{\pi \times p}$$

$$= \frac{4 \times 400 \times 1000}{120 \times \pi} = 4244.1318 \text{ mm}^2$$

$$D^2 - 100^2 = 4244.1318 \text{ mm}^2$$

$$\therefore \quad D^2 = 4244.1318 + 10000 = 14244.1318 \text{ mm}^2$$

$$\therefore \quad D = 119.348 \text{ mm} \approx 119 \text{ mm } \textbf{Ans}$$

Problem 1.2: A homogeneous 800 kg bar *AB* is supported at either end by a cable as shown in Figure. Calculate the smallest area of each cable if the stress is not to exceed 90 MPa in bronze and 120 MPa in steel.

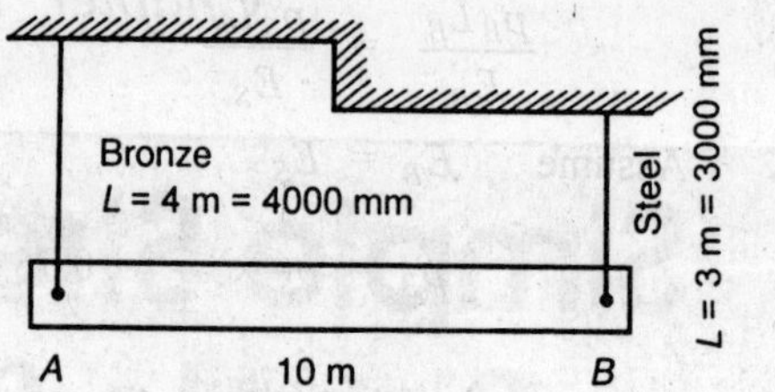

Fig. P. 1.2 (a)

Solution:

$$\text{Weight of bar} = 800 \text{ kg} = 800 \times 10 \text{ N} = 8 \times 10^3 \text{ N}$$

$$p_{\text{bronze}} = 90 \times 10^6 \; Pa = 90 \times 10^6 \text{ N/m}^2$$

$$= \frac{90 \times 10^6}{(1000)^2} \text{ N/mm}^2$$

$$= 90 \text{ N/mm}^2$$

$$p_{\text{steel}} = 120 \text{ MPa} = 120 \text{ N/mm}^2$$

F.B.D. P_B = Force in bronze

P_S = Force in steel

$\Sigma F_y = 0$

$$P_B + P_S = W$$

$$= 8 \times 10^3 \text{ N} \quad \ldots(1)$$

$\Sigma M_A = 0,$

$$P_S \times 10 = W \times 5$$

$$\therefore \quad P_S = \frac{8 \times 10^3 \times 5}{10}$$

$$= 4 \times 10^3 \text{ N} \quad \ldots(2)$$

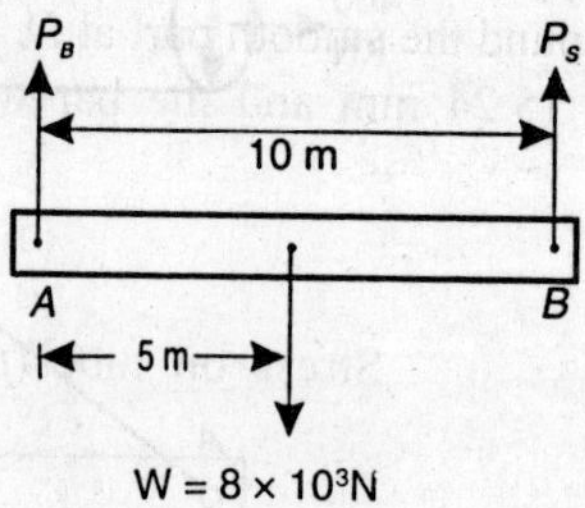

Fig. P. 1.2 (b)

Putting the value of P_S in eq (1), we get

$$P_B = 8 \times 10^3 - 4 \times 10^3$$

$$= [8 - 4] \times 10^3$$

$$= 4 \times 10^3 \text{ N}$$

So each bar carry equal loads

$\therefore$ Stresses is not to exceed 90 N/mm² in bronze and 120 N/mm² in steel.

$\therefore$ From compatibility equation,

$$\Delta B = \Delta S$$

$$\frac{p_B L_B}{E_B} = \frac{p_S L_S}{E_S}$$

Assume $E_B = E_S$

$$\therefore \quad p_B = p_S \times \frac{3}{4} = 0.75\ p_S$$

$$\therefore \quad p_B = 0.75\ p_S \qquad \ldots(3)$$

$$\therefore \quad p_B = 0.75 \times 120 = 90$$

$$\therefore \quad \text{if } p_S = 120 \text{ N/mm}^2 \text{ then } p_B = 90 \text{ N/mm}^2$$

$$\therefore \quad A_B = \frac{4 \times 10^3}{90} = 44.5 \text{ mm}$$

$$A_S = \frac{4 \times 10^3}{120} = 33.33 \text{ mm}$$

Problem 1.3: The homogeneous bar shown in Fig. P 1.3 (a) is supported by a smooth pin at *C* and a cable that runs from *A* to *B* around the smooth part at *D*. Find the stress in the cable if its diameter is 15.24 mm and the bar weights 27240 N.

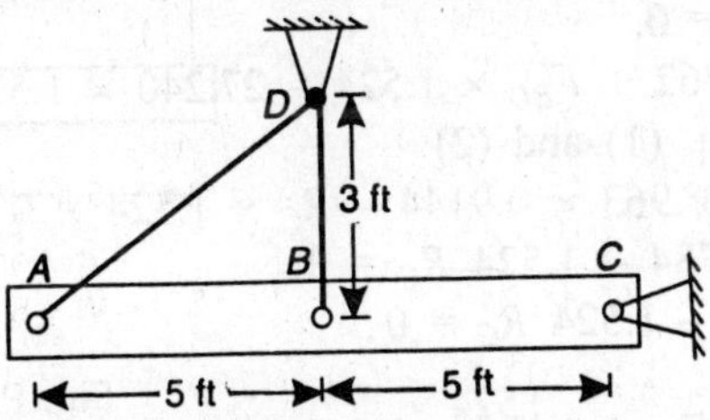

Fig. P. 1.3 (a)

Solution:

$$F.B.D. \quad \tan\theta = \frac{0.9144}{1.524}$$

$$\therefore \quad \theta = 30^\circ.963$$

$\therefore$ Horizontal component of $F_{AD} = F_{AD} \cos 30^\circ.963$

Vertical component of $F_{AD} = F_{AD} \sin 30^\circ.963$

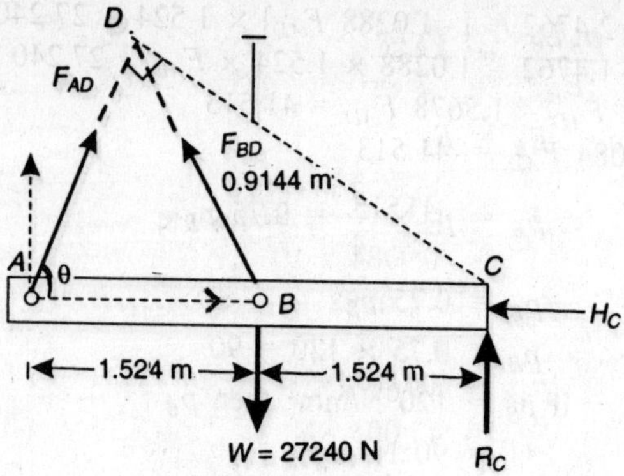

Fig. P. 1.3 (b)

$$AD = \sqrt{(1.524)^2 + (0.9144)^2} = 1.7772 \text{ m}$$

$$DC = \sqrt{(3.098)^2 - (1.7772)^2} = 2.4762 \text{ m}$$

$\therefore$ $\Sigma Fx = 0,$

$$F_{AD} \cos 30°.963 = Hc \qquad ...(1)$$

$\Sigma Fy = 0$

$$F_{AD} \sin 30°.963 + F_{BD} + Rc = 0 \qquad ...(2)$$

$\Sigma M_D = 0,$

$$H_C \times 0.9144 - R_C \times 1.524 = 0 \qquad ...(3)$$

$\Sigma M_C = 0,$

$$F_{AD} \times 2.4762 + F_{BD} \times 1.524 - 27.240 \times 1.524 = 0 \qquad ...(4)$$

$\therefore$ From eq. (1) and (2)

$F_{AD} \cos 30°.963 \times 0.9144 - R_C \times 1.524 = 0$

$\therefore F_{AD}\ 0.784 - 1.524\ R_C = 0$

$0.784\ F_{AD} - 1.524\ R_C = 0$

$$\therefore \qquad F_{AD} = \frac{1.524}{0.784} R_C = 1.9438\ R_C \qquad ...(5)$$

$$R_C = \frac{1}{1.9438} F_{AD} = 0.5144\ F_{AD} \qquad ...(6)$$

Putting this value in eq (2)

$F_{AD} \sin 30°.963 + F_{BD} + 0.5144\ F_{AD} = 0$

$0.5144\ F_{AD} + 0.5144 \times F_{AD} + F_{BD} = 0$

$1.0288\ F_{AD} + F_{BD} = 0$

$F_{BD} = -1.0288\ F_{AD}$

$\therefore$ Putting this value in eq (4)

$$F_{AD} \times 2.4762 + [-1.0288\ F_{AD}] \times 1.524 - 27.240 \times 1.524 = 0$$
$$F_{AD} \times 1.4762 - 1.0288 \times 1.524 \times F_{AD} - 27.240 \times 1.524 = 0$$
$$2.4762\ F_{AD} - 1.5678\ F_{AD} = 41.513$$
$$0.9084\ F_{AD} = 41.513$$
$$\therefore \quad F_{AD} = \frac{41.513}{0.9084} = 45.69 \text{ kN}$$
$$\therefore \quad F_{BD} = -1.0288 \times 45.69$$
$$= 47.005 \text{ kN}$$
$$\therefore \quad F_{AD} = 45.69 \text{ kN}$$
$$F_{BD} = 47.005 \text{ kN}$$
$$\text{Area} = \frac{\pi}{4} \times 15.24^2$$
$$= 182.41 \text{ mm}^2$$
$$\therefore \text{Stress in } AD = \frac{45.69 \times 1000}{182.41}$$
$$= 250.47 \text{ N/mm}^2$$
$$\text{Stress in } BD, = \frac{47.005 \times 1000}{182.41}$$
$$= 257.638 \text{ N/mm}^2$$

Problem 1.4: A rod is composed of an aluminium section rigidly attached between steel and bronze sections, as shown in Fig. P1.4 (a). Axial loads are applied at the positions indicated. If $P = 13620$ N and the cross-sectional area of the rod is 322.58 mm², determine the stress in each section.

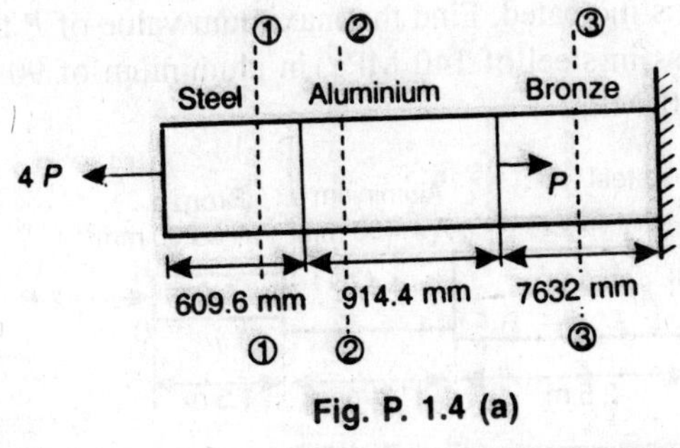

Fig. P. 1.4 (a)

Solution:

F.B.D. of section (1) - (1)

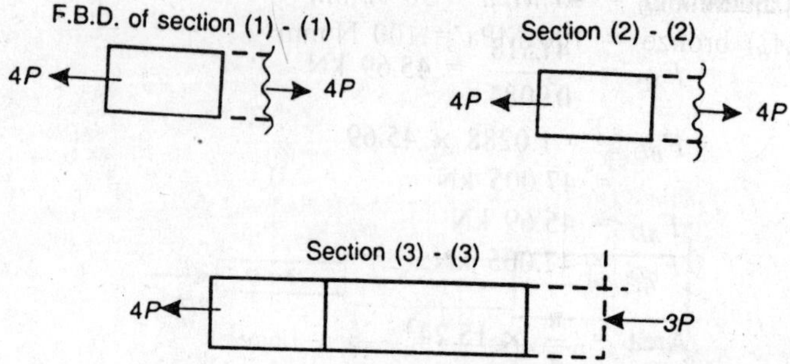

Fig. P. 1.4 (b)

Force at steel section

$$(P_S) = 4P = 4 \times 13620 = 54480 \text{ N } (T)$$

Force at aluminium section

$$(P_A) = 54480 \text{ N } (T)$$

Force at bronze section

$$(P_B) = -3P = -3 \times 13620 = -40860 = 40860 \ (C)$$

$$\therefore \quad \text{Stress in steel} = \frac{54480}{322.58} = 168.89 \text{ N/mm}^2 \ (T)$$

$$\text{Stress in aluminium} = 168.89 \text{ N/mm}^2 \ (T)$$

$$\text{Stress in bronze} = \frac{40860}{322.58} = -126.67 \text{ N/mm}^2 \text{ compression}$$

Problem 1.5: An aluminium rod is rigidly attached between a steel rod and a bronze rod as shown in Fig. P 1.5 (a). Arial loads are applied at the positions indicated. Find the maximum value of P that will not exceed a stress in steel of 140 MPa, in aluminium of 90 MPa, or in bronze of 100 MPa.

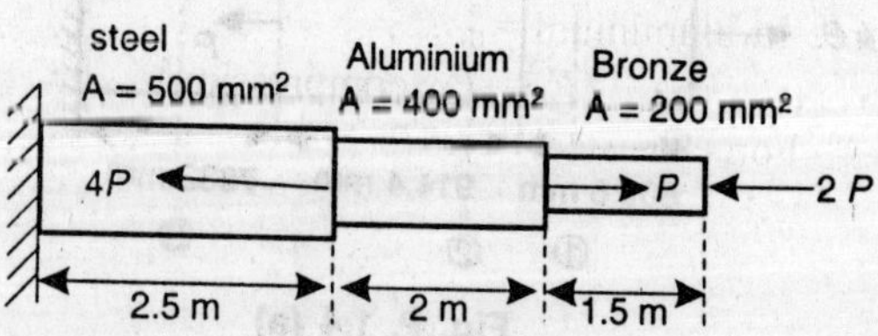

Fig. P. 1.5 (a)

Solution:

Permissible value of stress in,

steel = 140 MPa = 140 N/mm^2,

aluminium = 90 MPa = 90 N/mm^2,

(A_B) bronze = 100 MPa = 100 N/mm^2

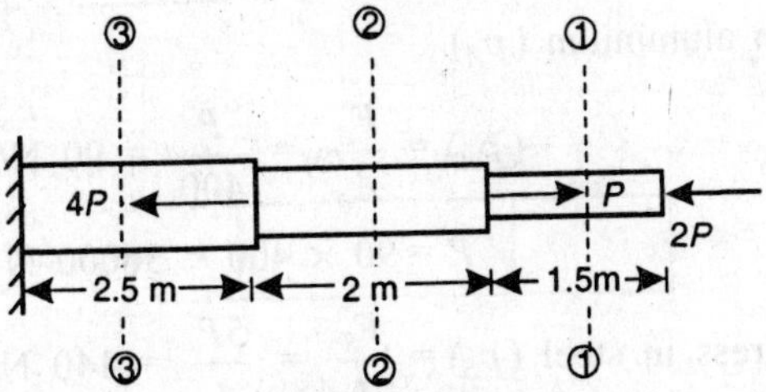

Fig. P. 1.5 (b)

F.B.D.

(1) Section (1) - (1)

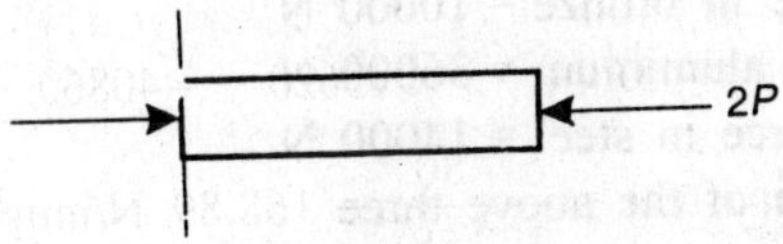

Fig. P. 1.5 (c)

Force in bronze = $-2P$

$F_B = 2P$ (compression)

(2) Section (2) - (2)

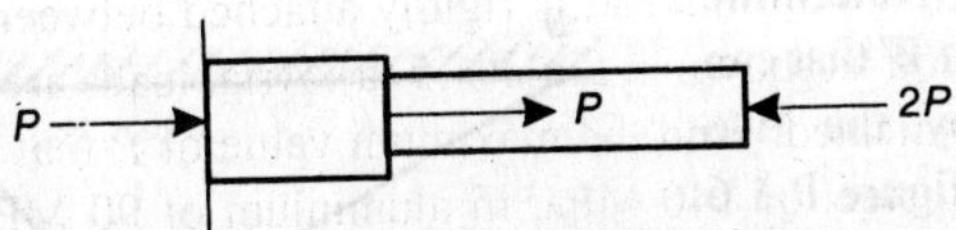

Fig. P. 1.5 (d)

$\therefore$ Force in aluminium = $-P$

$F_A = P$ (compression)

(3) Force in steel = $-5P$

Fig. P. 1.5 (e)

$$F_S = 5P \text{ (compression)}$$

$\therefore$ Stress in bronze $(p_B) = \dfrac{F_B}{A_B} = \dfrac{2P}{200} = 100$ N/mm^2

$\therefore$ $2P = 200 \times 100$

$\therefore$ $P = \dfrac{200 \times 100}{2} = 10000$ N

Stress in aluminium (p_A)

$$(p_A) = \frac{F_A}{A_A} = \frac{P}{400} = 90 \text{ N/mm}^2$$

$\therefore$ $P = 90 \times 400 = 36000$ N

Stress in steel $(p_S) = \dfrac{F_S}{A_S} = \dfrac{5P}{A_S} = 140$ N/mm^2

$\therefore$ $5P = 140 \times A_S = 140 \times 500$

$\therefore$ $P = \dfrac{140 \times 500}{5} = 140 \times 100 = 14000$ N

$\therefore$ **Force in bronze = 10000 N**
Force in aluminium = 36000 N
Force in steel = 14000 N

Take minimum of the above three

= 10000 N = 10 kN

$\therefore$ Minimum value of P that will not exceed a stress in steel of 140 MPa, in aluminimum of 90 MPa or in bronze of 100 MPa is 10 kN.

Problem 1.6: Determine the largest weight W that can be supported by the two wires shown in figure P 1.6 (a). The stress in either wire is not to exceed 211.11 N/mm^2. The cross-sectional areas of wires AB and AC are 258.064 mm^2 and 322.58 mm^2 respectively.

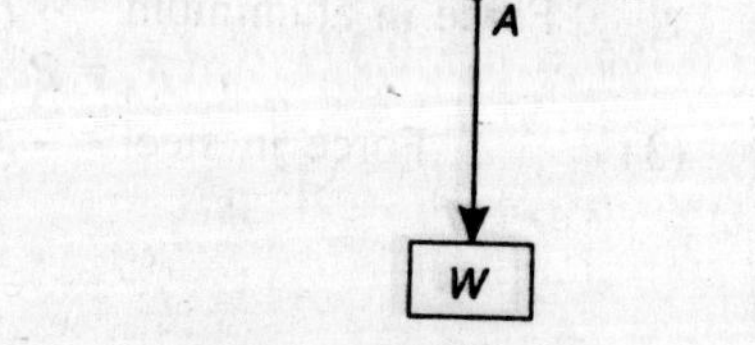

Fig. P. 1.6 (a)

Solution:

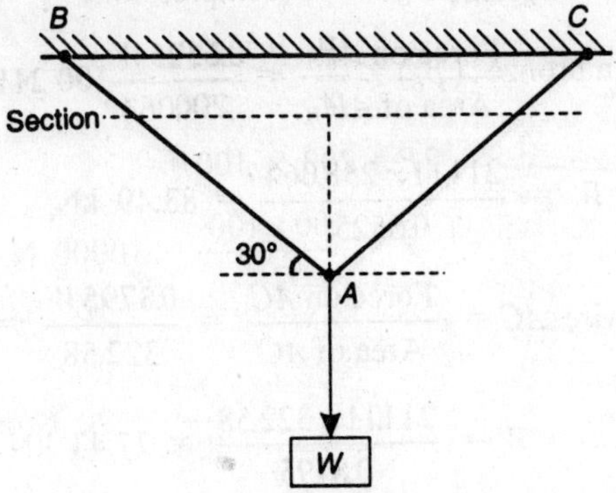

Fig. P. 1.6 (b)

F.B.D.

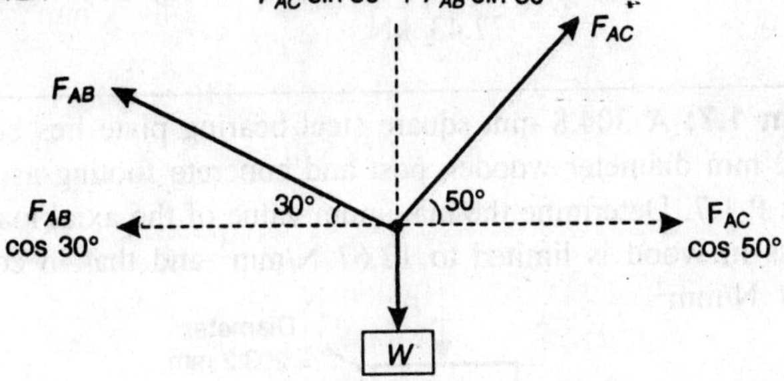

Fig. P. 1.6 (c)

$\Sigma F_x = 0,$

$$F_{AB} \cos 30^\circ = F_{AC} \cos 50^\circ$$

$$F_{AB} = F_{AC} \frac{\cos 50^\circ}{\cos 30^\circ} = 0.7422 \ F_{AC}$$

$$\therefore \quad F_{AB} = 0.742 \ F_{AC} \quad \ldots(1)$$

$\Sigma F_y = 0,$

$$F_{AC} \sin 50^\circ + F_{AB} \sin 30^\circ = W \quad \ldots(2)$$

Putting value of F_{AB} from equation (1) we get

$$F_{AC} \sin 50^\circ + 0.742 \ F_{AC} \times \sin 30^\circ = W$$

$$0.766 \ F_{AC} + 0.371 \ F_{AC} = W$$

$$1.137 \ F_{AC} = W$$

$$\therefore \quad F_{AC} = \frac{W}{1.137} = 0.8795 \ W \quad \ldots(3)$$

$\therefore$ From equation (1)

$$F_{AB} = 0.742 \ F_{AC} = 0.742 \times 0.8795 \ W$$

$$= 0.6525 \ W \quad \ldots(4)$$

$\therefore$ Stress on wire AB,

$$= \frac{\text{Force on } AB}{\text{Area of } AB} = \frac{0.6525\,W}{258.064} = 211.11$$

$$\therefore \qquad W = \frac{211.11 \times 258.064}{0.6525} = 83.49 \text{ kN}$$

$$\text{Stress on wire } AC = \frac{\text{Force on } AC}{\text{Area of } AC} = \frac{0.8795\,W}{322.58} = 211.11$$

$$\therefore \qquad W = \frac{211.11 \times 322.58}{0.8795} = 77.43 \text{ kN}$$

Take minimum value of $W = 77.43$ kN

$\therefore$ Largest weight W that can be supported by two wires $= 77.43$ kN.

Problem 1.7: A 304.8 mm square steel bearing plate lies between an 203.2 mm diameter wooden post and concrete footing as shown in figure P 1.7. Determine the maximum value of the axial load P of the stress in wood is limited to 12.67 N/mm² and that in concrete to 4.574 N/mm².

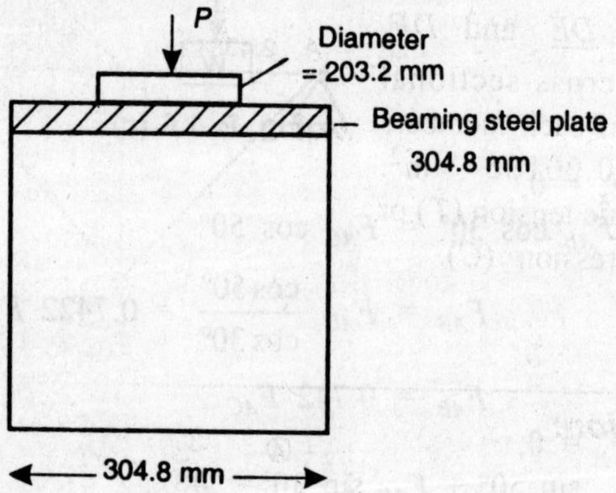

Fig. P. 1.7

Solution:

Stress in wooden post

$$(p_W) = \frac{\text{Force}}{\text{Cross-sectional area}} \quad \left[\begin{aligned} A_W &= \frac{\pi}{4}(203.2)^2 = 32429.3 \text{ mm}^2 \\ A_C &= 304.8 \times 304.8 = 92903.4 \text{ kN} \end{aligned}\right]$$

$$12.67 = \frac{P}{32429.3}$$

$$\therefore \quad P = 410.879 \text{ kN}$$

∴ Force on steel bearing plate = 410.879 kN

∴ Stress in concrete blower

$$(p_C) = \frac{\text{Force}}{\text{cross - sectional area}}$$

$$= \frac{410.879 \times 1000}{92903.04} = 4.4226 \text{ N/mm}^2$$

$$4.4226 < 4.574 \text{ (Permissible in concrete)}$$

∴ Maximum value of the arial load P is the stress on wood is limited to 12.67 N/mm² and that on concrete to 4.574 N/mm² is 410.879 kN.

Problem 1.8: For the truss shown in Figure P 1.8 (a), calculate the stresses in members $\underline{CF}$, $\underline{DE}$ and DF. The cross-sectional area of each members in 0.00166 m². Indicate tension (T) or compression (C).

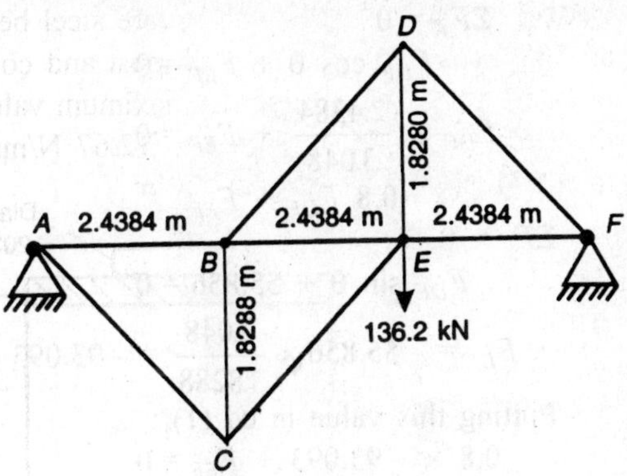

Fig. P. 1.8 (a)

Solution:

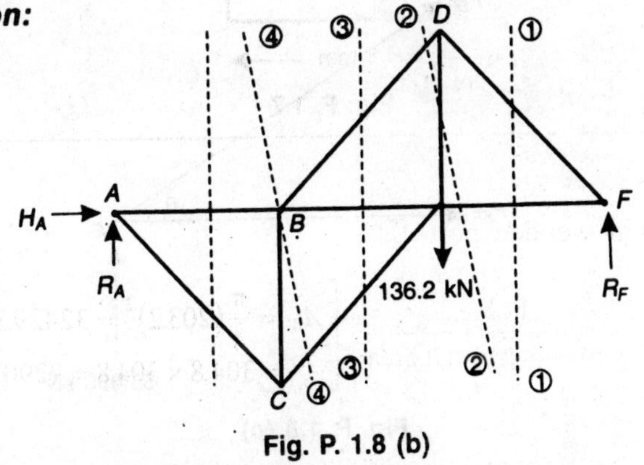

Fig. P. 1.8 (b)

$$\tan \theta = \frac{1.8288}{2.4384} \quad \text{or } \theta = 36.869°$$

For reaction using equilibrium equation

$$\Sigma F_x = 0,$$

$$\therefore \quad H_A = 0$$

$$\Sigma F_y = 0,$$

$$R_A + R_F = 136.2 \qquad ...(1)$$

$$\Sigma M_A = 0,$$

$$R_F \times [2.4384 \times 3] = 136.2 \times [2.4384 \times 2]$$

$$\therefore \quad R_F = \frac{136.2}{2.4384} = 55.856 \text{ kN}$$

$$\therefore \quad R_A = 136.2 - 55.856 = 80.344 \text{ kN}$$

For calculation of forces in member *CE*, *DE* and *DF* using method of section →

F.B.D. of the portion of the section (1) - (1)

$$\therefore \quad \Sigma F_x = 0$$

$$F_{DF} \cos \theta + F_{EF} = 0$$

$$F_{DF} \times \frac{2.4384}{3.048} + F_{EF} = 0$$

$$0.8 \; F_{DF} + F_{EF} \qquad ...(1)$$

$$\Sigma F_y = 0,$$

$$F_{DF} \sin \theta + 55.856 = 0$$

$$\therefore F_{DF} = -55.856 \times \frac{3.048}{1.8288} = -93.093 \text{ kN} = 93.093 \text{ kN } (C) \; ...(2)$$

Putting this value in eq (1),

$$0.8 \times -93.093 + F_{EF} = 0$$

$$\therefore \; F_{EF} = 74.97 \text{ kN } (T)$$

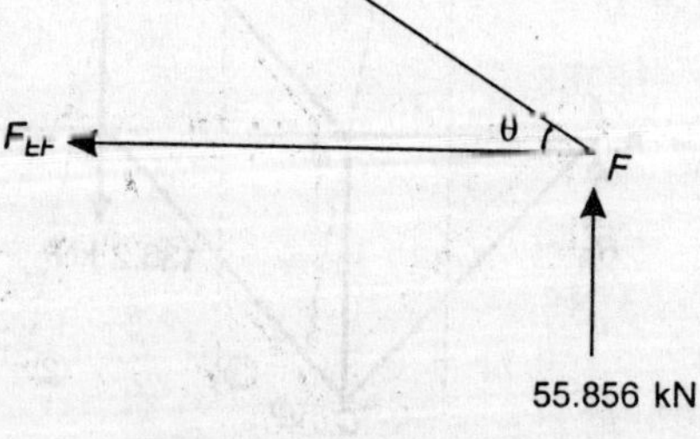

Fig. P. 1.8 (c)

F.B.D. of section (2) - (2)

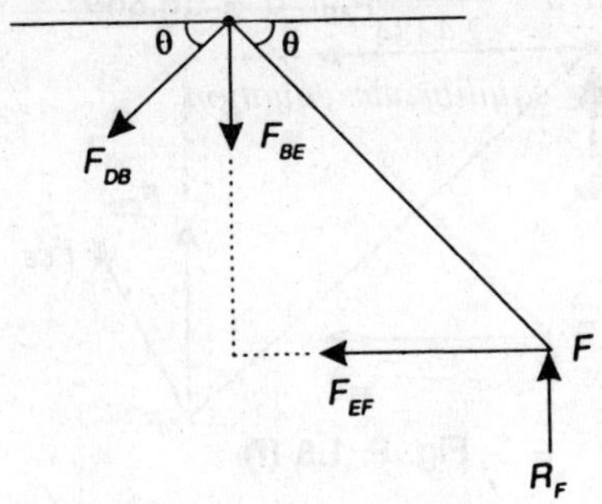

Fig. P. 1.8 (d)

$\Sigma M_F = 0,$

$F_{DE} \times 2.4384 + F_{DB} \times 3.048 = 0$

$\Sigma M_E = 0,$

$F_{DB} \times \cos\theta \times 1.8288 + R_F \times 2.4384 = 0$

$$\therefore\ F_{DB} \times \frac{2.4384}{3.048} \times 1.8286 + 55.856 \times 2.438 = 0$$

$F_{DB} = -93.093 \text{ kN} = 93.093 \text{ kN } (C)$

$\therefore\ F_{DE} \times 2.4384 = 93.093 \times 3.048$

$\therefore\ F_{DE} = 116.366 \text{ kN } (T)$

F.D.B. of right side of section (3) - (3)

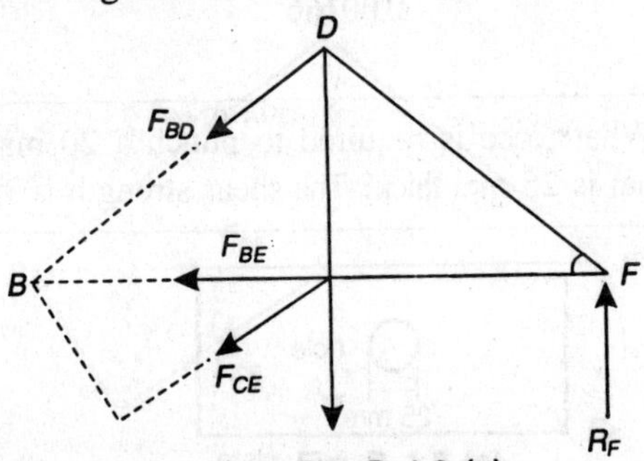

Fig. P. 1.8 (e)

$\therefore\ \Sigma M_B = 0$

$$-F_{CE} \times \frac{1.8288}{3.048} - 136.2 \times 2.4384 + R_F \times 2 \times 2.4384 = 0$$

$$-F_{CE} \times \frac{1.8288}{3.048} - 332.11 + 55.856 \times 2 \times 2.4384 = 0$$

$-0.6\ F_{CE} - 59.71 = 0$

$\therefore\qquad F_{CE} = -59.71/0.6 = -99.51 \text{ kN} = 99.51 \text{ kN } (C)$

F.B.D. of left portion of section (4) - (4)

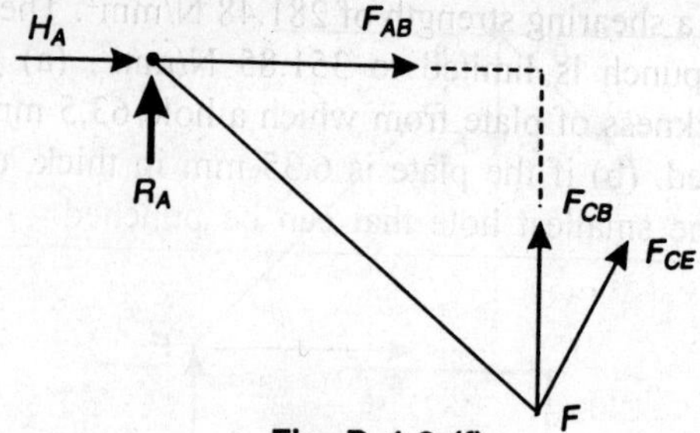

Fig. P. 1.8 (f)

$$\Sigma M_A = 0$$

$$F_{CB} \times 2.4384 + F_{CE} \times 3.098 = 0$$

$$F_{CB} \times 2.4384 + [-32.65 \times 3.098] = 0$$

$$F_{CB} = 40.812 \text{ kN } (T)$$

stress on member $CE = \dfrac{99.51}{0.00166} = 59945.78 \text{ kN/m}^2 \; (C)$

stress in member $DE = \dfrac{116.366}{0.00166} = 70100 \text{ kN/m}^2 \; (T)$

stress in member $DF = \dfrac{93.093}{0.00166} = 56080.12 \text{ kN/m}^2 \; (C)$

Problem 1.9: What force is required to punch a 20 mm diameter hole in a plate that is 25 mm thick? The shear strength is 350 MN/m².

Solution:

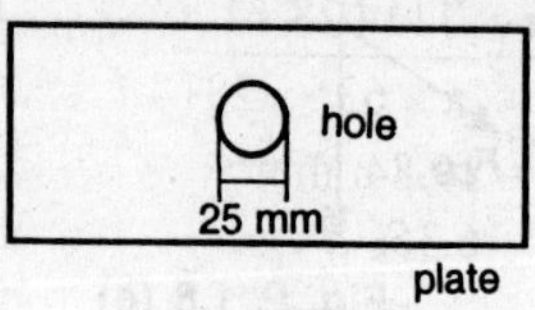

Fig. P. 1.9

$\therefore$ Resisting area of the hole $= \pi d \times t$

$= \pi \times 20 \times 25 = 1570.796 \text{ mm}^2$

$\therefore$ Shear strength $= \dfrac{\text{Force}}{\text{resisting area}} = 350 \text{ N/mm}^2$

$\therefore$ Force $= 350 \times 1570.796 = 549.77 \approx 550 \text{ kN}$

Problem 1.10: As in Fig. P. 1.10 a hole is to be punched out of a plate having a shearing strength of 281.48 N/mm². The compressive stress in the punch is limited to 351.85 N/mm². (a) Compute the maximum thickness of plate from which a hole 63.5 mm in diameter can be punched. (b) if the plate is 6.35 mm in thick, determine the diameter of the smallest hole that can be punched.

Solution:

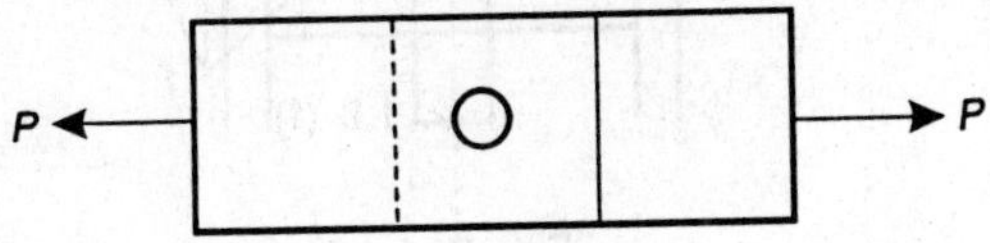

Fig. P. 1.10

(a) Compressive stress

$$= \frac{\text{Force}}{\text{Area of rivet}} = \frac{P}{\frac{\pi}{4}d^2} = \frac{P}{\frac{\pi}{4}(63.5)^2}$$

$$\therefore \quad \text{Force} = \frac{\pi}{4} \times (63.5)^2 \times 351.85 = 1114.28 \text{ kN}$$

$$\text{Shearing stress} = \frac{\text{Force}}{\text{Resisting area}} = \frac{\text{Force}}{\pi d \times t}$$

$$\therefore \quad t = \frac{\text{Force}}{\pi d \times 281.48}$$

$$= \frac{1114.28 \times 1000}{\pi \times 63.5 \times 281.48}$$

$$= 19.84 \text{ mm}$$

(b) If $t = 6.35$, $d = ?$

$$\therefore \quad d = \frac{1114.28 \times 1000}{\pi \times 6.35 \times 281.48}$$

$$= 198.437 \text{ mm}$$

Problem 1.11: Find the smallest diameter bolt that can be used in the device if P = 400 kN. The shearing strength for the bolt is 300 MPa.

Solution:

$$\text{Area} = \frac{\pi}{4} d^2,$$

$$P = 400 \text{ kN} = 4 \times 10^5 \text{ N}$$

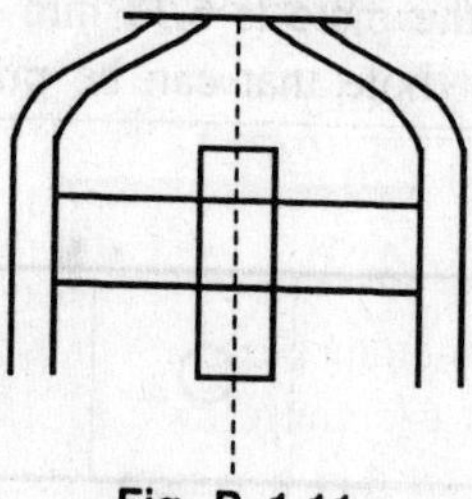

Fig. P. 1.11

$\therefore$ shearing strength for the bolt is = 300 MPa = 300 N/mm^2

$$\therefore \quad d^2 = \frac{400 \times 1000 \times 4}{\pi \times 300} = 1697.65$$

$$\therefore \quad d = 41.2025 \text{ mm}$$

Problem 1.12: A 200 mm diameter pulley is prevented from rotating relative to a 60 mm diameter shaft by a 70 mm long key, as shown in Fig. P. 1.12. If a torque T = 2.2 kN-m is applied to the shaft, determine the width b if the allowable shearing stress in the key is 60 MPa.

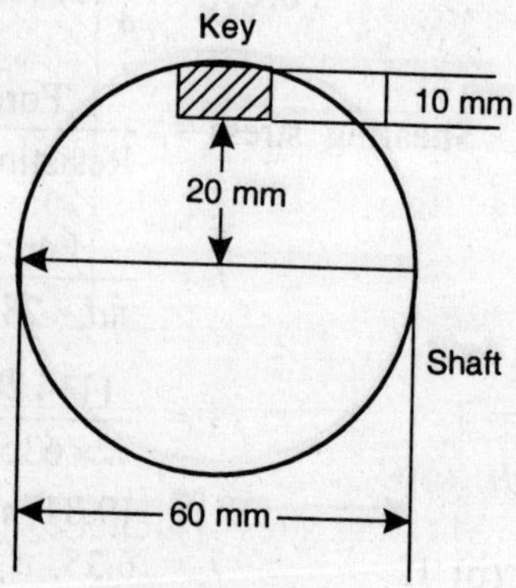

Fig. P. 1.12

Solution: Given :

Shaft diameter = 60 mm

distance of the key from the centre of the shaft (r) = 20 mm

Torque (T) = 2.2 × 10^6 N-mm

Length of the shaft = 70 mm

Allowable shearing stress

$= 60 \text{ N/mm}^2$

b = width of the key = ?

∴ Torque transmitted by the key (T)

$= q_K . r . l . b$

∴ $2.2 \times 10^6 = 60 \times 20 \times 70 \times b$

∴ $$b = \frac{2.2 \times 10^6}{60 \times 20 \times 70} = 26.19 \text{ mm}$$

Problem 1.13: A rectangular piece of wood 50 mm × 100 mm in cross-section, is used as a compression block as shown. Determine the maximum arial force P that can be safely applied to the block if the compressive stress in the wood is limited to 20 MN/m² and the shearing stress parallel to the grain is limited to 5 MN/m². The grain makes an angle of 20° with the horizontal, as shown.

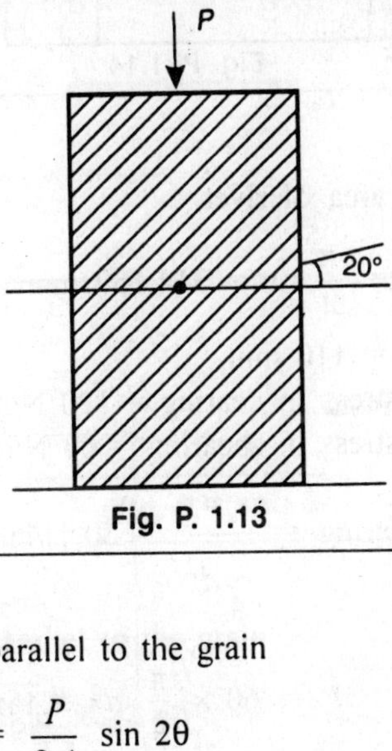

Fig. P. 1.13

Solution:

Shearing stress parallel to the grain

$$= \frac{P}{2A} \sin 2\theta$$

$= 5 \text{ N/mm}^2$ [A = Area of cross section]

∴ $$P = \frac{5 \times 2A}{\sin 2\theta} = \frac{10A}{\sin 2\theta}$$

$$= \frac{10 \times 50 \times 100}{\sin 40^\circ} = 77.78 \text{ kN}$$

Compressive stress

$$= \frac{\text{Force}}{\text{Area of cross-section}}$$

$\therefore$ Force $= 20 \times 50 \times 100 = 100$ kN

$\therefore$ Maximum arial force P that can be safely applied to the block $= 77.78$ kN.

Problem 1.14: Assume that a 20 mm diameter rivet joint the plates that are each 110 mm wide. The allowable stresses are 120 MPa for bearing in the plate material and 60 MPa for shearing of the rivet. Determine (a) the minimum thickness of each plate; and (b) the largest tangible stress in the plates.

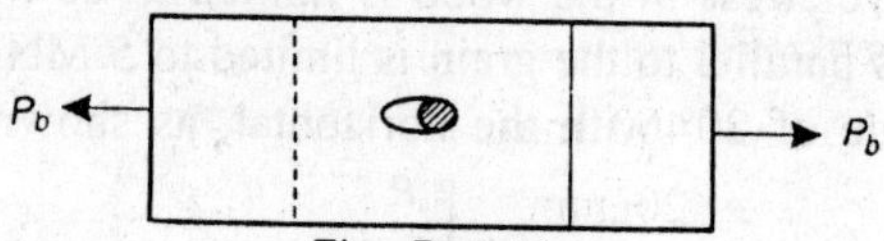

Fig. P. 1.14

Solution:

Cross-sectional area of rivet

$$= \frac{\pi}{4} 20^2 = 314.159 \text{ mm}^2$$

width of plate $= 110$ mm

The allowable stress in bearing $= 120$ N/mm^2

The allowable stress in shearing $= 60$ N/mm^2

$$\text{stress in shearing} = \frac{P}{\frac{\pi}{4} d^2} = 60 \text{ N/mm}^2$$

$$\therefore \quad P = 60 \times \frac{\pi}{4} d^2 = 15\pi \times d^2$$

$$= 15\,\pi \times (20)^2$$

$$= 18.849 \text{ kN} \quad \ldots(1)$$

$$\text{stress on Bearing} = \frac{P}{dt\,(\text{Projected area})} = 120$$

$$\therefore \quad P = 120\, dt \quad \ldots(2)$$

$\therefore$ from equation (1) and (2)

$$120\, dt = 18.849 \times 1000$$

$\therefore \quad t = 7.85$ mm

(b) The largest average tangible stress on the plate

$$= \frac{18849}{110 \times 7.85} = 26.7 \text{ N/mm}^2$$

Problem 1.15: A cylindrical steel pressure vessel 400 mm in diameter with a wall thickness of 20 mm, is subjected to an internal pressure of 4.5 MN/m². (a) Calculate the tangential and longitudinal stresses in the steel. (b) To what value may the internal pressure be increased if the stress in the steel is limited to 120 MN/m²?

Solution: Given

$$d = 400 \text{ mm}$$

$$t = 20 \text{ mm}$$

$$p_1 = 4.5 \text{ MN/m}^2 = 4.5 \text{ N/mm}^2$$

(a) $$\text{tangential stress} = \frac{pd}{2t} = \frac{4.5 \times 400}{2 \times 20} = 4.5 \times 10 = 45 \text{ N/mm}^2$$

$$\text{Longitudinal stress} = \frac{pd}{4t} = \frac{4.5 \times 400}{4 \times 20} = 22.5 \text{ N/mm}^2$$

(b) To what value of internal pressure removed if the stress in the steel as limited to 120 MN/m² ≈ 120 N/mm².

$$p_i = \frac{2t \times 6}{d} = \frac{2 \times 20 \times 120}{400} = 12 \text{ N/mm}^2$$

$$p_i = \frac{4t \times 6}{d} = \frac{4 \times 20 \times 120}{400} = 24 \text{ N/mm}^2$$

Take minimum of above two value 12 N/mm².

Problem 1.16: A cylindrical pressure vessel is fabricated from steel plating that has a thickness of 20 mm The diameter of the pressure vessel is 450 mm and its length is 2.0 m. Determine the maximum internal pressure that can be applied if the longitudinal stress is limited to 140 MPa, and the circumferential stress is limited to 60 MPa.

Solution: Given

$$t = 20 \text{ mm}$$
$$d = 450 \text{ mm}$$
$$L = 2.0 \text{ m} = 2000 \text{ mm}$$

$$\text{Longitudinal stress} = \frac{pd}{4t} = 140 \text{ N/mm}^2$$

$$\therefore \quad p_i = \frac{140 \times 4t}{d}$$

$$= \frac{140 \times 4 \times 20}{450} = 24.9 \text{ N/mm}^2$$

$$\text{Circumferential stress} = \frac{pd}{2t} = 60 \text{ N/mm}^2$$

$$p_i = \frac{60 \times 2 \times t}{d} = \frac{60 \times 2 \times 20}{450} = 5.34 \text{ N/mm}^2$$

Take minimum value of above two

$$p_i = 5.34 \text{ N/mm}^2$$

Problem 1.17: At what angular velocity will the stress in a rotating steel ring equal 150 MPa if its mean radius is 220 mm. The density of steel is 7.85 Mg/m³.

Solution:

$$\text{Stress } (\sigma) = 150 \text{ MP} = 150 \text{ N/mm}^2$$
$$\text{Mean radius } (r) = 220 \text{ mm}$$
$$\text{density } (\rho) = 7.85 \text{ Mg/m}^3$$
$$= \frac{7.85 \times 10^6}{10^9} \text{ g/mm}^3$$
$$= 7.85 \times 10^{-5} \text{ N/mm}^3$$

We know that,

$$\sigma \text{ (stress)} = \rho V^2 = \rho \times r^2 w^2$$

$$\therefore \quad w = \sqrt{\frac{\sigma}{\rho r^2}} = \sqrt{\frac{150 \times 10^5}{7.85 \times (220)^2}} = 6.28 \text{ rad/sec}$$

Problem 1.18: Calculate the modulus of rigidity and bulk modulus of a cylindrical bar of diameter 25 mm and of length 1.2 m if the longitudinal strain in a bar during a tensile test is four times the lateral strain. Find the change in volume when the bar is subjected to a hydrostatic pressure of 120 N/mm². Take $E = 1.2 \times 10^5$ N/mm².

Solution:

Volume of bar, $V = \frac{\pi}{4} d^2 L = \frac{\pi}{4} (25)^2 \times 1200 = 589049 \text{ mm}^3.$

(a) Value of m

Longitudinal strain = 4 × Lateral strain.

$$\therefore \quad \frac{1}{m} = \frac{\text{lateral strain}}{\text{longitudinal strain}} = \frac{1}{4} = 0.25$$

$$\therefore \quad m = 4$$

(b) Value of N

we know, $N = \frac{mE}{2(m+1)}$

$$= \frac{4 \times 1.2 \times 10^5}{2(4+1)} = 0.48 \times 10^5 \text{ N/mm}^2$$

(c) Value of K

we know, $K = \frac{mE}{3(m-1)}$

$$= \frac{4 \times 1.2 \times 10^5}{3(4-2)} = 0.8 \times 10^5 \text{ N/mm}^2$$

(d) Change in volume of bar

we know $K = \frac{P}{e_v}$

$$\therefore \quad e_v = \frac{\delta V}{V} = \frac{p}{K} = \frac{120}{0.8 \times 10^5}$$

$$\therefore \quad \delta V = \frac{120}{0.8 \times 10^5} \times 589049 = 884 \text{ mm}^3$$

Problem 1.19: A metallic bar 250 mm × 100 mm × 50 mm is loaded as shown in Fig. P 1.19. Find the change in volume. Take $E = 200$ kN/mm^2 and Poisson's ratio = 0.25. Also find the change that should be made in the 4000 kN load, in order that there should be no change in the volume of the bar.

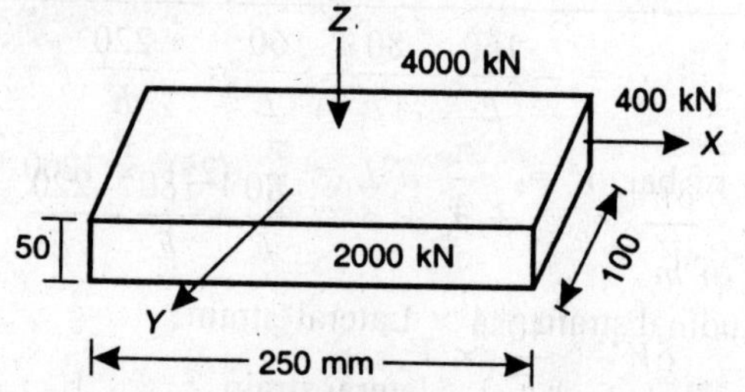

Fig. P. 1.19

Solution:

Volume of bar $V = 250 \times 100 \times 50 = 1250 \times 10^3$ mm^3;

$$\frac{1}{m} = 0.25; \; m = 4$$

(a) Change in volume

$$p_x = \frac{P_x}{A_x} = \frac{400 \times 10^3}{100 \times 50} = 80 \text{ N/mm}^2 \text{ (tension)}$$

$$p_y = \frac{P_y}{A_y} = \frac{2000 \times 10^3}{250 \times 50} = 160 \text{ N/mm}^2 \text{ (tension)}$$

$$p_z = \frac{P_z}{A_z} = \frac{4000 \times 10^3}{250 \times 100} = 160 \text{ N/mm}^2 \text{ (comp.)}$$

Taking tension as positive and compression as negative,

We know $$e_x = \frac{p_x}{E} - \frac{p_y}{mE} + \frac{p_z}{mE}$$

$$= \frac{80}{E} - \frac{160}{4E} + \frac{160}{4E} = \frac{80}{E}$$

$$e_y = \frac{p_y}{E} + \frac{p_z}{mE} - \frac{p_x}{mE}$$

$$= \frac{160}{E} + \frac{160}{4E} - \frac{80}{4E} = -\frac{180}{E}$$

$$e_z = -\frac{P_z}{E} - \frac{P_x}{mE} - \frac{P_y}{mE}$$

$$= -\frac{160}{E} - \frac{80}{4E} - \frac{160}{4E} = -\frac{220}{E}$$

$$\therefore \quad \frac{\delta V}{V} = e_x + e_y + e_z = \frac{80}{E} + \frac{180}{E} - \frac{220}{E} = \frac{40}{E}$$

$$\therefore \quad \delta V = \frac{40}{E} \times V$$

$$= \frac{40}{200 \times 1000} \times 1250 \times 10^3 = 250 \text{ mm}^3$$

(b) Change in 4000 kN load

Let P_z kN be the load, in place of 4000 kN load.

$$\therefore \quad p_z = \frac{P_z \times 10^3}{250 \times 100} = 0.04\ P_z \text{ N/mm}^2 \text{ (comp.)}$$

We know $e_x = \frac{p_x}{E} - \frac{p_y}{mE} + \frac{p_z}{mE}$

$$= \frac{80}{E} - \frac{160}{4E} + \frac{0.04\,P_z}{4E} = \frac{40}{E} + \frac{0.01\,P_z}{E}$$

$$e_y = \frac{p_y}{E} + \frac{p_z}{mE} - \frac{p_x}{mE}$$

$$= \frac{160}{E} + \frac{0.04\,P_z}{4E} - \frac{80}{4E} = \frac{140}{E} + \frac{0.01\,P_z}{E}$$

also $e_z = -\frac{p_z}{E} - \frac{p_x}{mE} - \frac{p_y}{mE}$

$$= -\frac{0.04\,P_z}{E} - \frac{80}{4E} - \frac{160}{4E}$$

$$= -\frac{0.04\,P_z}{E} - \frac{60}{E}$$

In order that $\frac{\delta V}{V} = e_x + e_y + e_z$ should be equal to zero,

We know $e_x + e_y + e_z = 0$

$$\therefore \left(\frac{40}{E} + \frac{0.01\,P_z}{E}\right) + \left(\frac{140}{E} + \frac{0.01\,P_z}{E}\right) - \left(\frac{0.04\,P_z}{E} + \frac{60}{E}\right) = 0$$

$$\therefore \qquad P_z = \frac{120}{0.02} = 6000 \text{ kN}$$

Hence P_z should be increased from its present value of 4000 kN to a new value of 6000 kN, in order that volumetric strain is zero.

Problem 1.20: A bar of steel is 40 mm in diameter and 500 mm long. A tensile load of 120 kN is found to stretch the bar by 0.24 mm. The same bar, when subjected to a torque of 1.3 kN-m is found to twist through 1.92°. Find the values of the four elastic constants.

Solution:

(a) Value of E

In case of bar, $\Delta = \frac{PL}{AE} = \frac{4\,PL}{\pi d^2 E}$

$$\therefore \qquad E = \frac{4\,PL}{\pi d^2 \Delta} = \frac{4 \times 120 \times 10^3 \times 500}{\pi\,(40)^2 \times 0.24}$$

$$= 1.989 \times 10^5 \text{ N/mm}^2$$

(b) Value of N

In case of shaft,

$$N = \frac{TL}{J\theta} = \frac{32\,TL}{\pi d^4\,\theta}$$

Where $\theta = \frac{\pi}{180} \times 1.92 = 0.0335$ radian,

and $T = 1.3$ kN-m $= 1.3 \times 10^6$ N-mm

$$\therefore \qquad N = \frac{32 \times 1.3 \times 10^6 \times 500}{\pi\,(40)^4 \times 0.0335} = 0.772 \times 10^5 \text{ N/mm}^2$$

(c) Value of $\frac{1}{m}$

We know $E = 2N\left(1+\frac{1}{m}\right)$

$$\therefore \quad \frac{1}{m} = \frac{E}{2N} - 1 = \frac{1.989 \times 10^5}{2 \times 0.772 \times 10^5} - 1 = 0.288$$

(d) Value of K

We know $K = \dfrac{E}{3\left(1-\frac{2}{m}\right)}$

$$= \frac{1.989 \times 10^5}{3\,(1 - 2 \times 0.288)} = 1.565 \times 10^5 \text{ N/mm}^2$$

Problem 1.21: Show that if E is assumed to be correct, an error of 1.5% in the determination of N will involve an error of about 6.5% in the calculation of Poisson's ratio when its correct value is 0.3.

Solution:

Let E, N and μ be the correct values of the constants and μ' be the calculated value of Poisson's ratio.

We know $E = 2\,N\,(1 + \mu)$ ...(i)

Due to an error of 1.5%, N is increased to 1.015 N, the calculated value μ' of Poisson's ratio is given by

$E = 2 \times 1.015\,N\,(1 + \mu')$ ...(ii)

Equating (i) and (ii), we get

$$2N\,(1 + \mu) = 2 \times 1.015\,N\,(1 + \mu')$$

$$\therefore \quad 1 + \mu = 1.015 + 1.015\,\mu'$$

$$\therefore \quad \mu' - \mu = -\,0.015 - 0.015\,\mu'$$

$$= -\,0.015\,(1 + \mu')$$

Hence % error is μ is

$$\frac{\mu'-\mu}{\mu} \times 100 = -0.015\,\frac{1+\mu'}{\mu} \times 100 \approx -\,1.5\,\frac{1+\mu}{\mu}$$

(Taking $\mu' \approx \mu$)

$$\therefore \frac{\mu'-\mu}{\mu} \times 100 \approx -1.5\left(\frac{1+0.3}{0.3}\right) \approx -6.5\%$$

> **Problem 1.22:** The values of E and N, determined experimentally were found to be 1.9×10^5 N/mm^2 and 0.75×10^5 N/mm^2. Calculate Poisson's ratio and the bulk modulus. If both the moduli are liable for an error of ± 1 percent, find the maximum percentage error in the derived value of Poisson's ratio.

Solution:

We have $\quad E = 2N(1 + \mu)$

$$\therefore \quad \mu = \frac{E}{2N} - 1 = \frac{1.9 \times 10^5}{2 \times 1.75 \times 10^5} - 1 = 0.267$$

Also, $$K = \frac{E}{3(1-2\mu)}$$

$$= \frac{1.9 \times 10^5}{3(1 - 2 \times 0.267)} = 1.357 \times 10^5 \text{ N/mm}^2$$

For maximum percentage error in the calculated value of μ, the error in the values of E and N should be of different sign. Let % error in E be + 1 and that in N be − 1.

$$\therefore \quad \mu = \frac{E}{2N} - 1$$

also $$\mu' = \frac{E'}{2N'} - 1,$$

where $\quad E'$ = incorrect value of $E = (1.9 \times 10^5) \times 1.01$

N' = incorrect value of $N = (0.75 \times 10^5) \times 0.99$

and $\quad \mu'$ = computed incorrect value of μ.

$$\therefore \quad \mu' = \frac{1.9 \times 10^5 \times 1.01}{2\left(0.75 \times 10^5 \times 0.99\right)} - 1 = 0.2923$$

$$\therefore \% \text{ error in } \mu = \frac{\mu' - \mu}{\mu} \times 100$$

$$= \frac{0.2923 - 0.267}{0.267} \times 100 = 9.46\%$$

Problem 1.23: A tensile load of 50 kN is acting on rod of 50 mm diameter and length of 5 m. Determine the length of a bore of 25 mm that can be made central in the rod, if the total extension is not to exceed by 25 percent under the same tensile load. Take $E = 2.05 \times 10^5$ N/mm².

Solution:

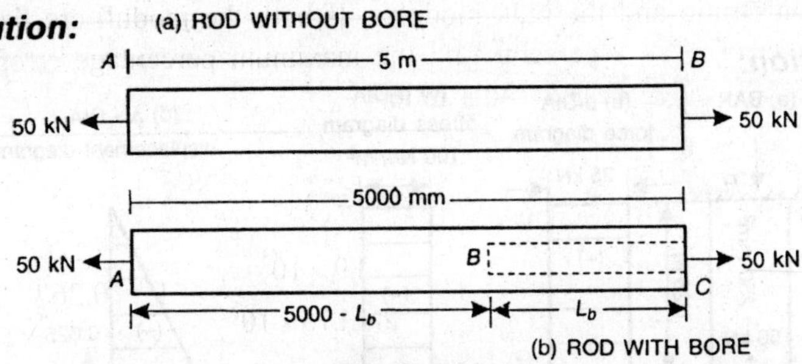

Fig. P. 1.23

Case I: Rod without bore

$$\text{Area of rod } A = \frac{\pi}{4}(50)^2 = 1963.5 \text{ mm}^2$$

$$\text{Length of rod } L = 5 \times 1000 = 5000 \text{ mm}$$

$$\therefore \quad \Delta = \frac{50 \times 10^3 \times 5000}{1963.5 \times 2.05 \times 10^5} = 0.621 \text{ mm}$$

Case II: Rod with bore

Let the length of the bore be L_b mm.

Hence length of unbored rod = $(5000 - L_b)$ mm.

Area of bored tube,

$$A_b = \frac{\pi}{4}(25)^2 = 490.87 \text{ mm}^2$$

Permissible total extension = (1 + 0.25) 0.621 = 0.7763 mm.

$$\text{But} \quad \Delta = \frac{P}{E}\left[\frac{L}{A} + \frac{L_b}{A_b}\right]$$

$$\therefore \quad 0.7763 = \frac{50 \times 10^3}{2.05 \times 10^5}\left[\frac{5000 - L_b}{1963.5} + \frac{L_b}{490.87}\right]$$

or $0.7763 = 0.2439\,[2.5465 - 5.0929 \times 10^{-4} L_b + 20.372 \times 10^{-4} L_b]$

or $15.2791 \times 10^{-4} L_b = 0.6363$

$$L_b = 416.5 \text{ mm}$$

Problem 1.24: For the steel bar shown in Fig. P. 1.24 determine the longitudinal force P and stress p at all cross-sections. Determine also the vertical displacement Δ of all cross-sections of the bar. Represent the result graphically by plotting P, p and Δ diagrams. Take $E = 2 \times 10^5$ N/mm². Also, locate the section of zero displacement.

Solution:

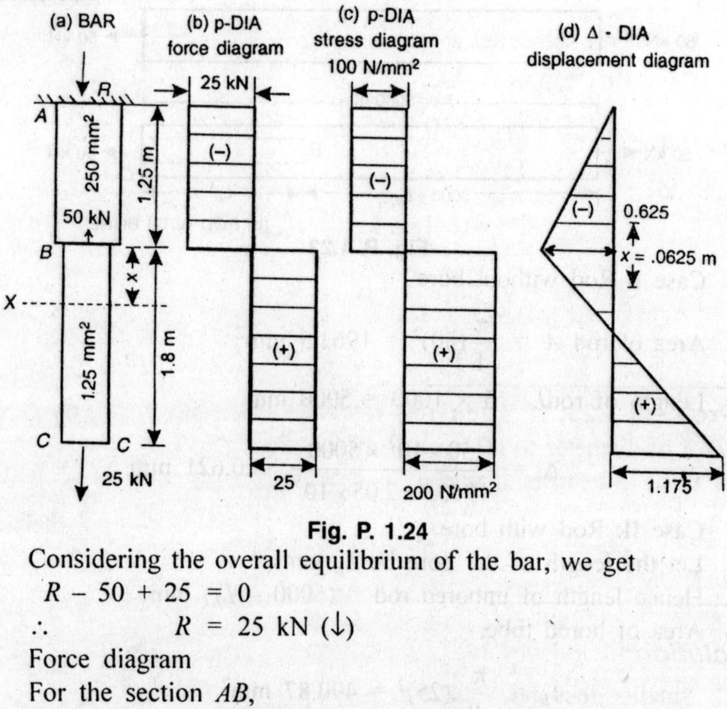

Fig. P. 1.24

Considering the overall equilibrium of the bar, we get

$$R - 50 + 25 = 0$$

$$\therefore \quad R = 25 \text{ kN } (\downarrow)$$

Force diagram

For the section AB,

force $\quad P = R = 25$ kN (comp.)

For the section BC,

force $\quad P = 50 - R = 50 - 25 = 25$ kN (tensile)

Stress diagram

For the section AB,

stress $\quad p = P/A = 25 \times 10^3/250$

$\quad = 100$ N/mm². (compressive)

For the section CD,

stress $\quad p = P/A = 25 \times 10^3/1250$

$\quad = 200$ N/mm² (tensile)

Δ Diagram

Point A is fixed in position. Hence $\Delta_A = 0$

Bar AB is subjected to a compressive force $P = 25$ kN

$$\therefore \quad \Delta_B = \frac{P_{AB} \cdot L_{AB}}{A_{AB} \cdot E}$$

$$= -\frac{(25 \times 10^3)(1.25 \times 1000)}{250 \times 2 \times 10^5} = -0.625 \text{ mm}$$

Bar BC is subjected a tensile force $P = 25$ kN

$$\therefore \quad \Delta_C = \Delta_{AB} + \Delta_{BC}$$

$$= \frac{P_{AB} \cdot L_{AB}}{A_{AB} \cdot E} + \frac{P_{BC} \cdot L_{BC}}{A_{BC} \cdot E}$$

$$= -\frac{(25 \times 10^3)(1.25 \times 1000)}{250 \times 2 \times 10^5} + \frac{(25 \times 10^3)(1.8 \times 1000)}{125 \times 2 \times 10^5}$$

$$= -0.625 + 1.8 = +1.175 \text{ mm}$$

Problem 1.25: If a tension test bar is found to taper from $(D + a)$ diameter to $(D - a)$ diameter, prove that the error involved in using the mean diameter to calculate the Young's modulus is $\left(\frac{10a}{D}\right)^2$ percent.

Solution:

Smaller diameter, $d_1 = D - a$ and larger diameter, $d_2 = D + a$

Let E_0 = modulus of elasticity calculated by considering mean diameter

E = actual Young's modulus of elasticity

We have
$$E = \frac{4\,P\,L}{\pi\, d_1\, d_2\, \Delta}$$

$$\therefore \quad E = \frac{4\,PL}{\pi\,(D-a)\,(D+a)\,\Delta} = \frac{K}{D^2 - a^2}$$

where
$$K = \frac{4\,PL}{\pi\Delta} \qquad \text{...(i)}$$

Also, based on the basis of mean diameter, we get

$$E_0 = \frac{PL}{A\Delta} = \frac{PL}{\frac{\pi}{4}D^2\Delta} = \frac{4PL}{\pi\Delta D^2} = \frac{K}{D^2} \quad \text{...(ii)}$$

$$\text{Hence \% error} = \frac{E - E_0}{E} \times 100 = \frac{\frac{K}{D^2 - a^2} - \frac{K}{D^2}}{\frac{K}{D^2 - a^2}} \times 100$$

$$= \frac{D^2 - (D^2 - a^2)}{D^2} \times 100 = \frac{100\,a^2}{D^2}$$

$$= \left(\frac{10\,a}{D}\right)^2 \text{ percent}$$

Problem 1.26: Two horizontal rigid bars *AB* and *CD* are connected by two wires of aluminium and copper as shown in Fig. P 1.26. Both the wires have length *L*. Vertical loads *P* are applied at mid points *E* and *F* of the two bars. Compute the increase Δ in the distance between points *E* and *F* when the loads are applied.

Solution:

The extensions Δ_{al} and Δ_{cu} of the two wires will be different. Hence the displacement Δ between points *E* and *F* is given by

$$\Delta = \frac{1}{2}(\Delta_{al} + \Delta_{cu}) \quad \text{...(i)}$$

Also, from equilibrium, by taking moments about plane *AC*, we get

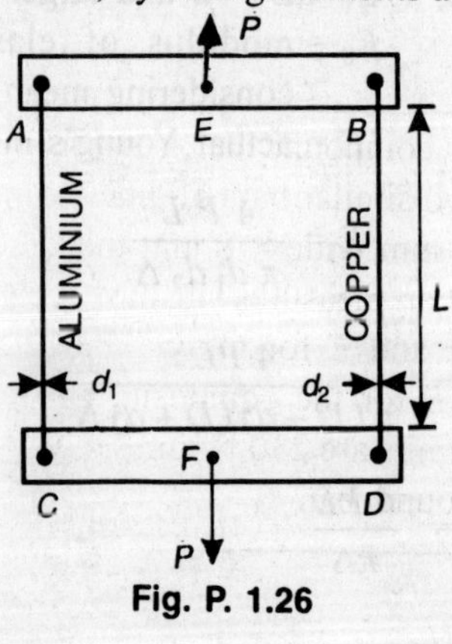

Fig. P. 1.26

$$P_2 \times AB = P \times AE$$

$$\therefore \quad P_2 = P\,\frac{AE}{AB} = \frac{1}{2}\,P \qquad \ldots\text{(ii)}$$

$$\therefore \quad P_1 = \frac{1}{2}\,P$$

$$\Delta_{al} = \frac{P_1 L}{A_1 E_1} = \frac{\left(\frac{P}{2}\right)L}{\frac{\pi}{4} d_1^2 E_1} = \frac{2\,PL}{\pi d_1^2 E_1}$$

$$\Delta_{cu} = \frac{P_2 L}{A_2 E_2} = \frac{\left(\frac{P}{2}\right)L}{\frac{\pi}{4} d_2^2 E_2} = \frac{2\,PL}{\pi d_2^2 E_2}$$

$$\therefore \quad \Delta = \frac{1}{2}\,\Delta_{al} + \Delta_{cu} = \frac{1}{2}\left[\frac{2\,PL}{\pi d_1^2 E_1} + \frac{2\,PL}{\pi d_2^2 E_2}\right]$$

or Total elongation $\Delta = \frac{PL}{\pi}\left[\frac{1}{d_1^2 E_1} + \frac{1}{d_2^2 E_2}\right]$, which is the required expression.

If, both the wires are of the same material, then $E_1 = E_2 = E$

$$\therefore \quad \Delta = \frac{PL}{\pi E}\left[\frac{1}{d_1^2} + \frac{1}{d_2^2}\right]$$

Problem 1.27: A compound bar is made by fastening one flat bar of steel between two similar bars of aluminium alloy. The dimensions of each bar are 40 mm wide × 8 mm thick, so that the cross-section of the composite bar measures 40 mm × 24 mm. If E for steel = 2.04×10^5 N/mm² and E for alloy = 0.612×10^5 N/mm², find the apparent value of E for the composite bar loaded in tension. If the respective elastic limits are 230 N/mm² and 50 N/mm², find the elastic limit of the compound bar.

Solution:

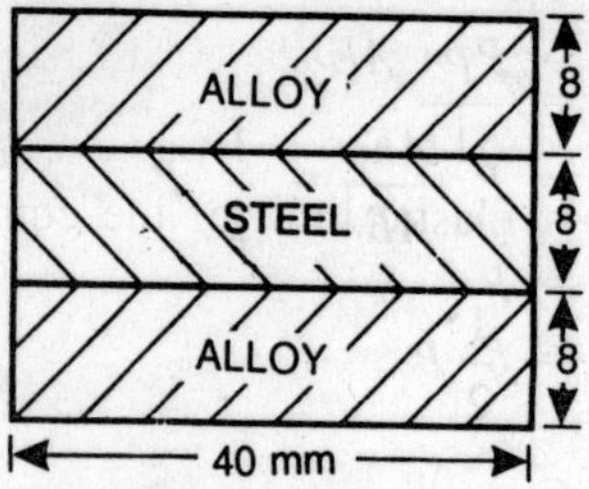

Fig. P. 1.27

Let Area of Steel $= A_s$

Area of Aluminium $= A_a$

$\therefore$ Area of steel $A_s = (40 \times 8) = 320 \text{ mm}^2$

and Area of Aluminium $= A_a = 2 \times 40 \times 8 = 640 \text{ mm}^2$

$\therefore$ Total area $A = A_s + A_a = 320 + 640 = 960 \text{ mm}^2$

We know
$$E = \frac{A_a E_a + A_s E_s}{A_a + A_s}$$

$$= \frac{640 \times 0.612 \times 10^5 + 320 \times 2.04 \times 10^5}{640 + 320}$$

$$= 1.008 \times 10^5 \text{ N/mm}^2$$

The *elastic limit* of the compound bar is the stress at which one of the members, either steel or aluminium alloy, is stressed to its elastic limit.

Let P be the load at elastic limit of the compound bar, and p_a and p_s be the corresponding stresses in the two materials at the elastic limit of the compound bar.

As the strains in the two materials are equal,

$$\frac{p_s}{E_s} = \frac{p_a}{E_a},$$

$$p_s = p_a \frac{E_s}{E_a}$$

$$p_a = \frac{2.04}{0.612} = 3.333 \, p_a \quad \ldots\text{(i)}$$

Also, $p_s A_s + p_a A_a = P$

$\therefore$ $3.333 \, p_a A_s + p_a A_a = P$

$\therefore$ $p_a [3.333 \times 320 + 640] = P$

$$p_a = \frac{P}{1706.7};$$

$$p_s = \frac{P}{512} \quad \text{...(ii)}$$

Let p = elastic limit of the compound bar.

$\therefore p\,(A_a + A_s) = P$

$\therefore \quad 960\,p = P \quad \text{...(iii)}$

But from eqn (ii)

$P = 1706.7\,p_a$

$\therefore \quad 960\,p = 1706.7\,p_a$

$$\therefore \quad p = \frac{1706.7}{960}\,p_a = 1.778\,p_a$$

$$= 1.778 \times 50 = 88.89 \text{ N/mm}^2 \quad \text{...(i)}$$

Also, from eq. (ii), $p = 512\,p_s$

Hence from (iii), $960\,p = 512\,p_s$

$$\text{or} \quad p = \frac{512}{960}\,p_s$$

$$= \frac{512}{960} \times 230 = 122.67 \text{ N/mm}^2 \text{ ...(ii)}$$

The elastic limit (P) is the lower of the two values of P given by eqn (i) and (ii)

Hence the elastic limit of the compound will be 88.89 N/mm^2

Problem 1.28: A railway is laid so that there is no stress in the rails at 8°C. Calculate (a) the stress on the rails at 50°C if there is no allowance for expansion, (b) the stress in the rails at 50°C if there is an expansion allowance of 8 mm per rail. (c) the expansion allowance if the stress in the rail is to be zero when the temperature is 50°C (d) the maximum temperature to have no stress in the rails if the expansion allowance is 12 mm per rail. The rails are 30 m long. Take $\alpha = 12 \times 10^{-6}$ per °C and $E = 2 \times 10^5$ N/mm^2.

Solution:

Change in temperature (rise) = 50 − 8 = 42°C

Case I: When No expansion allowance

Stress $p_t = \alpha t E$

$= 12 \times 10^{-6} \times 42 \times 2 \times 10^5 = 100.8 \text{ N/mm}^2$

Case II: Expansion allowance of 8 mm

$$\frac{p_t \times L}{E} = \Delta = (L\alpha t - a)$$

$$\therefore \quad p_t = E\alpha t - \frac{aE}{L}$$

$$= 100.8 - \frac{5 \times 2 \times 10^5}{30000} = 67.5 \text{ N/mm}^2$$

Case III: Value of a if p_t is to be zero

$$p_t = E\alpha t - \frac{aE}{L}$$

For p_t to be zero,

$$a = L\alpha t = 30000 \times 12 \times 10^{-6} \times 42$$
$$= 15.12 \text{ mm}$$

Case IV: Value of temperature for p_t to be zero

$$p_t = \frac{E}{L}(L\alpha t - a) = 0$$

$$\therefore \quad L\alpha t = a$$

$$\therefore \quad t = \frac{a}{L\alpha} = \frac{12}{30000 \times 12 \times 10^{-6}} = 33.33^\circ$$

$\therefore$ Allowable temperature $= 8 + 33.33 = 41.33^\circ$

Problem 1.29: A composite bar made up of aluminium and steel is held between two supports as shown in Fig. 1.29. The bars are stress free at a temperature of 42°C. What will be the stresses in the two bars when the temperature drops to 24°C, if (a) the supports are un-yielding, (b) the supports come nearer to each other by 0.1 mm. The cross-sectional area of steel bar is 160 mm^2 and that of aluminium bar is 240 mm^2.
Take E for aluminium as 0.7×10^5 N/mm^2 and that for steel as 2×10^5 N/mm^2. The coefficients of thermal expansion for aluminium and steel are 24×10^{-6} per °C and 12×10^{-6} per °C respectively.

Solution:

Using suffix s for steel and a for aluminium, the free contraction due to drop of temperature ($t = 42 - 24 = 18$°C) is given by:

$$\Delta = (L_s\alpha_s + L_a\alpha_a)t$$

$$= (500 \times 12\ 10^{-6} + 250 \times 24 \times 10^{-6})\ 18$$

$$= 0.216 \text{ mm} \qquad \ldots\text{(i)}$$

However, the above free contraction is checked by the supports, either fully or partially. Hence tensile stresses will be set up in both steel and aluminium bars. However, force in steel bar and aluminium bars will be the same. Hence

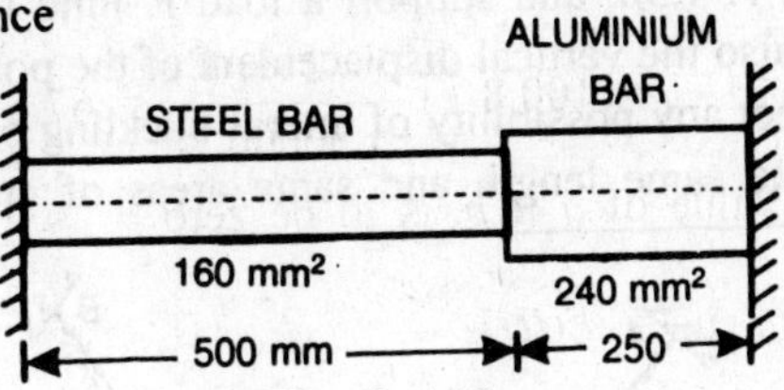

Fig. P. 1.29

$$A_s p_s = A_a p_a$$

or $\quad 160 p_s = 240\ p_a$

From which $p_s = 1.5\ p_a \qquad \ldots\text{(ii)}$

Let $\quad \Delta_s^p$ = elongation of steel bar due to temperature stress

$$= \frac{p_s}{E_s} L_s = \frac{500}{2 \times 10^5} \times p_s$$

$$= 250 \times 10^{-5}\ p_s \qquad \ldots\text{(iii)}$$

and $\quad \Delta_a^p$ = elongation of aluminium bar due to temperature stress

$$= \frac{p_a}{E_a} L_a = \frac{250}{0.7 \times 10^5} \times p_a$$

$$= 357.1 \times 10^{-5}\ p_a \qquad \ldots\text{(iv)}$$

Case (a): When supports non-yielding

$$\Delta_s^p + \Delta_a^p = \text{Free contraction } \Delta \qquad \ldots\text{(v)}$$

$\therefore \quad 250 \times 10^{-5}\ p_s + 357.1 \times 10^{-5}\ p_a = 0.216$

$\therefore$ From eqn (ii) $p_s = 1.5\ p_a$

$\therefore\ 250 \times 10^{-5} \times 1.5\ p_a + 357.1 \times 10^{-5}\ p_a = 0.216$

$\therefore \quad p_a = 29.50 \text{ N/mm}^2$ (tensile)

Hence $\quad p_s = 1.5 \times 29.5 = 44.26 \text{ N/mm}^2$ (tensile)

Case (b): When supports yielding by $a = 0.1$ mm

$$\Delta_s^p + \Delta_a^p = \Delta - a$$

$\therefore\ 250 \times 10^{-5}\ p_s + 357.1 \times 10^{-5}\ p_a = 0.216 - 0.1$

Substituting $p_s = 1.5\ p_a$ from eq. (ii), we get

$250 \times 10^{-5} \times 1.5\ p_a + 357.1 \times 10^{-5}\ p_a = 0.116$

$$\therefore \quad p_a = 15.84 \text{ N/mm}^2 \text{ (tensile)}$$
$$p_s = 1.5 \times 15.84 = 23.77 \text{ N/mm}^2 \text{ (tensile)}$$

Problem 1.30: Three identical pin-connected bars are arranged as shown in Fig. P. 1.30, and support a load P. Find the axial force in each bar and also the vertical displacement of the point of application of load. Neglect any possibility of lateral buckling of the bar. All the bars are of the same length and same areas of cross-section.

Solution:

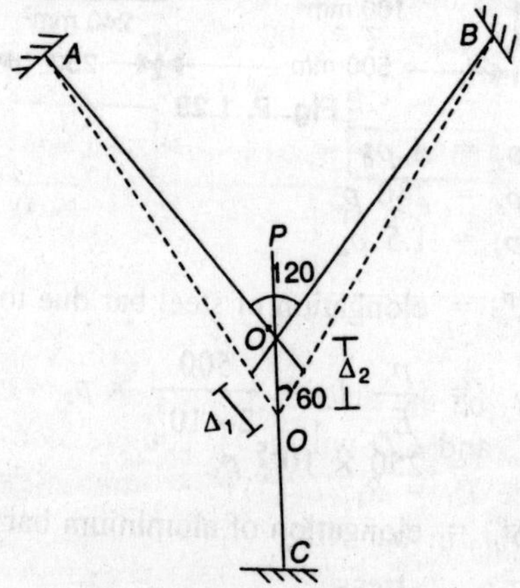

Fig. P. 1.30

Let force in $AO = BO = P_1$ (tensile)

force in $OC = P_2$ (compression)

$$\therefore \quad 2P_1 \cos 60° + P_2 = P$$

$$\therefore \quad P_1 + P_2 \qquad \text{...(i)}$$

The dotted lines show the deformed position of the structure. Assuming that the angles between the rods do not change appreciably,

$$\therefore \quad \Delta_t = \Delta_2 \cos 60° = \frac{1}{2} \Delta_2$$

$$\Rightarrow \quad \frac{P_1 L}{A E} = \frac{1}{2} \frac{P_2 L}{A E}$$

$$\therefore \quad P_2 = 2P_1 \qquad \text{...(ii)}$$

Substituting in (i), we get $P_1 + 2P_1 = P$

$$\therefore \quad P_1 = \frac{P}{3}$$

$$\therefore \quad P_2 = 2P_1 = \frac{2P}{3}$$

Also, $$\Delta_2 = \frac{P_2 L}{AE} = \frac{2PL}{3AE}$$

Problem 1.31: A circular bar *ABCD* is rigidly fixed at *A* and *D* and is subjected to axial forces as shown in Fig. 2.31. Determine the reactions, the forces in each portion of the bar and the displacement of point *B* and *C*. Take $E = 200 \text{ kN/mm}^2$.

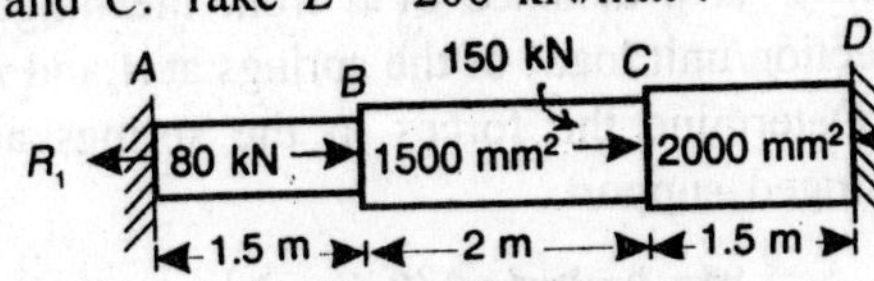

Fig. P. 1.31

Solution:

Let R_1 and R_2 be the end reactions, portion *AB* will be in tension while portions *BC* and *CD* will be in compression.

For *AB*, force $P_1 = R_1$; $L_1 = 1500$ mm; $A_1 = 1000 \text{ mm}^2$

Now, force P_{AB} in bar $AB = R_1 = 20000 - R_2$ (tensile)

Also, force P_{BC} in bar $BC = R_2$ (compressive)

$\therefore$ Extension of *AB*,

$$\Delta_{AB} = \frac{(20000 - R_2)\,1200}{113.1 \times 2 \times 10^5} = \frac{2000 - R_2}{18850} \text{ mm}$$

Contraction of *BC*,

$$\Delta_{BC} = \frac{R_2 \times 1800}{201.1 \times 2 \times 10^5} = \frac{R_2}{22344} \text{ mm}$$

In order that end *C* rests on the lower support to induce a reaction R_2, we have the compatibility equation:

$$\Delta_{AB} - \Delta_{AB} = a$$

$$\therefore \quad \frac{2000 - R_2}{18850} - \frac{R_2}{22344} = 1$$

$\therefore \quad 1.061 - 5.305 \times 10^{-5} R_2 - 4.475 \times 10^{-5} R_2 = 1$

$\therefore \quad R_2 = 6238 \text{ N} = 6.238 \text{ kN}$

$\therefore \quad R_1 = 20000 - 6238 = 13762 \text{ N} = 13.762 \text{ kN}$

$\therefore$ Stress in bar

$$AB = p_{AB} = \frac{R_1}{A_{AB}} = \frac{13762}{113.1} = 121.7 \text{ N/mm}^2 \text{ (tensile)}$$

Stress in bar

$$BC = p_{BC} = \frac{R_2}{A_{BC}} = \frac{6238}{201.1} = 31 \text{ N/mm}^2 \text{ (comp.)}$$

Problem 1.32: A rigid beam *ABC* is hinged at *D* and supported by two springs *A* and *B* as shows in fig. 1.32. The beam carries a vertical load *W* at a point *C* at a distance of *b* from the hinged end. The flexibilities (deflection/unit load) of the springs at *A* and *B* are d_1 and d_2 respectively. Determine the forces in the springs and also the reaction at the hinged support.

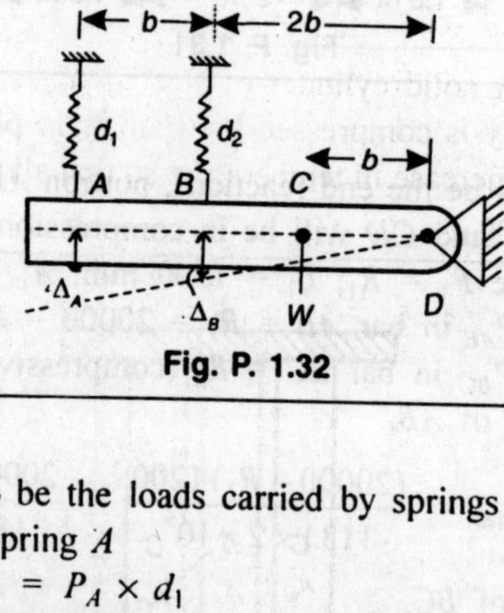

Fig. P. 1.32

Solution:

Let P_A and p_S be the loads carried by springs *A* and *B*.

Extension in spring *A*

$$\Delta_A = P_A \times d_1$$

Extension in spring *B*

$$\Delta_B = P_B \times d_2$$

From compatibility,

$$\frac{\Delta_A}{b + 2b} = \frac{\Delta_B}{2b}$$

$$\therefore \quad \Delta_A = \frac{3}{2}\Delta_B$$

Substituting the values of Δ_A and Δ_B,

$$P_A \times d_1 = \frac{3}{2} P_B \times d_2$$

$$\therefore \quad P_A = \frac{3}{2}\frac{d_2}{d_1} \times P_B \qquad ...(1)$$

Taking moments about hinge D, we get

$$P_A \times 3\,b + P_B \times 2\,b = W \times b$$

$$\therefore \quad 3\,P_A + 2\,P_B = W \qquad ...(2)$$

Substituting the value of P_A from (1) in (2), we get

$$3\left(\frac{3}{2}\frac{d_2}{d_1}P_B\right) + 2\,P_B = W$$

$$\therefore \quad P_B = \frac{2d_1}{4d_1 + 9d_2}W$$

$$P_A = \frac{3d_1}{4d_1 + 9d_2}W$$

Problem 1.33: A solid cylinder of steel is placed inside a copper tube. The assembly is compressed between rigid plates by forces P. Find the value of increase in temperature so that all the load is carried by the copper tube.

Solution:

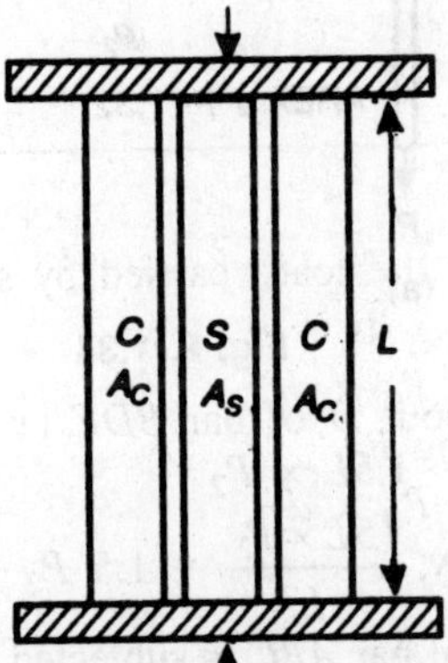

Fig. P. 1.33

Let the increase in temperature be t. Due to this increase, copper will expand more than steel.

Difference between the two expansions $= \Delta_t = L\,(\alpha_C - \alpha_S)\,t$

If the load P is to be carried entirely by the copper tube, the shortening (Δ_c) of the copper tube due to P should be equal to Δ_t.

$$\therefore \quad \Delta_c = \Delta_t$$

$$\therefore \quad \frac{PL}{A_C E_C} = L\,(\alpha_c - \alpha_s)\,t$$

$$\therefore \quad t = \frac{P}{A_C E_C(\alpha_c - \alpha_s)}$$

Problem 1.34: A vertical bar *ABC* is attached to a rigid bar *BDF*, as shown in Fig. 1.34 (a). Find the ratio of loads P_2 and P_1 so that the vertical deflection of point *C* is zero.

Solution:

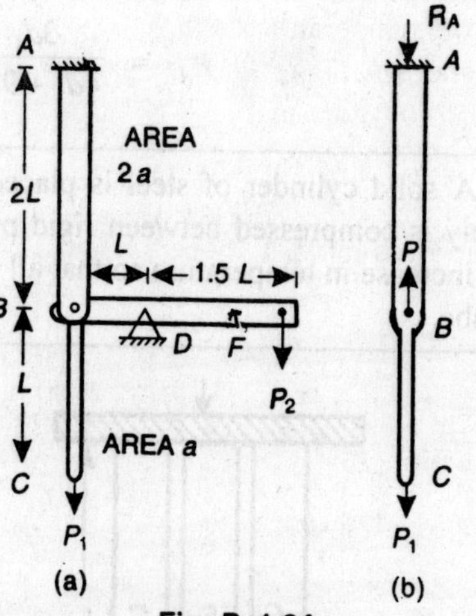

Fig. P. 1.34

Take moment about *D* of bar *BDF* i.e.,

$$P \times L = 1.5L \times P_2$$

$$\therefore \quad P = \frac{1.5L \times P_2}{L} = 1.5\ P_2$$

Hence the vertical bar *ABC* is subjected to forces as shown in Fig. 1.34 (b). The reaction $R_A = P - P_1$. Thus, the portion *AB* is subjected to a compressive force = $R_A\ (P - P_1)$ and portion *BC* is subjected to a tensile force P_1

$$\text{Hence} \quad \delta_C = \Sigma\,\frac{PL}{AE} = \frac{P_1(L)}{aE} - \frac{(P - P_1)(2L)}{2\,a\,E}$$

Substituting the value of *P* and nothing that S_C has to be zero, we have

$$\frac{P_1 L}{a\,E} - \frac{(1.5P_2 - P_1)2L}{2\,a\,E} = 0$$

$$\therefore\ P_1 - 1.5\,P_2 + P_1 = 0$$

or $$2P_1 = \frac{3}{2}\,P_2$$

$$\therefore\ \frac{P_2}{P_1} = \frac{4}{3}$$

Problem 1.35: If the mechanism of Fig. 1.35 is loaded with force $W = 50$ kN and the temperature increases by 25°F, calculate the axial stresses in *BC* and *DE*. Take coefficient of thermal expansion per degree *F* is 6.5×10^{-6} for steel and 13×10^{-6} for aluminium alloy.

Solution:

If temperature rises by 25°F then final position of the beam shown in Fig. P. 1.35.

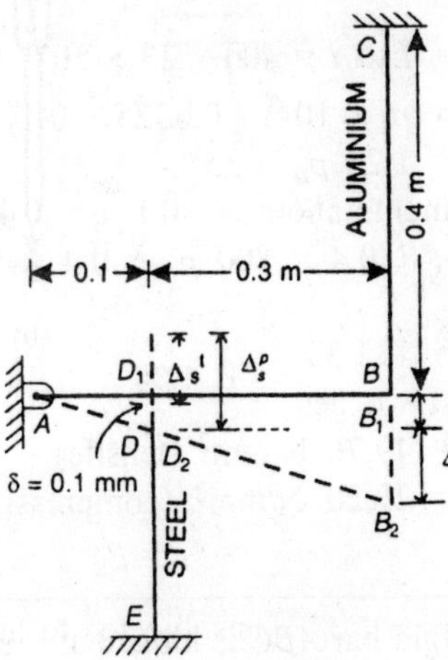

Fig. P. 1.35

Thus, $DD_1 = \Delta_s'$ = free expansion of steel bar due to temperature rise and $BB_1 = \Delta_a'$ = free expansion of aluminium bar due to temperature rise.

Now, when the load *W* is applied at point *B*, the rod *CB* extends further from B_1 to B_2 while the bar *DE* is compressed from D_1 to D_2 such that AD_2B_2 is the final position of the rigid bar.

Thus $B_1B_2 = \Delta_a^p$ = expansion of aluminium bar due to external load applied at B

and $D_1D_2 = \Delta_s^p$ = contraction of steel bar, due to external load applied at B

From the compatibilities equation,

$$\frac{\Delta_s^p - \Delta_s' + 0.1}{0.1} = \frac{\Delta_a' + \Delta_a^p}{0.4}$$

or $$4\left(\Delta_s^p - \Delta_s' + 0.1\right) = \Delta_a' + \Delta_a^p \quad \ldots\text{(i)}$$

We have $$\Delta_s^p = \frac{P_s L_s}{E_s} = \frac{p_s(200)}{2\times 10^5} = p_s \times 10^{-3}$$

$$\Delta_s' = L_s\alpha_s t = 200 \times 6.5 \times 10^{-6} \times 25 = 0.0325 \text{ mm}$$

$$\Delta_a^p = \frac{p_a L_a}{E_a} = \frac{P_a(400)}{0.8\times 10^5} = 5 \times 10^{-3}\ p_a$$

$$\Delta_a' = L_a\alpha_a t = 400 \times 13 \times 10^{-6} \times 25 = 0.13 \text{ mm}$$

Hence from (i), $4\,(p_s \times 10^{-3} - 0.0325 + 0.1) = 5 \times 10^{-3}\ p_a + 0.13$

$\therefore$ $p_s = 1.25\ p_a - 35$ ...(a)

Also, taking moments about A $0.1\ p_s + 0.4\ P_a = 0.4\ W$

or $0.1 \times 1500\ p_s + 0.4 \times 800\ p_a = 0.4 \times 50000$

which gives

$p_s + 2.133\ p_a = 133.33$ ...(b)

From (a) and (b) we get

$P_a = 49.76$ N/mm^2 (tensile)

and $p_s = 27.20$ N/mm^2 (compressive)

Problem 1.36: A rigid bar *ABC* is hinged at *A* and attached to brass bar *BF* (length 0.2 m, area 4000 mm^2) and steel bar *CD* (length 0.3 m and area 250 mm^2). The temperature of brass bar *BF* is lowered by 30 K and that of bar *CD* is raised by 30 K. Neglecting any possibility of lateral bucking, find the normal stresses in the brass and steel. Take $E = 0.9 \times 10^5$ N/mm^2 and $\alpha = 20 \times 10^{-6}$ per K for brass and $E = 2 \times 10^5$ N/mm^2 and $\alpha = 12 \times 10^{-6}$ per K for steel.

Solution:

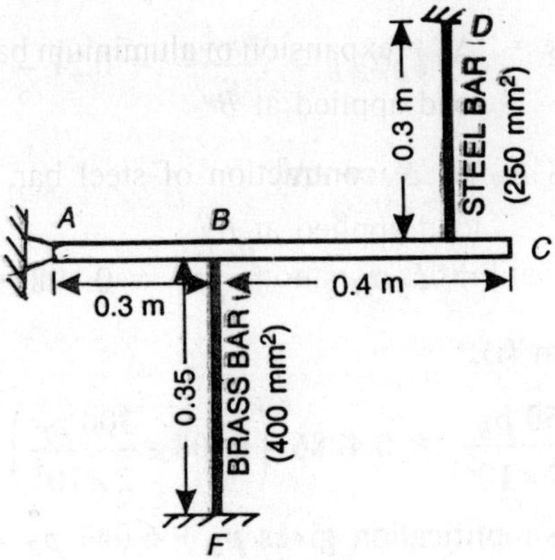

Fig. P. 1.36

Free expansion of steel bar due to temperature rise

$$= \Delta'_S = 300 \times 12 \times 10^{-6} \times 30 = 0.108 \text{ mm}$$

Free contraction of brass bar due to temperature fall

$$= \Delta'_B = 350 \times 20 \times 10^{-6} \times 30 = 0.21 \text{ mm}$$

For no temperature stresses to be developed,

$$\frac{\Delta'_B}{AB} = \frac{\Delta'_S}{AC}$$

$$\therefore \quad \Delta'_B = \Delta'_S \times \frac{AB}{AC}$$

$$= \Delta'_S \times \frac{300}{700} = 0.4286 \; \Delta'_S$$

$$= 0.4286 \times 0.108 = 0.0463 \text{ mm}$$

But $\Delta'_B = 0.21$ mm

Tensile stress developed in both brass and steel bar.

If Δ_B and Δ_S are the *final deformations* in the two bars at the *equilibrium position*, we have

$$\frac{\Delta_b}{AB} = \frac{\Delta_s}{AC}$$

$$\therefore \quad \Delta_B = \Delta_S \frac{AB}{AC} = \frac{0.3}{0.7} \Delta_S = 0.4286 \; \Delta_S \quad \text{...(i)}$$

Now $\Delta_S = \Delta'_B - \Delta^p_B$

$$= L_B\alpha_B t_B - \frac{p_B L_B}{E_B} = 0.21 - \frac{360\, p_B}{0.9 \times 10^5}$$

an $\Delta_S = \Delta'_S + \Delta^p_S$

$$= L_S\alpha_S t_S + \frac{p_S L_S}{E} = 0.108 + \frac{300\, p_S}{2 \times 10}$$

Hence from (i),

$$0.21 - \frac{350\, p_B}{0.9 \times 10^5} = 0.4286\left(0.108 + \frac{300\, p_S}{2 \times 10^5}\right)$$

which on simplification gives $p_S + 6.047\, p_B = 254.64$ ...(a)

Also, taking moments about A, $p_B \cdot A_B \times 350 - p_S\, A_S \times 700 = 0$

or $p_B \times 400 \times 350 = p_S \times 250 \times 700$

which gives $p_B = 1.25\, p_S$ ...(b)

Hence from (a) and (b), we get

$p_S = 29.74$ N/mm^2 (tension) and $p_B = 37.18$ N/mm^2 (tension)

Problem 1.37: A composite bar made up of aluminium bar and steel bar is firmly held between two unyielding supports as shown in Fig. 1.37. An axial load of 200 kN is applied at B at 20°C. Find the stresses in each material, when the temperature is 70°C. Take E for aluminium and steel as 0.7×10^5 N/mm^2 and 2×10^5 N/mm^2 respectively and coefficients of expansions for aluminium and steel as 24×10^{-6} per °C and 12×10^{-6} per °C respectively.

Solution:

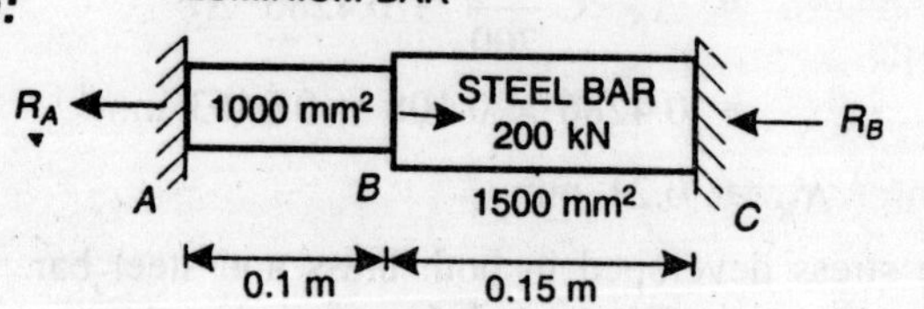

Fig. P. 1.37

Let us consider both the effects separately.

Case I: Effect of 200 kN load

Let R_A and R_B be the two reactions.

$\therefore \quad R_A = 200 - R_B$ ...(i)

Portion AB is subjected to a tensile load $= R_A = 200 - R_B$ while BC is subjected to a compressive load R_B $(= 200 - R_A)$.

From compatibility, extension (Δ_{AB}) of AB = contraction (Δ_{BC}) of BC.

$$\therefore \quad \frac{(200 - R_B) \times 1000 \times 100}{1000 \times 0.7 \times 10^5} = \frac{R_B \times 1000 \times 150}{1500 \times 2 \times 10^5}$$

$\therefore \quad R_B = 148.15$ kN

$\therefore$ force in AB, $\quad P_{AB} = 200 - R_B = 200 - 148.15$

$= 51.85$ kN (tension)

and force in BC, $\quad P_{BC} = R_B = 148.15$ kN (compression)

Case II: Effect of temperature rise:

$$t = 70 - 20 = 50°C$$

Since the temperature strain in restricted, both the bars will be subjected to equal temperature force P_t of compressive nature.

Free elongation of AB $= \Delta_{AB\,.\,t}$

$= 100 \times 24 \times 10^{-6} \times 50 = 0.12$ mm

Free elongation of BC $= \Delta_{BC\,.\,t} = 150 \times 12 \times 10^{-6} \times 50 = 0.09$ mm

$\therefore$ Total elongation,

$$\Delta_t = 0.21 + 0.09 = 0.21 \text{ mm}$$

Since this expansion is completely restricted, we have

$$\Delta_t = P_t \left[\frac{L_{AB}}{A_{AB} E_{AB}} + \frac{L_{BC}}{A_{BC} E_{BC}}\right]$$

$$0.21 = p_t \left[\frac{100}{1000 \times 0.7 \times 10^5} + \frac{150}{1500 \times 2 \times 10^5}\right]$$

$$= 0.1929 \times 10^{-5} P_t$$

From which $P_t = 1.08889 \times 10^5$ $N = 108.89$ kN (compression)

Case III

Final stresses

Final force in AB $= 108.89 - 51.85$

$= 57.04$ kN (compression)

and Final Force in BC $= 108.89 + 148.15$

$= 257.04$ kN (compression)

$\therefore$ Stress in AB, $\quad p_{AB} = \dfrac{57.04 \times 1000}{1000}$

$= 57.04$ N/mm^2 (compression)

Stress in BC, $\quad p_{BC} = \dfrac{257.04}{1500} \times 1000$

$= 171.36$ N/mm^2 (compression)

Problem 1.38: A flat bar of aluminium alloy, 30 mm wide and 8 mm thick is placed between two steel bars each 30 mm wide and 10 mm thick, to form a composite bar 30 mm × 28 mm, as shown in Fig. 1.38. The three bars are fastened together at their ends when the temperature is 12°C. Find the stress in each of the material when the temperature of the whole assembly is raised to 42°C. If, at the new temperature, a compressive load of 30 kN is applied to the composite bar, what are the final stresses in steel and aluminium? Take $E_s = 2 \times 10^5$ N/mm², $E_a = 0.7 \times 10^5$ N/mm²; $\alpha_s = 12 \times 10^{-6}$ per °C and $\alpha_a = 23 \times 10^{-6}$ per °C.

Solution:

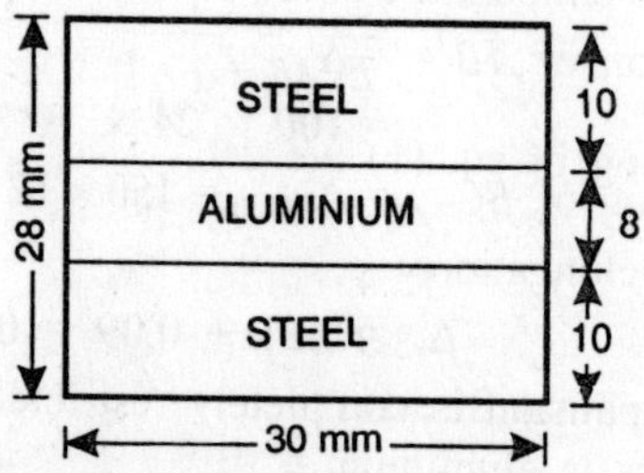

Fig. P. 1.38

Let A_s = Area of steel

A_a = Area of aluminium

$A_s = 20 \times 30 = 600$ mm²

and $A_a = 8 \times 30 = 240$ mm²

We shall consider both the effects separately.

Case I: Effect of temperature rise:

Due to temperature rise, steel will expand lesser than aluminium. Steel bar will be in tension and aluminium bar will be in compression.

In equilibrium,

$$p_s A_s = p_a A_a$$

or
$$p_s \times 600 = p_a \times 240$$

∴
$$p_a = 2.5\, p_s \qquad \ldots(1)$$

Also,
$$\Delta_s = \Delta_a$$

or
$$\alpha_s t l + \frac{p_s}{E_s} l = \alpha_a t l - \frac{p_a l}{E_a}$$

$$\therefore\ 12 \times 10^{-6} \times 30 + \frac{p_s}{2 \times 10^5} = 23 \times 10^{-6} \times 30 - \frac{p_a}{0.7 \times 10^5}$$

$$\text{or } 3.6 \times 10^{-4} + \frac{p_s}{2 \times 10^5} = 6.9 \times 10^{-4} - \frac{2.5\, p_s}{0.7 \times 10^5}$$

$$p_s = 42 \text{ N/mm}^2 \text{ (tension)}$$

and $$p_a = 105 \text{ N/mm}^2 \text{ (comp.)}$$

Case II: Effect of external load of 30 kN

$$f_s A_s + f_a A_a = P$$

$$\therefore f_s \times 600 + f_a \times 240 = 30000 \quad \text{...(i)}$$

We have $$\frac{f_s}{E_s} = \frac{f_a}{E_a}$$

$$\therefore \quad f_s = f_a \frac{E_s}{E_a} = \frac{2}{0.7} f_a = 2.857 f_a \quad \text{...(ii)}$$

Substituting in eq. (1) we get

$$600\,(2.854\, f_a) + 240\, f_a = 30000$$

$$\therefore \quad f_a = 15.4 \text{ N/mm}^2 \text{ (comp.)}$$

$$f_s = 43.9 \text{ N/mm}^2 \text{ (comp.)}$$

Case III: Final Stresses

Stress in aluminium = 105 + 15.4 = 120.4 N/mm² (comp.)

Stress in steel = – 42 + 43.9 = 1.9 N/mm² (tension)

Problem 1.39: A uniform bar of length L, cross-sectional area A, and unit mass ρ is suspended vertically from one end. Show that its total elongation is $\delta = \rho g L^2/2E$. If the total mass of the bar is M, show also that $\delta = MgL/2AE$.

Solution:

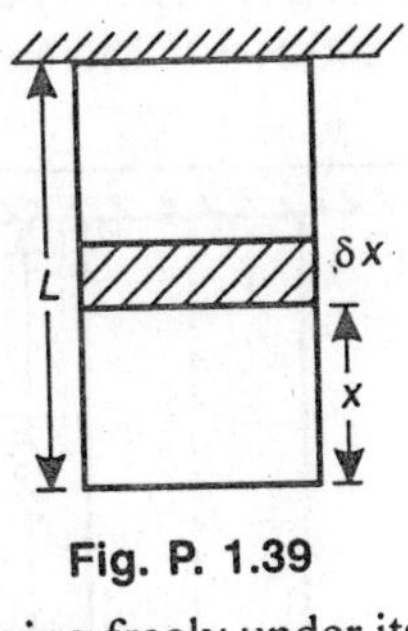

Fig. P. 1.39

Consider a bar AB hanging freely under its own weight. Considering a small section δx, its deformation $\delta\Delta$ will be given by

$$\delta\Delta = \frac{W_x \cdot \delta x}{AE}$$

Where W_x is weight of portion below the section = $Ax\lambda$.

Where λ is the specific weight or unit weight of the material i.e., $\lambda = \rho g$

$$\therefore \quad \delta\Delta = \frac{(Ax\lambda)\cdot\delta x}{AE} = \frac{x\,\lambda\,\delta x}{E}$$

$\therefore$ Total elongation of the rod

$$\Delta = \sum_{x=0}^{x=L} \delta\Delta = \int_0^L \frac{x\lambda}{E} \times dx$$

$$= \left[\frac{x^2\lambda}{2E}\right]_0^L = \frac{\lambda L^2}{2E}$$

$$\therefore \quad \Delta = \frac{\rho g L^2}{2E} \qquad [\because \lambda = \rho g]$$

If W = total weight of the bar, we have

$$\lambda = \frac{W}{AL}$$

$$\therefore \quad \Delta = \frac{WL}{2AE}$$

Problem 1.40: A steel rod having a cross-sectional area of 300 mm^2 and a length of 150 m is suspended vertically from one end. It supports a tensile load of 20 kN of the lower end. If the unit mass of steel is 7850 kg/m^3 and $E = 200 \times 10^3$ MN/m^2, find the total elongation of the rod.

Solution:

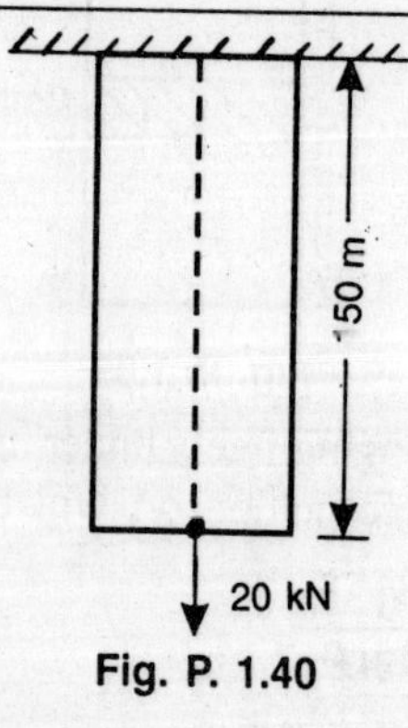

Fig. P. 1.40

$$
\begin{aligned}
A &= \text{cross-sectional area} = 300 \text{ mm}^2 \\
\text{(elongation) } \Delta &= ? \\
\text{Force} &= 20 \text{ kN} = 20 \times 10^3 \text{ N} \\
L &= 150 \times 1000 \text{ mm} \\
E &= 200 \times 10^3 \text{ N/mm}^2 \\
\rho &= 7850 \text{ kg/m}^3 \\
\lambda &= 7850 \times 9.8 \text{ N/m}^3 = 76930 \text{ N/m}^3 \\
&= 76930 \times 10^{-9} \text{ N/mm}^3
\end{aligned}
$$

Total elongation (Δ_T) = elongation due to self weight + elongation due to axial load

$$(\Delta_T) = \frac{\lambda L^2}{2E} + \frac{PL}{AE}$$

$$\therefore \quad \Delta_T = \frac{76930 \times 10^{-9} \times (150 \times 1000)^2}{200 \times 10^3} + \frac{20 \times 1000 \times 150 \times 1000}{300 \times 200 \times 10^3}$$

$$= \frac{76930 \times 10^{-9} \times 22500 \times 10^6}{200 \times 10^3} + \frac{2 \times 150}{3 \times 2}$$

$$= \frac{7693 \times 225}{200 \times 10^3} + \frac{150}{3} = 58.65 \text{ mm}$$

Problem 1.41: The rigid bar AB, attached to two vertical rods as shown in figure P 1.41 (a), is horizontal before the applied load P. Determine the vertical movement of P if its magnitude is 50 kN.

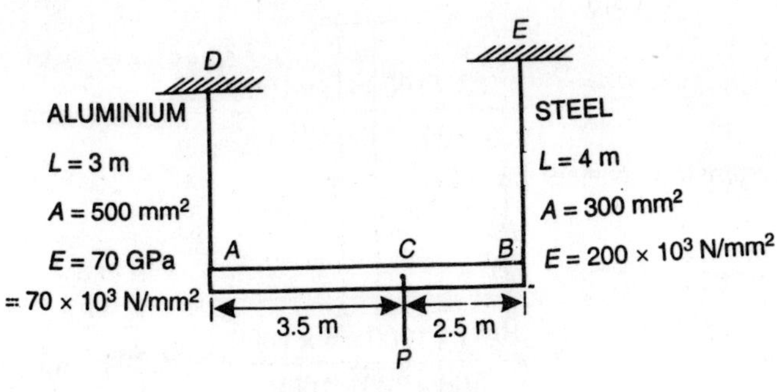

Fig. P. 1.41 (a)

Solution:

Since bar becomes horizontal before load is applied. Take a section (1) - (1),

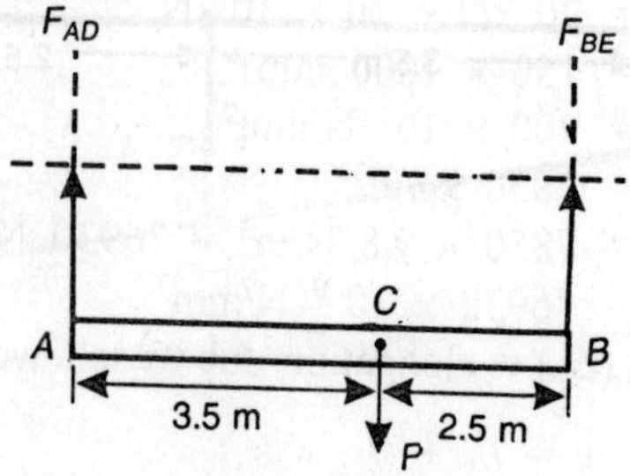

Fig. P. 1.41 (b)

$$\Sigma f_y = 0,$$

$$F_{AD} + F_{BE} = P \qquad ...(1)$$

$$\Sigma M_A = 0,$$

$$6F_{BE} = 3.5P \qquad ...(2)$$

$$\therefore \quad F_{BE} = \frac{3.5}{6} \times P$$

$$= \frac{3.5}{6} \times 50 \times 1000 = 583.34 \text{ N}$$

$$F_{AD} = 50 \times 1000 - F_{BE}$$

$$= 50000 - 583.34$$

$$= 49416.67 \text{ N} = 49.41 \text{ kN}$$

$\therefore$ Force in aluminium bar

$$= 0.583 \text{ kN}$$

$\therefore$ Extension in aluminium bar

$$(\Delta_a) = \frac{PL}{AE}$$

$$= \frac{0.583 \times 1000 \times 3 \times 1000}{500 \times 70 \times 10^3} = 0.0499 \text{ mm}$$

Extension in steel bar (Δ_S)

$$= \frac{PL}{AE}$$

$$= \frac{49.41 \times 1000 \times 4 \times 1000}{300 \times 200 \times 1000} = 3.294 \text{ mm}$$

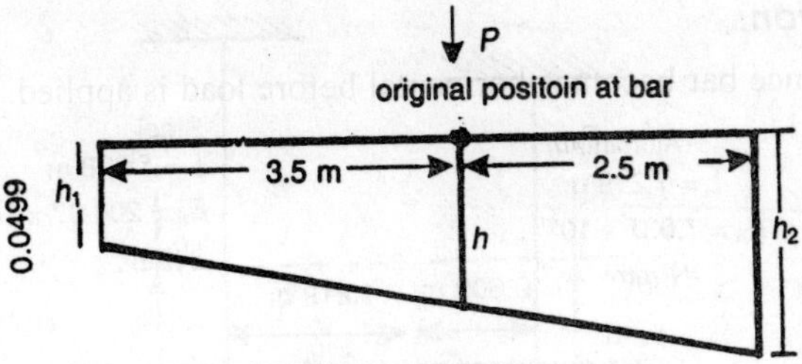

Fig. P. 1.41 (c)

$$h = h_1 + (h_2 - h_1)\ \frac{x}{L}$$

$$= 0.0499 + (3.294 - 0.0499)\ \frac{3.5}{6} = 1.942 \text{ mm}$$

∴ Vertical movement of

$$P = 1.942 \text{ mm}$$

Problem 1.42: A uniform concrete slab of total weight *W* is to be attached, as shown in Fig. to two rods whose lower ends are on the same level. Determine the ratio of the areas of the rods so that the slab will remain level.

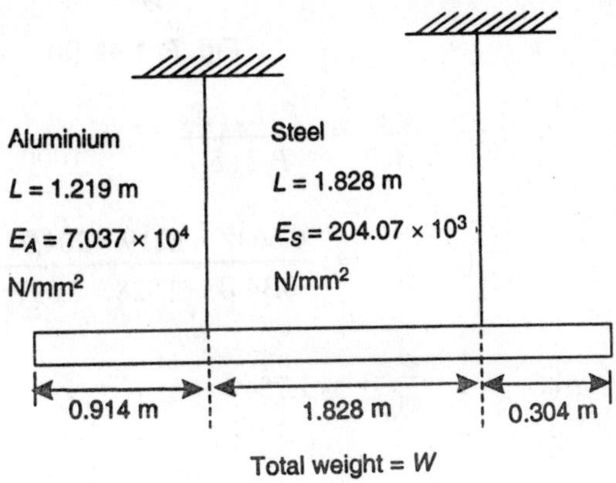

Fig. P. 1.42 (a)

Solution: From Equilibrium equation

$$\Sigma F_y = 0,$$

$$P_A + P_S = W \qquad \ldots(1)$$

$$\Sigma M_A = 0,$$

$$P_S \times 1.828 = W \times 0.609$$

$$P_S = \frac{0.609}{1.828}\ W = 0.34\ W \qquad \ldots(2)$$

$$P_A = 0.66\ W \qquad \ldots(3)$$

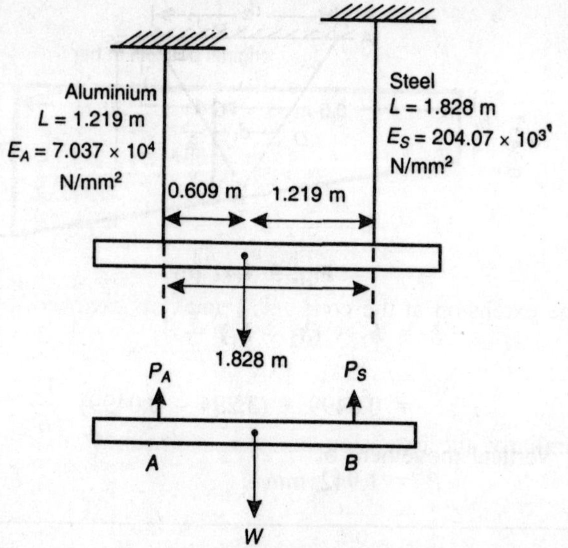

Fig. P. 1.42 (b)

$$\therefore \quad \frac{A_A}{A_S} = \frac{P_A L_A \cdot E_S}{P_S L_S E_A}$$

$$= \frac{0.66\, W \times 1.219 \times 204.07 \times 10^3}{0.34\, W \times 1.828 \times 7.037 \times 10^4} = 3.75$$

$$\therefore \quad \frac{A_A}{A_S} = 3.75$$

Problem 1.43: A round bar of length L, which tapers uniformly from Diameter d_2 at one end to a smaller diameter d_1 at the other is suspended vertically from the large end. If W is the weight per unit volume, find the elongation of the rod curved by its own weight.

Solution:

d_1 = diameter at free end of the truncated cone shaped bar
d_2 = diameter at fixed end

The length of the bar is L and it is hanging freely.

Prolong AD and BC to meet at O. Let l^1 be the length of cone DCO and l be the length of the full cone ABO.

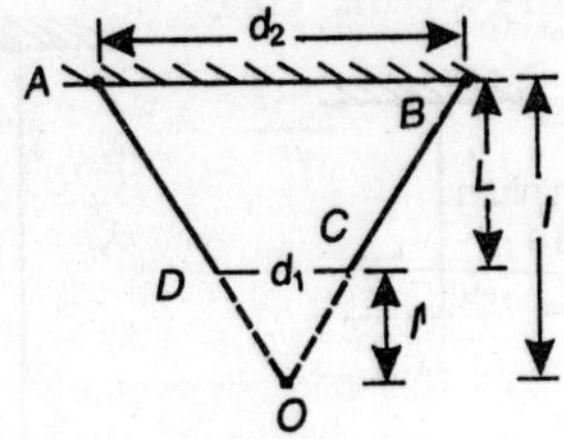

Fig. P. 1.43

The extension at the cone *ABO* under its own weight is given by

$$\Delta_C = \frac{\lambda l^2}{6\,E} \quad \text{...(1)}$$

Similarly, the extension of cone *DCO*, under its own weight is,

$$\Delta'_C = \frac{\lambda l'^2}{6\,E} \quad \text{...(2)}$$

The weight of cone *DCO*

$$= W' = \frac{1}{3}\left(\frac{\pi}{4}d_1^2\right) l'\lambda = \frac{\pi}{12}\, d_1^2\, l'\lambda$$

We know the extension of a tapering circular bar of length L, under an axial pull W' is given by

$$\Delta_{l'C} = \frac{4\,W'L}{\pi E d_1 d_2}$$

$$= \frac{4\,L}{\pi E d_1 d_2}\left[\frac{\pi}{12} d_1^2 l'\,\lambda\right]$$

$$\Delta_{l'C} = \frac{\lambda}{3}\,\frac{Ll'}{E}\cdot\frac{d_1}{d_2} \quad \text{...(1)}$$

Now extension of *ABCD* = Extension of *ABO* – extension of *DCO* – extension of *ABCD* under weight W'

or $\Delta_{te} = \Delta_C - \Delta'_C - \Delta'_{lC}$

$$\Delta_{te} = \frac{\lambda l^2}{6E} - \frac{\lambda l'^2}{6E} - \frac{\lambda}{3}\,\frac{Ll'}{E} \times \frac{d_1}{d_2} \quad \text{...(2)}$$

In the above expression l and l' are not directly known and hence these are to be eliminated.

From the geometry of the cone,

$$\frac{d_2}{d_1} = \frac{l}{l'} = \frac{L+l'}{l'} = \frac{L}{l'} + 1$$

Hence $\frac{L}{l'} = \frac{d_2 - d_1}{d_1}$,

from which $l' = \frac{d_1^2}{d_2 - d_1}$ ...(i)

and $l = \frac{d_2}{d_1} \times l'$

$$= \frac{d_2}{d_1} \times \frac{d_1 L}{d_2 - d_1} = \frac{d_2 L}{d_2 - d_1} \quad \text{...(ii)}$$

Substituting these values in (2), we get

$$\Delta_{te} = \frac{\lambda}{6E}\left(\frac{d_2 L}{d_2 - d_1}\right)^2 - \frac{\lambda}{6\,E} \times$$

$$\left(\frac{d_1 L}{d_2 - d_1}\right)^2 - \frac{\lambda}{3Ed_2 d_2 - d_1}$$

$$= \frac{\lambda L^2}{6E}\left[\frac{d_2^3 - d_1^2 d_2 - 2d_1^2(d_2 - d_1)}{d_2(d_2 - d_1)^2}\right]$$

or $$\Delta_{te} = \frac{\lambda L^2}{6E}\left[\frac{d_2^3 + 2d_1^3 - 3d_1^2 d_2}{d_2(d_2 - d_1)^2}\right]$$

Problem 1.44: A steel bar 50 mm in diameter and 2m long is surrounded by a shell of cast iron 5 mm thick. Compute the load that will compress the combined bar a total of 0.8 mm in the length of 2m. For steel E = 200 GPa, and for cast iron, E = 100 GPa.

Solution:

Total compression $(\Delta) = 0.8$ mm

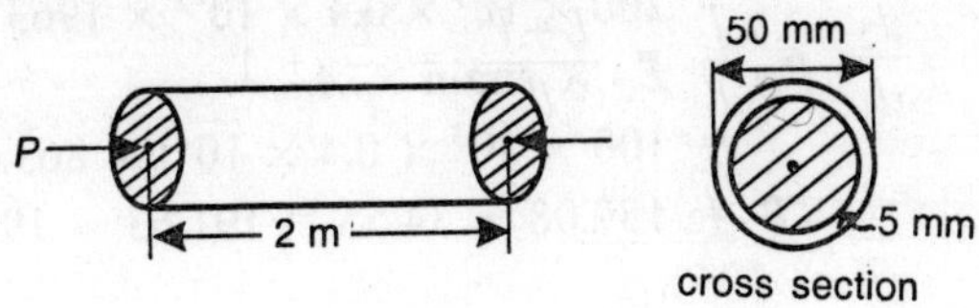

Fig. P. 1.44

$$E_S = 200 \times 10^3 \text{ N/mm}^2$$

$$E_C = 100 \times 10^3 \text{ N/mm}^2$$

$$A_S = \frac{\pi}{4} 50^2 = 1963.5 \text{ mm}^2$$

$$A_C = \frac{\pi}{4} (60)^2 - \frac{\pi}{4} 50^2 = 863.93 \text{ mm}^2$$

From condition equation i.e., $\Sigma F_x = 0$

$$P_S + P_C = P \quad ...(1)$$

From compatibilities equation

$$\Delta_1 \overset{?}{=} \Delta_2 \quad ...(2)$$

$$\therefore \quad \frac{P_S L}{A_S E_S} = \frac{P_C L}{A_C E_C}$$

$$\therefore \quad P_S = P_C \times \frac{A_S E_S}{A_C E_C}$$

$$= P_C \times \frac{1963.5 \times 200 \times 10^3}{863.93 \times 100 \times 10^3} = 4.53 \; P_C$$

$$P_S = 4.53 \; P_C$$

$\because$ Compress the combined bar 0.8 mm in the length of 2 m

$$\therefore \quad \text{Strain} = \frac{\text{Change in length}}{\text{Original length}}$$

$$= \frac{0.8}{2000} = 0.4 \times 10^{-3}$$

$$\text{Stress} = E \times \text{strain}$$

$$\frac{P_S}{A_S} = E_S \times \text{strain}$$

$$P_S = E_S \times \text{strain} \times A_S$$
$$= 200 \times 10^3 \times 0.4 \times 10^{-3} \times 1963.5 = 157.08 \text{ kN}$$
$$P_C = E_C \times \text{strain} \times A_C$$
$$= 100 \times 10^3 \times 0.4 \times 10^{-3} \times 863.93 = 34.55 \text{ kN}$$
$$\therefore \quad P = 157.08 + 34.55 = 191.63 \approx 192 \text{ kN}$$

Problem 1.45: A reinforced concrete column 200 mm in diameter is designed to carry an axial compressive load of 300 kN. Determine the required area of the reinforcing steel of the allowable stresses are 6 MPa and 120 MPa for the concrete and steel, respectively. Use E_{CO} = 14 GPa and E_{St} = 200 GPa.

Solution:

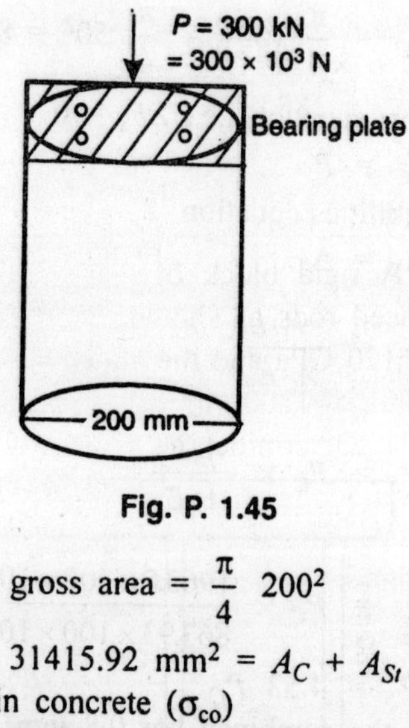

Fig. P. 1.45

$$A_g = \text{gross area} = \frac{\pi}{4}\,200^2$$
$$= 31415.92 \text{ mm}^2 = A_C + A_{St}$$

Allowable stress in concrete (σ_{co})

$$= 6 \text{ MPa} = 6 \text{ N/mm}^2$$
$$\sigma_{St} = 120 \text{ MPa} = 120 \text{ N/mm}^2$$
$$E_{CO} = 14 \text{ GPa} = 14 \times 10^3 \text{ N/mm}^2$$
$$E_{St} = 200 \text{ GPa}$$
$$= 200 \times 10^3 \text{ N/mm}^2$$

From condition equation,

$$P_{CO} + P_{St} = P = 300 \times 10^3 \text{ N} \qquad \ldots(1)$$

From compatibility equation,

$$\Delta_{St} = \Delta_{CO}$$

$$\therefore \quad \frac{P_{St} L}{A_{St} \cdot E_{St}} = \frac{P_{CO} L}{A_{CO} \cdot E_{CO}}$$

$$\therefore \quad \frac{\sigma_{St}}{E_{St}} = \frac{\sigma_{CO}}{E_{CO}}$$

$$\therefore \quad \sigma_{St} = \frac{E_{St}}{E_{CO}} \times \sigma_{CO} = \frac{200 \times 10^3}{14 \times 10^3} \times \sigma_{CO}$$

$$\sigma_{St} = 14.25\ \sigma_{CO} \qquad \ldots(2)$$

$\therefore$ If $\sigma_{St} = 6\ \text{MPa} = 6\ \text{N/mm}^2$

$$\therefore \quad \sigma_{St} = 14.28 \times 6 = 85.68\ \text{N/mm}^2$$

$\therefore$ From equation (1)

$$\sigma_{St} \times A_{St} + \sigma_{CO} \times A_{CO} = 300 \times 10^3$$

$$85.68\ A_{St} + 6 \times [31415.92 - A_{St}] = 300 \times 10^3$$

$$85.68\ A_{St} + 188495.52 - 6A_{St} = 300 \times 10^3$$

$$\therefore \quad A_{St} = 1399.40 \approx 1400\ \text{mm}^2$$

Problem 1.46: A rigid block of mass M is supported by three symmetrically spaced rods as shown. Each copper rod has an area of 900 mm^2; E = 120 GPa and the allowable stress is 70 MPa. The steel rod has an area of 1200 mm^2; E = 200 GPa; and the allowable stress is 140 MPa. Determine the largest mass on which can be supported.

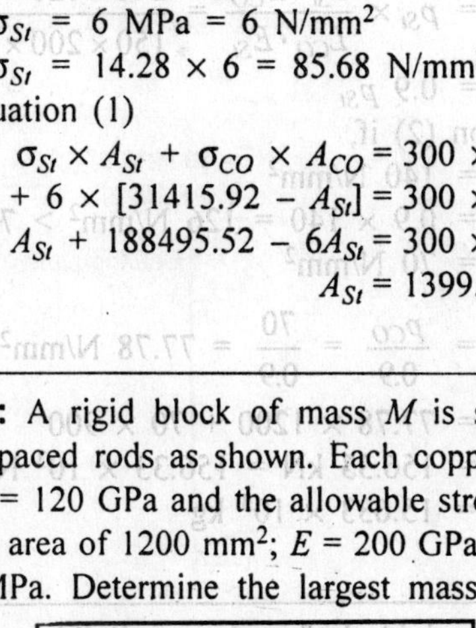

Fig. P. 1.46

Solution: Given :

$$A_{CO} = 900\ \text{mm}^2$$

$$E_{CO} = 120\ \text{MPa} = 120 \times 10^3\ \text{N/mm}^2$$

$$\sigma_{CO} = 70\ \text{MPa} = 70\ \text{N/mm}^2$$

$$L_{CO} = 160\ \text{mm}$$

$$L_{St} = 240\ \text{mm}$$

$$A_{St} = 1200\ \text{mm}^2$$

$$E_{St} = 200 \text{ MPa} = 200 \times 10^3 \text{ N/mm}^2$$
$$\sigma_{St} = 140 \text{ MPa} = 140 \text{ N/mm}^2$$

(1) From equation of equilibrium,

$$2\, p_{CO} \times A_{CO} + p_{St} \times A_{St} = M \quad \ldots(1)$$

(2) From equation of compatibility

$$\Delta_{CO} = \Delta_{St}$$

$$\frac{p_{CO} L_{CO}}{E_{CO}} = \frac{p_{St} \cdot L_{St}}{E_{St}}$$

$$p_{CO} = p_{St} \times \frac{L_{St} \cdot E_{CO}}{L_{CO} \cdot E_{St}} = \frac{240 \times 120 \times 10^3}{150 \times 200 \times 10^3}\, p_{St} = 0.9\, p_{St}$$

$$\therefore \quad p_{CO} = 0.9\, p_{St} \quad \ldots(2)$$

$\therefore$ From equation (2) if,

$$p_{St} = 140 \text{ N/mm}^2$$
$$\therefore \quad p_{CO} = 0.9 \times 140 = 126 \text{ N/mm}^2 > 70 \text{ N/mm}^2$$
$$\therefore \quad p_{CO} = 70 \text{ N/mm}^2$$
$$\therefore \quad p_{St} = \frac{p_{CO}}{0.9} = \frac{70}{0.9} = 77.78 \text{ N/mm}^2$$
$$\therefore \quad M = 77.78 \times 1200 + 70 \times 900$$
$$= 156.33 \text{ kN} = 156.33 \times 10^3 \text{ N}$$
$$= 15.633 \times 10^3 \text{ kg}$$

Problem 1.47: The rigid platform has negligible mass and rests on two steel bars, each = 50.0 mm long. The centre bar is aluminium and 249.90 mm long. Compute the stress in the aluminium bar after the centre load $P = 400$ kN has been applied. For each steel bar, the area is 1200 mm^2 and $E = 200$ MPa. For the aluminium bar, the area is 2400 mm^2 and $E = 70$ MPa.

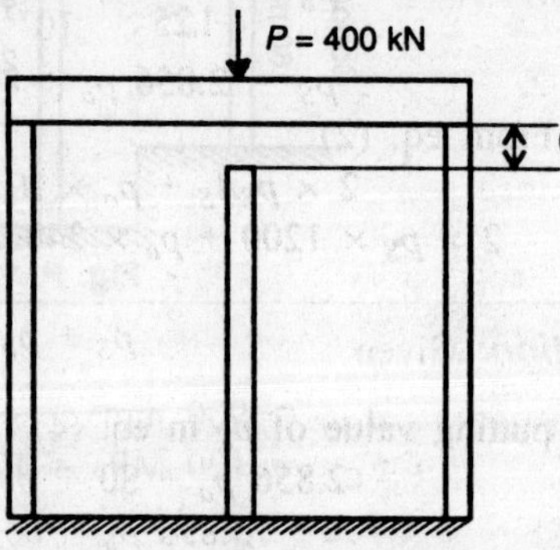

Fig. P. 1.47 (a)

Solution:

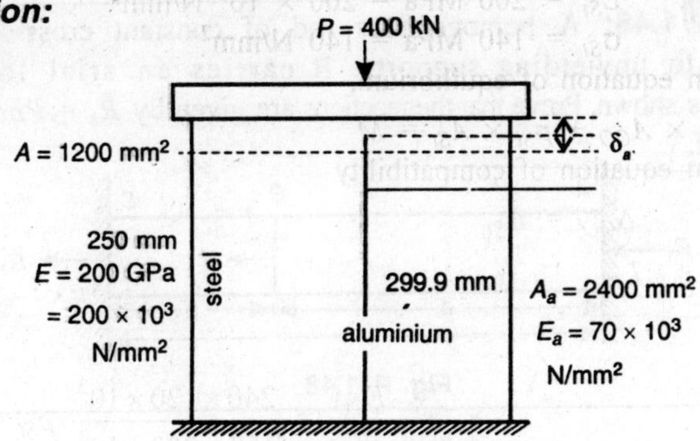

Fig. P. 1.47 (b)

$$A_a = 2400 \text{ mm}^2$$
$$E_a = 70 \times 10^3 \text{ N/mm}^2$$
$$\delta_S = 0.1 + \delta_a$$
$$\delta_S = \delta_a + 0.1 \quad \text{...(1)}$$

From equation of equilibrium,

$$= P_S + P_a = P = 400 \times 1000 \text{ N} \quad \text{...(2)}$$

From eq. (1)

$$\frac{p_s \cdot L_s}{E_s} = \frac{p_a \cdot L_a}{E_a} + 0.1$$

$$\frac{p_s \times 250}{200 \times 10^3} = \frac{p_C \times 249.9}{70 \times 10^3} + 0.1$$

$$1.25 \times 10^{-3}\ p_S = 3.57 \times 10^{-3}\ p_a + 0.1$$

$$\therefore \quad p_S = \frac{3.57}{1.25} \times \frac{10^{-3}}{10^{-3}}\ p_a + \frac{0.1}{1.25 \times 10^{-3}} = 2.856\ p_{at}\ 80$$

$$\therefore \quad p_S = 2.856\ p_a + 80 \quad \text{...(3)}$$

From eq. (2)

$$2 \times p_S A_S + p_a \times A_a = 400 \times 1000$$
$$2 \times p_S \times 1200 + p_a \times 2400 = 400 \times 1000$$

$$p_S + p_a = \frac{400 \times 1000}{2400} = 166.67 \quad \text{...(4)}$$

putting value of p_S in eq. (4) from eq. (3)

$$2.856\ p_a + 80 + p_a = 166.67$$
$$3.856\ p_a + 80 = 166.67$$
$$\therefore \quad p_a = 22.47 \text{ N/mm}^2 \approx 22.5 \text{ MPa}$$

Problem 1.48: A homogeneous rod of constant cross-section attached to unyielding supports. It carries an arial load P applied as shown. Prove that the reactions are given by $R_1 = Pb/L$ and $R_2 = Pa/L$

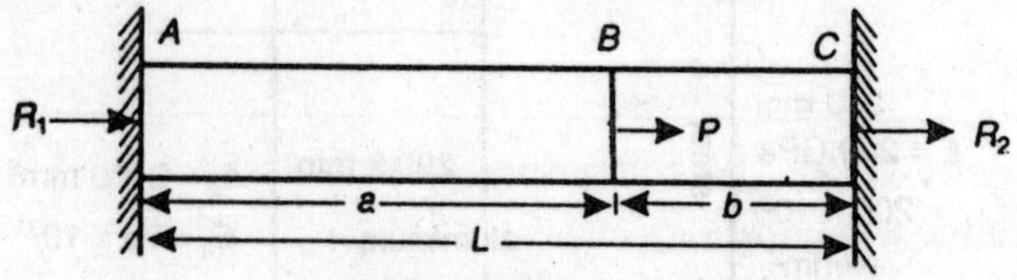

Fig. P. 1.48

Solution:

From equilibrium equation,

$$\therefore \quad \Sigma F_x = 0$$

$$R_1 + R_2 = P \qquad \ldots(1)$$

$\therefore$ Extension in AB = Compression in BC

$$\frac{R_1 a}{AE} = \frac{R_2 b}{AE}$$

$$R_1 = R_2 \frac{b}{a} \qquad \ldots(2)$$

$$\therefore R_2 \frac{b}{a} + R_2 = P$$

$$R_2 \left[\frac{b+a}{a}\right] = P$$

$$\therefore \quad R_2 = \frac{P_a}{a+b} = \frac{P_a}{L}$$

$$R_1' = P - \frac{P_a}{L} = P\left[\frac{L-a}{L}\right] = \frac{P_b}{L}$$

$$\frac{R_1 \times 0.6}{AE} = \frac{(25 - R_1) \times 1.2}{AE} + \frac{R_2 \times 0.9}{AE}$$

or $\quad 0.6\ R_1 = 30 - 1.2\ R_1 + 0.9\ R_2$

$\therefore \quad R_2 = 116.67$ kN

$R_1 = -41.66$ kN $= 41.66$ kN (compressive)

$\therefore \quad P_1 = R_1 = 41.66$ (compressive)

$$P_2 = 66.67 \text{ (compressive)}$$

$$P_3 = R_2 = 116.67 \text{ kN (compressive)}$$

$$\therefore \quad \text{Stress in } BC = \frac{66.67 \times 1000}{500} = 133.34 \text{ N/mm}^2$$

Problem 1.49: The composite rod in Fig. P 1.49 (a) is stress-free before the arial loads P_1 and P_2 are applied. Assuming that the walls are rigid, calculate the stress in each material if $P_1 = 150$ kN and $P_2 = 90$ kN.

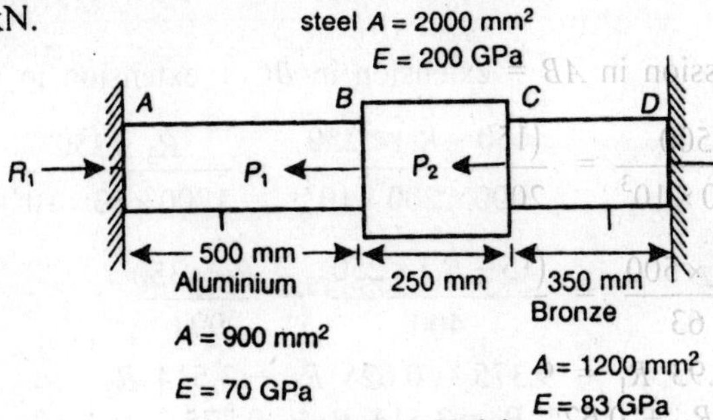

Fig. P. 1.49 (a)

Solution:

Fig. P. 1.49 (b)

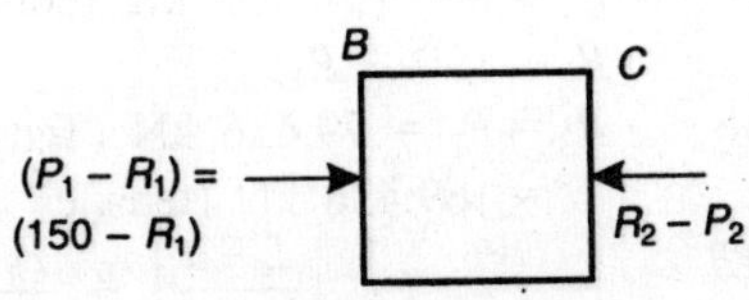

Fig. P. 1.49 (c)

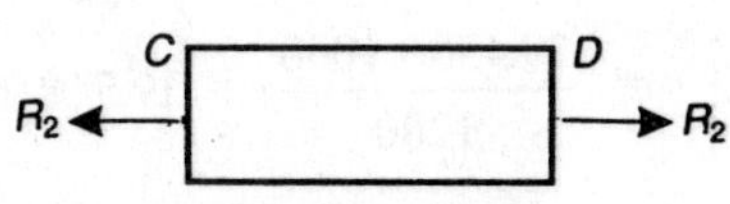

Fig. P. 1.49 (d)

aluminium $A = 900\ mm^2$
$E = 70$ MPa
bronze $A = 1200\ mm^2$
$E = 83$ MPa

$\therefore$ From Fig. P. 1.49 (b)

$$P_1 - R_1 = R_2 - P_2$$
$$R_1 + R_2 = P_1 + P_2$$
$$= 150 + 90$$
$$R_1 + R_2 = 240 \quad \ldots(1)$$

Compression in AB = extension in BC + extension in CD

$$\frac{R_1 \times 500}{900 \times 70 \times 10^3} = \frac{(150 - R_1) \times 250}{2000 \times 200 \times 10^3} + \frac{R_2 \times 350}{1200 \times 83 \times 10^3}$$

$$\frac{R_1 \times 500}{63} = \frac{(15 - R_1) \times 250}{400} + \frac{R_2 \times 350}{99.6}$$

$$7.93\ R_1 = 9.375 - 0.625\ R_1 + 3.514\ R_2$$
$$7.93\ R_1 + 0.625\ R_1 = 3.514\ R_2 + 9.375$$
$$8.555\ R_1 = 3.514\ R_2 + 9.375$$
$$\therefore \quad R_1 = 0.41\ R_2 + 1.095 \quad \ldots(2)$$

Putting this value in equation (1)

$$0.41\ R_2 + 1.095 + R_2 = 240$$
$$1.41\ R_2 = 238.905$$
$$\therefore \quad R_2 = 169.436 \text{ kN}$$
$$\therefore \quad R_1 = 70.563 \text{ kN}$$
$$\therefore \quad P_1 = R_1 = 70.563 \text{ kN (compressive)}$$
$$P_2 = 150 - R_1$$
$$P_3 = R_2 = 79.436 \text{ kN (Tensile)}$$
$$= 169.436 \text{ kN (tensile)}$$

$\therefore$ Stress in AB

$$= \frac{70.563 \times 1000}{900}$$
$$= 78.90 \text{ kN/mm}^2 \text{ (compressive)}$$

$$\text{Stress in } BC = \frac{79.436 \times 1000}{1200} = 39.718 \text{ N/mm}^2 \text{ (tensile)}$$

$$\text{Stress in } CD = \frac{169.436 \times 1000}{1200} = 141.196 \text{ N/mm}^2 \text{ (tensile)}$$

Problem 1.50: As shown in Fig. 1.50 (a) a rigid bar with negligible mass is pinned at O and attached to two vertical rods. Assuming that the rods were initially stress-free, what maximum load P can be applied without exceeding stresses of 150 MPa in the steel rod and 70 MPa in the bronze rod.

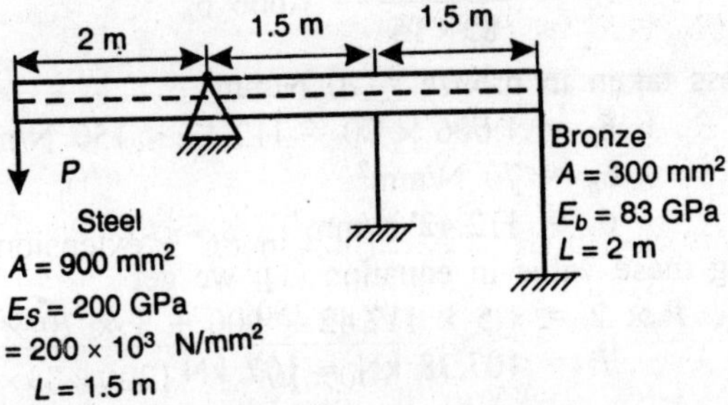

Fig. P. 1.50 (a)

Solution:

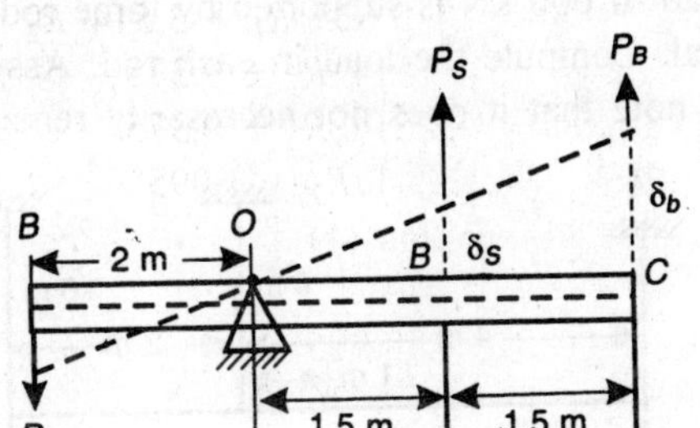

Fig. P. 1.50 (b)

From equilibrium equation,

Left side moment of force = Right side moment of the force

$$\therefore \quad P \times 2 = P_s \times 1.5 + P_B \times 3$$

$$= 6_s \times A_s \times 1.5 + 6_b \times A_b \times 3 \qquad \ldots(1)$$

One additional equation from compatibility,

$$\frac{\delta_s}{1.5} = \frac{\delta_b}{3}$$

$$2\delta_s = \delta_b$$

$$2 \times \frac{6_s \times L_s}{E_s} = \frac{6_b \times L_b}{E_b}$$

$$2 \times \frac{6_s \times 1.5}{200 \times 10^3} = \frac{\partial_b \times 2}{83 \times 10^3}$$

$$6_s = \frac{200 \times 6_b}{83 \times 1.5} = 1.606\ 6_b \qquad \ldots(2)$$

If stress taken in bronze = 70 N/mm²

$\therefore \quad 6_s = 1.606 \times 70 = 112.42 < 150$ N/mm²

$\therefore \quad 6_b = 70$ N/mm²

$6_s = 112.42$ N/mm²

Putting these value in equation (1), we get

$$P \times 2 = 1.5 \times 112.42 \times 900 + 3 \times 70 \times 300$$

$$\therefore \quad P = 107.38 \text{ kN} \approx 107 \text{ kN}$$

Problem 1.51: Shown in figure is a section through a balcony. The total uniform load of 600 kN is supported by three rods of the same area and material. Compute the load in each rod. Assume the floor to be rigid, but note that it does not necessarily remain horizontal.

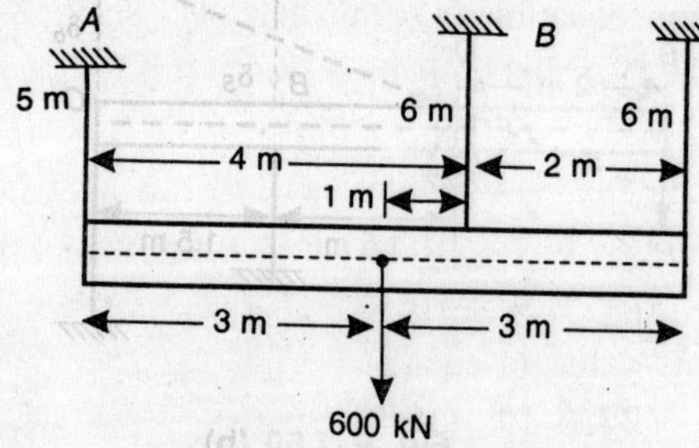

Fig. P. 1.51 (a)

Solution: Deflected position

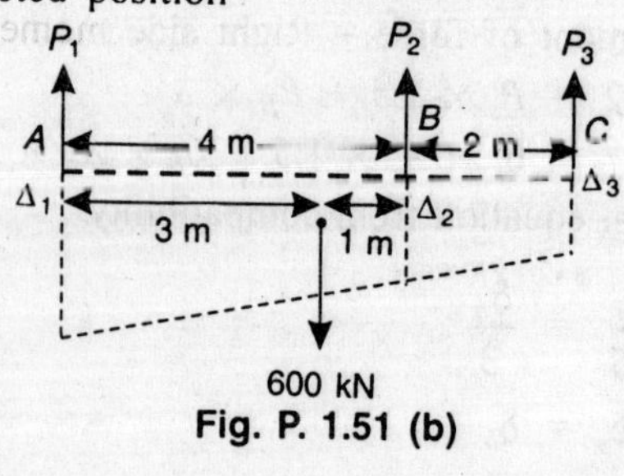

Fig. P. 1.51 (b)

From equilibrium equation,

$$P_1 + P_2 + P_3 = 600 \text{ kN} = 600 \times 1000 \text{ N} \quad \ldots(1)$$

$\therefore$ From deflection equation,

$$\therefore \quad \Delta_2 = \frac{\Delta_1 + \Delta_3}{2}$$

$$2\Delta_2 = \Delta_1 + \Delta_3 \quad \ldots(2)$$

$$2 \times \frac{6 \times 6}{E} = \frac{6_1 \times 5}{E} + \frac{6_3 \times 6}{E}$$

$$126_2 = 56_1 + 66_3$$

or $$12\,P_2 = 5P_1 + 6P_3 \quad \ldots(3)$$

Taking moment about C

$$6P_1 + 2P_2 = 600 \times 3 \times 1000 = 18 \times 10^5 \quad \ldots3(A)$$

Rearrange equation (3),

$$5P_1 + 6P_3 - 12\,P_2 = 0$$

$$6\left[\frac{5}{6}P_1 + P_3 - 2P_2\right] = 0$$

$$0.834\,P_1 + P_3 - 2P_2 = 0 \quad \ldots(4)$$

Substracting equation (4) from eq. (1)

$$P_1 + P_2 + P_3 - 0.834\,P_1 - P_3 + 2P_2 = 600 \times 1000$$

$$0.133\,P_1 + 3P_2 = 600 \times 1000 \quad \ldots(5)$$

Rearranging equation 3(A)

$$P_2 + 3P_1 = 9 \times 10^5$$

$$P_2 = 9 \times 10^5 - 3P_1 \quad \ldots(6)$$

Putting this value in eq. (5) we get

$$0.166\,P_1 + 27 \times 10^5 - 9P_1 = 600 \times 1000$$

$$-8.834\,P_1 = 600 \times 1000 - 27 \times 10^5 = -2100000$$

$$\therefore \quad P_1 = 237.825 \text{ kN}$$

$$P_2 = 9 \times 10^5 - 3 \times 237.825 \times 10^3$$

$$= 186.525 \text{ kN}$$

$$P_3 = 175.65 \text{ kN}$$

$$\therefore \quad P_1 = 237.825 \text{ kN}$$

$$P_2 = 186.525 \text{ kN}$$

$$P_3 = 175.65 \text{ kN}$$

Problem 1.52: Three rods, each of area 250 mm^2 jointly support a 7.5 kN load, as shown in figure. Assuming that there was no stack or stress in the rods before the load was applied, find the stresses in each rod. Use E_{st} = 200 MPa and E_{br} = 83 MPa

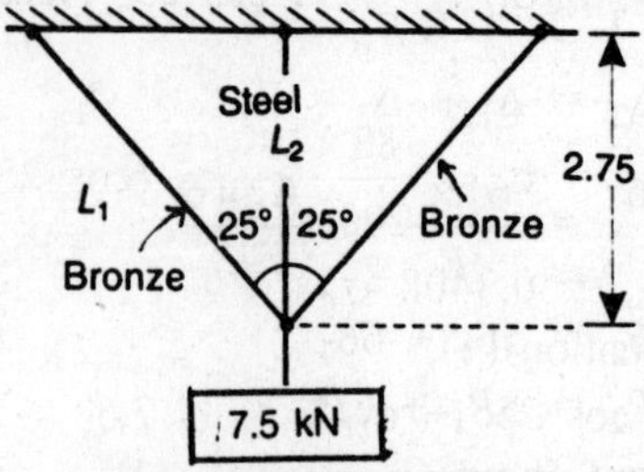

Fig. P. 1.52 (a)

Solution:

Area of each rod = 250 mm^2

Load = 7.5 kN = 7.5 × 1000 = 7500 N

F.B.D. of joint A

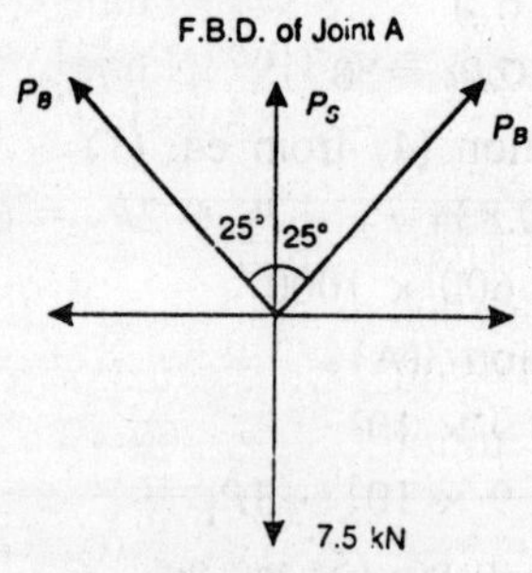

Fig. P. 1.52 (b)

$$\Sigma F_x = 0,$$

$$P_{br} \times \sin 25° = P_{br} \sin 50° = 0$$

$$\Sigma F_y = 0,$$

$$2P_{br} \cos 25° + P_{st} = 7.5 \qquad \ldots(1)$$

other equation from deflection,

$$\cos 0 = \frac{\Delta_1}{\Delta_2} \qquad \ldots(2)$$

$$\therefore \quad \Delta_1 = \Delta_2 \cos \theta$$

$$\frac{\sigma_{br} \cdot L_{br}}{E_{br}} = \frac{\sigma_{st} \cdot L_{st}}{E_{st}} \cos 25°$$

$$\sigma_{br} = \sigma_{st} \cdot \frac{E_{br}}{E_{st}} \cdot \frac{L_{st}}{L_{br}} \cdot \cos 25^\circ$$

$$= \sigma_{st} \cdot \frac{E_{br}}{E_{st}} \cdot \cos 25^\circ \cdot \cos 25^\circ$$

$$= \sigma_{st} \cdot \frac{83 \times 10^3}{200 \times 10^3} \cdot (\cos 25^\circ)^2$$

$$\sigma_{br} = 0.3408\ \sigma_{st} \qquad \ldots(3)$$

Recruiting equation (1),

$2\ \sigma_{br} \cdot A_{br} \cdot \cos 25^\circ + \sigma_{br} \cdot A_{st} = 7.5$

Putting value of σ_{br} from equation (3)

$2 \times 0.3408\ \sigma_{st} \times A_{br} \times \cos 25^\circ + \sigma_{st} \times A_{st} = 7.5 \times 1000$

$2 \times 0.3408\ \sigma_{st} \times 250 \times \cos 25^\circ + \sigma_{st} \times 250 = 7.5 \times 10^3$

or $2 \times 0.3408\ \sigma_{st} \times \cos 25^\circ + \sigma_{st} = 30$

$\therefore \quad \sigma_{st} = 18.544\ \text{N/mm}^2$

$\therefore \quad \sigma_{br} = 0.3408 \times 18.544 = 6.3197\ \text{N/mm}^2$

$\therefore$ Stress in steel $(\sigma_{st}) = 18.544\ \text{N/mm}^2$

Stress in bronze $(\sigma_{br}) = 6.3197\ \text{N/mm}^2$

Problem 1.53: A steel rod is stretched between two rigid walls and carries a tensile load of 5000 N at 20°C. If the allowable stress is not to exceed 130 MPa at – 20°C, what is the minimum diameter of the rod assume $\alpha = 11.7\ \mu\text{m/m}^\circ\text{C}$ $E = 200$ GPa.

Solution:

Tensile load = 5000 N at 20°C

$\Delta T = 20^\circ - [-20^\circ] = 40^\circ\text{C}$

Allowable stress $\ngtr$ 130 MPa = 130 N/mm²

Minimum diameter of rod (d) = ?

$\alpha = 11.7\ \mu\text{m/m}^\circ\text{C} = 11.7 \times 10^{-6}/^\circ\text{C}$

$E = 200\ \text{MPa} = 200 \times 10^3\ \text{N/mm}^2$

$\therefore$ Temperature fall and load is tensile

$\therefore$ Contraction due to fall of temp. (δ_{Ct}) = elongation due to tensile force (δ_P)

$$\therefore \quad \alpha L \Delta T = \frac{PL}{AE}$$

$$\therefore \quad \alpha \Delta T = \frac{P}{AE}$$

$$11.7 \times 10^{-6} \times 40 = \frac{5000}{A \times 200 \times 10^3}$$

$$\therefore \quad A = \frac{5000}{11.7 \times 10^{-6} \times 4 \times 200 \times 10^3}$$

$$= \frac{5000 \times 10 \times 1000}{11.7 \times 4 \times 200 \times 10^3} = \frac{5000 \times 10}{11.7 \times 4 \times 2}$$

$$\therefore \quad A = 534.18 \text{ mm}^2$$

$$\therefore \quad \frac{P}{A} = 93.60 \text{ N/mm}^2 < 130 \text{ N/mm}^2$$

$$\therefore \quad \frac{\pi}{4} d^2 = 534.18$$

$$\therefore \quad d = \sqrt{\frac{534.18 \times 4}{\pi}} = 26.079 \text{ mm}$$

Problem 1.54: Steel railroad rails 10 m long are laid with a clearance of 3 mm at a temperature of 15°C. At what temperature will the rails just touch? What stress would be induced in the rails at that temperature if there were no initial clearance. Assume $\alpha = 11.7$ µm/m°C and E = 200 GPa.

Solution:

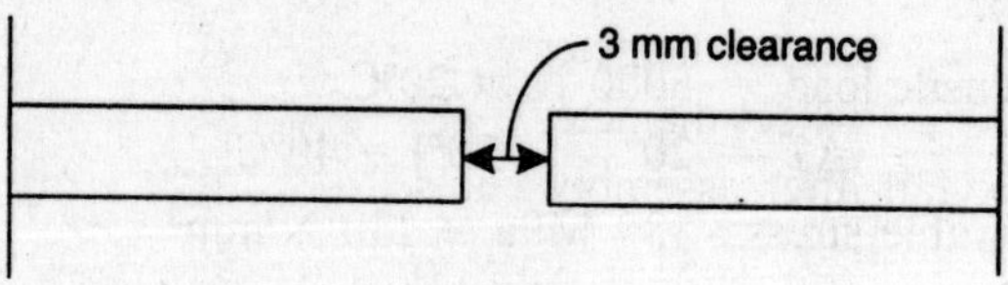

Fig. P. 1.54

Length of rail road = 10 m

Clearance = 3 mm of a temperature = 15°C (tt_1)

$\therefore \quad \Delta T = t_2 - t_1$

α = 11.7 µm/m°C = 11.7×10^{-6}/°C

E = 200 MPa = 200×10^3 N/mm²

Temperature at which rail just touch i.e. Stress = 0

$$\therefore \quad p = 0 = \frac{E}{L}(L\alpha\Delta T - 3)$$

$$\therefore \quad L\alpha\Delta T = 3$$

$$t_2 - 15 = \frac{3}{L \times \alpha} = \frac{3}{10 \times 1000 \times 11.7 \times 10^{-6}}$$

$$= \frac{3}{10^4 \times 11.7} \times 10^6 = \frac{300}{11.7} = 25.64$$

$$\therefore \quad t_2 = 25.64 + 15 = 40^\circ.64\ C \approx 41^\circ C$$

Stress induced in the rails at that temperature if there were no initial clearance

$$= \alpha + E = 11.7 \times \frac{10^{-6}}{10} \times 26 \times 200 \times 10^3 = 60.84\ \text{N/mm}^2$$

Problem 1.55: At temperature of 80°C, a steel tyre 12 mm thick and 90 mm wide that is to be shrunk onto a locomotive driving wheel 2 m, in diameter just fits over the wheel, which is at a temperature at 25°C. Determine the contact pressure between the tyre and the wheel after the assembly cools to 25°C. Neglect the deformation of the wheel caused by the pressure of the tyre. Assume $\alpha = 11.7\ \mu\text{m/m}^\circ\text{C}$, $E = 200$ GPa.

Solution: Given

tyre thickness at 80°C $t = 12$ mm

width = 90 mm

diameter = d

wheel diameter = D = 2m

Temp. fall after assembly cools to 25°C,

$$\alpha = 11.7 \times 10^{-6}/^\circ C$$

$$E = 200\ \text{GPa} = 200 \times 10^3\ \text{N/mm}^2$$

$$\Delta T = 80 - 25 = 55^\circ C$$

$\therefore$ Tensile strain imposed in the tone will be

$$= \frac{D - d}{d} = \alpha \times \Delta t$$

$$\therefore \quad \frac{2 - d}{d} = 11.7 \times 10^{-6} \times 55^\circ C = 643.5 \times 10^{-6}$$

$$2 - d = 643.5 \times 10^{-6}\, d$$

$$643.5 \times 10^{-6}\, d + d = 2$$

$$\therefore \quad d = 1.998 \text{ m}$$

$\therefore$ Contact pressure between the time and wheel

$$= \left(\frac{D-d}{d}\right) \times E$$

$$= \frac{2-1.998}{1.998} \times 200 \times 10^3$$

$$= 200.2 \text{ N/mm}^2$$

Problem 1.56: A bronze bar 3 m long with a cross-sectional area 320 mm^2 is placed between two rigid walls as shown in figure. At a temperature of – 20°C the gap Δ = 2.5 mm. Find the temperature at which the compressive stress in the bar will be 35 MPa. Use $\alpha = 18.0 \times 10^{-5}$ m/m°C) and E = 80 GPa.

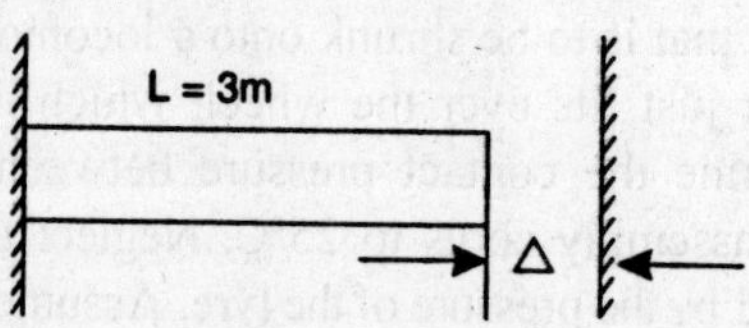

Fig. P. 1.56 (a)

Solution: Given

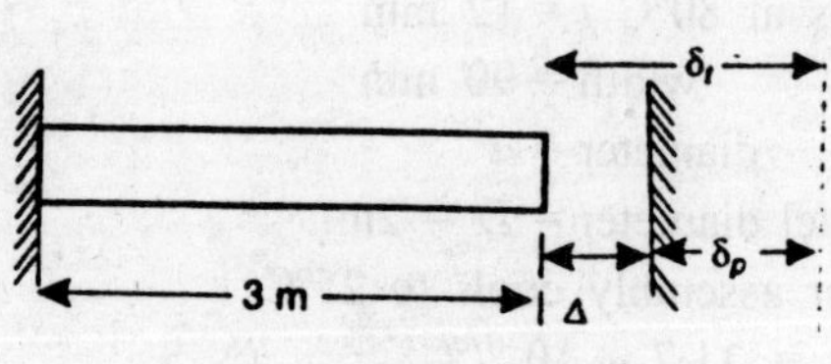

Fig. P. 1.56 (b)

Length = 3 m

Cross-sectional area = 320 mm^2

At a temperature – 15°C, the gap Δ = 2.5 mm

i.e. $t_1 = -15$°C,

$\Delta = 2.5$ mm

Find the temp. at which the compressive stress in the bar will be 35 MPa = 35 N/mm²,

$$\alpha = 18.0 \times 10^{-6}/^\circ C$$

$$E = 80 \text{ GPa} = 80 \times 10^3 \text{ N/mm}^2$$

∴ From question we obtain relation,

$$\delta_t = + \delta_P$$

$$\alpha L \Delta T = \Delta + \frac{PL}{AE} = \Delta + 6 \frac{L}{E}$$

$$18.0 \times 10^{-6} \times 3 \times 1000 \times (t_2 - t_1) = 2.5 + \frac{35 \times 3 \times 1000}{80 \times 10^3}$$

$$\therefore \quad t_2 - t_1 = 70.60 \approx 71$$

$$\therefore \quad t_2 = 71 + t_1 = 71 - 15 = 56^\circ C$$

Chapter 2

Shear and Moment in Beams

Problem 2.1: Write shear and moment equations for the beams in the following problems. In each problem, let x be the distance measured from the left end of the beam. Also, draw shear and moment diagrams. specifying values at all change of loading positions and at all points of zero shear. Neglect the mass of the beam in each problem. Beam loaded as shown in Fig. P. 2.1 (a).

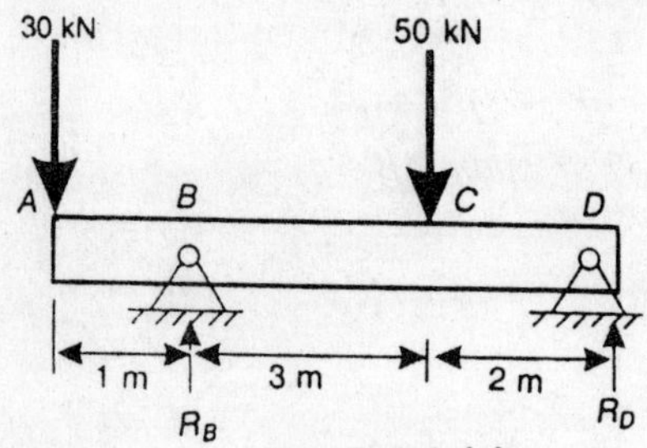

Fig. P. 2.1 (a)

Solution:

First reaction from equilibrium equation

$$\Sigma F_y = 0,$$

$$R_B + R_D = 30 + 50 = 80 \quad \ldots(i)$$

$$\Sigma M_B = 0,$$

$$R_D \times 5 - 50 \times 3 = -30 \times 1$$

$$R_D \times 5 = -30 + 150 = 120$$

$$\therefore \quad R_D = \frac{120}{5} = 24 \text{ kN}$$

$$\therefore \quad R_B = 80 - 24 = 56 \text{ kN}$$

$$\therefore \quad R_B = 56 \text{ kN}$$

$$R_D = 24 \text{ kN}$$

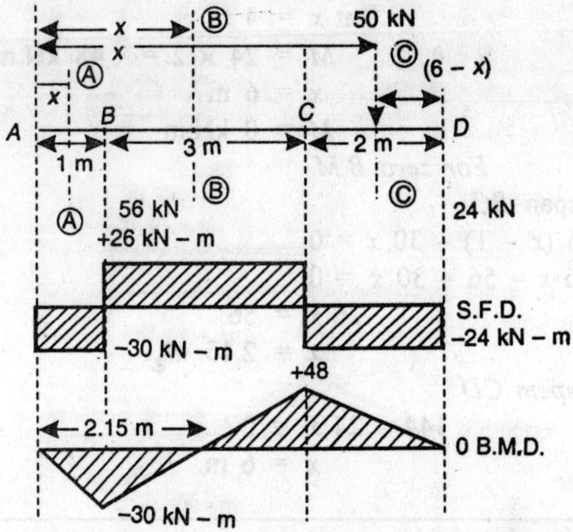

Fig. P. 2.1 (b)

For span AB

$$V_x = -30 \text{ kN}$$

$$M_x = -30\,x$$

$\therefore$ at $x = 1$ m

$$M = -30 \times 1$$

$$= 30 \text{ kN.m}$$

For span BC

$$V_x = +56 - 30 = +26 \text{ kN}$$

$$M_x = +56\,(x - 1) - 30\,x$$

$$= 56\,(x - 1) - 30\,x$$

x varies 1m to 4m

$\therefore$ At $x = 1$ m

$$M = -30 \text{ kN.m}$$

$$\text{At} = 4 \text{ m}$$

$$M = 56 \times 3 = -120$$

$$= +48 \text{ kN.m}$$

For span CD

$$V = -24 \text{ kN}$$

$$M_x = +24(6 - x)$$

x varies 4 m to 6 m

$\therefore$ at $x = 4$ m

$$M = 24 \times 2 = +48 \text{ kN.m}$$

at $x = 6$ m

$$M = 0 \text{ kN.m}$$

For zero B.M.

For span BC

$$56(x - 1) - 30x = 0$$

$$56x - 56 - 30x = 0$$

$$26x = 56$$

$\therefore$ $x = 2.15$ m

For spem CD

$$144 - 24x = 0$$

$$x = 6 \text{ m}$$

Problem 2.2: Beam loaded as shown in Fig. 2.2 (a). Find the B.M. and S.F.

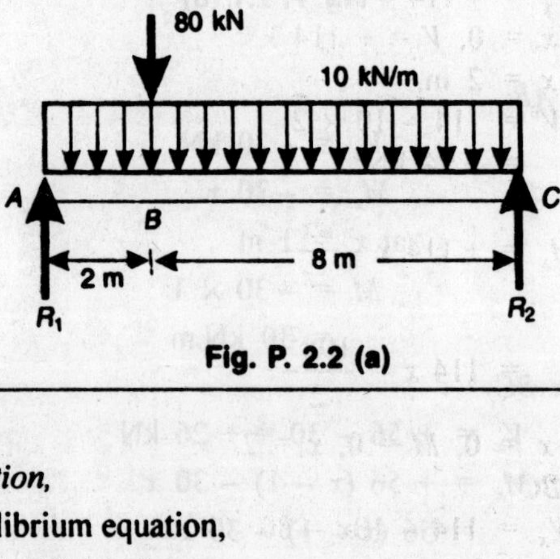

Fig. P. 2.2 (a)

Solution:

For relation,

Use equilibrium equation,

$$\Sigma F_y = 0$$

$$R_A + R_C = 80 + 10 \times 10 = 180 \quad \text{...(i)}$$

$$\Sigma M_A = 0$$

$$R_C \times 10 - 80 \times 2 - 10 \times 10 \times 5 = 0$$

$\therefore$ $$R_c = \frac{660}{10} = 66 \text{ kN}$$

$\therefore$ $$R_A = 180 - 66 = 114 \text{ kN}$$

For portion AB

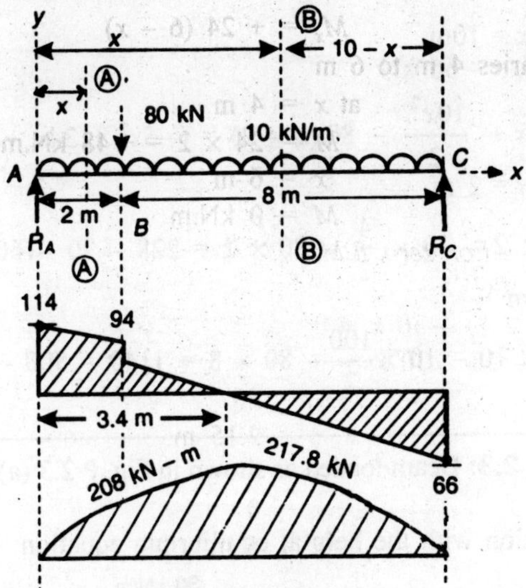

Fig. P. 2.2 (b)

$$V_x = +114 - wx$$

∴ at $x = 0$, $V = +114$ kN

at $x = 2$ m,

$$V = 114 - 10 \times 2$$
$$= +94 \text{ kN}$$

$$M_x = +114\,x - \frac{wx^2}{z}$$

$$= 114\,x - \frac{10x^2}{2}$$

$$x = 0,\ M = 0,\ x = 2,\ M = +208 \text{ kN.m}$$

For portion BC

$$V_x = 114 - 10x - 80$$

x varies 2 m to 10 m

∴ at $x = 2$ m

$$V_x = 114 - 20 - 80$$
$$= +14 \text{ kN}$$

at $x = 10$ m

$$V = -66 \text{ kN}$$

From other equation

$$V_x = -R_c + 10\,(10 - x) = -66 + 10\,(10 - x)$$

at $x = 10\text{m}$

$V_x = -66$ kN

$$M_x = 114\,x - \frac{10x^2}{2} - 80.(x-2) = (-5x^2 + 34x + 160)\ \text{kN.m}$$

$\therefore$ at $x = 2$ m

$$M = 114 \times 2 - 10 \times \frac{4}{2} - 80 \times 2 = 228 - 20 - 160 = +48\ \text{kN.m}$$

at $x = 10$ m

$$M = 114 \times 10 - 10 \times \frac{100}{2} - 80 \times 8 = 1140 - 500 - 640 = 0$$

Problem 2.3: Beam loaded as shown in Fig. P 2.3 (a). Find the B.M. and S.F.

Find reaction with the help at equilibrium equation

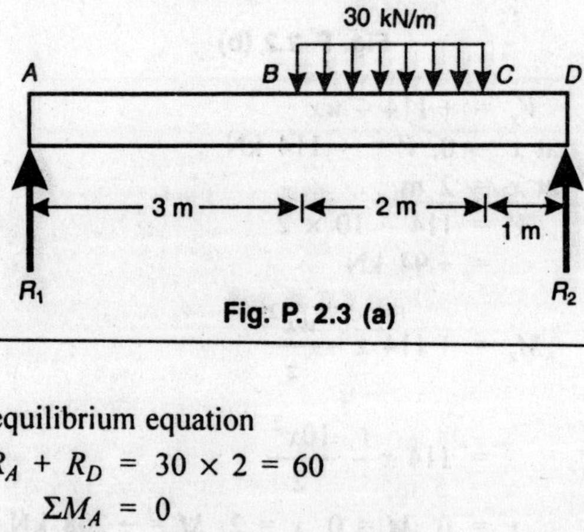

Fig. P. 2.3 (a)

Solution:

From equilibrium equation

$$R_A + R_D = 30 \times 2 = 60 \qquad \text{...(i)}$$

$$\Sigma M_A = 0$$

$$R_D \times 6 = 30 \times 2 \times 4 = 240$$

$$\therefore \quad R_D = 40\ \text{kN}$$

$$R_A = 20\ \text{kN}$$

For portion AB

$$V = +20\ \text{kN}$$

$$M_x = +20\,x$$

at $x = 0$, $M = 0$

at $x = 3$ m, $M = +60$ kN.m

For portion BC

$$V_x = +20 - 30\,x$$

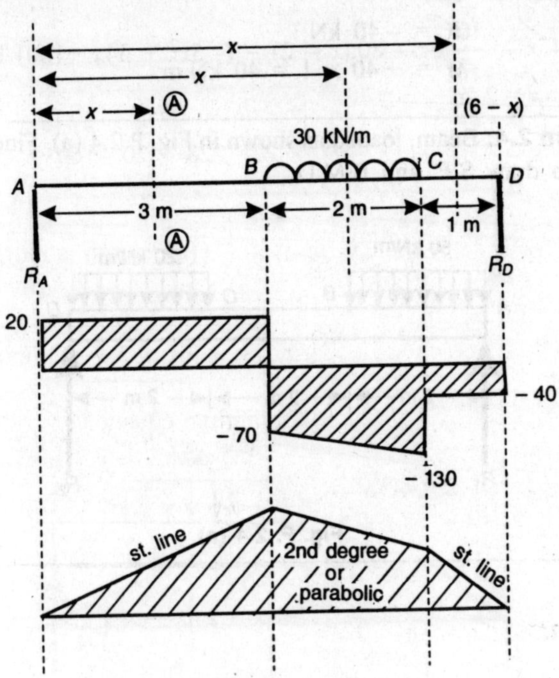

Fig. P. 2.3 (b)

at $\quad x = 3$ m, $V = -70$ kN

$x = 5$ m,

$V = -130$ kN .

$$M_x = +20 \times x - w \frac{(x-3)^2}{2}$$

$$= 20\,x - 30\frac{(x-3)^2}{2} = 20\,x - 15\,(x-3)^2$$

at $\quad x = 3$ m

$M = 60 - 15 \times 0 = 60$ kN.m

at $\quad x = 5$ m

$M = 100 - 15 \times 4 = 100 - 60 = 40$ kN.m

For portion CD,

$$V = -40 \text{ kN}$$

$$M = +40 \times 1 = 40 \text{ kN.m.}$$

Problem 2.4: Beam, loaded as shown in Fig. P 2.4 (a). Find reaction and also draw S.F. and B.M.D.

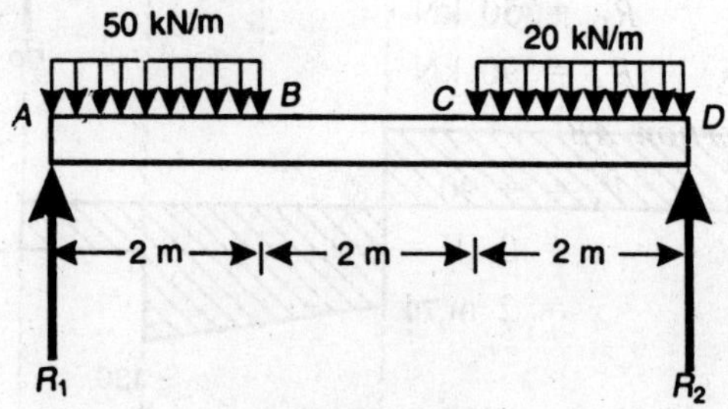

Fig. P. 2.4 (a)

Solution:

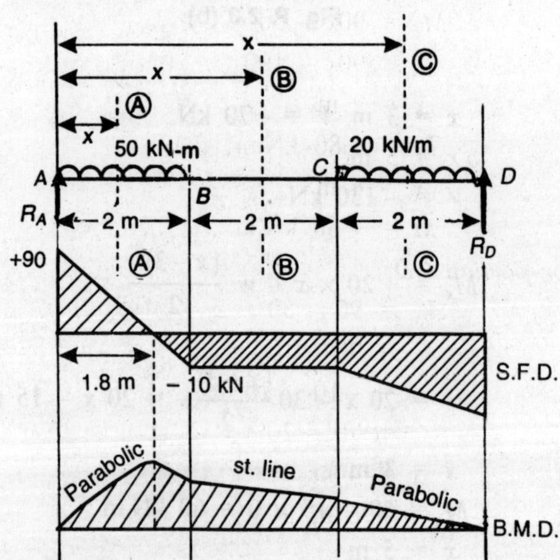

Fig. P. 2.4 (b)

First calculate reaction

$$R_A + R_D = 50 \times 2 + 20 \times 2 = 100 + 40 = 140 \quad \ldots\text{(i)}$$

$$\Sigma M_A = 0,$$

$$R_D \times 6 = 50 \times 2 + 1 + 20 \times 2 \times 5 = 100 + 200 = 300$$

$$\therefore \quad R_D = \frac{300}{6} = 50 \text{ kN}$$

$$\therefore \quad R_D = 50 \text{ kN}$$

$$R_A = 90 \text{ kN}$$

For portion AB

$$V_x = +90 - 50x$$

At $x = 0, V = +90$ kN

At $x = 2$ m, $V = -10$ kN

$$M_x = +90 \times x - \frac{50x^2}{2} = 90x - 25x^2$$

at $x = 0, M = 0$

at $x = 2, M = +80$ kN.m

For portion BC

$$V = 90 - 50 \times 2 = 90 - 100 = -10 \text{ kN}$$

$$M_x = 90x - 50 \times 2 \times (x - 1)$$

$$= 90x - 100(x - 1)$$

at $x = 2$ m

$$M = +80 \text{ kN.m}$$

at $x = 4$ m

$$M = +60 \text{ kN.m}$$

For portion CD

$$V_x = 90 - 50 \times 2 - 20(x - 4)$$

$$= 90 - 100 - 20(x - 4)$$

$$= -10 - 20(x - 4) = -10 - 20x + 80$$

$$= 70 - 20x$$

At $x = 4$ m, $V = -10$ kN

At $x = 6$ m, $V = -50$ kN

$$M_x = 90x - 50 \times 2(x - 1) - 20\frac{(x-4)^2}{2}$$

$$= 90\,x - 100\,(x - 1) - 10\,(x - 4)^2$$

at $x = 4$ m, $M = 90 \times 4 - 100 \times 3 - 10 \times 0$

$= 360 - 300 = +\,60$ kN.m

at $x = 6$ m, $M = 90 \times 6 - 100 \times 5 - 10 \times 2^2$

$= 540 - 500 - 40 = 0$ kN.m

Problem 2.5: Cantilever beam loaded as shown in Fig. P 2.5 (a). Find the reaction S.F.D. and B.M.D.

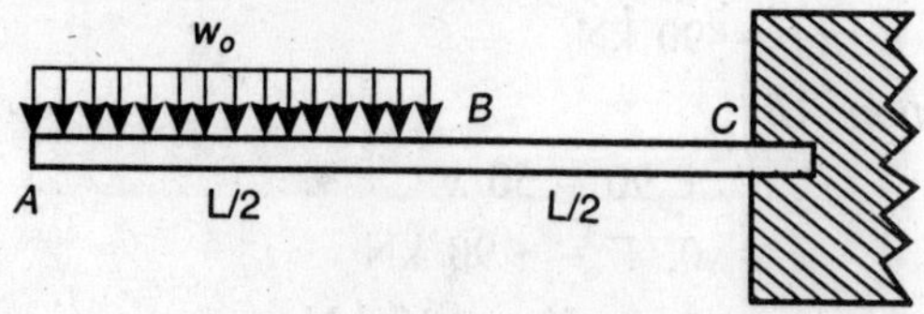

Fig. P. 2.5 (a)

Solution:

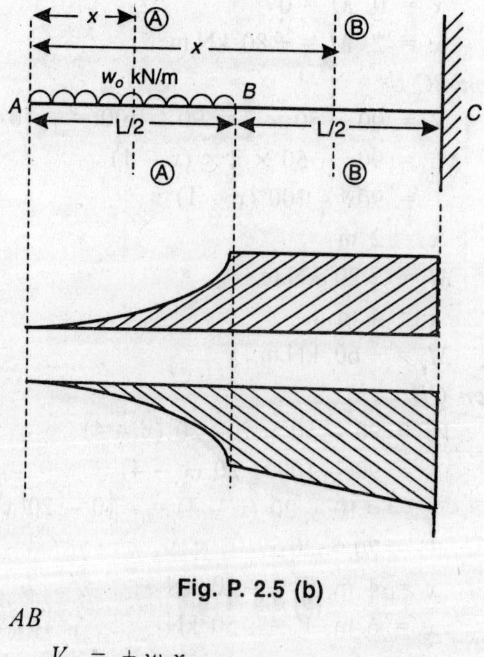

Fig. P. 2.5 (b)

Portion AB

$$V_x = +\,w_0 x$$

at $x = 0$, $V = 0$

at $$x = \frac{L}{2},\ V = -\frac{w_0 L}{8}$$

$$M_x = -\frac{w_0 x^2}{2}$$

at $$x = 0,\ M = 0$$

at $$x = \frac{L}{2},\ M = -\frac{w_0 L^2}{8}$$

For portion BC

$$V = +\frac{w_0 L}{2} \text{ (Constant)}$$

$$M_x = -w_0 \frac{L}{2} \times \left(x - \frac{L}{4}\right)$$

at $$x = \frac{L}{2},\ M = -w_0 \frac{L}{2} \times \frac{L}{4} = -\frac{w_0 L^2}{8}$$

at $$x = L$$

$$M = \frac{-3 w_0 L^2}{8}$$

Problem 2.6: Cantilever beam carrying the uniformly varying load shown in Fig. 2.6 (a). Find the S.F., B.M. and S.F.D. & B.M.D.

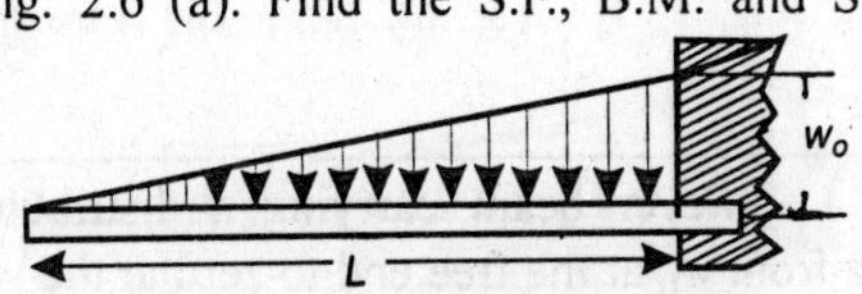

Fig. P. 2.6 (a)

Solution:

$$\frac{w_x}{x} = \frac{w_0}{L}$$

$$\therefore \quad w_x = \frac{w_0 x}{L}$$

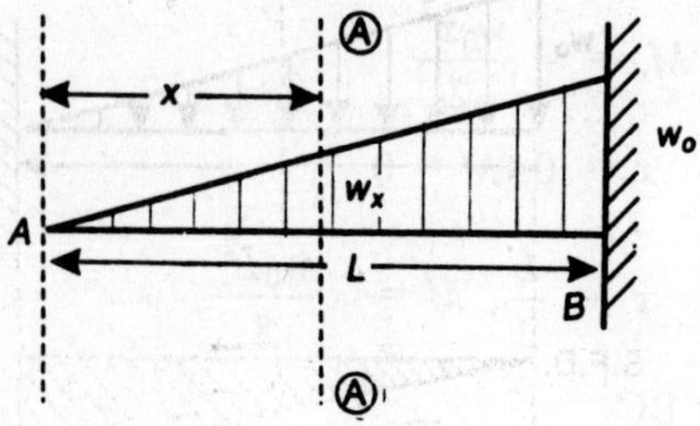

Fig. P. 2.6 (b)

$$V_x = -\frac{1}{2}w_x.x = -\frac{1}{2}\frac{w_0.x}{2} \times x = -\frac{w_0 x^2}{2L}$$

At $x = 0, V = 0$

At $x = L, V = -\frac{w_0 L^2}{2L} = -\frac{w_0 L}{2}$

$$M_x = -\frac{1}{2} \times w_x.x \times \frac{x}{3} = -\frac{1}{2} \times \frac{w_0.x}{L}.x \times \frac{x}{3} = -\frac{w_0 x^3}{6L}$$

$\therefore$ at $x = 0, m = 0,$

at $x = L, m = -\frac{w_0 L^3}{6}$

Problem 2.7: Cantilever beam carrying a distributed load with intensity varying from w_0 at the free end to zero at the wall, as shown in Fig. P. 2.7 (a). Find the B.M. and S.F. and draw the corresponding S.F.D. & B.M.D.

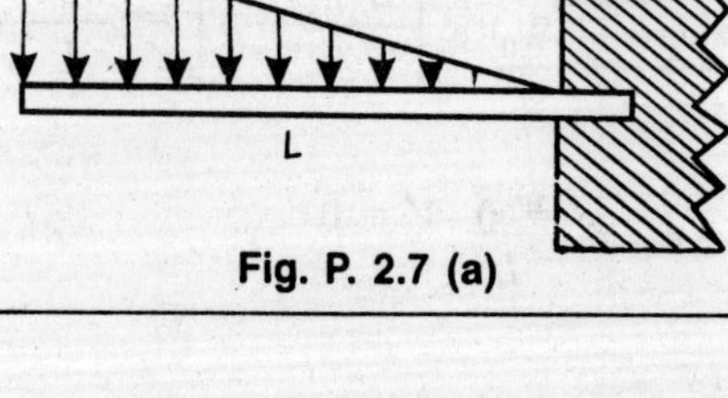

Fig. P. 2.7 (a)

Solution:

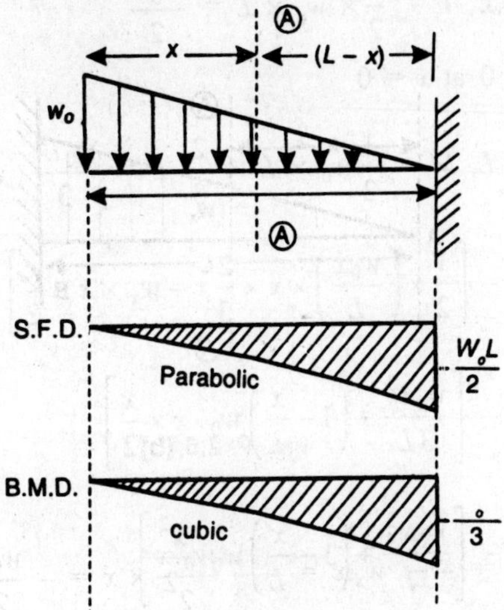

Fig. P. 2.7 (b)

$$\frac{w_0}{L} = \frac{w_x}{L-x}$$

$$\therefore \qquad w_x = \frac{(L-x)w_0}{L}$$

$$= \left(1-\frac{x}{2}\right)w_0 \qquad \ldots\text{(i)}$$

$$w_0 - w_x = w_0 - \left(1-\frac{x}{2}\right)w_0 = w_0\left[1-\left(1-\frac{x}{L}\right)\right]$$

$$= w_0\left[1-\frac{L-x}{L}\right] = w_0\left[\frac{L-L+x}{L}\right]$$

$$= w_0\left[\frac{L-L+x}{L}\right] = \frac{w_0 x}{L}$$

S.F.

at $\qquad x = 0.\ V = 0$

$$\text{at} \quad x = L,\ V = \frac{1}{2} \times w_0 \times L = \frac{w_0 L}{2}$$

$$M = 0 \text{ at } x = 0$$

$$\text{at} \quad x = L,\ M = \frac{1}{2} w_0 \times L \times \frac{2}{3} L = \frac{w_0 L^2}{3}$$

$$M_x = -\left[\frac{1}{2} \times \left(\frac{w_0 x}{L}\right) \times x \times \frac{2}{3} x + w_x \times x \times \frac{x}{2}\right]$$

$$= -\left[\frac{w_0 x^3}{3L} + \left(1 - \frac{x}{L}\right) w_0 \cdot x \times \frac{x}{2}\right]$$

$$= -\left[\frac{w_0 x^3}{3L} + \left(1 - \frac{x}{L}\right) w_0 \cdot \frac{x^2}{2}\right]$$

$$= -\left[\frac{w_0 x^3}{3L} + \frac{w_0 x^2}{2} - \frac{w_0 x^3}{2L}\right]$$

$$= -\left[\frac{w_0 x^3}{3L} - \frac{w_0 x^3}{2L} + \frac{w_0 x^2}{2}\right]$$

$$= -\left[\frac{2w_0 x^3 - 3w_0 x^3}{6L} + \frac{w_0 x^2}{2}\right] = -\left[-\frac{w_0 x^3}{6L} + \frac{w_0 x^2}{2}\right]$$

$$= \frac{w_0 x^3}{6L} - \frac{w_0 x^2}{2} \quad \text{proved.}$$

Problem 2.8: Cantilever beam loaded as shown in Fig. 2.8 (a). Find the S.F. and B.M. and S.F.D. and B.M.D.

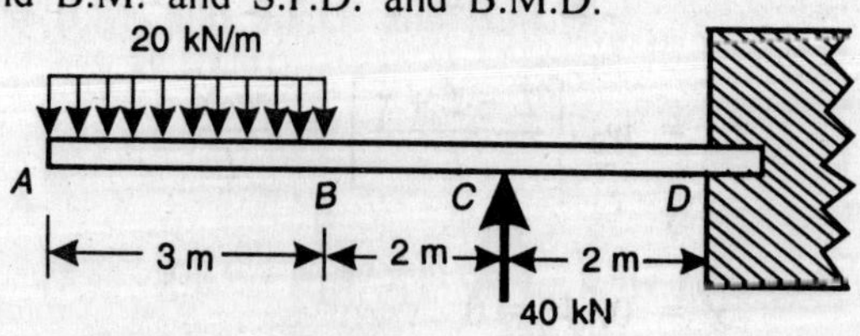

Fig. P. 2.8 (a)

Solution:

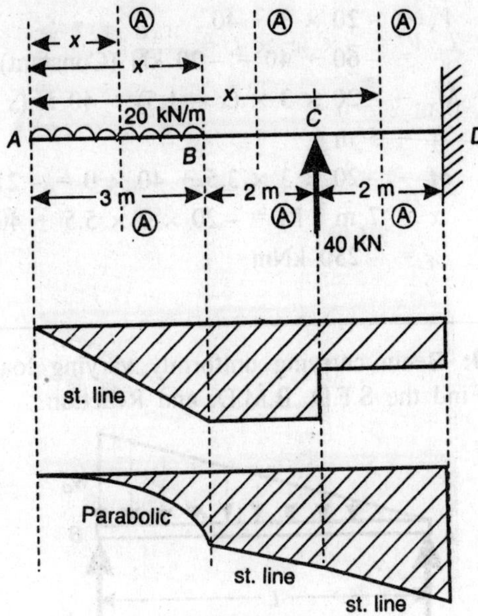

Fig. P. 2.8 (b)

For portion AB

$$V_x = -20\,x$$

at $x = 0,\ V = 0$

at $x = 3$ m, $V = -60$ kN

$$M_x = -20\frac{x^2}{2},\ \text{at } x = 0,\ M = 0$$

at $x = 3$ m,

$$M = -20 \times \frac{9}{2} = -20 \times 4.5 = -90 \text{ kN.m}$$

For portion BC

$$V = -20 \times 3$$

$$= -60 \text{ kN (Constant)}$$

$$M_x = -20 \times 3\ (x - 1.5)$$

at $x = 3,\ M = -20 \times 3 \times 1.5 = -90$ kN.m

at $x = 5$ m, $M = -210$ kN.m

For portion CD

$$V_x = -20 \times 3 + 40$$
$$= -60 + 40 = -20 \text{ kN (Constant)}$$
$$M_x = -20 \times 3 \times (x - 1.5) + 40 \times (x - 5)$$

at $x = 5$ m

$$M = -20 \times 3 \times 3.5 + 40 \times 0 = -210 \text{ kN.m}$$

at $x = 7$ m, $M = -20 \times 3 \times 5.5 + 40 \times 2$

$$= -250 \text{ kNm}$$

Problem 2.9: Beam carrying uniformly varying load shown in Fig. 2.9 (a). Find the S.F.D. B.M.D. and Reaction.

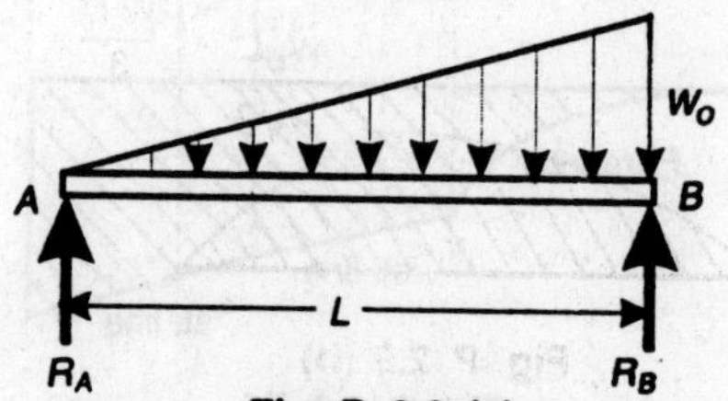

Fig. P. 2.9 (a)

Solution:

From equlibrium equation,

$$\Sigma F_y = 0$$

$$R_A + R_B = \frac{1}{2} w_0 \times L$$

or, $$R_A + R_B = \frac{w_0 L}{2}$$

$$\Sigma M_B = 0,$$

$$R_A \times L = \frac{1}{2} w_0 \times L \times \frac{L}{3} = \frac{w_0 L^2}{6}$$

$$\therefore \quad R_A = \frac{w_0 L^2}{6L} = \frac{w_0 L}{6}$$

$$\therefore \quad R_B = \frac{w_0 L}{2} - \frac{w_0 L}{6} = \frac{3w_0 L - w_0 L}{6}$$

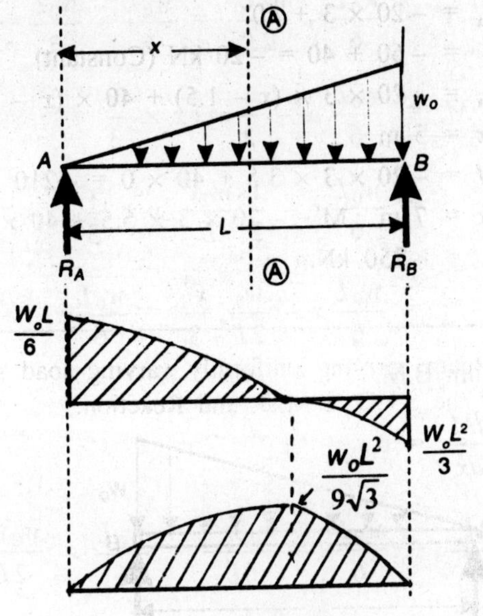

Fig. P. 2.9 (b)

$$= \frac{2w_0 L}{6} = \frac{w_0 L}{3}$$

$$\therefore \quad R_A = \frac{w_0 L}{6}, \quad R_B = \frac{w_0 L}{3}$$

$$\frac{w_x}{x} = \frac{w_0}{L}$$

$$\therefore \quad w_x = \frac{w_0 x}{L} \qquad \text{...(i)}$$

$$V_x = \frac{w_0 L}{6} - \frac{1}{2} w_x \,.\, x = \frac{w_0 L}{6} - \frac{1}{2} \times \frac{w_0 x}{L} \times x$$

$$= \frac{w_0 L}{6} - \frac{1}{2L} w_0 x^2 \qquad \text{...(2)}$$

at $x = 0,\ V = \dfrac{w_0 L}{6}$

$\therefore \quad V$ at $x = L$

$$V = \frac{w_0 L}{6} - \frac{w_0 L^2}{2L} = \frac{w_0 L}{6} - \frac{w_0 L}{2} = -\frac{w_0 L}{3} \quad ...(3)$$

$$M_x = \frac{w_0 L}{6} .x - \frac{1}{2} w_x . x \times \frac{x}{3}$$

$$= \frac{w_0 L}{6} x - \frac{1}{2} \frac{w_0 x}{L} . x \times \frac{x}{3}$$

$$= \frac{w_0 L}{6} x - \frac{w_0}{2L} . \frac{x^3}{3} = \frac{w_0 L}{6} x - \frac{w_0 x^3}{6L}$$

For maximum B.M.

$$\frac{dM}{dx} = 0$$

$$\frac{dM_x}{dx} = \frac{w_0 L}{6} - \frac{w_0 3x^2}{6L} = \frac{w_0 L}{6} - \frac{w_0 x^2}{2L} = 0$$

$$\frac{w_0 x^2}{2L} = \frac{w_0 L}{6}$$

$$x^2 = \frac{2L^2}{6} = \frac{L^2}{3}$$

$$\therefore \quad x = \sqrt{\frac{L^2}{3}} = \frac{L}{\sqrt{3}} = 0.577\ L$$

$$\therefore \quad M \text{ at } x = 0.577\ L$$

$$= \frac{w_0 L}{6} \times 0.577\ L - \frac{w_0}{6L} \times (0.577L)^3$$

$$= \frac{w_0 L}{6} \times 0.577\ L - \frac{w_0 L^3}{6L} \times (0.577)^3$$

$$= \frac{w_0 L^2}{6} \times 0.577 - \frac{w_0 L^2}{6} \times (0.577)^3$$

$$= \frac{w_0 L^2}{6} [0.577 - 0.192] = \frac{w_0 L^2}{6} \times 0.384$$

$$= \frac{w_0 L^2}{9\sqrt{3}} \quad \text{Proved.}$$

Problem 2.10: Cantilever beam loaded as shown in Fig. 2.10 (a). Find the S.F.D., B.M.D. and reactions.

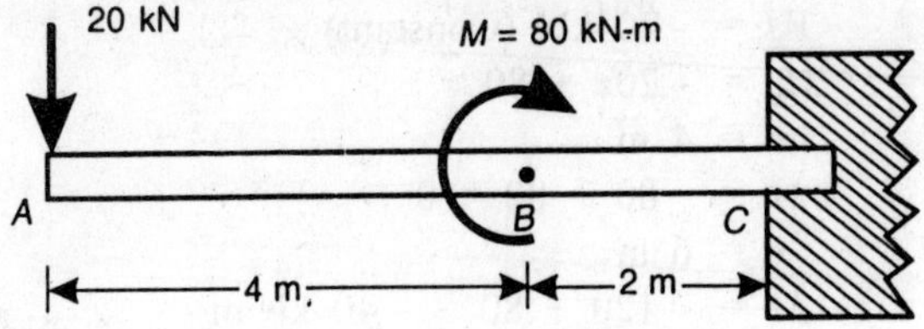

Fig. P. 2.10 (a)

Solution:

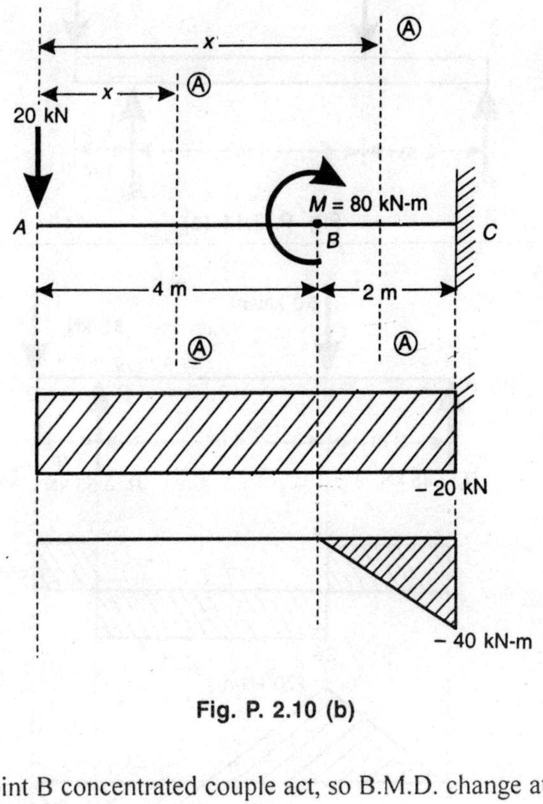

Fig. P. 2.10 (b)

At point B concentrated couple act, so B.M.D. change at that point.

For portion AB

$$V_x = -20 \text{ kN (Constant)}$$

$$M_x = -20\,x, \text{ at } x = 0, M = 0$$

at $x = 4$ m, $M = -20 \times 4 = -80$ kN.m

For portion BC

$$V_x = -20 \text{ kN (Constant)}$$

$$M_x = -20x + 80$$

at $x = 4$ m

$$M = -80 + 80 = 0$$

at $x = 6$ m

$$M = -120 + 80 = -40 \text{ kN.m}$$

Problem 2.11: Beam loaded as shown in Fig. 2.11 (a). Find the S.F.D., B.M.D. and reaction.

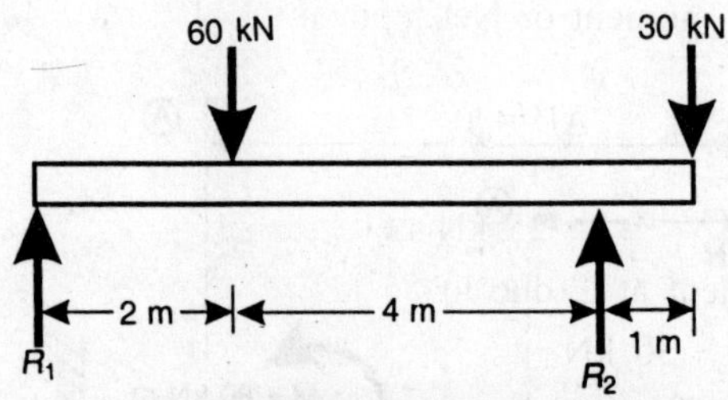

Fig. P. 2.11 (a)

Solution:

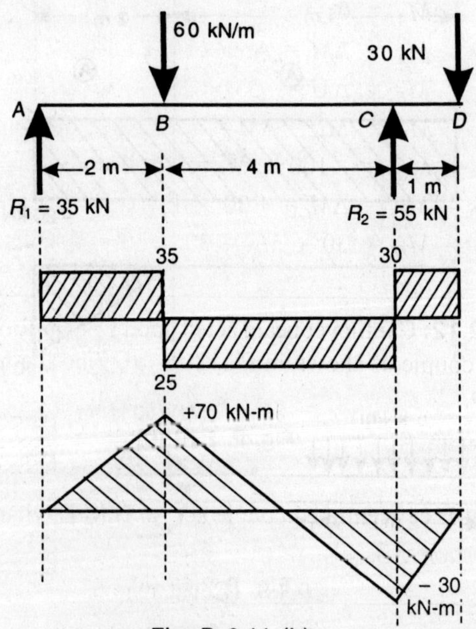

Fig. P. 2.11 (b)

From equalibrium equation,

$$\Sigma F_y = 0$$

$$R_1 + R_2 = 60 + 30 = 90 \quad \text{...(i)}$$

$$\Sigma M_A = 0,$$

$$R_2 \times 6 = 60 \times 2 + 30 \times 7 = 120 + 210 = 330$$

$$\therefore \quad R_2 = 55 \text{ kN}$$

$$\therefore \quad R_1 = 35 \text{ kN}$$

For S.F.D

We know that,

$$\Delta V = V_2 - V_1 = \text{area of load diagram}$$

$$\therefore \quad V_A = +35 \text{ kN}$$

$V_B = V_A = +35$ kN Left moment at point B due to 60 kN (↓)Right moment or Net moment at at B as $+35 - 60 = -25$ kN,

$$V_C - V_B = \Delta V = 0$$

$$\therefore \quad V_c = +V_B$$

$$= -25 \text{ kN Left moment at } C$$

Right moment at C due to 55 kN reaction,

$55 - 25 = +30$ kN

**Forcheck*: Total shear force diagram must be zero because difference in B.M. at both end is zero this case.

For B.M.D $\quad M_A = 0$

$$M_B - M_A = \Delta M = \text{Area of S.F.D}$$

$$M_B = \Delta M = 35 \times 2 = +70 \text{ kN.m}$$

$$M_C = M_B = \Delta M = -25 \times 4 = -100$$

$$\therefore \quad M_C = -100 + M_B = -100 + 70 = -30 \text{ kN.m}$$

$$M_D - M_C = \Delta M = +30 \times 1 = +30 \text{ kN.m}$$

$$\therefore \quad M_D = 30 + M_C = 30 - 30 = 0 \text{ kN.m}$$

Problem 2.12: Cantilever beam acted upon by a uniformly distributed load and a couple as shown in Fig. 2.12 (a). Find the reaction, S.F.D. and B.M.D.

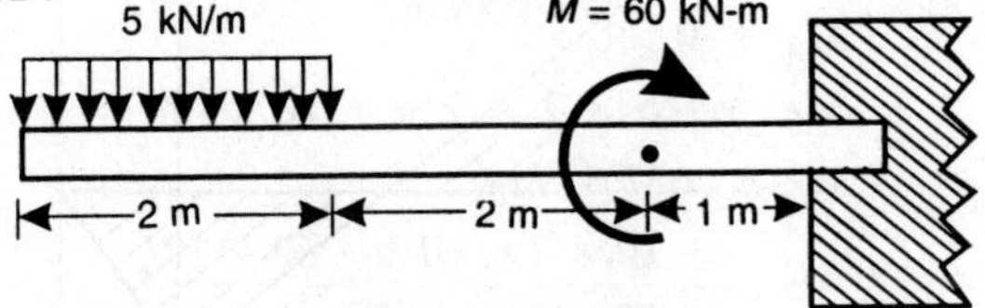

Fig. P. 2.12 (a)

Solution:

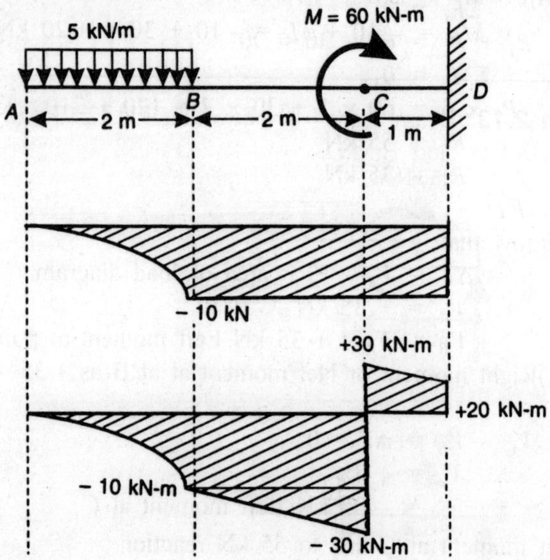

Fig. P. 2.12 (b)

For S.F.D.

We know that

$$V_2 - V_1 = \Delta V = \text{area of load diagram.}$$

$$V_B - V_A = \Delta V = -5 \times 2 = -10 \text{ kN}$$

$$\therefore \quad V_B = -10 \text{ kN}$$

$$V_C - V_B = 0$$

$$\therefore \quad V_C = V_B = -10 \text{ kN}$$

$$V_0 - V_C = 0$$

$$\therefore \quad V_D = +V_C = -10 \text{ kN}$$

For B.M.D

We know that,

$$M_2 - M_1 = \Delta M = \text{area of S.F.D}$$

$$M_A = 0$$

$$M_B = -5 \times 2 \times 1 = -10 \text{ kN.m}$$

$$M_C = \Delta M + M_B$$

$$= -10 \times 2 - 10 = -20 - 10$$

$$= -30 \text{ kN.m Left moment of C}$$

Right moment at C = + 60 − 30 = + 30 kN.m

$$M_D - M_C = \Delta M = -10 \times 1 = -10 \text{ kN.m}$$

$$\therefore \quad M_D = -10 + M_C = -10 + 30 = +20 \text{ kN.m}$$

Problem 2.13: Beam loaded as shown in Fig. 2.13 (a). Find the reactions, S.F.D. and B.M.D.

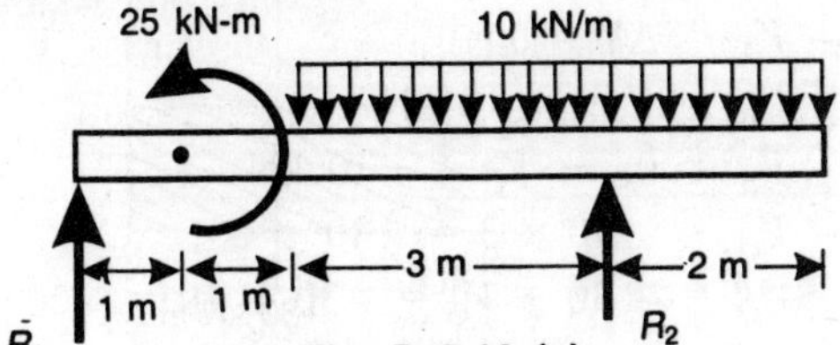

Fig. P. 2.13 (a)

Solution:

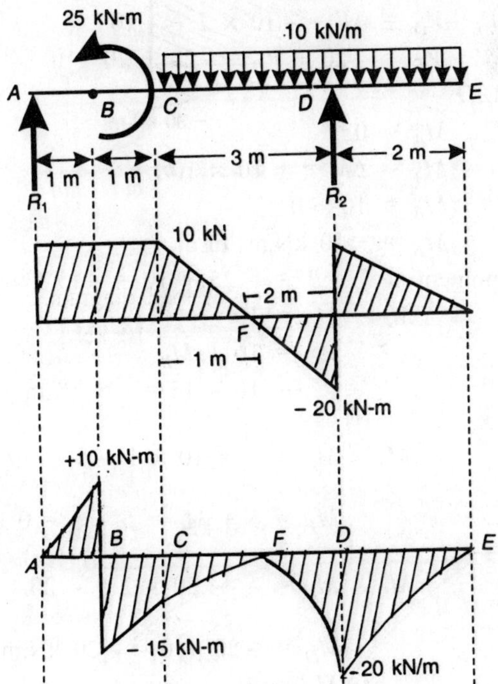

Fig. P. 2.13 (b)

For reaction, use equilibrium equation,

$$R_1 + R_2 = 10 \times 5 = 50 \qquad \text{...(i)}$$

$$\Sigma M_A = 0,$$

$$R_2 \times 5 + 25 = 10 \times 5 \times 4.5 = 225$$

$$\therefore \quad R_2 = 40 \text{ kN}$$

$$\therefore \quad R_1 = 50 - 40 = 10 \text{ kN}$$

$$\therefore \quad R_1 = 10 \text{ kN}$$

$$R_2 = 40 \text{ kN}$$

For S.F.D $V_A = +10$ kN

$$V_B - V_A = \Delta V = 0$$

$$\therefore \quad V_B = V_A = +10 \text{ kN}$$

$$V_C - V_B = 0, \Rightarrow V_C = +10 \text{ kN}$$

$$V_D - V_C = -10 \times 3 = -30$$

$$\therefore \quad V_D = -30 + V_C = -30 + 10 = -20 \text{ kN left S.F.}$$

Right S.F. at D.

$$= +40 - 20 = +20 \text{ kN}$$

$$V_E - V_D = \Delta V = -10 \times 2 = -20$$

$$V_E = -20 + V_D = -20 + 20 = 0$$

For B.M.D

$$M_A = 0$$

$$M_B - M_A = \Delta M = +10 \times 1$$

$$\therefore \quad M_B = 10 + 0$$

$$\therefore \quad M_B = +10 \text{ kN.m left}$$

Right moment at $B = -25 + 10 = -15$ kN.m

$$M_C - M_B = 10$$

$$M_C = 10 + M_B$$

$$= 10 - 15 = -5 \text{ kN.m}$$

$$M_F - M_F = \frac{1}{2} \times 10 \times 1 = 5$$

$$M_F = 5 + M_C = 5 - 5 = 0 \text{ kN.m}$$

$$M_D - M_F = -\frac{1}{2} \, 20 \times 2 = -20$$

$$M_D = -20 + 0 = -20 \text{ kN.m}$$

$$M_E = 0$$

Problem 2.14: Beam loaded as shown in Fig. 2.14 (a). Find the reactions S.F.D. and B.M.D.

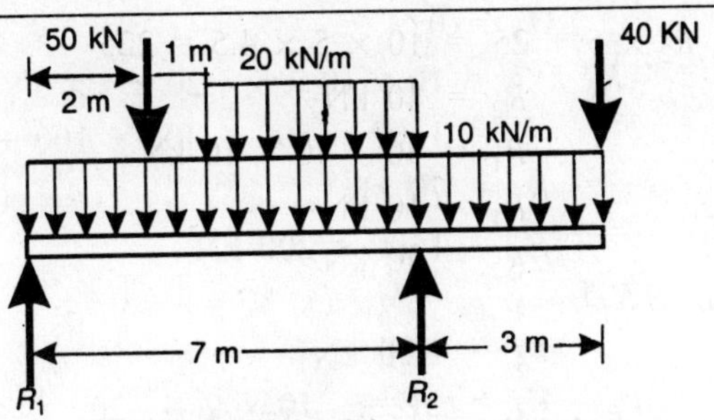

Fig. P. 2.14 (a)

Solution:

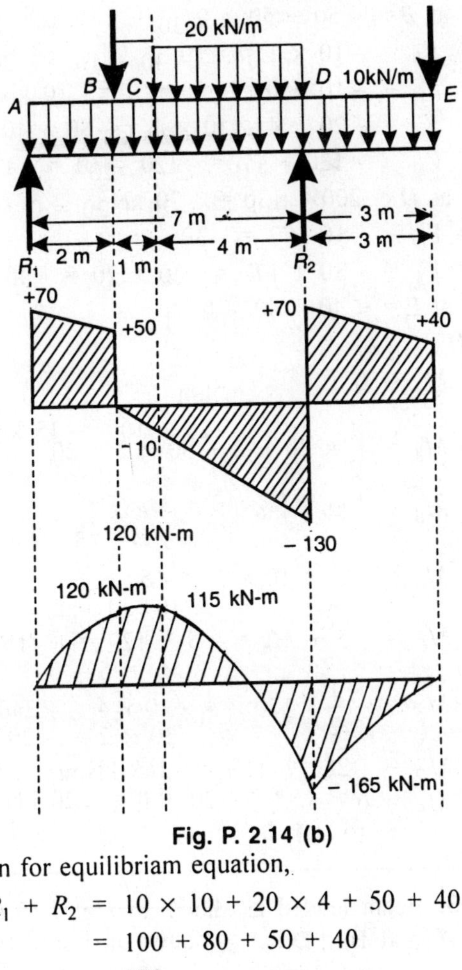

Fig. P. 2.14 (b)

Reaction for equilibriam equation,

$$R_1 + R_2 = 10 \times 10 + 20 \times 4 + 50 + 40$$
$$= 100 + 80 + 50 + 40$$
$$= 270 \qquad \ldots(1)$$

$$\Sigma M_A = 0,$$

$$R_2 \times 7 = 10 \times 10 \times 5 + 20 \times 4 \times 5 + 50 \times 2 + 10 \times 40$$

$$= 1400 = 500 + 400 + 100 + 400 = 1400$$

$$R_1 = 70 \text{ kN}$$

$$\therefore \quad R_2 = 1400 = 200 \text{ kN}$$

For S.F.D

$$V_A = +70 \text{ kN}$$

$$V_B - V_A = \Delta V = -10 \times 2 = -20$$

$$\therefore \quad V_B = -20 + V_A = -20 + 70 = +50 \text{ kN} - \text{left S.F. at B}$$

$$\text{Right S.F. at } B = -50 + 50 = 0$$

$$V_C - V_B = -10 \times 1 = -10$$

$$V_C = -10 + 0 = -10 + 0 = -10 \text{ kN}$$

$$V_D - V_C = -20 \times 4 - 10 \times 4 = -80 - 40 = -120 \text{ kN}$$

$$\therefore \quad V_D = -120 + V_C = -120 - 10 = -130 \text{ kN}$$

$$\text{Right S.F. at } D = 200 - 130 = +70 \text{ kN}$$

$$V_E - V_D = -10 \times 3 = -30$$

$$V_E = -30 + VD = -30 + 70 = +40 \text{ kN}$$

$$\text{Right S.F. at E} = +40 - 40 = 0$$

For B.M.D

$$M_A = 0$$

$$M_B - M_A = \frac{1}{2} \times 20 \times 2 + 50 \times 2 = 20 + 100 = 120$$

$$\therefore \quad M_B = 120 \text{ kN.m}$$

$$M_C - M_B = -\frac{1}{2} \times 10 \times 1 = -5$$

$$\therefore \quad M_C = -5 + M_B = -5 + 120 = +115 \text{ kN.m}$$

$$M_D - M_C = -\frac{1}{2} \times 120 \times 4 - 10 \times 4 = -240 - 40 = -280$$

$$M_D = -280 + 115 = -165 \text{ kN.m}$$

$$M_E = 0$$

Problem 2.15: Beam loaded as shown in Fig. 2.15 (a). Find the reactions, S.F.D. and B.M.D.

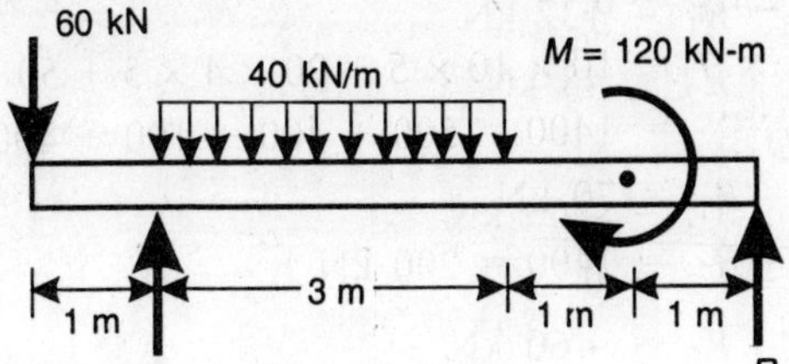

Fig. P. 2.15 (a)

Solution:

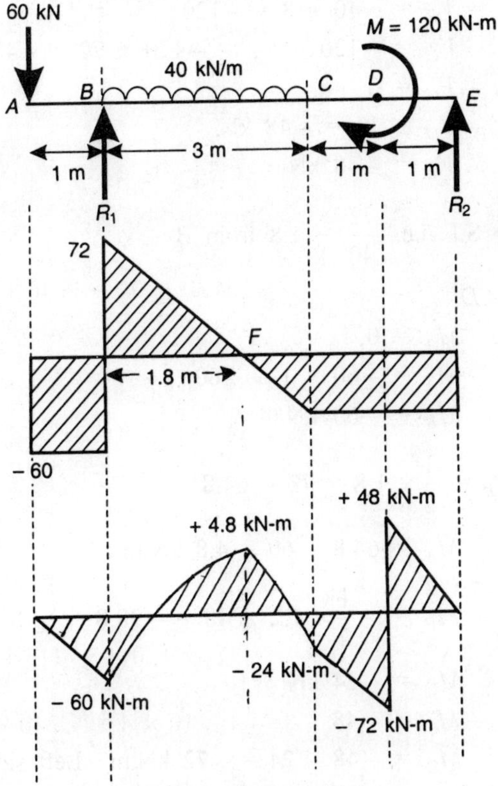

Fig. P. 2.15 (b)

From equilibrium equation,

$$R_1 + R_2 = 40 \times 3 + 60 = 120 + 60 = 180 \qquad \text{...(i)}$$

$$\Sigma M_{R_2} = 0$$

$$R_1 \times 5 + 120 = 60 \times 6 + 40 \times 3 \times 3.5 = 780$$

$\therefore \quad R_1 = 132$ kN

$\therefore \quad R_2 = 48$ kN

For S.F.D

$$V_A = -60 \text{ kN}$$

$$V_B - V_A = 0$$

$$\therefore \quad V_B = -60 \text{ kN}$$

Right shear force at B = 132 – 60

$$V_B = +72 \text{ kN}$$

$$V_C - V_B = -40 \times 3 = -120$$

$$V_C = -120 + V_B = -120 + 72 = -48 \text{ kN}$$

$$V_D - V_C = 0$$

$$\therefore \quad V_D = V_C = -48 \text{ kN}$$

$$V_E = -48 \text{ kN}$$

For zero S.F. i.e. $\frac{72}{40} = 1.8$ from B

For B.M.D

$$M_A = 0,$$

$$M_B - M_A = -60 \times 1 = -60$$

$$M_B = -60 \text{ kN.m}$$

$$M_F - M_B = \frac{1}{2} \times 1.8 \times 72 = 64.8$$

$$M_F = 64.8 - 60 = 4.8 \text{ kN.m}$$

$$M_C - M_F = -\frac{1}{2} \times 1.2 \times 48 = -28.8$$

$$M_C = -24 \text{ kN}$$

$$M_D - M_C = -48$$

$$M_D = -48 - 24 = -72 \text{ kN.m} \quad \text{Left side moment}$$

$\therefore$ Right moment at $D = 120 - 72 = +48$ kN.m

$$M_E - M_D = +48$$

$$M_E = 48 - 48 = 0$$

Problem 2.16: Beam loaded as shown in Fig. 2.16 (a). Find the reaction S.F.D. and B.M.D.

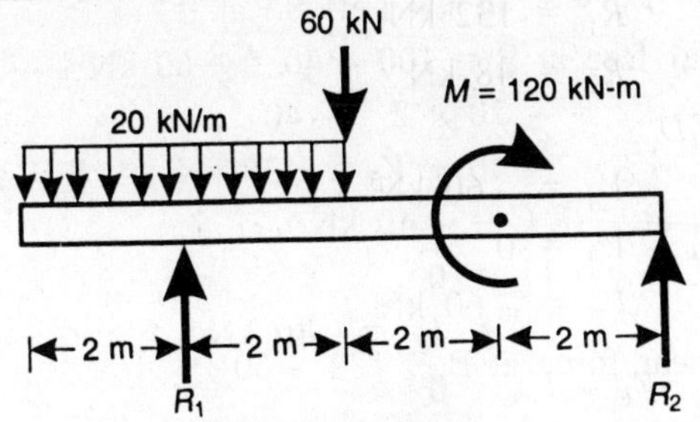

Fig. P. 2.16 (a)

Solution:

From equilibrium equation,

$$R_1 + R_2 = 20 \times 4 + 60 = 80 + 60 = 140 \text{ kN} \quad \ldots\text{(i)}$$

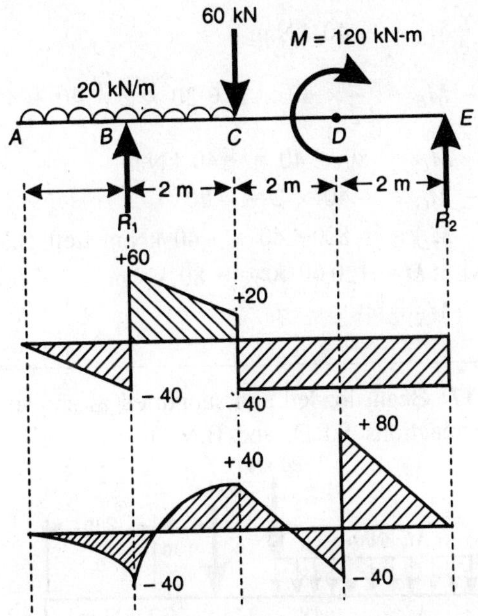

Fig. P. 2.16 (b)

$$\Sigma M_{R_2} = 0, \quad R_1 \times 6 + 120 = 20 \times 4 \times 6 + 60 \times 4$$

$$\therefore \quad R_1 = 100 \text{ kN}$$

$$R_2 = 140 - 100 = 40 \text{ kN}$$

For S.F.D

$$V_A = 0$$

$$V_B - V_A = -20 \times 2 = -40$$

$V_B = -40$ kN Left S.F.at B

Right shear fore at B = $100 - 40 = +60$ kN

$V_C - V_B = -20 \times 2 = -40$

$V_C = -40 + V_B = +20$ kN left S.F. at C

Right S.F. at $C = -40$ kN

$V_D - V_C = 0$

$\therefore \quad V_D = V_C = -40$ kN

$V_E - V_D = 0$

$V_E = -40$ kN

For B.M.D

$M_A = 0$

$$M_B - M_A = -\frac{1}{2} \times 40 \times 2 = -40 \text{ V}$$

$\therefore \quad M_B = -40$ kN.m

$$M_C - M_B = \frac{1}{2} \times 40 \times 2 + 20 \times 2 = 40 + 40 = 80$$

$M_C = 80 - 40 = +40$ kN.m

$M_D - M_C = -40 \times 2 = -80$

$M_D = -80 + 40 = -40$ kN.m Left B.M.at D

Right B.M at $D = 120 - 40 = +80$ kN.m

$M_E = 0$

Problem 2.17: Beam loaded and supported as shown in Fig. 2.17 (a). Find the reactions S.F.D. and B.M.D.

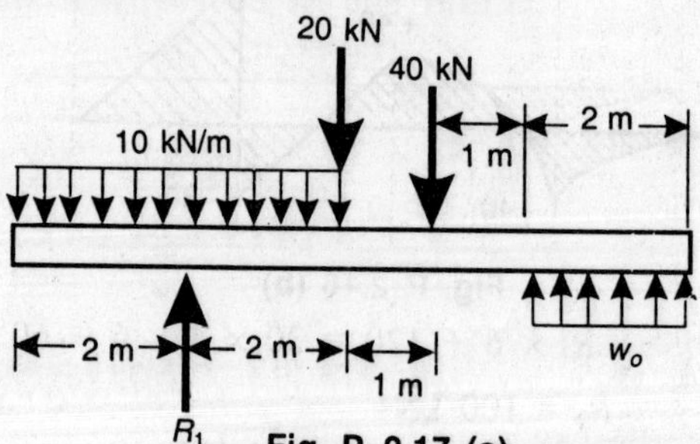

Fig. P. 2.17 (a)

Solution:

For reaction, using equilibrium equation

$\Sigma F_y = 0$

$$R_1 + 2\,w_0 = 10 \times 4 + 20 + 40$$

$$= 40 + 20 + 40 = 100 \qquad \ldots\text{(i)}$$

$$\Sigma M_B = 0,$$

$$2\,w_0 \times 5 - 20 \times 2 - 40 \times 3 - 2 \times 10 \times 1 = -10 \times 2 \times 1$$

$$\therefore 10\,w_0 - 40 - 120 - 20 + 20 = 0$$

$$\therefore \quad 10\,w_0 = 160$$

$$\therefore \quad w_0 = \frac{160}{10} = 16 \text{ kN/m}$$

$$\therefore \quad R_1 = 100 - 32 = 68 \text{ kN.}$$

For shear force

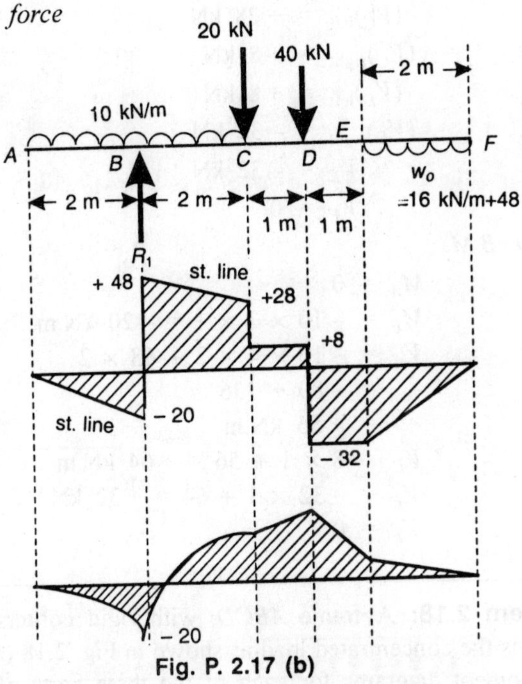

Fig. P. 2.17 (b)

$$V_A = 0$$

$$V_B - V_A = -10 \times 2 = -20$$

$$\therefore \quad (V_B)_{\text{left}} = -20 \text{ kN}$$

$$(V_B)_{\text{Right}} = 68 - 20 = +48 \text{ kN}$$

$$V_C - V_B = -10 \times 2 = -20$$

$$V_c = -20 + V_B = -20 + 48 = 28 \text{ kN}$$

$$\therefore \quad (V_C)_{\text{left}} = +28 \text{ kN}$$

$$(V_c)_{\text{right}} = +28 - 20 = +8 \text{ kN}$$
$$V_D - V_C = 0$$
$$\therefore \quad (V_D)_{\text{left}} = V_c = +8 \text{ kN}$$
$$(V_D)_{\text{right}} = +8 - 40 = -32 \text{ kN}$$
$$V_E = (V_D)_{\text{right}} = -32 \text{ kN}$$
$$V_F - V_E = +16 \times 2 = +32$$
$$V_F = +32 - 32 = 0$$

$\therefore$ Conclusion $V_A = 0$

$$(V_B)_{\text{left}} = -20 \text{ kN}$$
$$(V_B)_{\text{right}} = +48 \text{ kN}$$
$$(V_C)_{\text{left}} = +28 \text{ kN}$$
$$(V_C)_{\text{right}} = +8 \text{ kN}$$
$$(V_D)_{\text{left}} = +8 \text{ kN}$$
$$(V_D)_{\text{right}} = -32 \text{ kN}$$
$$V_E = -32 \text{ kN}$$
$$V_F = 0$$

For B.M

$$M_A = 0$$
$$M_B = -10 \times 2 \times 1 = -20 \text{ kN.m}$$
$$V_C = -10 \times 4 \times 2 + 68 \times 2$$
$$= -80 + 136$$
$$= +56 \text{ kN.m}$$
$$V_D = 8 \times 1 + 56 = +64 \text{ kN.m}$$
$$V_E = -32 \times 1 + 64 = +32 \text{ kN}$$
$$V_F = 0$$

Problem 2.18: A frame *ABCD*, with rigid corners at *B* and *C*, supports the concentrated load as shown in Fig. 2.18 (a) (Draw shear and moment diagrams for each of the three parts of the frame.)

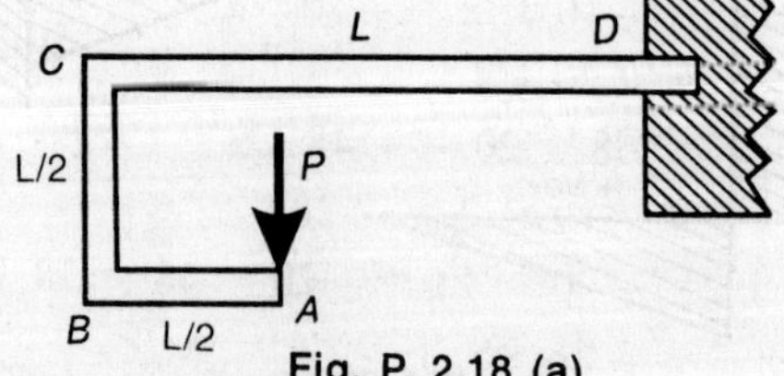

Fig. P. 2.18 (a)

Solution:

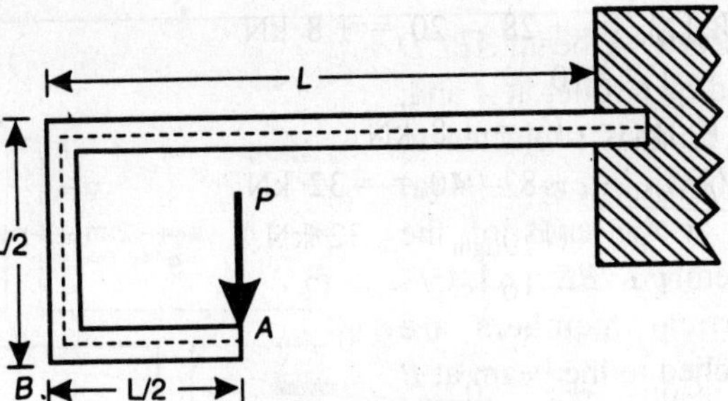

Fig. P. 2.18 (b)

First of all brack all part and draw free body diagram of each part, then draw S.F.D and B.M.D

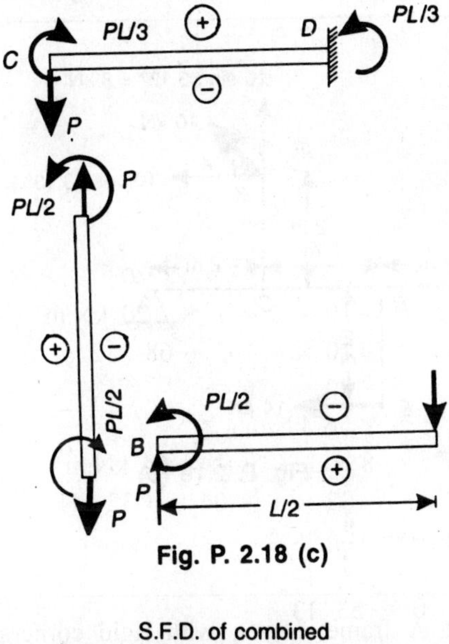

Fig. P. 2.18 (c)

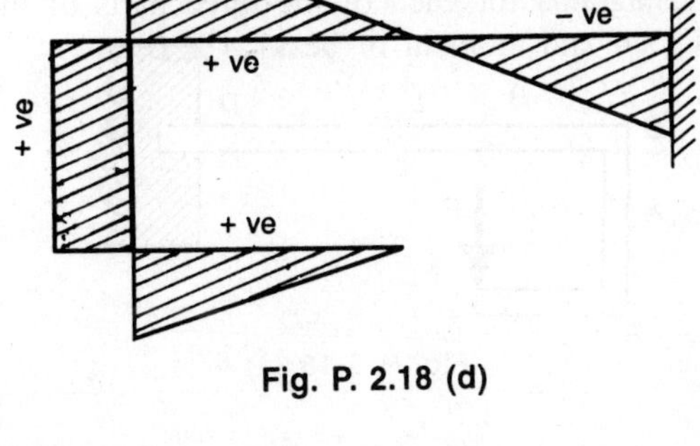

Fig. P. 2.18 (d)

Problem 2.19: A beam *ABCD* is supported by a roller at *A* and, *a* hinge at *D*. It is subjected to the loads shown in Fig. 2.19 (a) which act at the ends of the vertical members *BE* and *CF*. These vertical members are rigidly attached to the beam at *B* and *C*. (Draw shear and moment diagrams for the beam *ABCD* only)

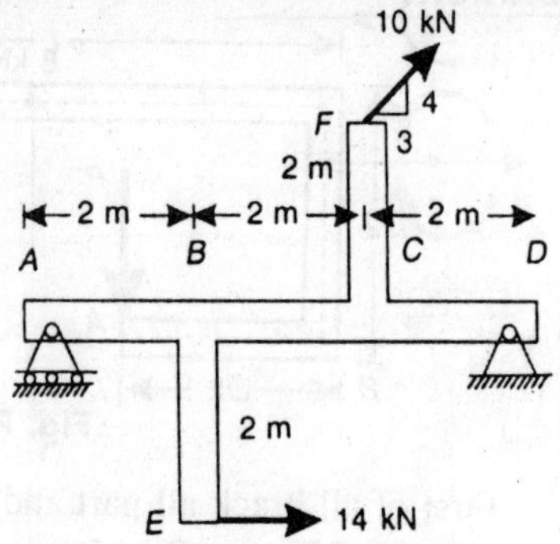

Fig. P. 2.19 (a)

Solution:

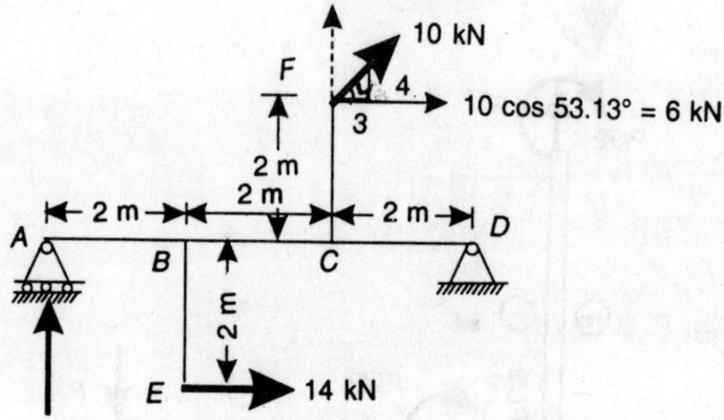

Fig. P. 2.19 (b)

$$\tan\theta = \frac{4}{3}$$

$\therefore \qquad \theta = 53°.13$

F. B. diagram of joint *BC*

Finally at B

∴ Final load and moment of beam *ABCD*.

$$\Sigma f_y = 0$$

$$R_A + R_D - 8 \text{ kN} = 0$$

$$\Sigma M_A = 0$$

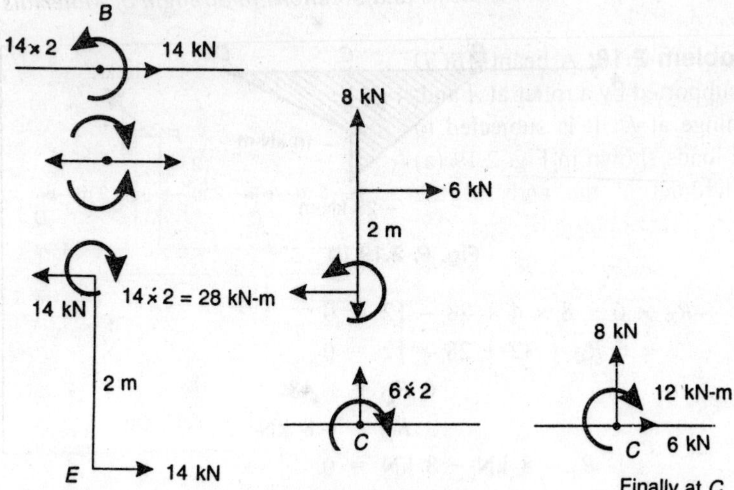

Fig. P. 2.19 (c)

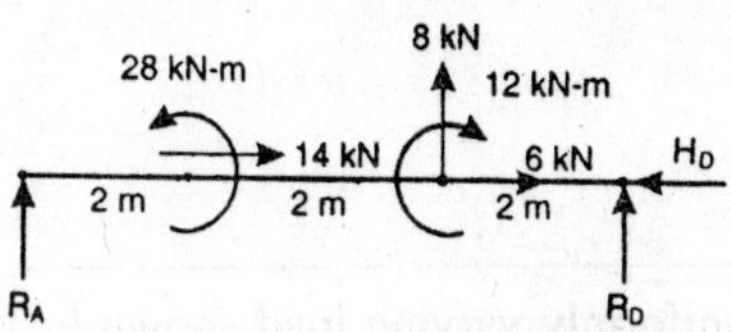

Fig. P. 2.19 (d)

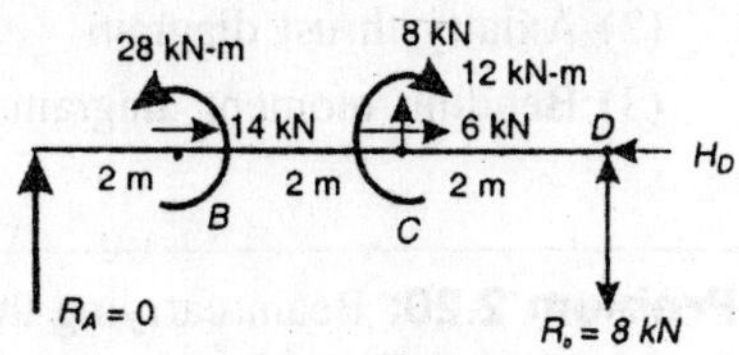

Fig. P. 2.19 (e)

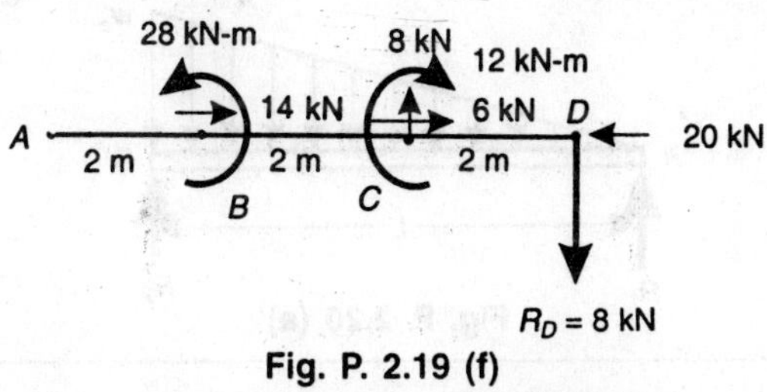

Fig. P. 2.19 (f)

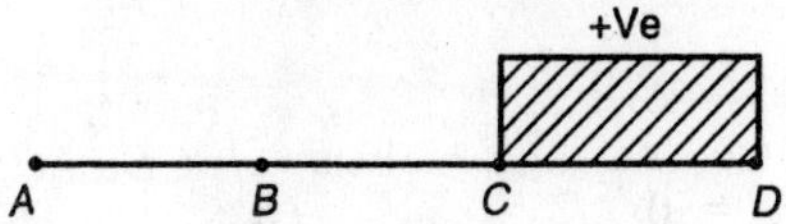

Fig. P. 2.19 (g)

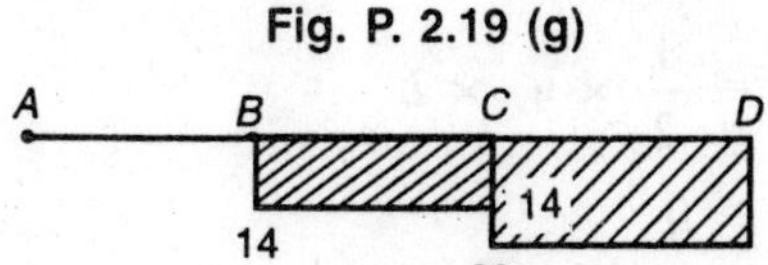

Fig. P. 2.19 (h)

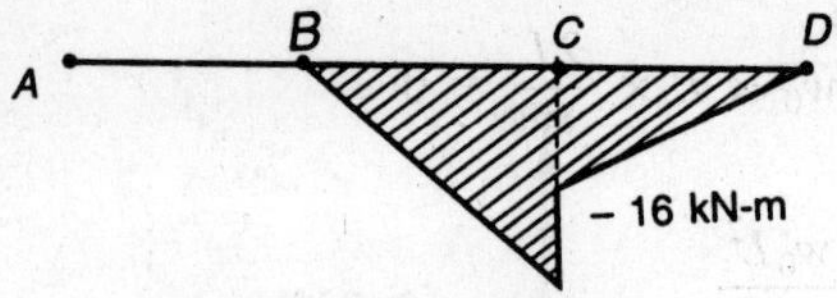

Fig. P. 2.19 (i)

$$R_D \times 6 + 8 \times 4 + 28 - 12 = 0$$

$$6\,R_D + 32 + 28 - 12 = 0$$

$$6\,R_D = -48$$

$$R_D = -8 \text{ kN}$$

$$\therefore \quad R_A - 8 \text{ kN} + 8 \text{ kN} = 0$$

$$\therefore \quad R_A = 0$$

(1) S.F.D

(2) Axially thrust diagram

(3) Bending moment diagram

Problem 2.20: Beam carrying the uniformly varying load shown in Fig. 2.20. (a). Find the maximum B.M.

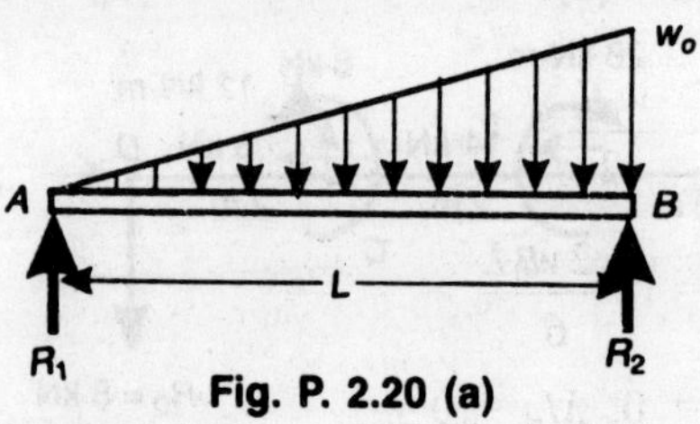

Fig. P. 2.20 (a)

Solution:

$$\Sigma F_y = 0$$

$$R_1 + R_2 = \frac{1}{2} \times w_0 \times L \qquad \ldots(i)$$

$$\Sigma M_A = 0$$

$$R_2 \times L - \frac{1}{2} \times w_0 \times L \times \frac{2L}{3} = 0$$

$$R_2 \times L = \frac{w_0 L^2}{3}$$

$$\therefore \quad R_2 = \frac{w_0 L}{3}$$

$$\therefore \quad R_1 = \frac{w_0 L}{2} - \frac{w_0 L}{3} = \frac{3w_0 L - 2w_0 L}{6} = \frac{w_0 L}{6}$$

$$\therefore \quad R_1 = \frac{w_0 L}{6}$$

$$R_2 = \frac{w_0 L}{3}$$

$$V_A = +\frac{w_0 L}{6}$$

$$\therefore \quad V_B - V_A = \text{Area at load diagram} = -\frac{1}{2} \times w_0 L$$

$$\therefore \quad V_B = -\frac{w_0 L}{2} + V_A = -\frac{w_0 L}{2} + \frac{w_0 L}{6} = \frac{-3w_0 L + w_0 L}{6}$$

$$= \frac{-2w_0 L}{6} = -\frac{w_0 L}{3}$$

$$M_A = 0, \; M_B = 0$$

$$M_x = \frac{w_0 L}{6} \times x - \frac{1}{2} w_x \times x$$

But from similar triangle

$$\frac{w_0}{L} = \frac{w_x}{x}$$

$$\therefore \quad w_x = \frac{w_0}{L} x$$

$$\therefore \quad M_x = \frac{w_0 L}{6} x - \frac{1}{2} \times \frac{w_0}{L} x \times x \times \frac{x}{3}$$

$$= \frac{w_0 L}{6} x - \frac{w_0 x^3}{6L}$$

B.M. maximum at S.F. = 0

$$M_x = \frac{w_0 L}{6} x - \frac{w_0}{6L} x^3$$

$$M_{max} = \frac{w_0 L}{6} \times \frac{L}{\sqrt{3}} - \frac{w_0}{6L} \times \left(\frac{L}{\sqrt{3}}\right)^3$$

$$= \frac{w_0 L^2}{6\sqrt{3}} - \frac{w_0 L^3}{6L \times 5.2}$$

$$= \frac{w_0 L^2}{10.392} - \frac{w_0 L^2}{31.2} = w_0 L^2 \left[\frac{1}{10.392} - \frac{1}{31.2}\right]$$

$$= w_0 L^2 \times 0.06417 = 0.06417\ w_0 L^2 = \frac{w_0 L^2}{9\sqrt{3}}$$

Problem 2.21: Shear diagram as shown in Fig. 2.21 (a). Find the B.M. diagram.

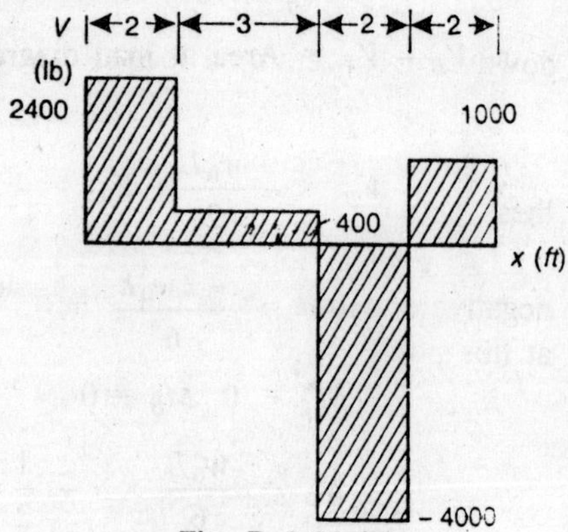

Fig. P. 2.21 (a)

Solution:

Start from origin 0,

Take a section just right of *O* then left side shear force be 2400 kN upward so reaction exist here. Between 0 and B S.F.D straight so no loading exist between them.

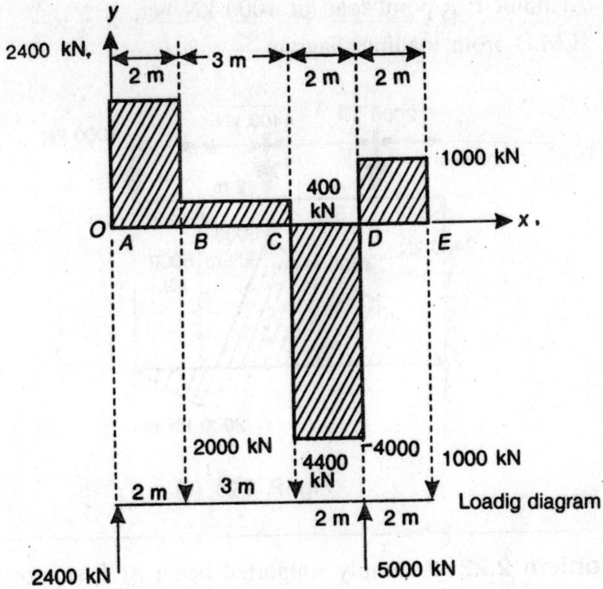

Fig. P. 2.21 (b)

At point B shear force reduces from 2400 kN to 400 kN. So a downward concentrated load out at point. ↓ 2000 kN

B

Between B and C.S.F.D. follow straight line so no loading between these point.

At c shear force first decreases to zero and increases to 4000 in negative direction, so at this point a concentrated load of 4400 kN act at this point.

↓ 4400 kN

C

Between C and D Shear force Constant.

At point D shear force first each zero value then increases to 1000 kN in positive direction, So at this point a reaction of magnitude 5000 kN ats.

D

↑ 5000 kN

Between point D and E shear force us constant, because no loading present here.

At point E a point load of 1000 kN act,

B.M.D from loading diagram

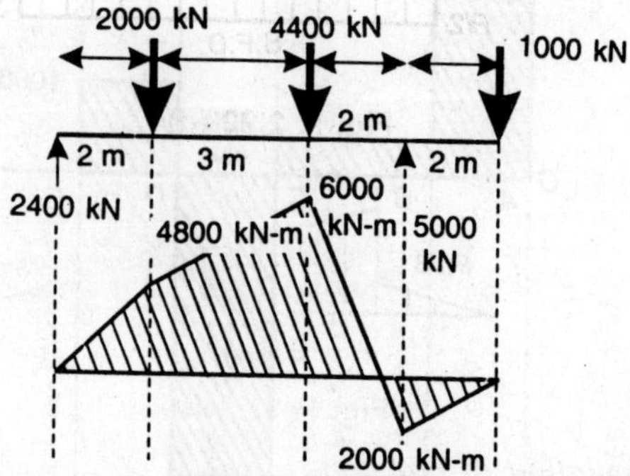

Fig. P. 2.21 (c)

Problem 2.22: A simply supported beam *ACB* supports a vertical load *P* by means of a bracket *CDE*, as shown in Fig 2.22 (a). Draw the *S.F.* and *B.M.* diagrams for the beam.

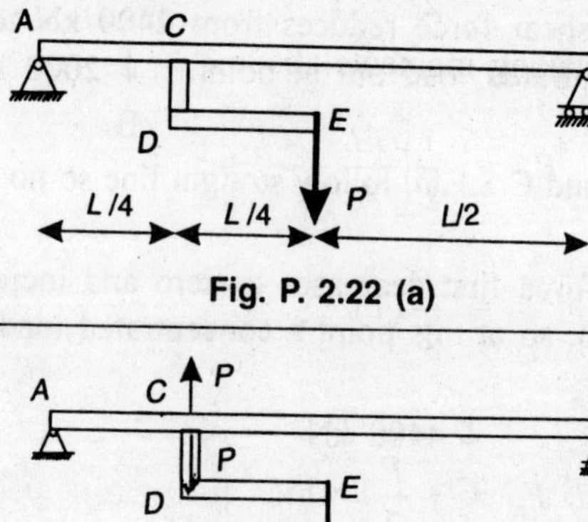

Fig. P. 2.22 (a)

Solution:

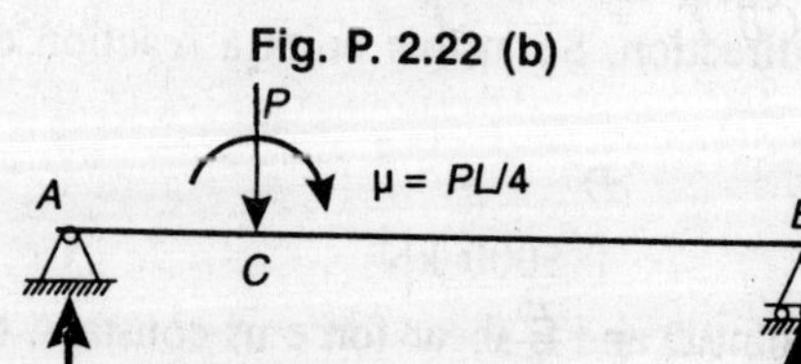

Fig. P. 2.22 (b)

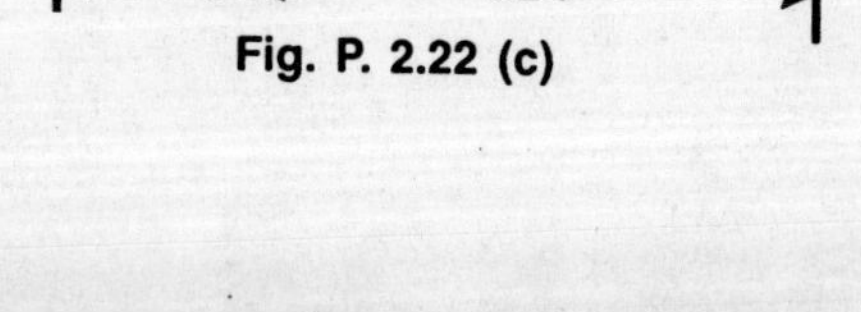

Fig. P. 2.22 (c)

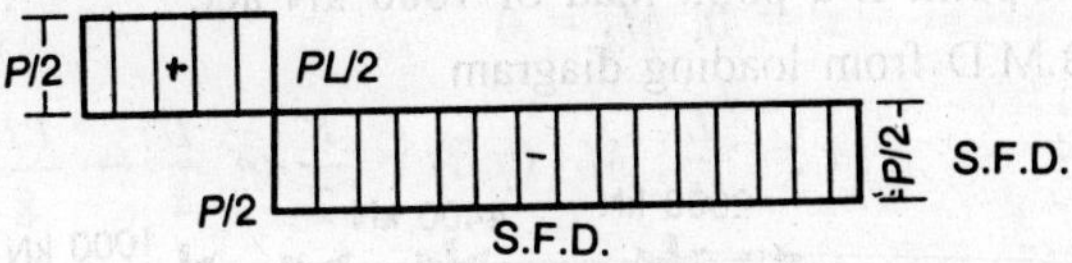

Fig. P. 2.22 (d)

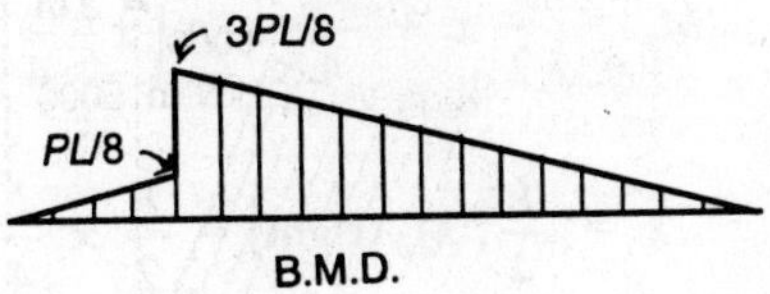

Fig. P. 2.22 (e)

For calculation of reaction

Taking moments about B,

$$R_A = \frac{1}{L}\left[P \times \frac{3L}{4} - \frac{PL}{4}\right]$$

$$= \frac{P}{2}(\uparrow)$$

Similarly, taking moments about A,

$$R_B = \frac{1}{L}\left[\frac{PL}{4} + \frac{PL}{4}\right]$$

$$= \frac{P}{2}(\uparrow)$$

S.F.D. :

For $AC,\ F_x = +\dfrac{P}{2}$, which is constant between A and C.

For $CB\ F_x = \dfrac{P}{2} - P = -\dfrac{P}{2}$, which is constant between C and B.

B.M.D.:

For $AC,\ M_x = \dfrac{P}{2}x$, which varies lineraly.

At $\quad x = 0, M_A = 0$

At $\quad x = \frac{L}{4}, M_C \text{ (left)} = \frac{P}{2} \times \frac{L}{4} = \frac{PL}{8}$

For portion CB

$$M_x = \frac{P}{2}x + \frac{PL}{4} - p\left(x - \frac{L}{4}\right), \text{ which varies linerly.}$$

At $\quad x = \frac{L}{4}, M_c \text{ (right)} = \frac{P}{2} \times \frac{L}{4} + \frac{PL}{4} - p\left(\frac{L}{4} - \frac{L}{4}\right)$

$$= \frac{3PL}{8}$$

At $\quad x = L,$

$$M_B = \frac{PL}{2} + \frac{PL}{4} - p\left(L - \frac{L}{4}\right) = 0, \text{ as expected.}$$

Problem 2.23: Draw the *S.F.* and *B.M.* diagrams for the *ACB* shown in Fig. 2.23 (a).

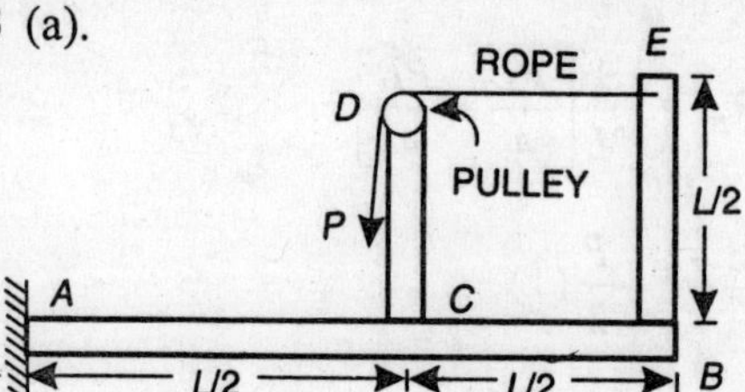

Fig. P. 2.23 (a)

Solution:

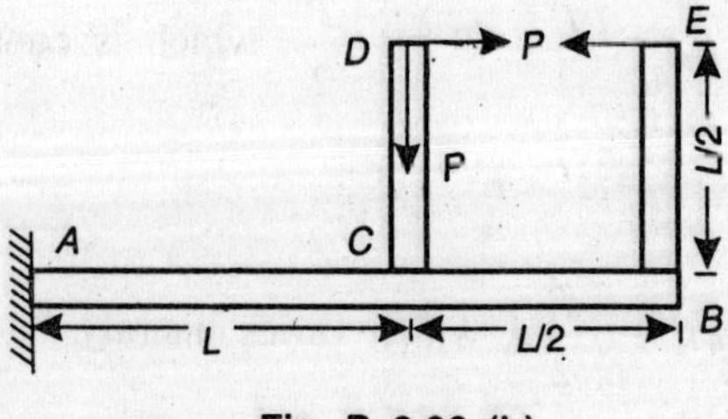

Fig. P. 2.23 (b)

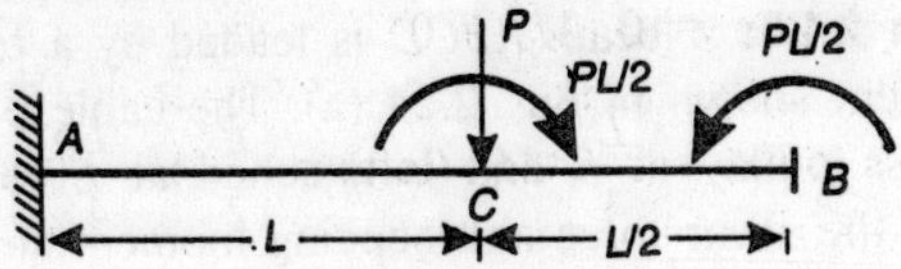

Fig. P. 2.23 (c)

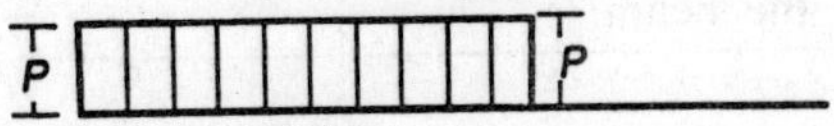

S.F.D.

Fig. P. 2.23 (d)

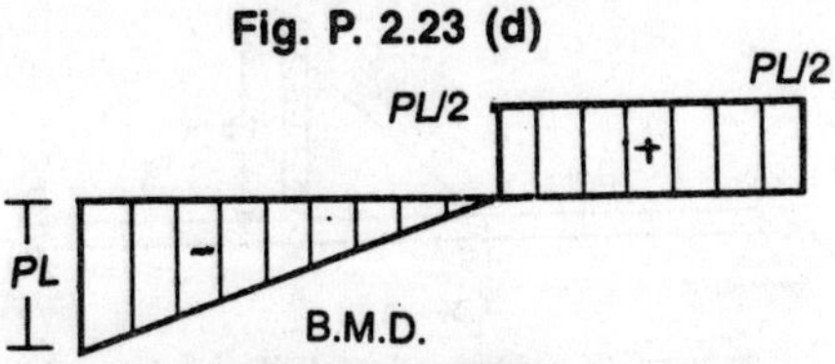

Fig. P. 2.23 (e)

Free diagram of the contilever *ABC* shown in Fig. P. 2.23 (b) final force and moment shown in figure (c)

Reaction : Since $\Sigma V = 0$, $R_A = P$ (↑)

Reacting moment at $A = P.L.$ in anticlockwise direction.

S.F.D.:

For BC, $F_x = 0$

For CA, $F_x = +P$,

which is constant from C to A.

B.M.D:

For BC, $M_x = \dfrac{PL}{2}$ which is constant from BC,

For CA, $M_x = \dfrac{PL}{2} - \dfrac{PL}{2} - p\left(x - \dfrac{L}{2}\right)$

which varies linearly.

At $x = \dfrac{L}{2}$

$$M_C \text{ (left)} = \frac{PL}{2} - \frac{PL}{2} - p\left(\frac{L}{2} - \frac{L}{2}\right) = 0$$

At $x = \dfrac{3L}{2}$

$$M_A = \frac{PL}{2} - \frac{PL}{2} - P\left(\frac{3L}{2} - \frac{L}{2}\right) = -PL$$

This S.F. and B.M diagrams are shown in Fig. 2.23 (*d*) and (*e*) respectively.

Problem 2.24: *A* beam *ABCD* is loaded by a force $W = 8$ kN by arrangement shown in Fig. 2.24 (a). The cable passes over a small frictionless pulley at *B* and is attached at *E* to the vertical arm. Compute the shear force and bending moment at the section, which is just to the left of the vertical arm. Also, sketch the S.F. and B.M. diagram for the beam.

Solution:

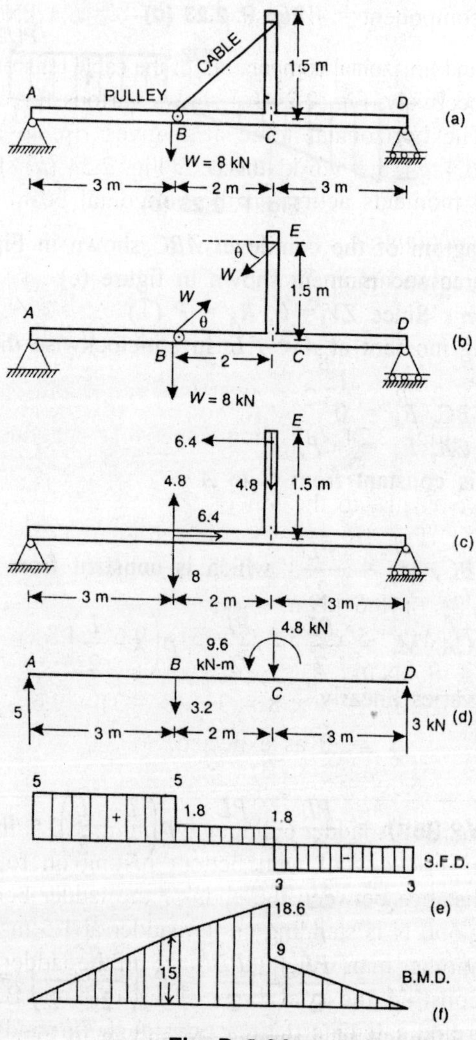

Fig. P. 2.24

Force body diagram of beam ABC and given in Fig. P. 2.24.

From geometry, we have $BE = \sqrt{(2)^2 + (1.5)^2} = 2.5$ m.

Hence $\cos\theta = \frac{2}{2.5}$ and $\sin\theta = \frac{1.5}{2.5}$ .At B,

Vertical component of tension in cable $= W\sin\theta = 8 \times \frac{1.5}{2.5} = 4.8$ kN and horizontal component $= W\cos\theta = 8 \times \frac{2}{2.5} = 6.4$ kN. Similarly, at E, the vertical and horizontal components of the cable tension are 4.8 kN and 6.4 kN respectively. Fig. 2.24 (c) shows various forces marked on the elements. The horizonatal force at E gives rise to an anticlockwise moment = 6.4 × 1.5 = 9.6 kN.m at C. Fig. 2.24 (*d*) shows the vertical forcess and moments acting on the horizontal beam *ABCD*.

For Reactions :

Taking moments at D,

$$R_A = \frac{1}{8}(3.2 \times 5 + 4.8 \times 3 + 9.6) = 5 \text{ kN } (\uparrow)$$

$$\therefore \quad R_B = 3.2 + 4.8 - 5 = 3 \text{ kN } (\uparrow)$$

S.F.D.: Fig. 2.24 (*e*) shows the S.F. diagram, which is self explanatory.

Hence S.F. just to the left of C = 1.8 kN

B.M.D.: For AB, $M_x = 5x$, which gives $M_B = 5 \times 3 = 15$ kN.m

For BC, $M_x = 5x - 3.2(x - 3)$, which gives M_c (left) $= 5 \times 5 - 3.2(5 - 3) = 18.6$ kN.m

For CD, $M_x = 5x - 3.2(x - 3) - 9.6 - 4.8(x - 5)$, which gives M_C (right) = 9 kN.m

and $M_D = 5 \times 8 - 3.2(8 - 3) - 9.6 - 4.8(8 - 5)$
$= 0$ as expected.

Problem 2.25: A ladder 6 m long weighing 250 N/metre run, against a smooth vertical wall with bottom resting on rough ground, the vertical distance between the ends of the ladder being 4 m. A man weighting 650 N is standing on the ladder at 1.5 m from the bottom end and another man weighting 700 N on the ladder at 3 m from the bottom. Construct the axial thrust, shear force and B.M. diagrams for the ladder and calculate the reactions at the ground and at the wall.

Solution:

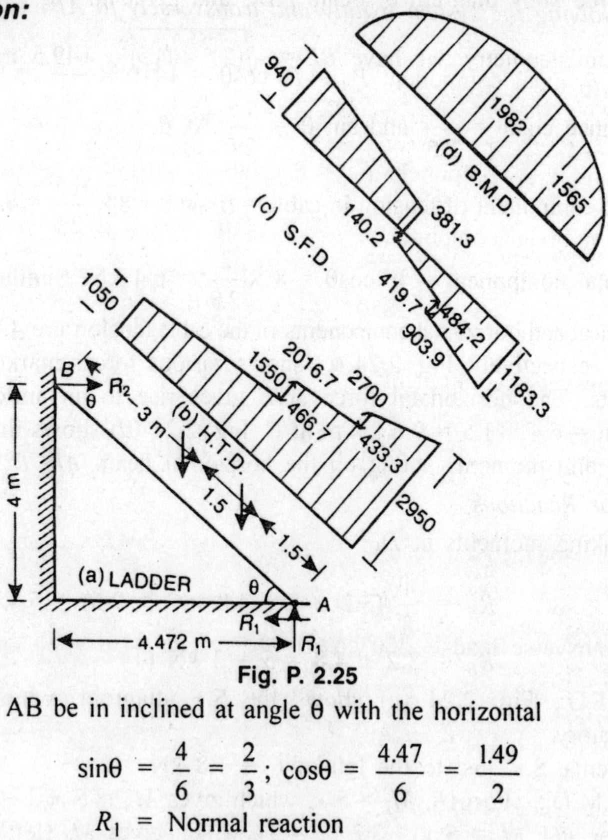

Fig. P. 2.25

Let AB be in inclined at angle θ with the horizontal

$$\sin\theta = \frac{4}{6} = \frac{2}{3};\ \cos\theta = \frac{4.47}{6} = \frac{1.49}{2}$$

Let R_1 = Normal reaction on rough ground

R_2 = Normal reaction on wall

R_f = Frictional resistance of ground.

Resolving force vertically

$$R_1 = 700 + 650 + 250 \times 6 = 2850 \text{ N}$$

Resolving force horizontally

$$R_2 = R_f$$

Taking moments about A, Then we get

$$R_2 \times 4 = 700 \times 3 \cos\theta + 350 \times 1.5 \cos\theta + 650 \times 6 \times \frac{4.472}{2}$$

$$= \left(2100 \times \frac{1.49}{2}\right) + \left(975 \times \frac{1.49}{2}\right) + \left(\frac{1500 \times 4.472}{2}\right)$$

or, $R_2 = 1410 \text{ N} = R_f$

Resolving the system axially and transversely to AB

At B, axial component $P_B = R_2 \cos\theta = 1410 \times \frac{1.49}{2} = 1050 \text{ N}$

Transverse component $F_B = R_2 \sin\theta = 1410 \times \frac{2}{3} = 940 \text{ N}$

At A, axial component $P_A = R_1 \sin\theta + R_f \cos\theta$

$$= \left(2850 \times \frac{2}{3}\right) + \left(1410 \times \frac{1.49}{2}\right) = 2950 \text{ N}$$

Transverse component $F_A = R_1 \cos\theta - R_f \sin\theta$

$$= \left(2850 \times \frac{1.49}{2}\right) - \left(1410 \times \frac{2}{3}\right) = 1183.3 \text{ N}$$

For ladder, axial load $= 250 \sin\theta = 250 \times \frac{2}{3} = 166.7 \text{ N/m}$

Transverse load $= 250 \cos\theta = 250 \times \frac{1.49}{2} = 186.25 \text{ N/m}$

At C, axial load $= 700 \sin\theta = 700 \times \frac{2}{3} = 466.7 \text{ N}$

Transverse component

$$= 700 \cos\theta = 700 \times \frac{1.49}{2} = 521.5 \text{ N}$$

At D, axial load $= 650 \sin\theta = 650 \times \frac{2}{3} = 433.3 \text{ N}$

Transverse load $= 650 \cos\theta = 650 \times \frac{1.49}{2} = 484.2 \text{ N}$

Max B.M. occurs where S.F. change its sign.

B.M. at point C = $M_c = +940 \times 3 - \left(186.25 \times 3 \times \frac{3}{2}\right)$

$$= +1982 \text{ N.m}$$

$$\text{B.M. at point } D = M_D = (+1183.3 \times 1.5) - \left(186.25 \times 1.5 \times \frac{1.5}{2}\right)$$

$$= +1565 \text{ N.m}$$

Problem 2.26: A beam *ABCD* has an internal hinge at B and is loaded as shown in Fig 2.26 (a). Determine the reactions at *A*, *C* and *D* and plot the *S.F.* and *B.M.* diagrams, indicating principal values.

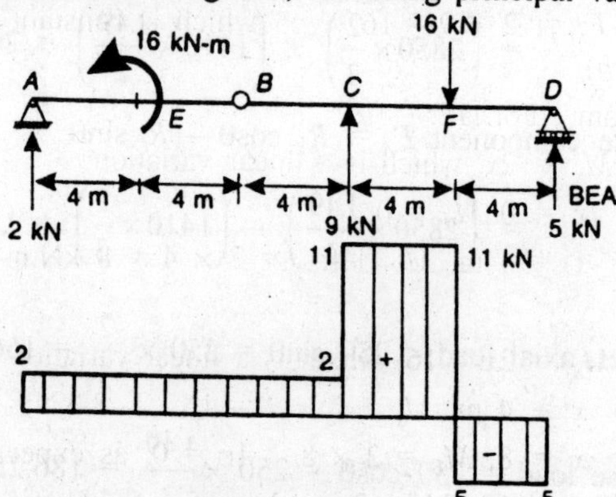

Fig. P. 2.26 (a)

Solution: Considering L.H.S.

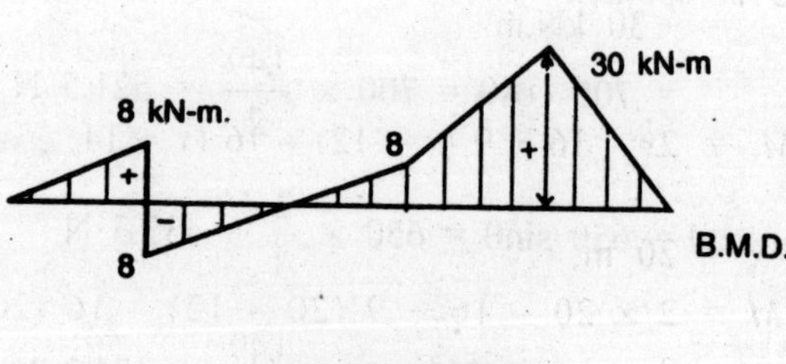

Fig. P. 2.26 (b)

$$M_B = 0 = R_A \times 8 - 16$$

$$\therefore \quad R_A = \frac{16}{8} = 2 \text{ kN } (\uparrow)$$

Considering R.H.S.,

$$M_B = 0 = R_D.\ 12 + R_C.\ 4 - 16 \times 6$$

or $R_C + 3\,R_D = 24$ ...(i)

For the whole beam

$R_A + R_C + R_D = 16$

$\therefore \quad R_C + R_D = 16 - 2 = 14$ ...(ii)

From (i) and (ii), we get

$$R_D = 5 \text{ kN } (\uparrow)$$

S.F. Diagram : For *AC* ;

$F_x = 2\text{kN}$, which is constant from *A* to C

For *CF* ; $F_x = 2 + 9 = 11$ kN, constant from *C* to *F*

For *FD*; $F_x = 2 + 9 - 16 = -5$ which is constant from *F* to *D* in Fig. 2.26 (*b*)

B.M. Diagram : For *AE*:

$M_x = 2x$, which is a linear variation

at $x = 0\ M_A = 0.$

At $x = 4$ m, M_E (left) $= 2 \times 4 = 8$ kN.m

For *EC*:

$M_x = 2x - 16$, which is a linear variation.

At $x = 4$ m, $M_x = 2 \times 4 - 16 = -8$ kN

At $x = 8$, $M_B = 2 \times 8 - 16 = 0$, as expected

At $x = 12$, $M_x = 2 \times 12 - 16 = 8$ kN.m

For *CF*:

$M_x = 2x - 16 + 9\,(x - 12)$, which is a linear variation.

At $x = 14$ m, $M_{max} = 2 \times 14 - 16 + 9\,(14 - 12)$

$= +\,30$ kN.m

For *FD*:

$M_x = 2x - 16 + 9\,(x - 12) - 16\,(x - 14)$, which is a linear variation.

At $x = 20$ m,

$M = 2 \times 20 - 16 + 9\,(20 - 12) - 16\,(20 - 14)$

$= 0$, as expected.

Problem 2.27: A beam *AB*, 8 m long and supported at *A*, has a simple support of 1 m length between C and D. Assuming uniformly distributed reaction between *C* and *D*, draw the S.F. and B.M. diagrams for the loading shown in Fig. 2.27 (a)

Solution:

Replace reaction between CD into a concentrated reaction, taking moments about A then we get,

$$R_2 \times (6.5) = \frac{1}{2} \times 1 \times 4.5 \left(\frac{2}{3} \times 4.5\right) + 1 \times 8$$

From which $R_2 = 2.2692$ kN

Taking moments about R_2

$$R_1 \times 6.5 + 1 \times 1.5 = \frac{1}{2} \times 4.5 \times 1 \left(2 + \frac{1}{3} \times 4.5\right)$$

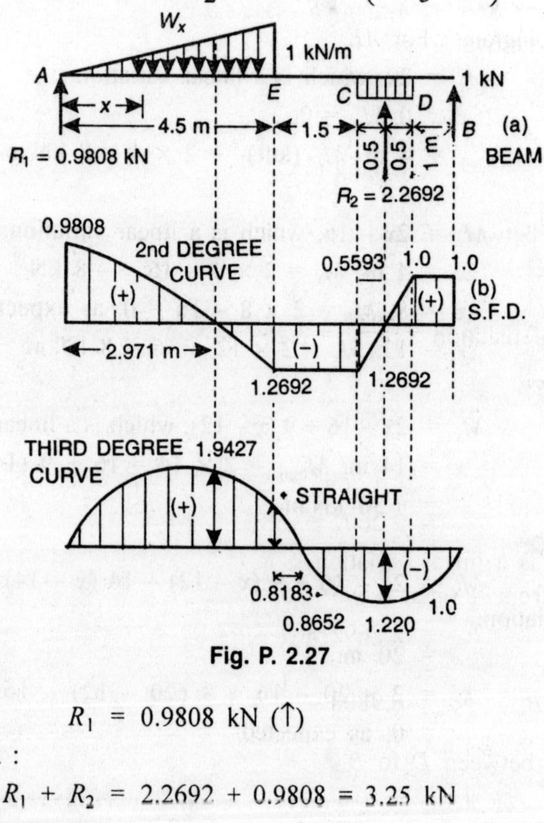

Fig. P. 2.27

$$\therefore \quad R_1 = 0.9808 \text{ kN } (\uparrow)$$

Check :

$$R_1 + R_2 = 2.2692 + 0.9808 = 3.25 \text{ kN}$$

$$\text{Total load} = 1 + \frac{1}{2} \times 1 \times 4.5 = 3.25 \text{ kN}$$

At any distance x from A, rate of loading in

$$AE = \frac{1}{4.5}x = \frac{x}{4.5}$$

S.F. diagram :

For AE :

$$F_x = 0.9808 - \frac{1}{2} \times x \times \frac{x}{4.5} = 0.9808 - \frac{x^2}{9} \text{ which}$$

is a second degree a curve.

At $x = 0$, $F_A = 0.9808$ kN

At $x = 4.5$ m, $F_E = -1.2692$ kN.

S.F. is zero at

$$x = (0.9808 \times 9)^{1/2} = 2.971 \text{ m}$$

For EC

$$F_x = 0.9808 - \frac{1}{2} \times 1 \times 4.5 = -1.2692 \text{ kN, which}$$

is constant between E and C.

For CD

$$\text{Rate of reaction} = \frac{R_2}{1} = 2.2692 \text{ kN.m}$$

Hence $$F_x = 0.9808 - \frac{1}{2} \times 1 \times 4.5 + 2.2692\ (x - 6)$$

$$= -1.2692 + 2.2692\ (x - 6)$$

$$= 2.2692\ x - 14.8844$$

which is a linear variation. S.F. is zero at $x = 14.8844/2.2692 =$ 6.5593 m from A or at $6.593 - 6 = 0.5593$ m from C. At $x = 7$ m,

$$F_D = 2.2692 \times 7 - 14.8844 = +1 \text{ kN}$$

For DB: $F_x = 0.9808 - \frac{1}{2} \times 1 \times 4.5 + 2.2692 = 1$ kN, which

is constant between D to B.

B.M. Diagram : For AE, $M_x = 0.9808\ x - \frac{1}{2} \times x \times \frac{x}{4.5} \times \frac{1}{3} x$

$= 0.9808\ x - \frac{x^3}{27}$ which is a third degree curve. B.M. is maximum at

$x = 2.971$ m, where S.F. is zere

$$M_{max} = 0.9808 \times 2.971 - \frac{(2.971)^3}{27} = 1.9427 \text{ kN.m}$$

$$x = 4.5,\ M_E = 0.9808 \times 4.5 - \frac{(4.5)^3}{27} = +\ 1.0386 \text{ kN.m}$$

For EC

Measuring x from E, $M_x = 0.9808\,(x + 4.5) - \frac{1}{2} \times 1 \times 4.5\,(x + 4.5)$

$$= 1.0386 - 1.2692\,x$$

This is a linear variation. The B.M. is zero at

$$x = \frac{1.0386}{1.2692} = 0.8183 \text{ m from } E$$

At $x = 1.5$ m, $M_C = 1.0386 - 1.2692 \times 1.5$

$$= -\,0.8652 \text{ kN.m}$$

For CD

Measuring x from C,

$$M_x = 0.9808\,(6 + x) - \frac{1}{2} \times 4.5 \times 1\,(3 + x) + \frac{2.2692\,x^2}{2},$$

Which is a parabolic variation. M_{max} (at $x = 0.5593$ m) $=$

$$0.9808\,(6 + 0.593) - \frac{1}{2} \times 4.5 \times 1 \times (3 + 0.5593) + \frac{2.2692}{2}\,(0.5593)^2$$

$= -\ 1.2201$ kN.m.

At $x = 1$, $M_D = 0.9808\,(6 + 1) - \frac{1}{2}\,4.5\,(3+1) + \frac{2.2692(1)^2}{2}$

$$= -\ 1.0 \text{ kN.m}$$

From BD

Measuring x from B, $M_x = -\,1 \times x$, which is a linear variation, giving

$$M_B = 0 \text{ and } M_x = -\,1 \times 1 = -1 \text{ kN.m}$$

Problem 2.28: *AB* is a vertical post of a crane Fig. 2.28 (a). The sockets at *A* and *B* offer no constrant against flexure. The horizontal arm *CD* is hinged to *AB* at *C* and supported by strut *FE* which is freely hinged at its two extremities to *AB* and *CD*. Construct the bending moment diagrams for *AB* and *CD*.

Solution:

$$FE = \sqrt{3^2 + 2^2} = 3.606$$

$$\therefore \quad \sin\theta = \frac{2}{3.606} = 0.5547$$

$$\cos\theta = \frac{3}{3.606} = 0.8321$$

Using equilibrium of the whole frame

$$H \times 6 = 20 \times 6 + 2 \times 6 \times 3$$

$$\therefore \quad H = 26 \text{ kN}$$

Let T be thrust in EF.

For the arm CD,

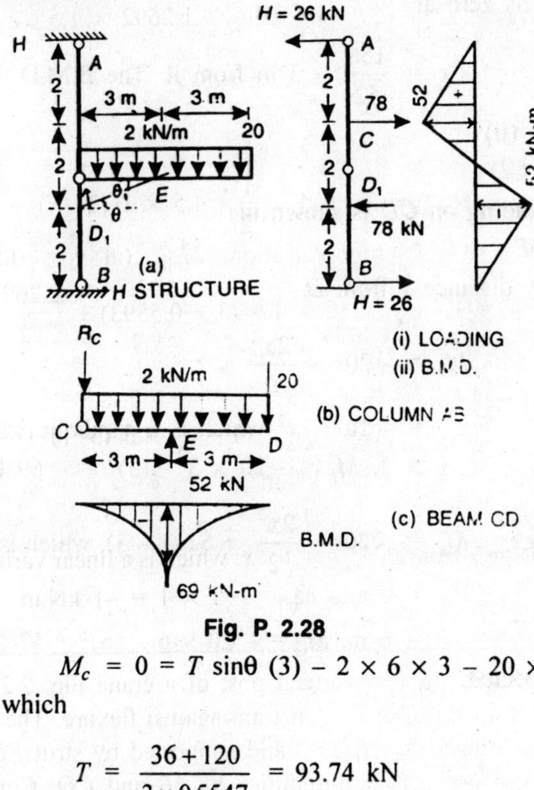

Fig. P. 2.28

$$M_c = 0 = T \sin\theta\,(3) - 2 \times 6 \times 3 - 20 \times 6$$

From which

$$T = \frac{36 + 120}{3 \times 0.5547} = 93.74 \text{ kN}$$

The vertical reaction at $E = T \sin\theta = 93.74 \times 0.5547$

$= 52$ kN

Horizontal pull in $CE = P = T \cos\theta = 93.74 \times 0.8321$

$= 78$ kN $(\rightarrow)$

Also, horizontal load at F

$= T \cos\theta = 78$ kN $(\leftarrow)$

Post AB : The various forces acting on the vertical post AB are shown in Fig. 2.28 (b) Evidently.

$$M_A = M_B = 0$$

$$M_C = +26 \times 2 = +52 \text{ kN.m}$$

$$M_F = +26 \times 4 - 78 \times 2 = -52 \text{ kN.m}$$

For CF,

$$M_x = 26\,x - 78\,(x - 2) = 156 - 52\,x$$

which is zero at

$$x = \frac{156}{52} = 3 \text{ m from } A.$$ The B.M.D. is shown in Fig. 2.28 b (ii)

Beam CD

The loading on CD is shown in

For DE

At any distance x from D.

$$M_x = -20x - \frac{2x^2}{2}$$

$= -20x - x^2$ which is a parabolic variation.

At $x = 3$, $M_E = -20 \times 3 - (3)^2 = -69$ kN.m.

For EC: $M_x = -20x - \dfrac{2x^2}{2} + 52\,(x - 3)$, which is a parabolic variation.

At $x = 6$ m, $M_C = -20 \times 6 - (6)^2 + 52\,(6 - 3) = 0$, as expected.

Chapter 3

Moments of Inertia

Problem 3.1: Determine the moment of inertia of a triangle of base b and altitude h with respect to an axis through the apex parallel to the base.

Solution:

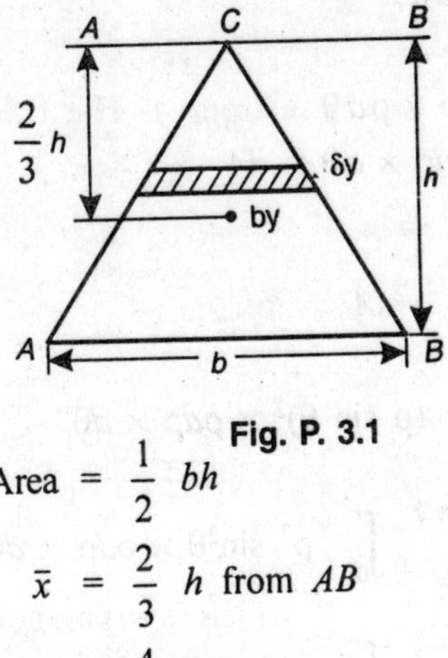

Fig. P. 3.1

$$\text{Area} = \frac{1}{2}bh$$

$$\bar{x} = \frac{2}{3}h \text{ from } AB$$

$$\bar{x}^2 = \frac{4}{9}h^2$$

$$I_{AB} = I_G + A\bar{x}^2$$

Where I_G = moment of enertia about centroidal axis

$$\therefore \quad I_G = \frac{1}{36}bh^3$$

$$\therefore \quad I_{AB} = \frac{1}{36}bh^3 + \frac{1}{2}bh \times \frac{4}{9}h^2$$

$$= \frac{bh^3}{36} + \frac{2bh^3}{9} = \frac{bh^3 + 8bh^3}{36} = \frac{9bh^3}{364} = \frac{bh^3}{4}$$

Problem 3.2: Determine the moment of inertia of the quarter circle w.r.t. given axis.

Solution:

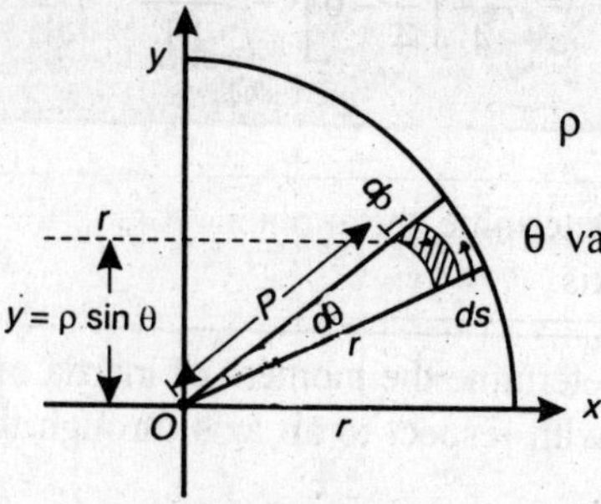

ρ varies from 0 to r

θ varies from 0 to $\frac{\pi}{2}$

Fig. P. 3.2

$$\therefore \quad ds = \rho d\theta$$

∴ Area of segment

$$= d\rho \times \rho d\theta$$

$$= \rho d\rho \times d\theta = dA$$

We know that,

$$I_x = \int y^2 dA$$

$$= \int (\rho \sin \theta)^2 \times \rho d\rho \times d\theta$$

$$= \int_0^{\pi/2} \int_0^r \rho^2 \sin^2\theta \times \rho d\rho \times d\theta$$

$$= \int_0^{\pi/2} \int_0^r \rho^3 \, d\rho \times \sin^2\theta \, d\theta$$

$$= \frac{r^4}{4} \int_0^{\pi/2} \sin^2\theta \, d\theta$$

$$= \frac{r^4}{4} \int_0^{\pi/2} \frac{1 - \cos 2\theta}{2} \, d\theta$$

$$= \frac{r^4}{4} \left[\frac{\theta}{2} - \frac{\sin 2\theta}{4}\right]_0^{\pi/2}$$

$$= \frac{r^4}{4}\left[\frac{1}{2} \times \frac{\pi}{2} - \frac{\sin 80^\circ}{4}\right]$$

$$= \frac{r^4}{4}\left[\frac{\pi}{4} - 0\right] = \frac{\pi r^4}{16}$$

Problem 3.3: Determine the moment of inertia of the semi-circle w.r.t. to given axis.

Solution:

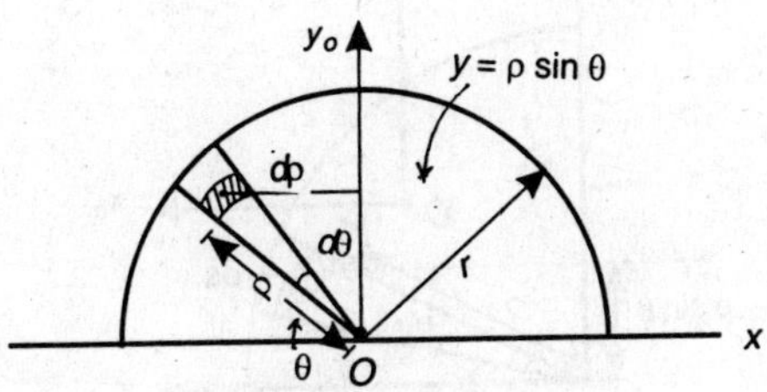

Fig. P. 3.3

$$I_x = \int y^2 dA$$

$$= \int_0^r \int_0^\pi \rho^2 \times \sin^2\theta \times \rho d\theta \times d\rho$$

$$= \frac{r^4}{4} \int_0^\pi \sin^2\theta \times d\theta$$

$$= \frac{r^4}{4} \int_0^\pi \frac{1 - \cos 2\theta}{2} d\theta$$

$$= \frac{r^4}{4} \left[\frac{1}{2}\theta - \frac{\sin 2\theta}{4}\right]_0^\pi$$

$$= \frac{r^4}{4} \left[\frac{\pi}{2} - \frac{\sin 360}{4}\right]$$

$$= \frac{r^4}{4} \times \left[\frac{\pi}{2} - 0\right]$$

$$= \frac{r^4}{4} \times \pi = \frac{\pi r^4}{8}$$

$$I_G = 0.11\, r^4$$

Problem 3.4: Determine the moment of inertia after the quarter circle with respect to a centroidal x-axis.

Solution:

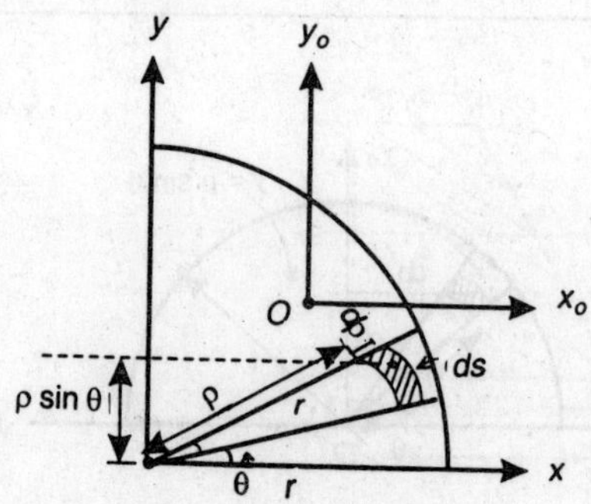

Fig. P. 3.4

$$I_x = \int y^2 d\,A$$

$$= \int_0^{\pi/2}\int_0^r \rho^2 \sin^2\theta \,.\, \rho\, d\rho \,.\, d\theta$$

$$= \frac{r^4}{4}\int_0^{\pi/2} \frac{1-\cos 2\theta}{2}\, d\theta$$

$$= \frac{\pi r^4}{16}$$

$$\therefore \quad I_x = I_G + A\bar{h}^2$$

$$\therefore \quad I_G = I_x - A\bar{h}^2$$

$$\frac{\pi r4}{16} - \frac{\pi r^2}{4} \times \left(\frac{4r}{3\pi}\right)^2$$

$$= \frac{\pi r^4}{16} - \frac{\pi r^2}{4} \times \frac{16r^2}{9\pi^2}$$

$$= 0.0549\, r^4$$

$$\approx 0.055\, r^4$$

Problem 3.5: Determine the position of the centroid of the plane figure shown in Fig. P 3.5.

Solution:

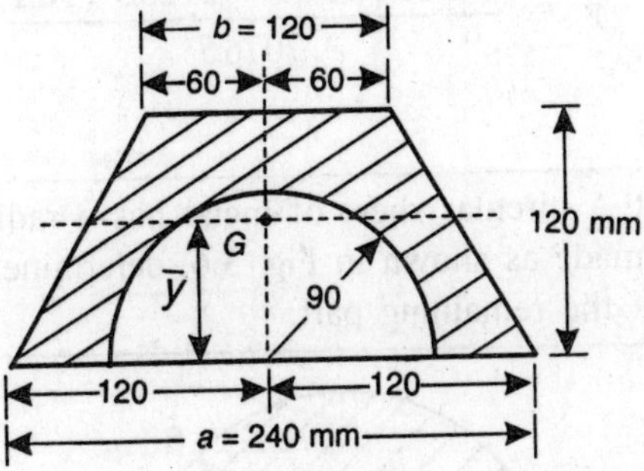

Fig. P. 3.5

Let the distance of the centroid G be $\bar{y}$ from the base.

Area of the trapezium

$$= A_1 = \frac{120 + 240}{2} \times 110$$

$$= 19800 \text{ mm}^2$$

The distance of its centroid from the bar

$$= y_1 = \frac{a + 2b}{a + b} \times \frac{h}{3}$$

$$= \frac{240 + 2 \times 120}{240 + 120} \times \frac{120}{3} = 53.33 \text{ mm}$$

Area of semi-circle

$$= A_2 = \frac{\pi r^2}{2} = \frac{\pi (90)^2}{2}$$

$$= 12723.5 \text{ mm}^2$$

The distance of its centroid from the base

$$= y_2 = \frac{4r}{3\pi} = \frac{4 \times 90}{3\pi} = 38.2 \text{ mm}$$

Total area $= A = A_1 - A_2$

$= 19800 - 12723.5 = 7076.5 \text{ mm}^2$

Now taking moments of the areas about the base,

$$A\,\bar{y} = A_1 y_1 - A_2 y_2$$

$$\bar{y} = \frac{19800 \times 53.33 - 12723.5 \times 38.2}{7076.5} = 80.53 \text{ mm}$$

Problem 3.6: A circular sheet of metal has a radius R. If a hole of radius r is made as shown in Fig. 3.6, determine the position of the centroid of the remaining part.

Solution:

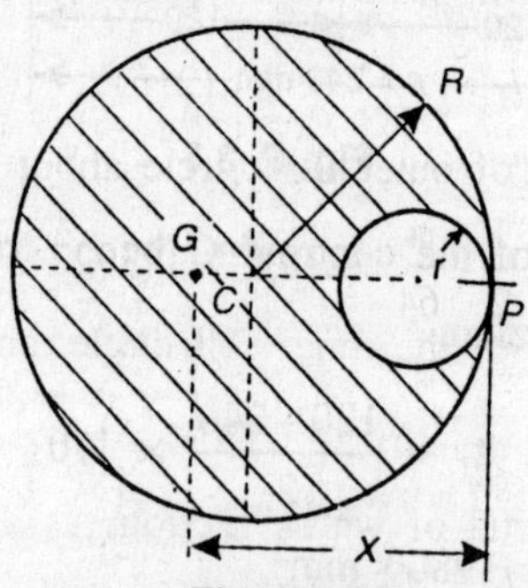

Fig. P. 3.6

Let G be the position of the centroid, distance $\bar{x}$ from P.

Area of circular sheet $= A_1 = \pi R^2$

Distance of the centroid from $P = x_1 = R$

Area of hole $= A_2 = \pi r^2$

Distance of the centroid of the hole $= x_2 = r$

Total area $A = (R^2 - r^2)$

Taking moments of areas about point P, we get

$$A\,\bar{x} = A_1 x_1 - A_2 x_2$$

$$\therefore \quad \bar{x} = \frac{A_1 x_1 - A_2 x_2}{A_1 + A_2} = \frac{\pi R^2 \times R - \pi r^2 \times r}{\pi R^2 - \pi r^2} = \frac{R^3 - r^3}{R^2 - r^2}$$

Problem 3.7: Find the moment of inertia of the shaded area shown in Fig. 3.7 about the axis *AB*.

Solution:

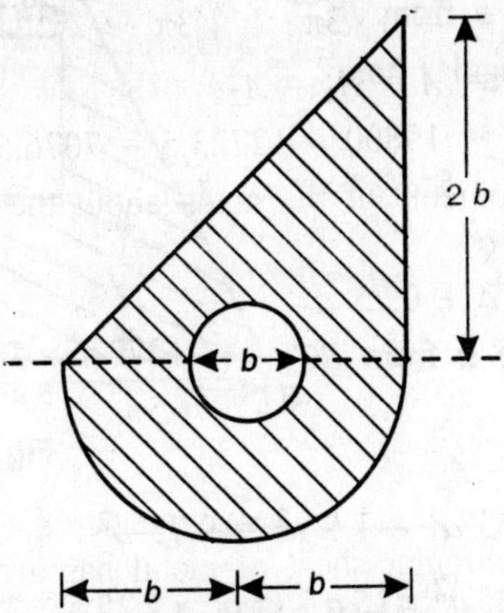

Fig. P. 3.7

Moment of inertia of the triangle, about its base *XX*

$$= I_1 = \frac{1}{12}(2\,b)\,(2\,b)^3 = 1.3333\ b^4$$

Moment of inertia of outer half circle about its diameter

$$I_2 = \frac{1}{2}\ \frac{\pi}{64}\ (2\,b)^4 = 0.3927\ b^4$$

Moment of inertia of the inner full circle, about *XX*

$$I_3 = \frac{\pi}{64}\ (b)^4 = 0.0491\ b^4$$

∴ Moment of inertia of whole section,

$$I = I_1 + I_2 - I_3$$
$$= 1.3333\ b^4 + 0.3927\ b^4 - 0.0491\ b^4 = 1.6769\ b^4$$

Problem 3.8: Find the moment of inertia of the shaded area shown in Fig. 3.8 about the base *BC* and also about the centroidal axis parallel to the base.

Solution:

Let the centroidal axis be at a distance $\bar{y}$ above *BC*

$$\text{Area of triangle} = A_1 = \frac{1}{2}\ (5\ \text{a})\ (4.5\ \text{a}) = 11.25\ a^2$$

and $y_1 = \frac{1}{3}$ (4.5 a)

= 1.5 a from BC

Area of rectangular hole

$= A_2$

$= a \times 1.5\ a$

$= 1.5\ a^2$

and $y_2 = 1.5\ a + 0.75\ a$

$= 2.25\ a$ from BC

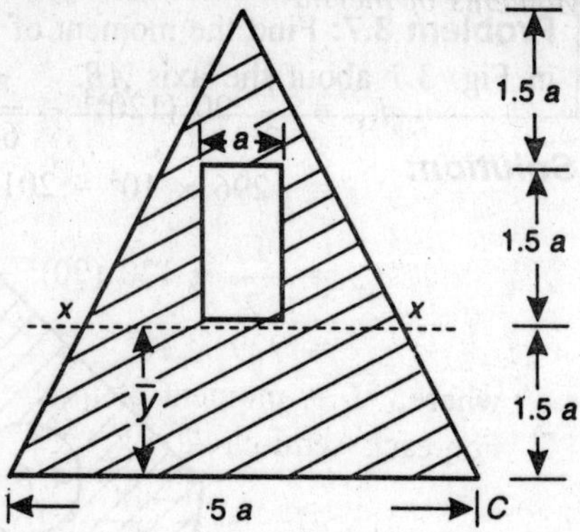

Fig. P. 3.8

∴ Total

$$A = A_1 - A_2$$

$$= 11.25\ a^2 - 1.5\ a^2 = 9.75\ a^2$$

$$\therefore \quad \bar{y} = \frac{11.25\,a^2 \times 1.5\,a - 1.5\,a^2 \times 2.25\,a}{9.75\,a^2} = 1.3846\ a$$

$$I_{BC} = \frac{1}{12}\ (5\ a)\ (4.5\ a)^3 - \left[\frac{1}{12}(a)(1.5\,a)^3 + 1.5\,a^2(2.25\,a)^2\right]$$

$$= 30.094\ a^4$$

Hence $I_{xx} = I_{BC} - A\bar{y}^2 = 30.094\ a^4 - 9.75\ a^2\ (1.3846\ a)^2$

$$= 11.402\ a^4$$

Problem 3.9: Fig. 3.9 shows the section of a beam. Find the M.I. of each semi-circle also M.I. of beam about y-axis

Solution:

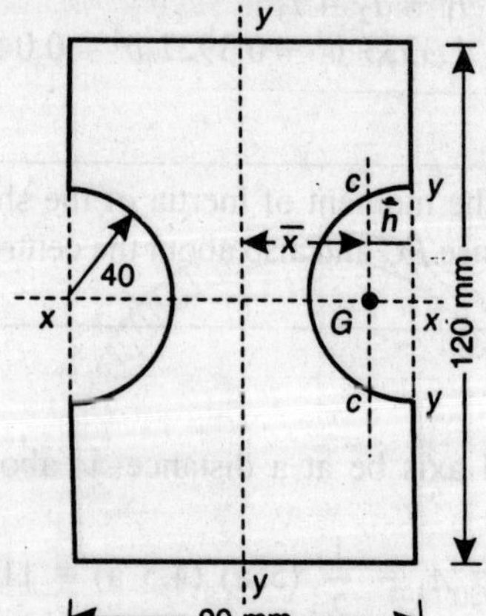

Fig. P. 3.9

$$I_{xx} = \frac{1}{2}\ 90\ (120)^3 - \frac{\pi}{64}\ (80)^4$$
$$= 1296 \times 10^4 - 201 \times 10^4 = 1095 \times 10^4 \text{ mm}^4$$

$$I_{yy} = \frac{1}{12} \times 120\ (90)^3 - 2\ I$$
$$= 729 \times 10^4 - 2\ I$$

where I = moment of inertia of each semi-circle about *yy*-axis

For each semi-circle

$$I_{yy} = \frac{1}{2}\ \frac{\pi}{64}\ (80)^4 = 100.5 \times 10^4 \text{ mm}^4$$

$$A = \frac{\pi}{2}\ (40)^2 = 2513 \text{ mm}^2;$$

$$\bar{h} = \frac{4r}{3\pi} = \frac{4 \times 40}{3\pi} = 17 \text{ mm}$$

$\therefore$
$$I_{cc} = I_{yy} - A\ h^2$$
$$= 100.5 \times 10^4 - 2513\ (17)^2 \gg 28 \times 10^4 \text{ mm}^4$$

Hence
$$I = I_{cc} + A\bar{x}$$
$$= 28 \times 10^4 + 2513\ (45 - 17)^2 = 225 \times 10^4 \text{ mm}^4$$

$\therefore$
$$I_{yy} = 729 \times 10^4 - 2\ I$$
$$= 729 \times 10^4 - 2 \times 225 \times 10^4 = 279 \times 10^4 \text{ mm}^4$$

Problem 3.10: Fig. 3.10 shows the section of a steel beam in the shape of an inverted semi-circular channel with flanges. Determine the position of the centroid and the second moment of the area of the section about the centroidal axis parallel to the base.

Solution:

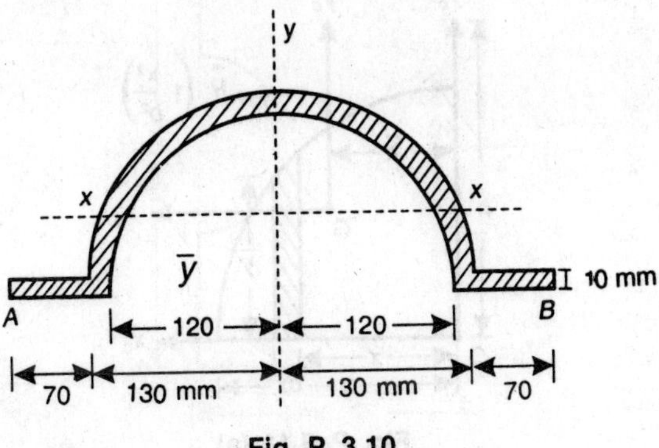

Fig. P. 3.10

$$\text{Total area } A = 2 \times 70 \times 10 + \frac{\pi}{2}(130^2 - 120^2)$$

$$= 1400 + 26546 - 22619 = 5327 \text{ mm}^2$$

Let the centroidal axis x–x be at a distance $\bar{y}$ above AB. Distance of centroid of a semicircle from its base

$$= \frac{4r}{3\pi} = 0.4244\, r$$

$$\therefore \quad 5327\, \bar{y} = 1400 \times 5 + 26546 \times (0.4244 \times 130) - 22619\,(0.4244 \times 120)$$

$$\text{From which } \bar{y} = 60 \text{ mm}$$

$$I_{AB} = \frac{1}{3} \times 140 \times 10^3 + \frac{1}{2}\frac{\pi}{4}(130^4 - 120^4)$$

$$= 3077 \times 10^4 \text{ mm}^4$$

$$\text{Now} \quad I_{xx} = I_{AB} - A\bar{y}^2 = 3077 \times 10^4 - 5327\,(60)^2$$

$$= 1160 \times 10^4 \text{ mm}^4$$

Problem 3.11: Determine the moment of inertia of the shaded parabolic area.
(a) Moment of inertia about x_o, y_o axes, and
(b) moments of inertia about the centroidal axes x_g, y_g.

Solution:

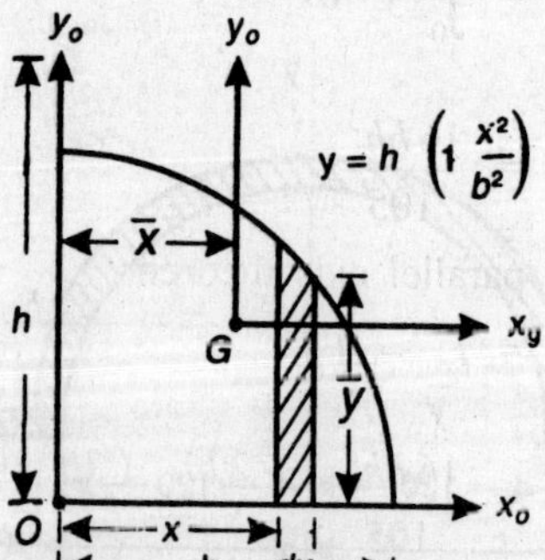

Fig. P. 3.11 (a)

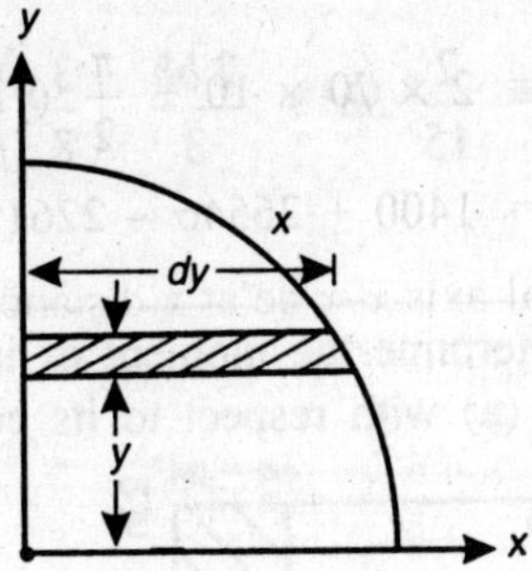

Fig. P. 3.11 (b)

$$A = \frac{2bh}{3}, \quad \bar{x} = \frac{3}{8}\, b \text{ and } \bar{y} = \frac{2h}{5}$$

$$I_{yo} = \int_0^b x^2 \times y\, dx$$

$$= \int_0^b x^2 h\left(1 - \frac{x^2}{b^2}\right) dx = \frac{2}{15}\, hb^3$$

In order to find I_{xo}, consider a strip of thickness dy at distance y from the x-axis.

The width of the strip

$$= x = b\left(1 - \frac{y}{h}\right)^{1/2}$$

$$\therefore \quad I_{xo} = \int_0^h y^2\, dx \times dy = \int_0^h y^2 \times b\left(1 - \frac{y}{h}\right)^{1/2} dy$$

$$= \frac{16\,bh^3}{105}$$

Now, from the parallel axis theorem,

$$I_{xg} = I_{xo} - A\,\bar{y}^2$$

$$= \frac{16\,bh^3}{105} - \frac{2\,bh}{3}\left(\frac{2h}{5}\right)^2 = \frac{8}{175}\, bh^3$$

$$I_{yg} = I_{yo} - A\,\bar{x}^2$$

$$= \frac{2}{15} bh^3 - \frac{2\,bh}{3}\left(\frac{3}{8}b\right)^2 = \frac{19}{480} hb^3$$

Problem 3.12: Determine the moment of inertia of the T section shown in Fig. 3.12 (a) with respect to its centroidal x_o axis.

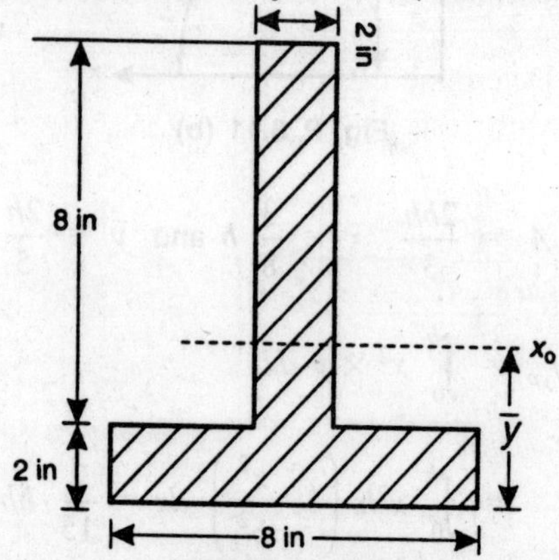

Fig. P. 3.12 (a)

Solution:

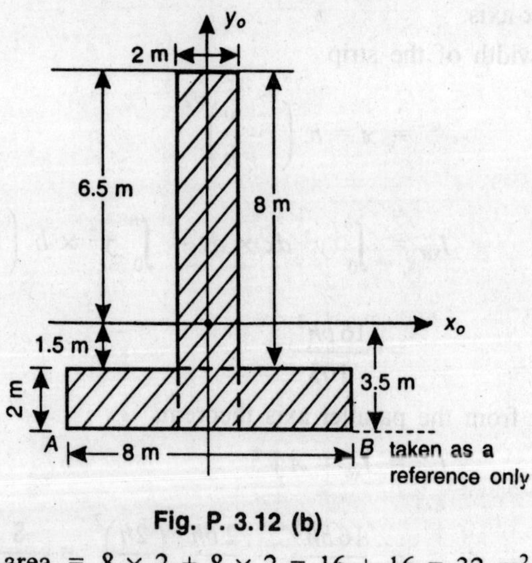

Fig. P. 3.12 (b)

$$\text{Total area} = 8 \times 2 + 8 \times 2 = 16 + 16 = 32 \text{ m}^2$$

To find the value of $\bar{y}$, take the moments of the areas about AB,

$$32\,\bar{y} = 2 \times 8 \times 6 + 2 \times 8 \times 1$$

$$\bar{y} = 3.5 \text{ m}$$

$$I_{AB} = \frac{1}{3} \times 2 \times 10^3 + \frac{1}{3} \times 6 \times 2^3 = 682.67 \text{ mm}^4$$

$$\therefore \quad I_G = I_{AB} - A \times \bar{y}^2$$

$$= 682.67 - 32 \times 3.5^2 = 290.67 \text{ mm}^4 \approx 91 \text{ mm}^4$$

Problem 3.13: Determine the moment of inertia of the area shown in Fig. 3.13 (a) with respect to its centroidal axes.

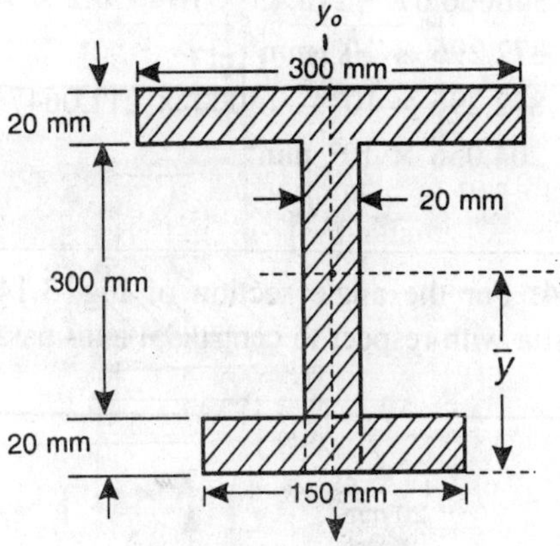

Fig. P. 3.13 (a)

Solution:

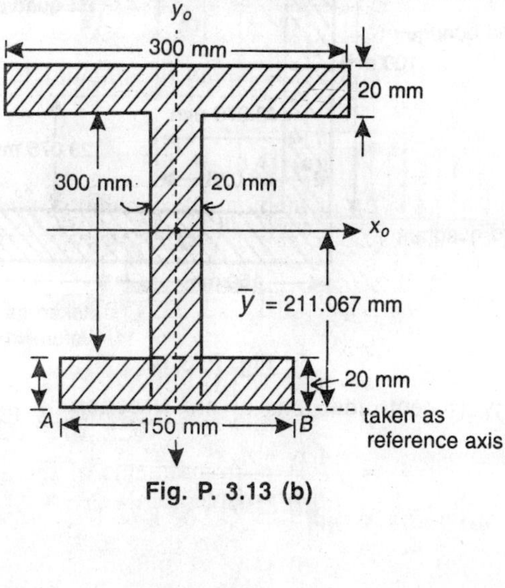

Fig. P. 3.13 (b)

Total area $= 300 \times 20 + 300 \times 20 + 150 \times 20 = 15000 \text{ mm}^2$

$$\bar{y} \times 15000 = 320 \times 20 + 330 + 320 \times 20 \times 160 + 150 \times 20 \times 160$$

$$\bar{y} = 211.067 \text{ mm}$$

$$I_{AB} = \frac{1}{3} \times 130 \times 20^3 + \frac{1}{3} \times 20 \times (320)^3 + \frac{1}{12} \times 300 \times 20^3 \times 20^3 + 300 \times 20 \times 330^2$$

$$= 346666.67 + 218.45 \times 10^6 + 0.2 \times 10^6 + 653.4 \times 10^6$$

$$= 872.296 \times 10^6 \text{ mm}^4$$

$$\therefore \quad I_G = 872.296 \times 10^6 - 15000 \times 211.0647^2$$

$$= 204.056 \times 10^6 \text{ mm}^4$$

Problem 3.14: For the angle section of Fig. 3.14, determine the product of inertia with respect to centroidal axes parallel to the x and y axes.

Solution:

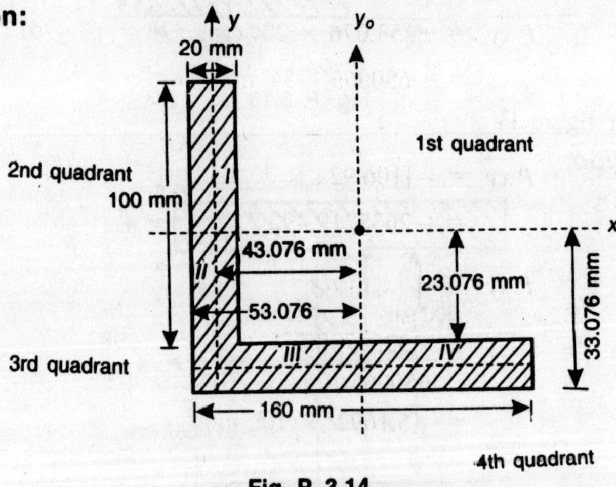

Fig. P. 3.14

$x_0 - y_0$ is centroidal axis of cross-section

$$\bar{x} = \frac{A_1x_1 + A_2x_2}{A_1 + A_2}$$

$$= \frac{160 \times 20 \times 80 + 100 \times 20 \times 10}{160 \times 20 + 100 \times 20}$$

$$= \frac{256000 + 20000}{5200} = 53.076 \text{ mm}$$

$$\bar{y} = \frac{160 \times 20 \times 10 + 20 \times 100 \times 70}{5200} = 33.076 \text{ mm}$$

We know that, product of inertia with respect to centroidal axes parallel to the x and y axes,

$$P\overline{xy} = \int x_o y_o \, dA$$

$\therefore$ For figure I

$$P\overline{xy} = -[20 \times 86.924 \times 43.076 \times 43.462]$$
$$= -3254728.558 \text{ mm}^4$$

For figure II

$$P\overline{xy} = +[13.076 \times 20 \times 43.076 \times 6.538]$$
$$= +73652.10983 \text{ mm}^4$$

For figure III

$$P\overline{xy} = +[53.076 \times 20 \times 26.538 \times 23.076]$$
$$= +650065.1754 \text{ mm}^4$$

For figure IV

$$P\overline{xy} = -[106.924 \times 20 \times 53.462 \times 23.076]$$
$$= -2638219.492$$

$$\therefore \quad P\overline{xy} = \int x_o y_o \, dA$$
$$= -3254728.558 + 73652.10983 + 650065.1754 + 2638219.992$$
$$= -5.1692. \times 10^6 \text{ mm}^4$$

Problem 3.15: Determine the product of inertia of the Z section shown in Fig. 3.15 (a) with respect to the centroidal x and y axes.

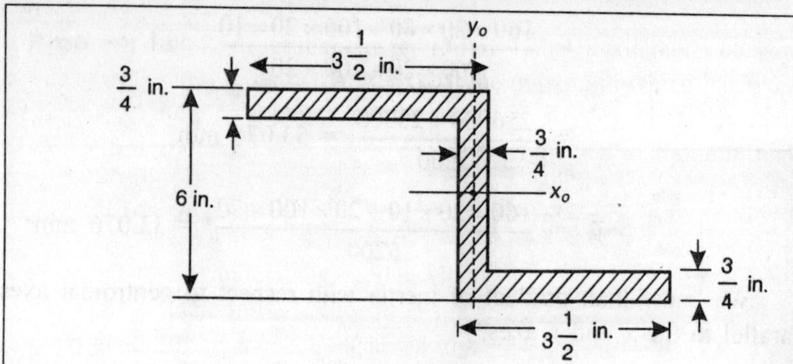

Fig. P. 3.15 (a)

Solution:

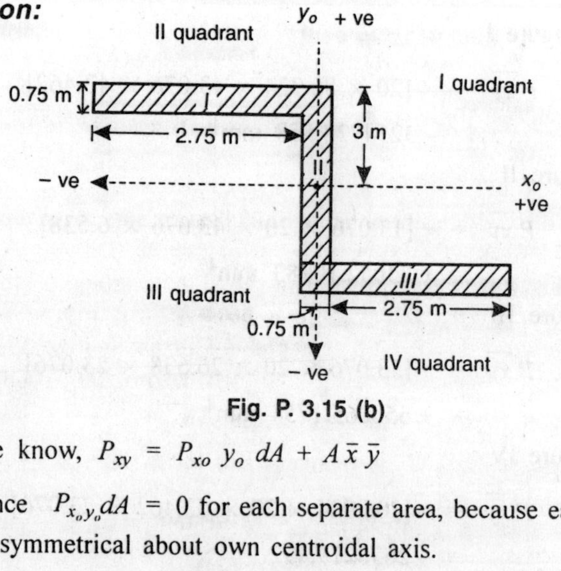

Fig. P. 3.15 (b)

We know, $P_{xy} = P_{xo}\ y_o\ dA + A\,\bar{x}\,\bar{y}$

Since $P_{x_o y_o} dA = 0$ for each separate area, because each separate area is symmetrical about own centroidal axis.

$$\therefore \quad P_{xy} = \Sigma A\,\bar{x}\,\bar{y}$$

For area I $P_{xy} = A\,\bar{x}\,\bar{y}$

$$= -2.75 \times 0.75 \times 1.75 \times 2.625 = -9.4746$$

For area II $P_{xy} = 0$

For area III $P_{xy} = P_{xy}$ of I area

$$= -9.4746 \text{ mm}^4$$

$$\therefore \quad \Sigma P_{xy} = -9.4746 + 9.4746$$

$$= -18.9492 \text{ mm}^4$$

Chapter 4

Bending Stress in Beams

Problem 4.1: A cantilever beam, 50 mm wide by 150 mm high and 6 m long, carries a load that varies uniformly from zero at the end to 1000 N/m at the wall. (a) Compute the magnitude and location of the maximum flexural stress. (b) Determine the type and magnitude of the stress in a fiber 20 mm from the top of the beam at a section 2 m from the free end.

Solution:

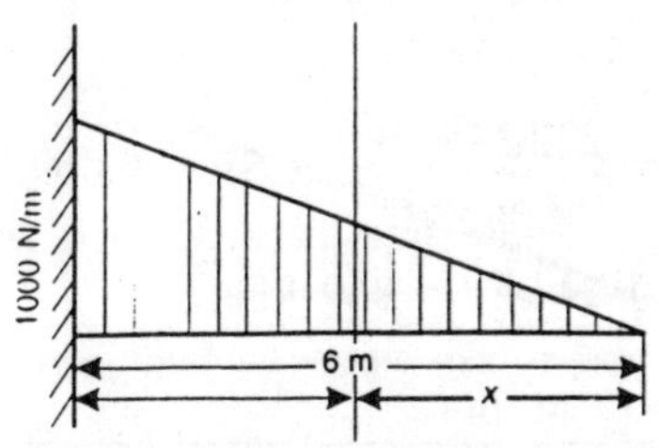

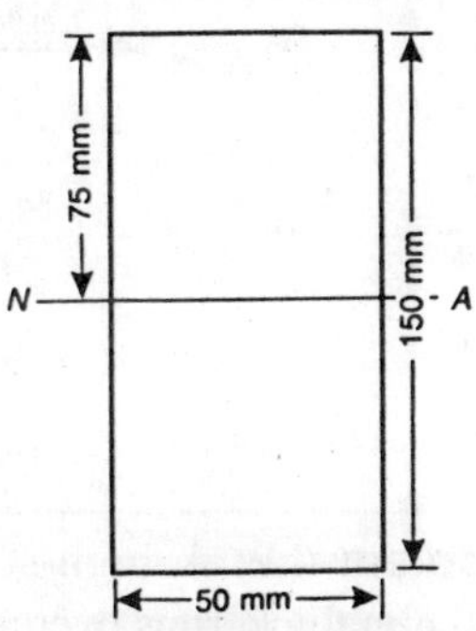

cross section of beam

Fig. P. 4.1

(a) Magnitude and location of the maximum flexural stress.

we know that,

$$f = \frac{M}{I} \cdot y$$

For maximum flexural stress

$$y = \frac{d}{2} = \frac{150}{2} = 75 \text{ mm}$$

$$I = \frac{1}{12}bd^3 = \frac{1}{12} \times 50 \times (150)^3 = 14.06 \times 10^6 \text{ mm}^4$$

For maximum B.M., take a section of a distance x from free end, hence B.M. at that section if,

$$M_x = \text{load} \times \text{C.G. of load}$$

$$= \left(\frac{1000x}{6}\right) \times \frac{x}{3} = \frac{1000x^2}{18} = \frac{500x^2}{9}$$

∴ Max. B.M. at $x = 6$ m

$$M_{max} = \frac{500 \times 6^2}{9} = \frac{500 \times 36}{9} = 2000 \text{ N.m}$$

$$\therefore \quad f_{max} = \frac{2000 \times 1000}{14.06 \times 10^6} \times 75 = \frac{150}{14.06} = 10.67 \text{ N/mm}^2$$

(b) Type and magnitude of the stress in a fibre 20 mm from the top of the beam at a section 2 m from the free end.

→ $y = 75 - 20 = 55$ mm from neutral axis

$$M_{at\ x=2} = \frac{500(2)^2}{9} = \frac{500 \times 4}{9} = 222.23 \text{ N.m}$$

$$\therefore \quad f = \frac{M}{I} \cdot y = \frac{222.23 \times 10^3}{14.06 \times 10^6} \times 55 = 0.869 \text{ N/mm}^2$$

$$= 869 \times 10^3 \text{ N/m}^2 = 869 \text{ kPa.}$$

Problem 4.2: A high-strength steel band saw, 20 mm wide by 0.80 mm thick, runs over pulleys 600 mm in diameter. What maximum flexural stress is developed ? What minimum diameter pulleys can be used without exceeding a flexural stress of 400 MPa? Assume E = 200 GPa.

Solution:

$$E = 200 \text{ GPa} = 200 \times 10^3 \text{ N/mm}^2$$

$$f_{max} = \frac{M}{I} \cdot y = \frac{E}{R} \cdot y$$

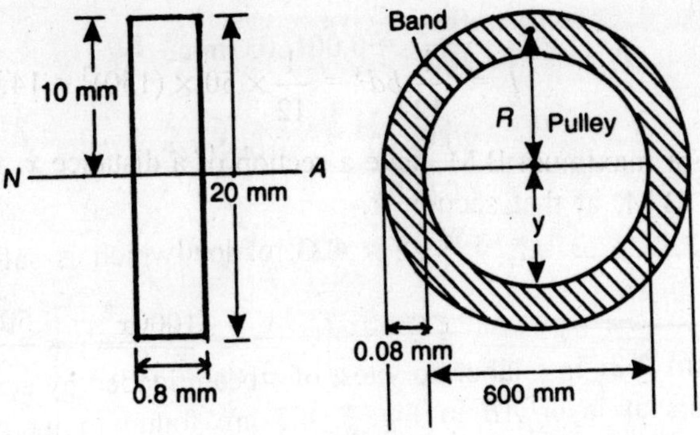

Fig. P. 4.2

$$R = \text{radius of curvature of the neutral fibers}$$

$$= 300 + \frac{0.8}{2}$$

$$= 300 + 0.4$$

$$= 300.4 \text{ mm}$$

$$y_{max} = 300 + 0.8$$

$$= 300.8 \text{ mm}$$

$$\therefore \quad f_{max} = \frac{200 \times 10^3 \times 300.8}{300.4}$$

$$= 200.266 \times 10^3 \text{ N/mm}^2$$

$$\approx 200.267 \times 10^3 \text{ N/mm}^2$$

Minimum diameter of pully can be used without exceeding a flexural stress of 400 MPa,

$$f = 400 \text{ N/mm}^2$$

$$f = \frac{E}{R} \times y$$

$$f = \frac{E}{\left[\frac{d}{2} + 0.4\right]} \quad \frac{d}{2}$$

$$\therefore \quad \frac{d}{2} + 0.4 = \frac{200 \times 10^3}{400} \cdot \frac{d}{2} = 500 \times \frac{d}{2} = 250\, d$$

$$\therefore \quad d = \frac{0.8}{499} = 0.001603 \text{ mm.}$$

$$\therefore \quad f_{max} = \frac{200 \times 10^3}{0.4008} \times \frac{0.001603}{2}$$

$$= 399.95 < 400 \text{ N/mm}^2 \text{ which is safe}$$

Problem 4.3: In a laboratory test of a beam loaded by end couples, the fibers at layer *AB* in Fig. P. 4.3 are found to increase 60×10^{-3} mm whereas those at *CD* decrease 100×10^{-3} mm in the 200-mm gage length. Using E = 70 GPa, determine the flexural stress in the top bottom fibers.

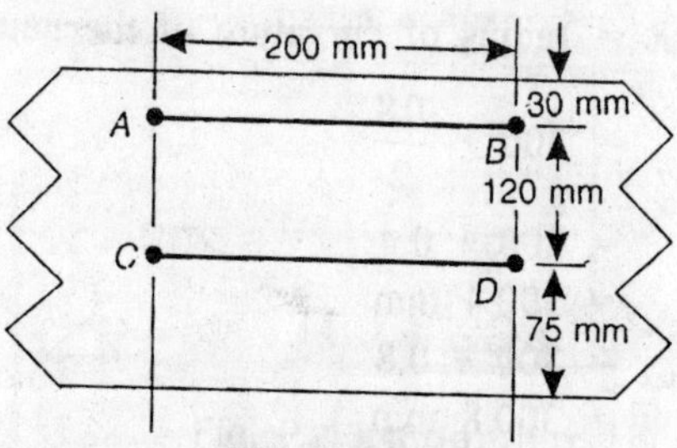

Fig. P. 4.3 (a)

Solution: Given:

Fibre *AB* increase 60×10^{-3} mm

Fibre *CD* decrease 100×10^{-3} mm

Total gauge length = 200 mm

$$E = 70 \text{ GPa} = 70 \times 10^{-3} \text{ N/mm}^2$$

Strain in fibre *AB*

$$= \frac{\text{Change in length}}{\text{Gauge length}} = \frac{60 \times 10^{-3}}{200} = 0.0003$$

Strain in fibre *CD*

$$= \frac{100 \times 10^{-3}}{200} = 0.0005$$

$\therefore$ Stress in fibre *AB*

$$= \text{Strain} \times \text{Modulus of elasticity}$$

$$= 0.0003 \times 70 \times 10^3$$
$$= 21 \text{ N/mm}^2$$

Stress in fibre

$$CD = 0.0005 \times 70 \times 10^3$$
$$= 35 \text{ N/mm}^2$$

∴ Stress distribution diagram in fibre *AB* and *CD* as,

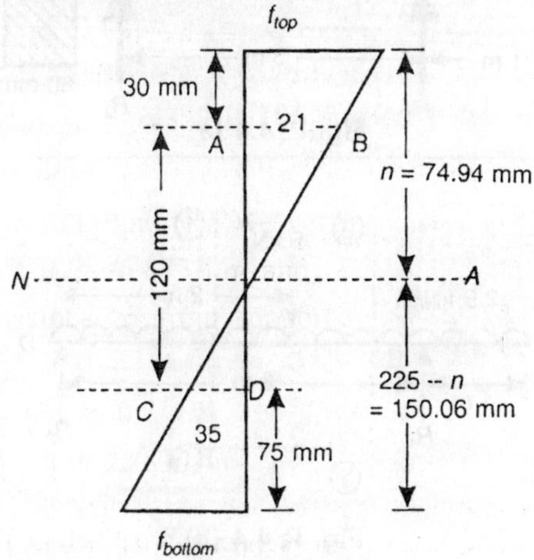

Fig. P. 4.3 (b)

∴ From geometry

$$\frac{21}{n-30} = \frac{35}{225-n-75} = \frac{35}{150-n}$$

$$\therefore [150 - n]\, 21 = [n - 30]\, 35$$

$$150 - n = [n - 30]\,\frac{35}{21} = [n - 30]\, 1.67 = 1.67\, n - 50.1$$

$$\therefore \quad 2.67\, n = 200.1$$

$$\therefore \quad n = \frac{200.1}{2.67} = 74.94 \text{ mm}$$

$$\therefore \quad f_{\text{top}} = \text{Stress of top fibre}$$

$$= \frac{21}{44.94} \times 74.94 = 35.01 \text{ N/mm}^2$$

$$f_{\text{bottom}} = \text{Stress at bottom fibre}$$

$$= \frac{35}{75.06} \times 115.06 = 69.97 \text{ N/mm}^2$$

Problem 4.4: Determine the minimum height h of the beam shown in Fig. 4.4 (a) if the flexural stress is not to exceed 20 MPa.

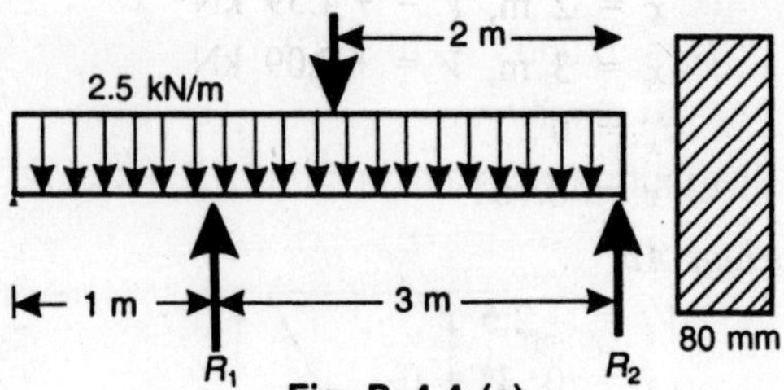

Fig. P. 4.4 (a)

Solution:

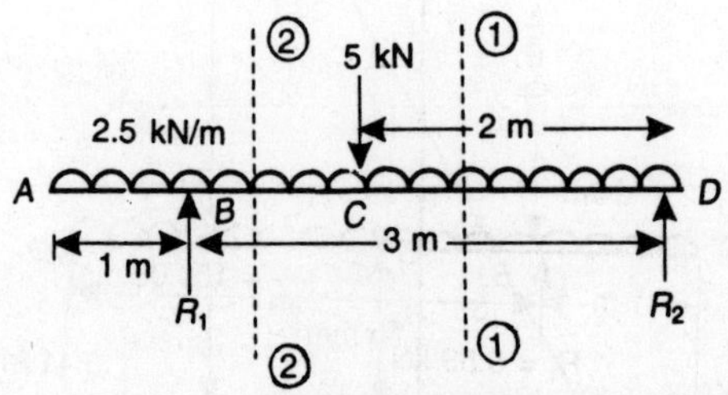

Fig. P. 4.4 (b)

Determine the minimum height h of the beam if the flexual stress is not to exceed 20 MP ≈ 20 N/mm²

First draw S.F.D of the loaded beam

From equilibrium equation,

$$R_1 + R_2 = 5 + 2.5 \times 4 = 15 \quad \text{...(i)}$$

$$\Sigma M_{R1} = 0,$$

$$R_2 \times 3 = 5 \times 1 + 2.5 \times 3 \times 1.5 = 16.25$$

$$\therefore \quad R_2 = 5.41 \text{ kN}$$

$$\therefore \quad R_1 = 9.59 \text{ kN}$$

For portien CD

$$V_x = 2.5\,x - 5.41$$

$$\therefore \text{ at } \quad x = 0,\ V = -5.41$$

$$\text{at} \quad x = 2 \text{ m},\ V = -0.41 \text{ m}$$

For zero S.F.

$$V_x = 0$$

$$\therefore \quad x = 2.164$$

For portion BC

$$V_x = 2.5x + 5 - 5.41$$

at $x = 2$ m, $V = +4.59$ kN

at $x = 3$ m, $V = +7.09$ kN

$$V_x = 0$$

at $x = 0.164$

For portion AB

$$V_x = -2.5\,x$$

at $x = 0,\ V = 0$

at $x = 1$ m, $V = -2.5$ kN

∴ *S.F.D*

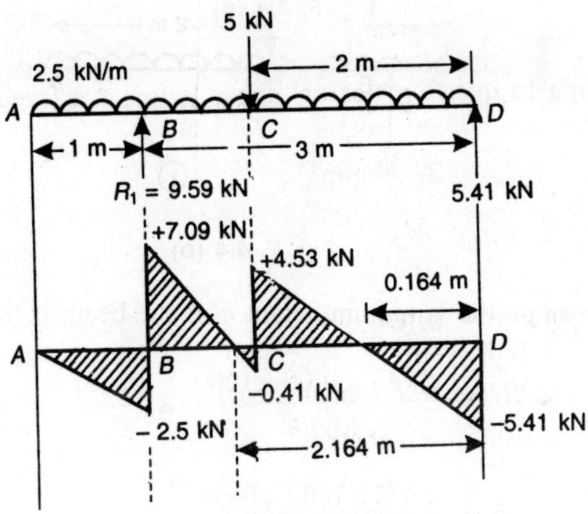

Fig. P. 4.4 (c)

Since shear force is zero at $x = 0.164$ m and $x = 2.164$ m,

∴ Hence B.M. at $x = 0.164$ m

$$M = -\frac{1}{2} \times 0.164 \times 5.41 = -0.443 \text{ kN.m} \qquad \text{...(i)}$$

B.M. at $x = 2.164$ m

$$M = \frac{1}{2} \times 4.59 \times 1.836 = -\frac{1}{2} \times 0.164 \times 5.41$$

$$= -\frac{1}{2} \times 0.41 \times 0.164 = 4.213 - 0.443 - 0.0336$$

$$= +3.736 \text{ kN.m} \qquad \text{...(ii)}$$

$$I = \frac{1}{12} \times 80 \times h^3 = 6.67\ h^3$$

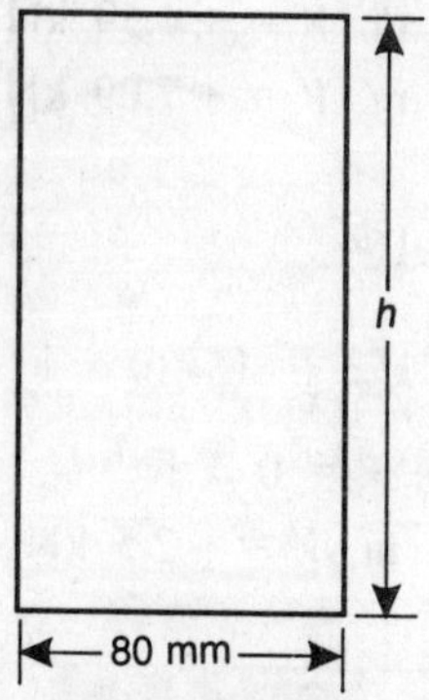

Fig. P. 4.4 (d)

According to question flexural stress is not to exceed 20 MPa ≈ 20 N/mm^2 i.e.

$$f_{max} = 20\ \text{N/mm}^2$$

$$= \frac{M}{I} \cdot y$$

From B.M at x = 0.164 m

$$20 = \frac{0.443 \times 1000 \times 1000}{6.67\ h^3} \times \frac{h}{2}$$

$$\therefore \quad h^2 = \frac{0.443 \times 1000 \times 1000}{6.67 \times 20 \times 2}$$

$$\therefore \quad h = 40.74\ \text{mm}$$

From B.M. at x = 2.164 m

$$20 = \frac{3.376 \times 10^6}{6.67\ h^3} \times \frac{h}{2}$$

$$h^2 = \frac{3.376 \times 10^6}{6.67 \times 20 \times 2}$$

$$\therefore \quad h = 112.48\ \text{mm}$$

∴ Minimum depth of the section is 40.74 mm.

Problem 4.5: A 50-mm diameter bar is used as a simply supported beam 3 m long. Determine the largest uniformly distributed load that can be applied over the right two-thirds of the beam if the flexural stress is limited to 50 MPa.

Solution:

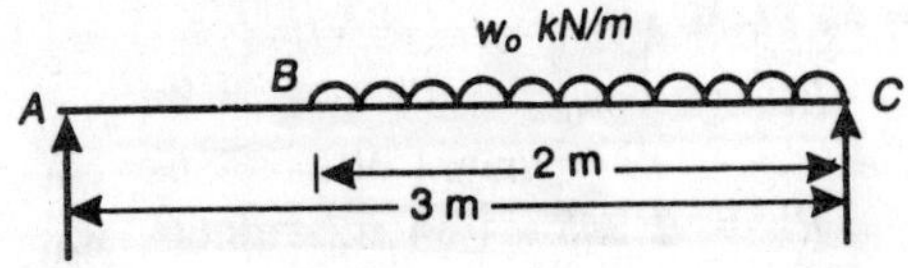

Fig. P. 4.5 (a)

Cross-section of the beam

$$I = \frac{\pi}{64} d^4$$

$$= \frac{\pi}{64} \times (50)^4$$

$$= 3.067 \times 10^5 \text{ mm}^4$$

$$y = \frac{d}{2} = 25 \text{ mm}$$

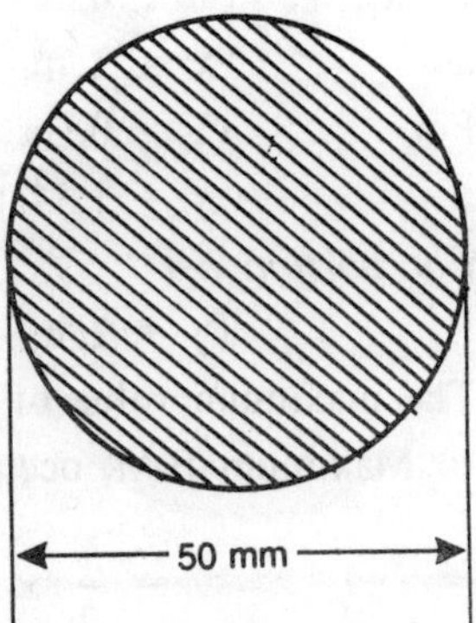

Fig. P. 4.5 (b)

Flexural stress (f) is limited to

$= 50$ MPa

$= 50$ N/mm^2

Shear Force diagram (S.F.D)

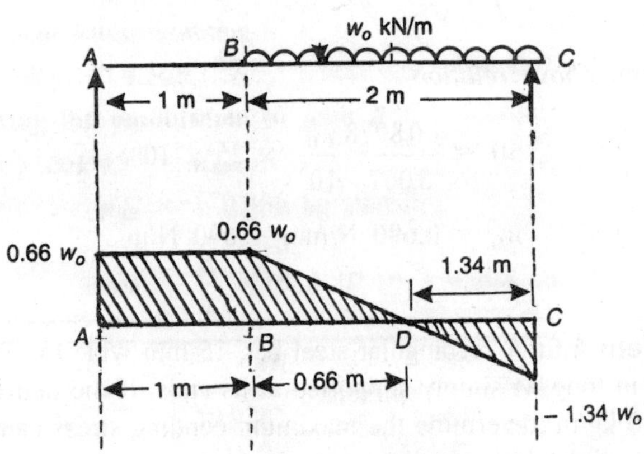

Fig. P. 4.5 (c)

Reaction:

From equilibrium equation

$$R_A + R_C = w_0 \times 2 \quad \text{...(i)}$$

$$\Sigma M_A = 0$$

$$R_C \times 3 = w_0 \times 2 \times 2$$

$$\therefore \quad R_C = \frac{4}{3} w_0 = 1.34\ w_0$$

$$\therefore \quad R_A = 2\ w_0 - 1.34\ w_0 = 0.66\ w_0$$

For portion BC

$$V_x = -1.34\ w_0 + w_0.x = w_0.x - 1.34\ w_0$$

$\therefore$ at $x = 0,\ V = -1.34\ w_0$

at $x = 2$ m, $V = 0.66\ w_0$

For $V_x = 0$, in BC

$$x = 1.34 \text{ m}$$

For portion AB

$$V_x = 2\ w_0 - 1.34\ w_0 = 0.66\ w_0 \text{ Constant}$$

The maximum value of w_0 to reduce the shear to zero at D,

$\therefore$ Maximum B.M. occurs of D is

$$= \frac{1}{2} \times - 1.34\ w_0 \times 1.34 = -0.8978\ w_0$$

Applying the flexural formula,

$$f = \frac{M}{I}.y \quad \text{...(i)}$$

From *above relation,*

$$50 = \frac{0.8978\ w_0}{3.067 \times 10^5} \times 25 \times 10^6$$

$$\therefore \quad w_0 = 0.690 \text{ N/mm} = 690 \text{ N/m.}$$

Problem 4.6: A rectangular steel bar, 15 mm wide by 30 mm high and 6 m long, is simply supported at its ends. If the density of steel is 7850 kg/m^3,determine the maximum bending stress caused by the weight of the bar.

Solution:

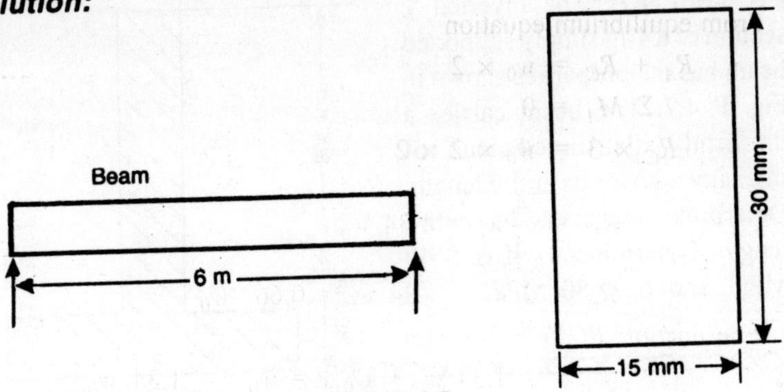

Fig. P. 4.6

Span of beam = 6 m = 6×10^3 mm
Width of beam = 15 mm
Depth of beam = 30 mm
Density of steel = 7850 kg/m^3

$\therefore$ Weight of the bar = Volume of bar × density of bar
= 0.015 × 0.030 × 6 × 7850
= 21.195 kg = 211.95 N

Maximum B.M. due to self weight of the beam

$$= \frac{wl^2}{8}, \text{ where } w = \frac{211.95}{6} = 35.325 \text{ N/m}$$

$$\therefore \quad M_{max} = \frac{35.325 \times 6^2}{8} = 158.9625 \text{ N.m}$$

$$Z = \frac{I}{y_{max}}, \quad I = \frac{1}{12} bd^3$$

$$= \frac{1}{12} \times 15 \times 30^3 = 33750 \text{ mm}^4$$

$$y_{max} = 15 \text{ mm}$$

$$\therefore \quad Z = \frac{33750}{15} = 2250 \text{ mm}^3$$

$$f_{max} = \frac{M}{I} \cdot y_{max} = \frac{M}{Z} = \frac{158.9625 \times 10^3}{2250}$$

$$= 70.65 \text{ N/mm}^2 = 70.65 \text{ MPa}$$

Problem 4.7: The inverted T section of a 4-m simply supported beam has the properties shown in Fig. P 4.7. The beam carries a uniformly distributed load of intensity w_0 over its entire length. Determine w_0 over. Its entire length. Determine w_0 if $\sigma_t \leq 40$ MPa and $\sigma_c < 80$ MPa.

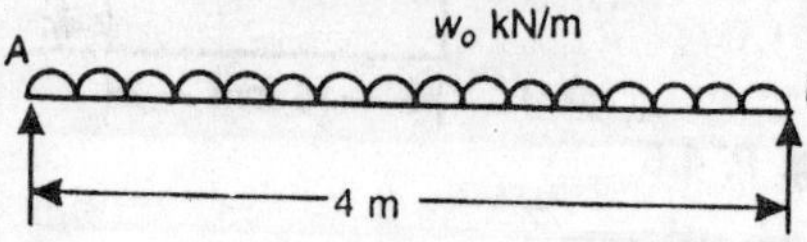

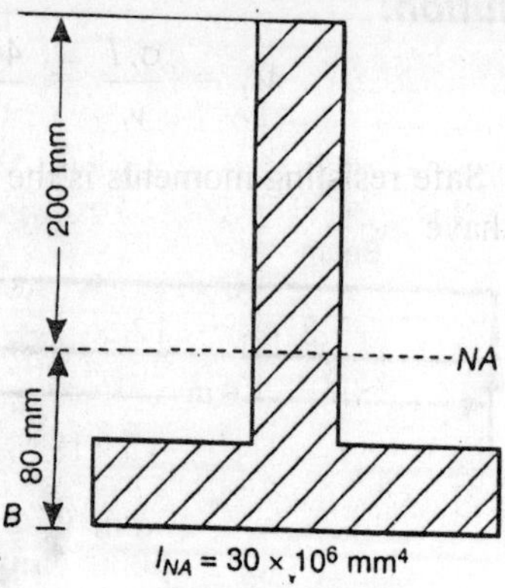

Fig. P. 4.7 (a)

Solution: Given :

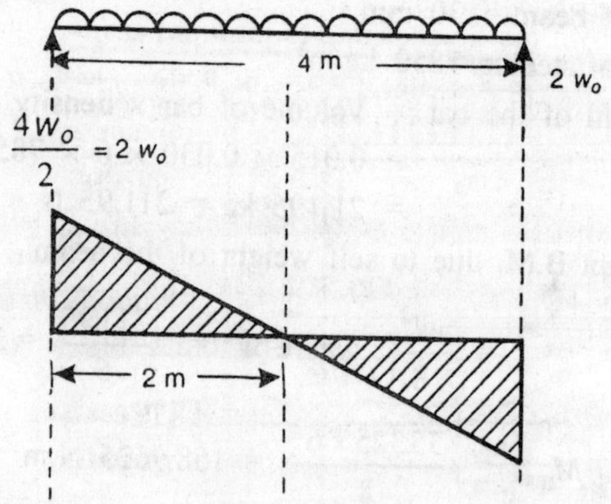

Fig. P. 4.7 (b)

$\sigma_t \leq 40$ MPa ≈ 40 N/mm^2

$\sigma_c \leq 80$ MPa ≈ 80 N/mm^2

since S.F. is zero at $x = 2$ m

B.M. maximum at $x = 2$ m

$$M_{max} = \frac{1}{2} \; 2\, w_0 \times 2 = +2\, w_0 = +2\, w_0 \text{ kN.m}$$

The B.M. at $x = 2$ m indicate concave upward i.e. upper portion in compression and lower portien in tension.

Hence, $$M_c = \frac{\sigma . I}{y_c} = \frac{80 \times 30 \times 10^6}{200} = 12 \times 10^6 \text{ N.mm}$$

$$M_t = \frac{\sigma . I}{y_t} = \frac{40 \times 30 \times 10^6}{80} = 15 \times 10^6 \text{ N.mm}$$

Safe resisting moments is the lower value equality this to the B.M., we have

$$2\ w_0 = 12 \times \frac{10^6}{10^3}$$

$$\therefore \quad w_0 = \frac{12 \times 10^6}{2 \times 10^3} = 6 \times 10^3 \text{ N.m}$$

$$\therefore \quad w_0 = 6000 \text{ N-m}$$

Problem 4.8: Find the maximum tensile and compressive flexure stresses for the cantilever beam shown in Fig. 4.8 (a).

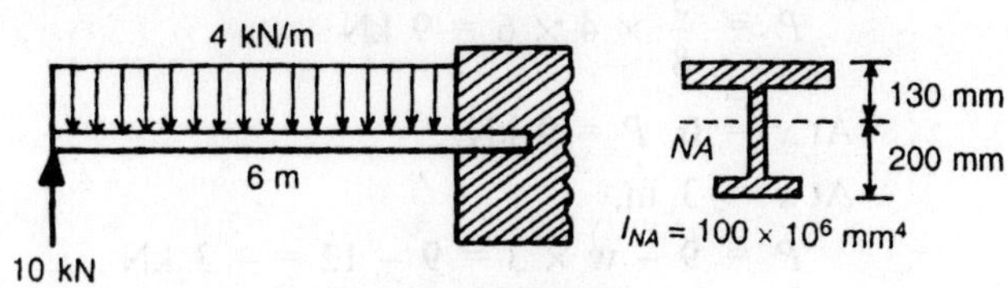

Fig. P. 4.8 (a)

Solution:

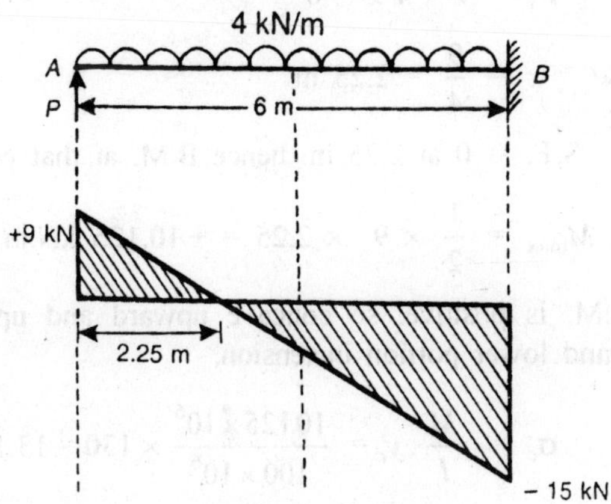

Fig. P. 4.8 (b)

S.F.D of beam

For Prop P

The downward deflection of A due to the reaction P (in above of the prop) is,

$$y_B = \frac{wL^4}{8EI} \text{ (downward)}$$

The upward deflection of B due to the reaction P (in the absence of load) is

$$y'_B = \frac{PL^3}{3EI}$$

Equating the two,

$$\frac{PL^3}{3EI} = \frac{wL^4}{8EI}$$

$$\therefore \quad P = \frac{3}{8} wL \qquad \ldots(i)$$

$$\therefore \quad P = \frac{3}{8} \times 4 \times 6 = 9 \text{ kN}$$

$$\therefore \quad \text{At } x = 0,\ P = 9 \text{ kN}$$

$$\text{At } x = 3 \text{ m},$$

$$P = 9 - w \times 3 = 9 - 12 = -3 \text{ kN}$$

For zero S.F.

$$P_x = 9 - 4x = 0$$

$$\therefore \quad x = \frac{9}{4} = 2.25 \text{ m}$$

since S.F. = 0 at 2.25 m, hence B.M. at that point

$$M_{max} = \frac{1}{2} \times 9 \times 2.25 = +10.125 \text{ kN.m}$$

Since B.M. is positive, so concave upward and upper portion compression and lower portion in tension,

$$\therefore \quad \sigma_c = \frac{M}{I} \cdot y_c = \frac{10.125 \times 10^6}{100 \times 10^6} \times 130 = 13.1625 \text{ N/mm}^2$$

$$\sigma_t = \frac{M}{I} \cdot y_t = \frac{10.125 \times 10^6}{100 \times 10^6} \times 200 = 20.25 \text{ N/mm}^2$$

Problem 4.9: A beam carries a concentrated load W and a total uniformly distributed load of 4 W as shown in Fig. P. 4.9. What safe value of W can be applied if σ_c = 100 MPa and $\sigma_c \leq 60$ MPa? Can a greater load be applied if the section is inverted ? Explain.

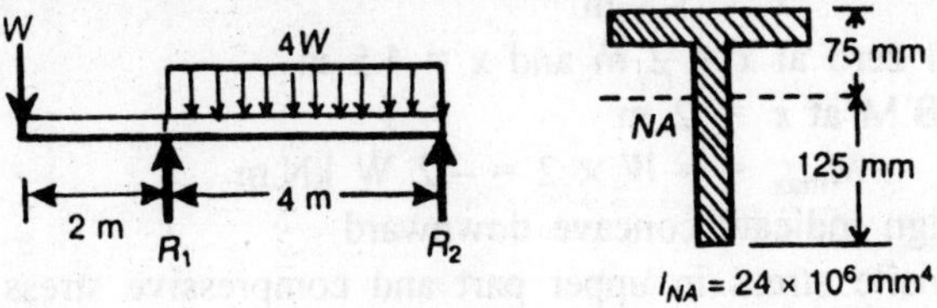

Fig. P. 4.9 (a)

Solution:

$$\Sigma F_y = 0$$

$$R_1 + R_2 = W + 4\ W = 5\ \text{W} \qquad \ldots\text{(i)}$$

$$\Sigma M_c = 0$$

$$R_1 \times 4 = W \times 6 + 4\ W \times 2 = 6\ \text{W} + 8\ \text{W} = 14\ \text{W}$$

$$\therefore \quad R_1 = \frac{14W}{4} = 3.5\ W$$

$$R_2 = 5\ W - 3.5\ W = 1.5\ W$$

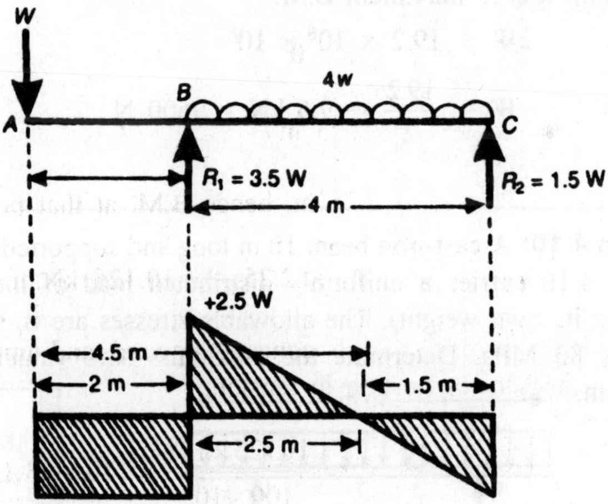

Fig. P. 4.9 (b)

For *Span AB*

$$F = -W \text{ Constant}$$

For *Span BC*

$$F_x = -1.5\ W + Wx$$

at $x = 0, F = -1.5\ W$

at $x = 4$ m, $F = -1.5\ W + 4\ W = +2.5\ W = +2.5\ W$

For zero S.F.,

$$F_x = 0 = -1.5\ W + Wx$$

$$\therefore \quad x = 1.5 \text{ m}$$

S.F. is zero at $x = 2$ m and $x = 4.5$ m

$\therefore$ B.M at $x = 2$ m

$$M_{max} = -W \times 2 = -2\ W \text{ kN.m}$$

–ve sign indicate concave downward

$\therefore$ Tensile stress in upper part and compressive stress in lower part,

$$\sigma_t = \frac{M}{I}.y_t$$

$$\therefore \quad M = \frac{\sigma_t}{y_t}.I = \frac{60}{75} \times 24 \times 10^6$$

$$= 19.2 \times 10^6 \text{ N.mm}$$

$$M = \frac{\sigma_c}{y_c}.I = \frac{100}{125} \times 24 \times 10^6$$

$$= 19.2 \times 10^6 \text{ N.m}$$

Equating this to maximum B.M.

$$2W = 19.2 \times 10^6 \times 10^6$$

$$\therefore \quad W = \frac{19.2}{2} = 9.6 \text{ kN} = 9600 \text{ N}$$

Problem 4.10: A cast-iron beam 10 m long and supported as shown in Fig P 4.10 carries a uniformly distributed load of intensity w_0 (including its own weight). The allowable stresses are $\sigma_t \leq 20$ MPa and $\sigma_t \leq 80$ MPa. Determine the maximum safe value of w_o if $x = 1.0$ m.

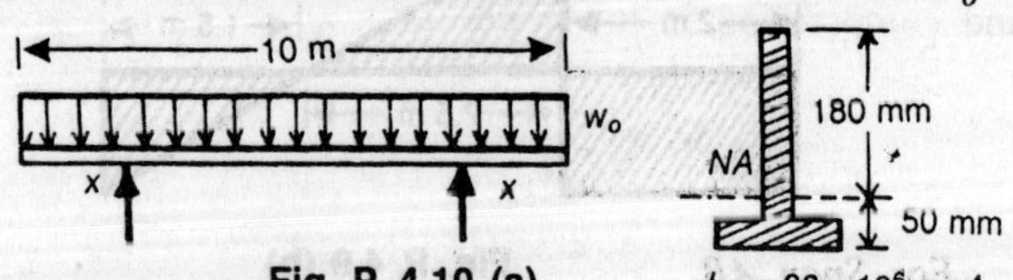

Fig. P. 4.10 (a)

Solution: Given :

$$\sigma_t \leq 20 \text{ MPa} = 20 \text{ N/mm}^2$$

$$\sigma_c \leq 80 \text{ MPa} = 80 \text{ N/mm}^2$$

$$\Sigma F_y = 0$$

$$R_1 + R_2 = w_0 \times 10 \quad \ldots\text{(i)}$$

At point B,

by symmetry

$$R_1 = R_2 = \frac{w_o \times L}{2} = \frac{10w_o}{2} = 5\, w_o$$

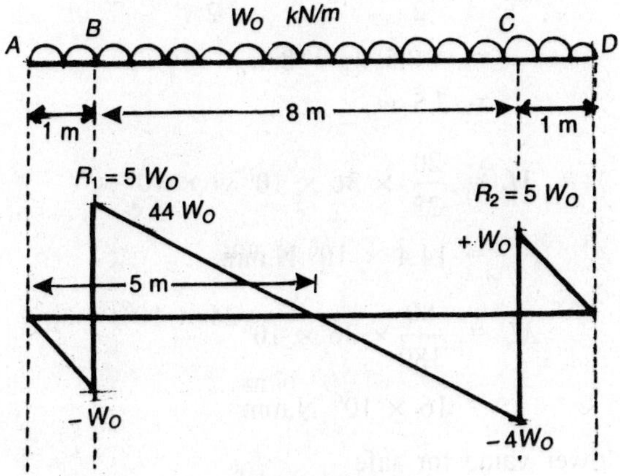

Fig. P. 4.10 (b)

∴ S.F. zero at $x = 1$ m and $x = 5$ from A

∴ At $x = 1$ m

$$\text{B.M} = -\frac{1}{2} \times w_o \times 1 = -0.5\, w_o$$

–ve B.M. indicated concave downward, so tension in upper portion and compression in lower portion,

$$M_t = \frac{\sigma_t . I}{y_t} = \frac{20}{180} \times 36 \times 10^6$$

$$= 4 \times 10^6 \text{ N.mm}$$

$$M_C = \frac{\sigma_c\, I}{y_c} = \frac{80}{50} \times 36 \times 10^6$$

$$= 57.6 \times 10^6 \text{ N.mm}$$

Take minimum value for safe B.M

$$\therefore \quad M = 4 \times 10^6 \text{ N.mm}$$

Equating this with B.M

$$0.5\, w_o \times 10^6 = 4 \times 10^6 \times 10^6$$

$$\therefore \quad W_o = 8 \text{ kN/m}$$

At $\quad x = 5$ m

$$M_{max} = -\frac{1}{2} \times w_o \times 1 + \frac{1}{2} \times 4\, w_o \times 4$$

$$= -0.5\, w_o + 8\, w_o$$

$$= 7.5\, w_o$$

$$\therefore \quad M_t = \frac{20}{25} \times 36 \times 10^6$$

$$= 14.4 \times 10^6 \text{ N.mm}$$

$$M_c = \frac{80}{180} \times 36 \times 10^6$$

$$= 16 \times 10^6 \text{ N.mm}$$

Take lower value for safe

$$\therefore \quad M_{max} = 14.4 \times 10^6 \text{ N.mm}$$

$$= 14.4 \text{ kN-m}$$

Equating these from above equation,

$$7.5\, w_o = 14.4$$

$$\therefore \quad w_o = \frac{14.4}{7.5} = 1.92 \text{ kN/m}$$

Problem 4.11: Compare the flexural strength of the following three beams of equal weight.

(i) *I*-section 100 mm × 200 mm having 10 mm flange thickness and 8 mm thickness.

(ii) A rectangular section having depth equal to twice the width

(iii) Solid circular section.

Solution:

(a) *I*-Section :

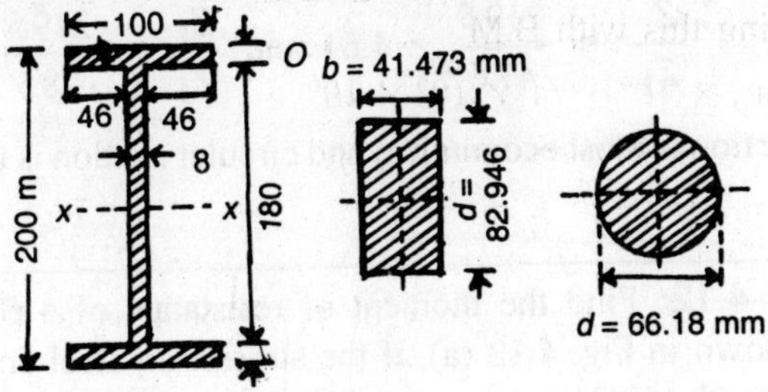

Fig. P. 4.11

$$\text{Area } A_1 = 2 \times 100 \times 10 + 180 \times 8 = 3440 \text{ mm}^2$$

$$\text{Moment of Inertia } I_1 = \frac{1}{12}[BD^3 - bd^3]$$

$$= \frac{1}{12}[100\ (200)^3 - 92\ (180)^3]$$

$$\approx 2195 \times 10^4 \text{ mm}^4$$

Section modulus Z_1

$$= \frac{2195 \times 10^4}{100} = 219.5 \times 10^3 \text{ mm}^3$$

(b) Rectangular Section

$$A_2 = bd = b\ (2b) = 2b^2$$

But $A_2 = A_1$

$\therefore$ $2b^2 = 3440$

From which

$b = 41.473$ mm

and $d = 82.946$ mm

$$\therefore \quad Z_2 = \frac{1}{6}bd^2 = \frac{1}{6}(41.473)(82.946)^2 = 47.6 \times 10^3 \text{ mm}^3$$

(c) Circular section

Let d be the diameter in mm

$$\therefore \quad A_3 = \frac{\pi}{4}d^2$$

But $A_3 = A_1$

$$\therefore \quad \frac{\pi}{4}d^2 = 3440, \text{ from which } d = 66.18 \text{ mm}$$

$$\therefore \quad Z_3 = \frac{\pi}{32} d^3 = \frac{\pi}{32}(66.18)^3 = 28.5 \times 10^3 \text{ mm}^3$$

$$\frac{Z_1}{Z_2} = \frac{219.5}{47.6} = 4.61 \text{ and } \frac{Z_1}{Z_3} = \frac{219.5}{28.5} = 7.70$$

$\therefore$ I section is most economical and circular section is uneconomical.

Problem 4.12: Find the moment of resistance of a flitched beam section shown in Fig. 4.12 (a), if the stresses in steel and timber are not to exceed 160 N/mm² and 8 N/mm² respectively. Take $E_s = 2 \times 10^5$ N/mm² and $E_w = 1 \times 10^1$ N/mm²

Solution:

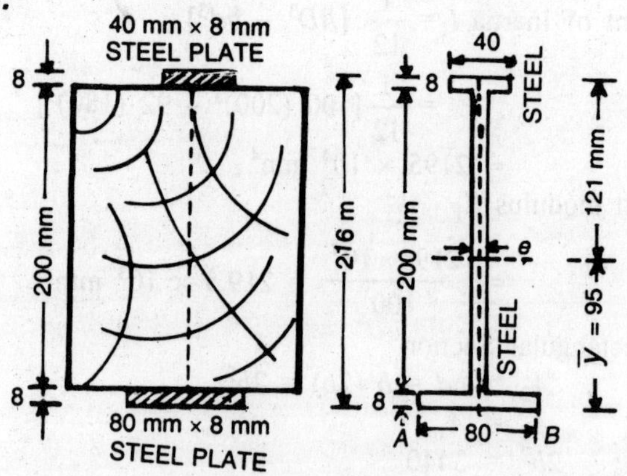

Fig. P. 4.12

$$\text{modular ratio} \quad (m) = \frac{E_S}{E_T} = \frac{2 \times 10^5}{1 \times 10^4} = 20$$

The original section is shown in Fig. 4.12 (a), which can be replaced by an equivalent section in terms of steel, as shown in Fig. 4.12 (b) The 200 × 160 mm timber section can be replaced by steel section having depth equal to 200 mm, but width $= \dfrac{160}{20} = 8$ mm

In order to find $\bar{y}$ of the equivalent section, take moments about base *AB* of *I* section

$$\therefore \quad \bar{y} = \frac{40 \times 8 \times 212 + 200 \times 8 \times 108 + 80 \times 8 \times 4}{40 \times 8 + 200 \times 8 + 80 \times 8}$$

$$= \frac{243200}{2560} = 95 \text{ mm}$$

$$I_{AB} = \frac{1}{3} \times 72\ (8)^3 + \frac{1}{3} \times 8\ (208)^3 +$$

$$\left\{\frac{1}{12} \times 40(8)^3 + 40 \times 8(212)^2\right\} = 3839 \times 10^4 \text{ mm}^4$$

Moment of inertia about cantroidal axis

$$\therefore \quad I_{xx} = 3839 \times 10^4 - 2560\ (95)2 = 1529 \times 10^4 \text{ mm}^4$$

Taking $f_{s,\ max}$ = 160 N/mm^2, extrene stress in timber

$$= \frac{1}{2} \times \frac{160}{121} \times 113 = 7.47 \text{ N/mm}^2$$

which is less than the permissible value of 8 N/mm^2

$$M_r = \frac{f}{y} \times I = \frac{160}{121} \times 1529 \times 10^4$$

$$= 20.22 \times 10^6 \text{ N.mm} = 20.2 \text{ kN.m}$$

Problem 4.13: A solid rod of bronze 20 mm in diameter is surrounded by a fitting steel cylinder of external diameter 28 mm. If the permissible bending strees in bronze and steel are 100 and 150 N/mm^2, find the moment of resistance of the composite section. The young's modulus for steel may be taken as 1.75 times that of bronze.

Solution:

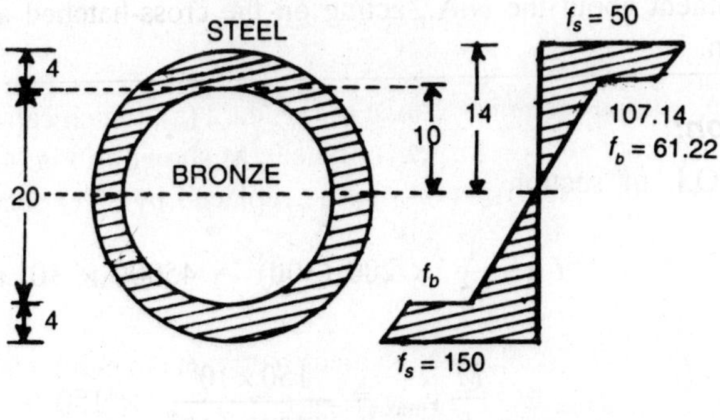

Fig. P. 4.13

modular ratio $(m) = \dfrac{E_S}{E_b} = 1.75$

Let the maximum stress in steel reach its permissible value of 150 N/mm^2. Hence corresponding maximum stress reached in bronze

$$= 150 \times \frac{10}{14} \times \frac{1}{m} = \frac{150 \times 10}{14 \times 1.75}$$

$= 61.22$ N/mm^2, which is less than the permissible value of 100 N/mm^2

For bronze portion

$$Z_b = \frac{\pi}{32}(20)^3 = 785.4 \text{ mm}^3$$

For steel portion

$$Z_S = \frac{\pi}{32}\frac{\left(28^4 - 20^4\right)}{28} = 1594.1 \text{ mm}^3$$

$$M_{rb} = f_b Z_b = 61.22 \times 785.4 = 48082 \text{ N.mm}$$

$$M_{rs} = f_s Z_s = 150 \times 1594.1 = 239120 \text{ N.mm}$$

$$M_r = 48082 + 239120 = 287202 \text{ N.m} \approx 0.287 \text{ kN.m}$$

Problem 4.14: Shows a rectangular section, 200 mm × 300 mm, subjected to a sagging B.M. of 150 kN.m acting about its horizontal axis. Find (a) the compressive force, and its moment about the N.A., acting on the shaded area of the section, and (b) the tensile force and its moment about the N.A., acting on the cross-hatched area of the section.

Solution:

M.O.I. of section

$$I = \frac{1}{12} \times 200\ (300)^3 = 45000 \times 10^4 \text{ mm}^4$$

$$\therefore \quad f_{max} = \frac{M}{I} y_{max} = \frac{150 \times 10^6}{45000 \times 10^4} \times 150 = 50 \text{ N/mm}^2$$

(a) Compressive force on the shaded area

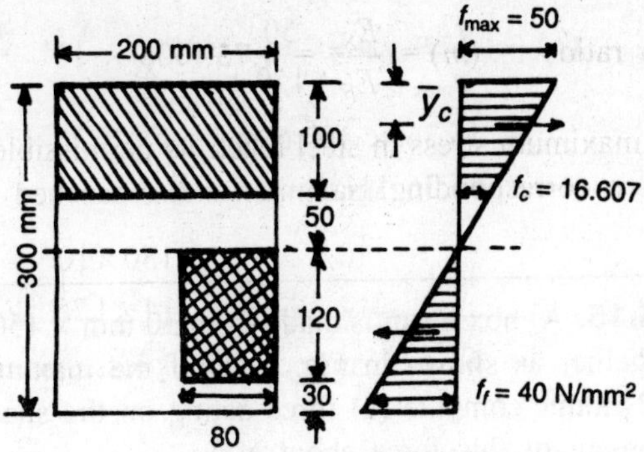

Fig. P. 4.14

$$P_C = \frac{f_{max}}{y_{max}} A\,\bar{y}$$

$$= \frac{50}{150} (120 \times 100)\ (50 + 50)$$

$$= 666667 \text{ kN} = 666.667 \text{ kN}$$

Moment of this force about N.A. is

$$M_C = \frac{f_{max}}{y_{max}} . I_A$$

and I_A = moment of inertia of the shaded area about the N.A.

$$= \frac{1}{12} \times 200\ (100)^3 + 200 \times 100\ (100)^2$$

$$= 21667 \times 10^4$$

$$M_C = \frac{50}{150} = 21667 \times 10^4 = 72.222 \times 10^6 \text{ N.mm}$$

$$= 72.222 \text{ kN.m}$$

(b) Tensile force on the cross-hatched area

$$P_T = \frac{f_{max}}{y_{max}} A\,\bar{y} = \frac{50}{150} (80 \times 120) \times 60$$

$$= 19200 \text{ N} = 192 \text{ kN}$$

$$I_A = \frac{1}{3} \times 80\ (120)^3 = 4608 \times 10^4 \text{ mm}^4$$

$$\therefore \quad M_T = \frac{f_{max}}{y_{max}} IA = \frac{50}{150} \times 4608 \times 10^4$$
$$= 15.36 \times 10^6 \text{ N.m}$$
$$= 15.36 \text{ kN.m}$$

Problem 4.15: A box beam is made from 50 mm × 150 mm pieces screwed together as shown in Fig. 4.15. If the maximum flexural stress is 8 N/mm², compute (a) force acting on the shaded portion, and (b) moment of this force ahout N.A.

Solution:

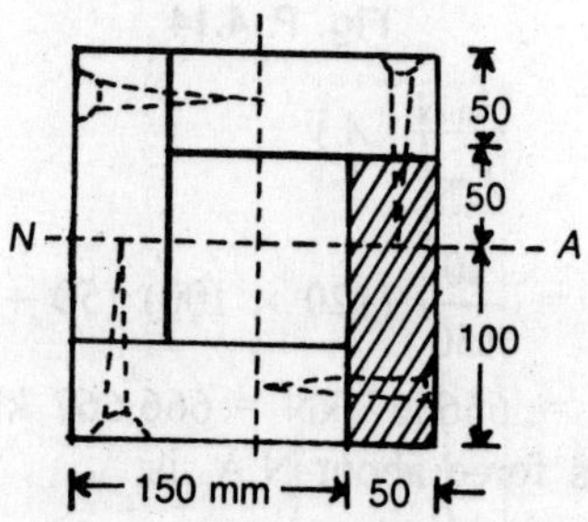

Fig. P. 4.15

Area of shaded portion,

$$A = 50 \times 150 = 7500 \text{ mm}^2$$

$$\bar{y} = 25 \text{ mm, from N.A}$$

$$\therefore \quad p = \frac{f_{max}}{y_{max}} . A\,\bar{y}$$

$$= \frac{8}{100} \times 7500\ (25) = 15000 \text{ N} = 15\text{kN}$$

Moment of this force about N.A is

$$M = \frac{f_{max}}{y_{max}} . I_A$$

where $I_A = \frac{1}{12} \times 50\ (150)^3 + 50 \times 150\ (25)^2$

$$= 1875 \times 10^4 \text{ mm}^4$$

$$M = \frac{8}{100} \times 1875 \times 10^4$$

$$= 1.5 \times 10^6 \text{ N.m} = 1.5 \text{ kN.m}$$

Problem 4.16: A uniformly tapering vertical post of height H having diameter D at the base and a diameter d at the top is fixed at the base. A horizontal force P is applied at the top of the post. Determine the maximum bending stresses for the post and state where it occurs.

Solution:

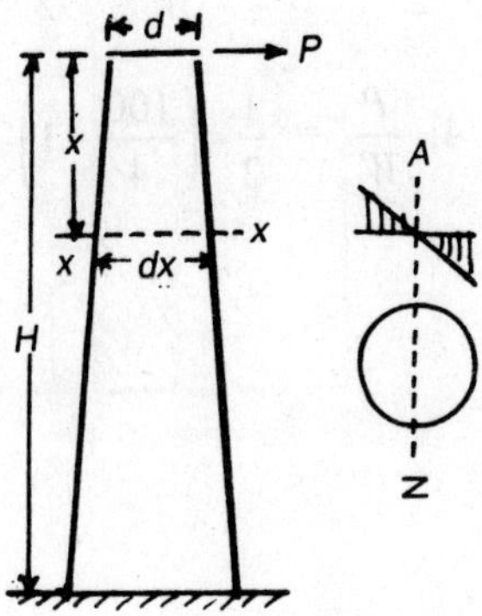

Fig. P. 4.16

Let us take a horizontal section at a distanc x from the top. Its diameter d_x at a distance x from the top of the post.

$$d_x = d + \frac{D-d}{H} x = d + kx$$

where $$k = \frac{D-d}{H}$$

$$\therefore \qquad Z = \frac{\pi d_x^3}{32} \text{ and } M = P.x$$

Stress $$f = \frac{M}{Z} = \frac{P.x}{\pi \dfrac{d_x^3}{32}} = \frac{32p}{\pi} \cdot \frac{x}{d_x^3} = \frac{32p}{\pi} \cdot \frac{x}{(d+kx)^3}$$

$$\text{Hence \% error} = \frac{f_2 - f_1}{f_2} \times 100$$

$$\therefore \quad \frac{k\left(P+\dfrac{W}{2}\right) - kP}{k\left(P+\dfrac{W}{2}\right)} \times 100 = e \quad \text{or} \quad \frac{\dfrac{W}{2}}{P+\dfrac{W}{2}} = \frac{e}{100}$$

$$\text{or} \quad \frac{W}{2P+W} = \frac{e}{100} \quad \text{or} \quad \frac{2P+W}{W} = \frac{100}{e}$$

$$\therefore \quad \frac{2P}{W} = \frac{100}{e} - 1 \quad \text{or} \quad \frac{P}{W} = \frac{1}{2}\left(\frac{100}{e} - 1\right) \quad \textbf{Ans}$$

$$\text{When} \quad e = 4, \ \frac{P}{W} = \frac{1}{2}\left(\frac{100}{4} - 1\right) = 12$$

Chapter 5

Shearing Stresses in Beams

Problem 5.1: A timber beam 80 mm wide by 160 mm high is subjected to a vertical shear $V = 40$ kN. Determine the shearing stress developed at layers 20 mm apart from top to bottom to the section.

Solution:

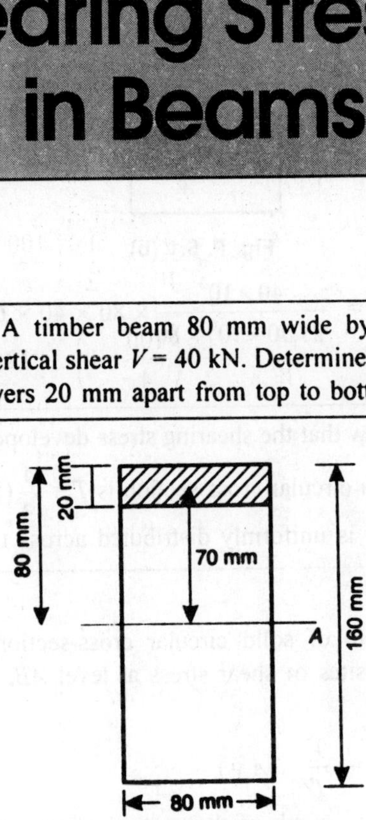

Fig. P. 5.1 (a)

$$V = 40 \text{ kN} = 40 \times 10^3 \text{ N}$$

$$I = \frac{1}{12} bd^3 = \frac{1}{12} \times 80 \times 160^3 = 27.30 \times 10^6 \text{ mm}^4$$

We know that, shearing stress us,

$$q = \frac{v}{Ib} . A\,\bar{y}$$

$$= \frac{40 \times 10^3}{27.30 \times 10^6 \times 80} \times 80 \times 20 \times 70 = 2.05 \text{ N/mm}^2$$

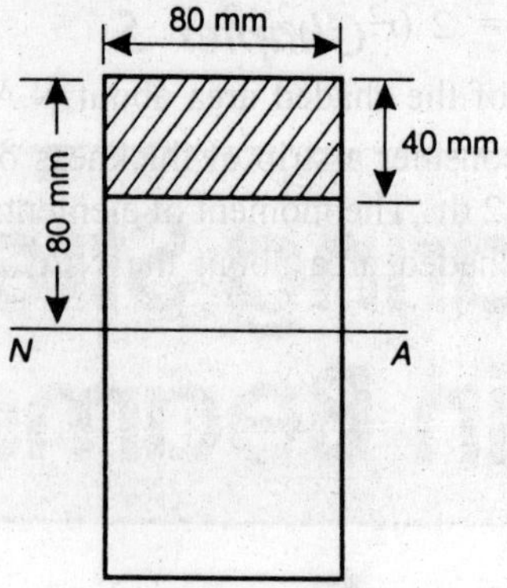

Fig. P. 5.1 (b)

$$q = \frac{40 \times 10^3}{27.30 \times 10^6 \times 80} \times 80 \times 40 \times 60 = 3.51 \text{ N/mm}^2$$

> **Problem 5.2:** Show that the shearing stress developed at the neutral axis of a beam with circular cross section is $T = \frac{4}{3}(V/\pi r^2)$. Assume that shearing stress is uniformly distributed across the neutral axis.

Solution:

Consider a beam of solid circular cross-section, of diameter d (radius r). The intensites of shear stress at level AB, at height y from N.A is given by

$$q = \frac{V}{IZ} \quad (A\bar{y})$$

where $\quad z$ = width of the section of the height

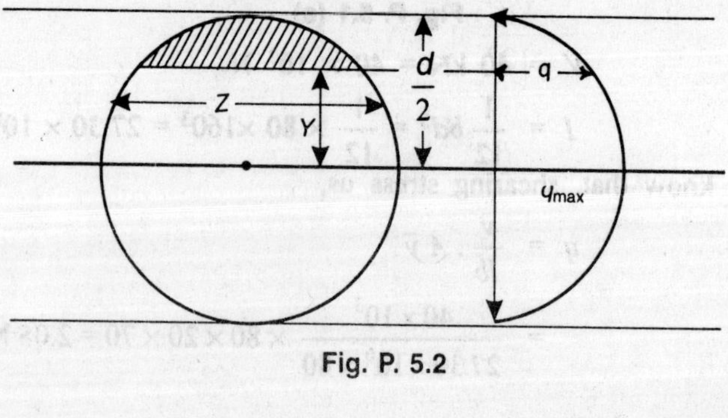

Fig. P. 5.2

$$y = 2\,(r^2 - y^2)^{1/2}$$

$A\bar{y}$ = moment of the shaded area about N.A To find the moment of the shaded area, consider a strip of thickness δy at the height y. The area da of the strip = 2 dy. The moment of elements area = $da.y.$ = Hence the moment of the shaded area about the N.A.

$$A\bar{y} = \int_{y=y}^{y=r} yz\,dy$$

But $\quad z = 2\sqrt{z^2 - y^2}$ or $z^2 = 4\,(r^2 - y^2)$

$\therefore \quad 2\,zdz = 4\,(-2y)\,dy = -8y\,dy$

or $\quad ydy = -\dfrac{1}{4}zdz$

when $\quad y = r,\ z = 0$; when $y = y,\ z = z$

$$\therefore \quad A\bar{y} = \int_0^z \left(-\frac{z^2}{4}\right) d^2 = \frac{1}{4}\int_0^z z^2 dz = \frac{z^3}{12} \qquad \text{...(i)}$$

$$\therefore \quad q = \frac{V}{Iz} A\bar{y} = \frac{V}{Iz}\cdot\frac{z^3}{12} = \frac{Vz^2}{12I}$$

$$\text{or} \quad q = \frac{V}{12I}\{4(r^2 - y^2)\} = \frac{V}{3I}(r^2 - y^2)$$

The variation of q is, therefore, parabolic

$$\text{at} \quad y = r,\ q = 0 \text{ and at } y = 0,\ q = d_{max} = \frac{V}{3I}\cdot r^2$$

$$= \frac{Vd^2}{12I} \text{ at N.A}$$

$$q_{max} = \frac{Fd^2}{12\dfrac{\pi d^4}{64}} = \frac{4}{3}\,\frac{V}{\dfrac{\pi r^2}{4}}$$

$$= \frac{4}{3}\,\frac{V}{\pi r^2} \quad \text{Proved.}$$

Problem 5.3: The cross section of a beam is an isosceles triangle with vertex uppermost, of altitude h and base b. If V is the vertical shear, show that the maximum shearing stress is 3 V/bh located at the midpoint of the altitude.

Solution:

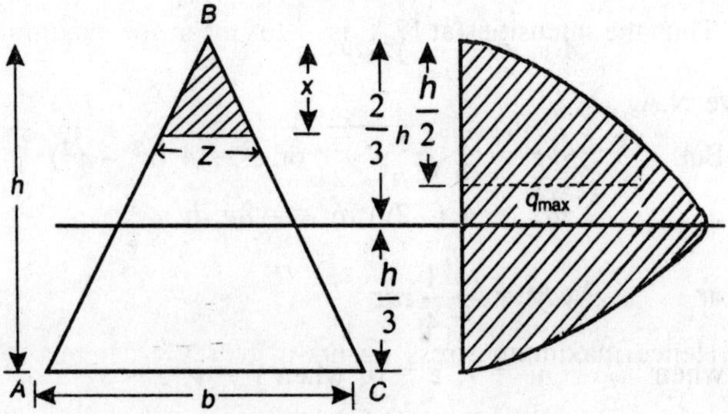

Fig. P. 5.3

$$I_{NA} = \frac{1}{36} bh^3$$

At any section x from the apex B,

$$q = \frac{V}{Iz}(A\,\bar{y})$$

where $\quad z = b\,\dfrac{x}{h}$

$$A\,\bar{y} = \left(\frac{1}{2}zx\right)\left(\frac{2}{3}h - \frac{2}{3}x\right) = \frac{1}{3}\; z\,x\,(h - x)$$

$$q = \frac{V}{Iz}\,\frac{1}{3}.\; zx\,(h - x) = \frac{V}{3I}\; x\,(h - x)$$

$$= \frac{V}{3I} x\,(h - x)$$

The variation is, therefore, parabolic

At $\quad x = 0,\; q = 0$

At N.A.,

$$q_z = \frac{V}{3I}\,\frac{2}{3}h.\left(\frac{1}{3}h\right) = \frac{2}{27}\,\frac{V}{I}.\,h^2 = \frac{2}{27}\,\frac{36V}{bh^3}.\,h^2$$

$$= \frac{72}{27} \cdot \frac{V}{bh} = \frac{8}{3} \frac{V}{bh} \quad \text{...(i)}$$

$$= \frac{4}{3} \frac{V}{area} = \frac{4}{3} q_{mean}$$

Thus the intensities at N.A. is $\frac{4}{3}$ of mean for maximum intensities above N.A.

$$\frac{dq}{dx} = 0 = h - 2x$$

$$\therefore \quad x = \frac{h}{2}$$

Hence maximum stress occurs at half the height

$$q_{max} = \frac{V}{3I} \frac{h^2}{4} = \frac{V}{12I} h^2 = \frac{3V}{bh}$$

Problem 5.4: In the beam section shown in Fig. P 5.4 (a) prove that the maximum horizontal shearing stress occurs at a layer $h/8$ above or below the N.A.

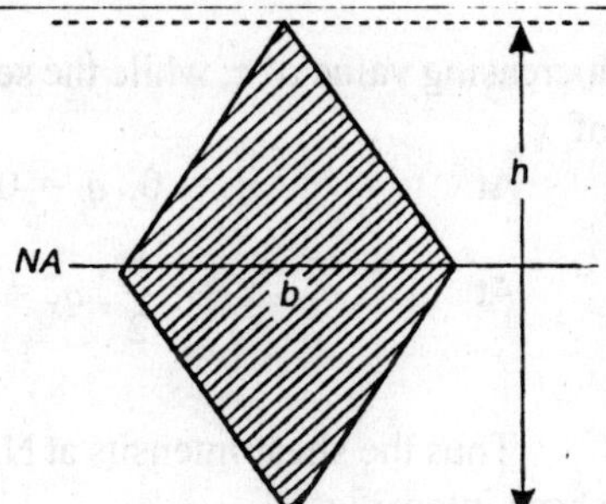

Fig. P. 5.4 (a)

Solution:

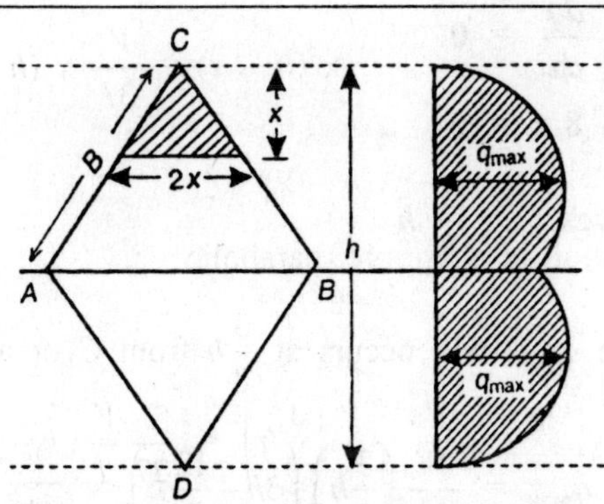

Fig. P. 5.4 (b)

Let B be the length of the side Let h = length of the diagonal = $\sqrt{2}$ B

$$I_{AB} = 2 \times \frac{1}{12} h \times \left(\frac{h}{2}\right)^3 = \frac{h^4}{48}$$

At a distance x from c,

$$q = \frac{V}{I \times 2x} \times \text{moment of shaded area about N.A.}$$

$$= \frac{V}{I \times 2x}\left\{x^2\left(\frac{h}{2} - \frac{2}{3}x\right)\right\}$$

$$= \frac{48Vx}{2h^4}\left\{\frac{3h - 4x}{6}\right\} = \frac{4Vx}{h^4}(3h - 4x)$$

The variations is parabolic

In the above expression, the first term $\frac{4Vx}{h^4}$ increases with the increasing value of x, while the second term $(3h - 4x)$ with the increasing of x

At $x = 0, q = 0$

At $x = \frac{h}{2}, q_x = \frac{2V}{h^3}(3h - 2h) = \frac{2V}{h^2} = \frac{V}{area} = q_{mean}$

Thus the shear intensits at N.A is the mean intensities. For maximum shear intensities

$$\frac{dq}{dx} = 0$$

$$\therefore \quad 3h - 8x = 0$$

Which gives, $x = \frac{3}{8}h$

Hence the q_{max}. will occurs at $\frac{3}{8}h$ from C or at $\frac{h}{8}$ from N.A.

$$\therefore \quad q_{max} = \frac{4V}{h^4}\left(\frac{3}{8}h\right)\left(3h - \frac{12}{8}h\right) = \frac{9V}{4h^2} = \frac{9}{8}\frac{V}{B^2}$$

$$q_{mean} = \frac{V}{area} = \frac{V}{B^2} = \frac{2V}{h^2}$$

$$\therefore \quad \frac{q_{max}}{q_{mean}} = \frac{9}{8} = 1.125$$

Problem 5.5: The T section shown in Fig. P. 5.5 (a) is the cross section of a beam formed by joining two rectangular pieces of wood together The beam is subjected to a maximum shearing force of 60 kN. Show that the NA is 34 mm from the top and that $I_{NA} = 10.57 \times 10^6 \text{ mm}^4$. Using these values, determine the shearing stress (a) at the neutral axis and (b) at the junction between the two pieces of wood.

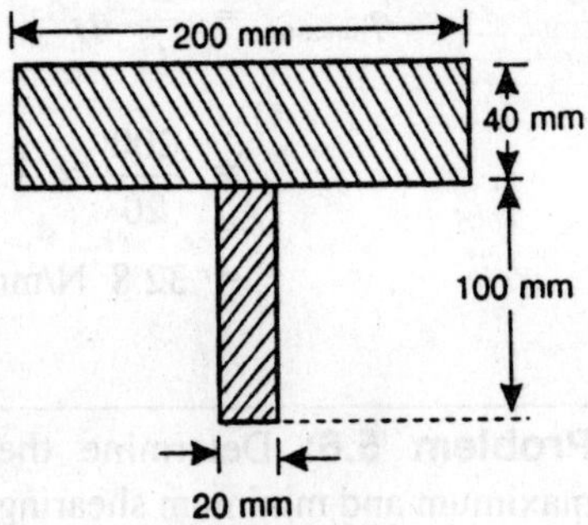

Fig. P. 5.5 (a)

Solution:

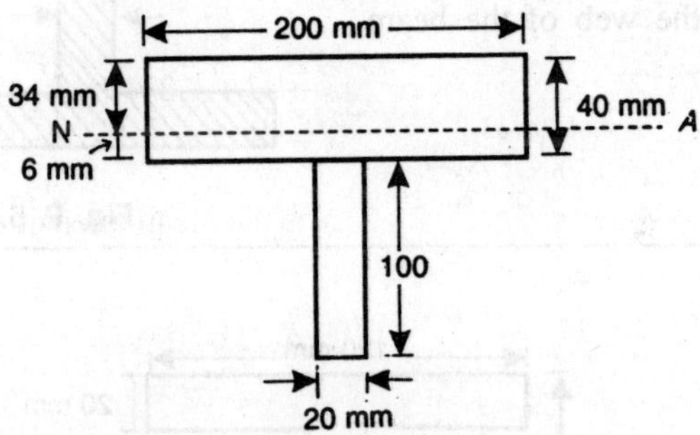

Fig. P. 5.5 (b)

$$V_{max} = 60 \text{ kN}$$
$$= 60 \times 10^3 \text{ N}$$
$$I_{NA} = 10.57 \times 10^6 \text{ mm}^4$$

(*a*) Shearing stresses at the N.A.

$$q_f = \frac{V}{Ib} \cdot A\bar{y}$$

$$= \frac{60 \times 10^3}{10.57 \times 10^6 \times 200} \times 200 \times 34 \times 17$$

$$= 3.28 \text{ N/mm}^2$$

$$= 3.28 \text{ MPa}$$

(*b*) Shearing stress at the junction between two pieces of wood,

$$q_{\text{junction}} = \frac{B}{t} \cdot q_f$$

$$= \frac{200}{20} \times 3.28$$

$$= 32.8 \text{ N/mm}^2$$

Problem 5.6: Determine the maximum and minimum shearing stress in the web of the wide flange section in Fig. P 5.6 (a) if $V = 100$ kN. Also, compute the percentage of vertical shear carried only by the web of the beam.

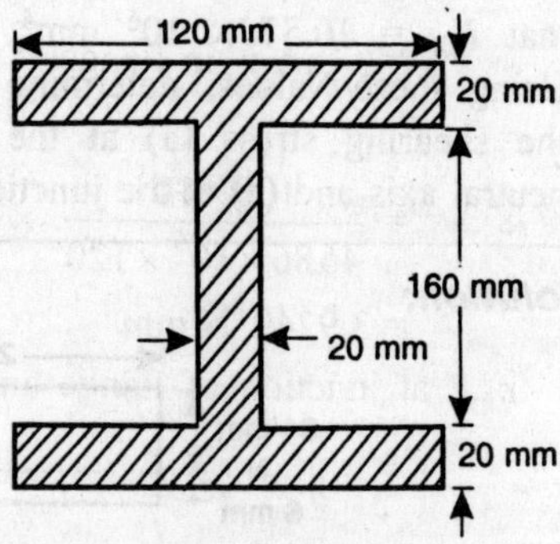

Fig. P. 5.6 (a)

Solution:

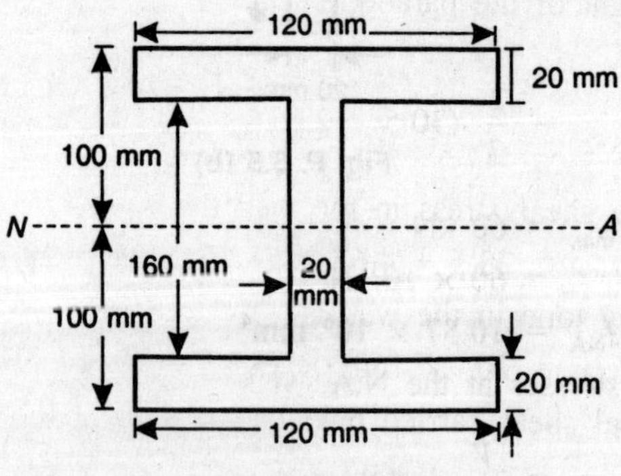

Fig. P. 5.6 (b)

$$V = 100 \text{ kN}$$

$$I_{NA} = \frac{1}{12}[BD^3 - bd^3]$$

$$= \frac{1}{12}[120 \times 200^3 - 100 \times 160^3]$$

$$= 45.86 \times 10^6 \text{ mm}^4$$

$$q_{\max} = \frac{V}{Ib}.A\bar{y} \text{ at N.A.}$$

$$= \frac{100 \times 10^3}{45.86 \times 10^6 \times 20} \times [120 \times 20 \times 90 + 80 \times 80 \times 20 \times 40]$$

$$= 30.527 \text{ N.mm}^2 = 30.5 \text{ MPa}$$

$$q = \frac{V}{Ib}.\text{A}\bar{y} \text{ in flange}$$

$$= \frac{100 \times 10^3}{45.86 \times 10^6 \times 120} \times 120 \times 20 \times 90$$

$$= 3.9249 \text{ N.mm}$$

$q_{\min}$ at junction of flange and web,

$$q_{\min} = \frac{B}{t} \times q_f = \frac{120}{20} \times 3.9249 = 23.5 \text{ MPa}$$

These stresses vary parabolically from top to bottom of the web,

The average height of the parabolie segment is $\frac{2}{3}$ of (30.5 – 23.5)

$$= \frac{2}{3}(30.5 - 23.5) = 4.67 \text{ MPa}$$

The average shear stress in the web the

$$q_{ave} = 23.5 + 4.67 = 28.17 \text{ MPa} = 28.17 \text{ MPa}$$

The shearing tone in the web is

$$V_{web} = 20 \times 160 \times 28.17 = 90144 \text{ N}$$

% of vertical shear carried only be the web at the beam,

$$= \frac{90144}{100 \times 1000} \times 100 = 90.14 \approx 90.2\ \%$$

Problem 5.7: Find the cross-sectional dimensions of the smallest square beam that can be loaded as shown in Fig P. 5.7 (a) if $\tau \leq 1.0$ MPa and $\sigma \leq 8$ MPa.

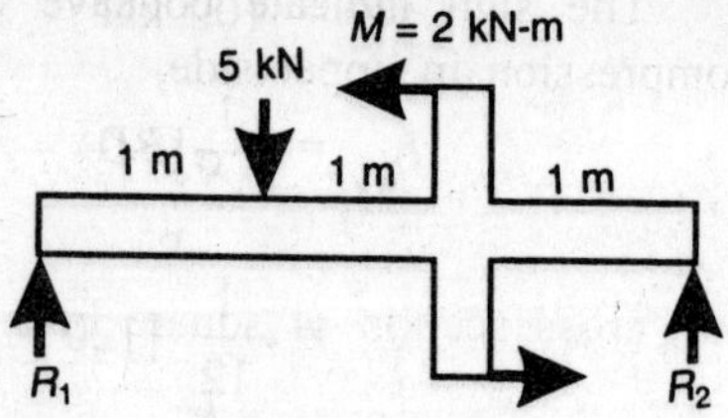

Fig. P. 5.7 (a)

Solution:

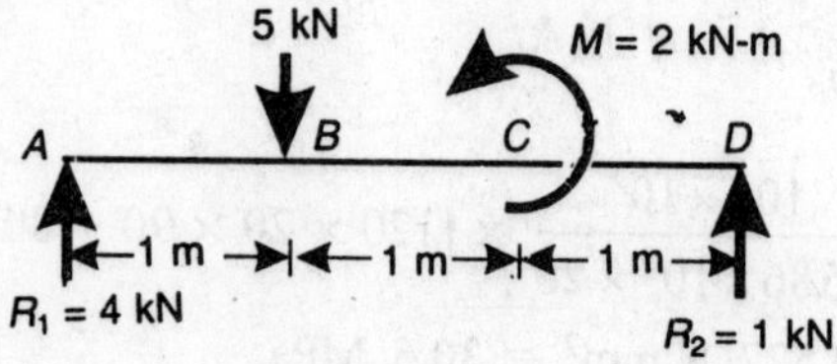

Fig. P. 5.7 (b)

$$\Sigma F_y = 0$$

$$R_1 + R_2 = 5 \text{ kN}$$

$$\Sigma M_A = 0$$

$$R_2 \times 3 + 2 = 5 \times 1$$

$$R_2 \times 3 = 5 - 2 = 3$$

$$\therefore \quad R_2 = 3/3 = 1 \text{ kN}$$

$$\therefore \quad R_1 = 5 - 1 = 4 \text{ kN}$$

S.F.D.

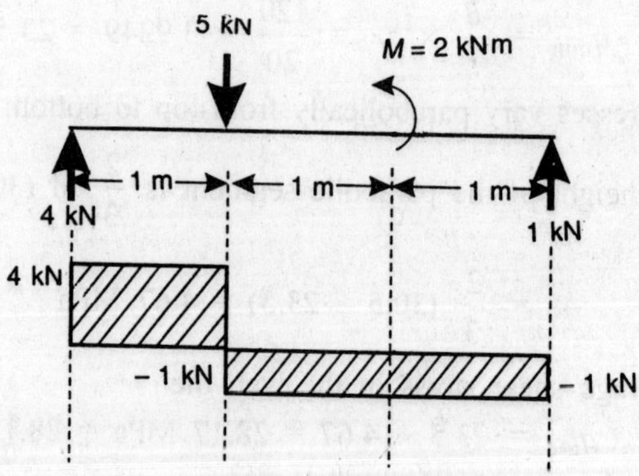

Fig. P. 5.7 (c)

Since shear froce diagram charges sign at

$$x = 1 \text{ m}$$

$$\therefore \quad M_{max} = 4 \times 1 = +4 \text{ kN.m}$$

The sign indicate concave upward tension in lower side and compression in upper side,

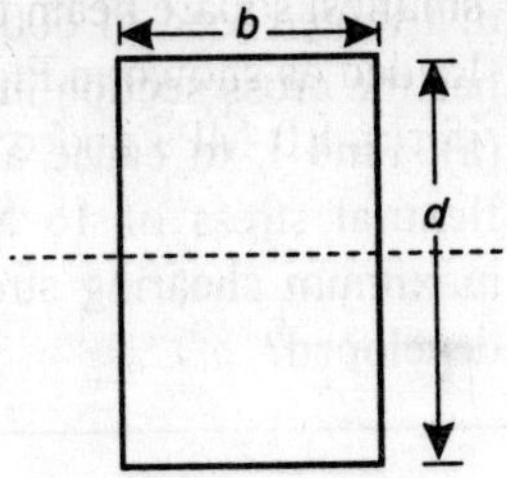

Fig. P. 5.7 (d)

$$M_C = \frac{\sigma_c \cdot I}{y_c}$$

cross section of square beam

$$y_c = \frac{b}{2}$$

$$I_{NA} = \frac{1}{12}\ bd^3 = \frac{1}{12}\ b^4$$

$$\therefore \quad M_C = \frac{8 \times \frac{1}{12} b^4}{\frac{b}{2}} = \frac{4b^3}{3}$$

Equating these from above moment equation,

$$4 \times 10^6 = \frac{4}{3} b^3$$

$$\therefore \quad b^3 = \frac{4 \times 10^6 \times 3}{4} = 3 \times 10^6$$

$$\therefore \quad b = \left[3 \times 10^6\right]^{\frac{1}{3}}$$

$$= 144.22 \text{ mm} = 144 \text{ mm}$$

From shear stress equation,

$$\tau_{max} = \frac{V}{Ib} .A\,\bar{y} = \frac{4 \times 1000}{\frac{1}{12} b^4} \times b \times \frac{b}{2} \times \frac{b}{4}$$

$$= \frac{4000}{\frac{b^4}{12}} \times \frac{b^3}{8} = \frac{4000 \times 12}{b^4} \times \frac{b^3}{8} = \frac{6000}{b}$$

$$\therefore \quad \frac{6000}{b} = 1$$

$$\therefore \quad b = 6000 \text{ mm}$$

$\therefore$ Take minimum value 144 mm.

Problem 5.8: A simply supported beam of length/carries a uniformly distributed load of 6000 N/m and has the cross section in Fig. P. 5.8 (a). Find I. to cause a maximum flexural stress of 16 MPa. What maximum shearing stress is then developed?

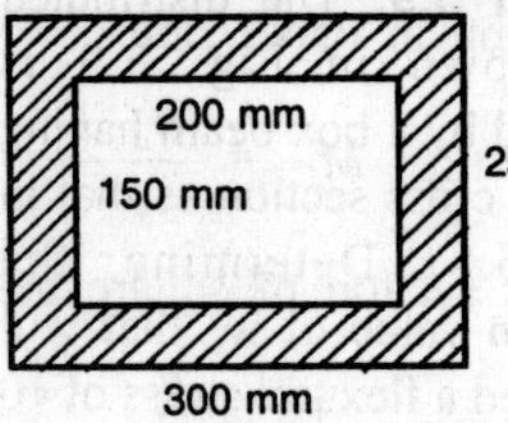

Fig. P. 5.8 (a)

Solution:

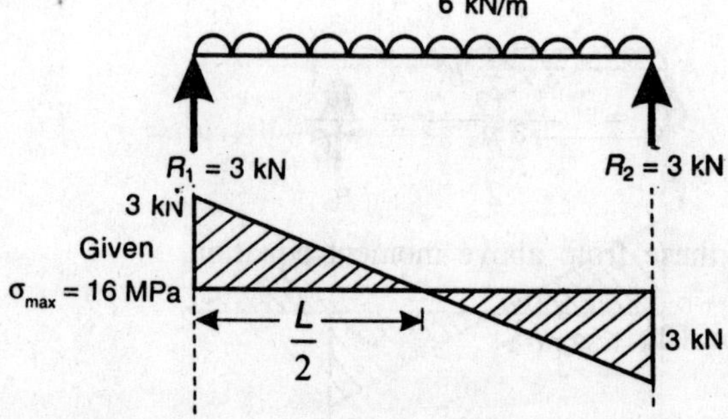

Fig. P. 5.8 (b)

$$\sigma \max = 16 \text{ MPa}$$

Shear force zero at x = L/2

$$M_{max} = \frac{1}{2} 3 \times \frac{L}{2} = + \frac{3}{4} L = + 0.75 L$$

The sign indicates concave upward, compression in upper portion and tension in lower portion,

$$I_{NA} = \frac{1}{12}[300 \times 125^3 - 200 \times 75^3]$$

$$= 41.8 \times 10^6 \text{ mm}^4$$

$$M - \frac{\sigma}{y} I - \frac{16 \times 41.8 \times 10^6}{125}$$

$$= 5.35 \times 10^6 \text{ N.mm}$$

Equating these from above moment,

$$0.75 L \times 10^6 = 5.35 \times 10^6$$

$$\therefore \quad L = 7.13 \text{ m} = 7.2 \text{ m}$$

Problem 5.9: The distributed load shown in Fig. 5.9 is supported by a box beam having the same cross section as that in Prob. 5.8. Determine the maximum value of w_o that will not exceed a flexural stress of 10 MPa or a shearing stress of 1.0 MPa.

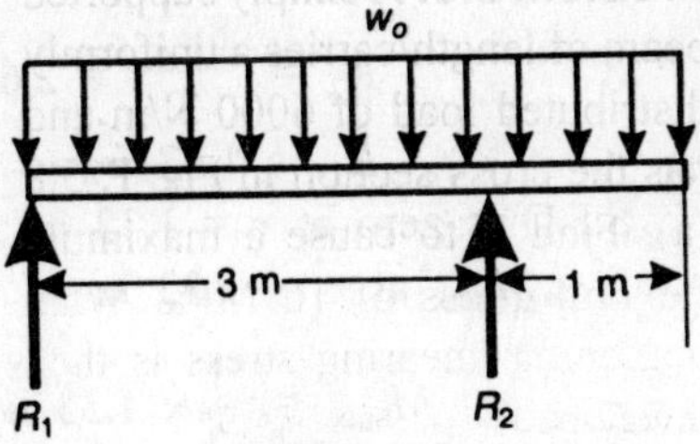

Fig. P. 5.9

Solution:

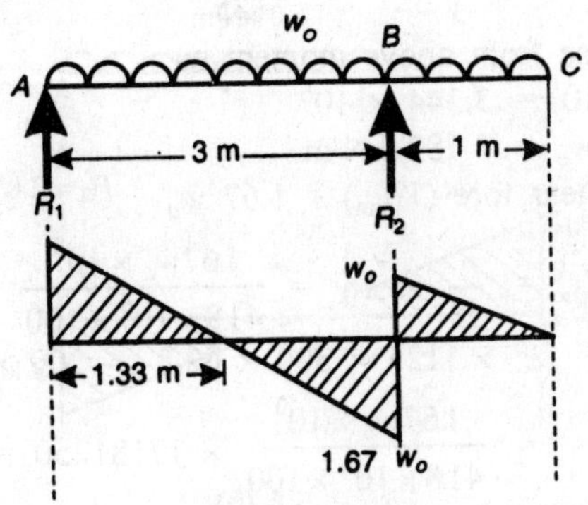

Fig. P. 5.9 (a)

$$\Sigma F_y = 0$$

$$R_1 + R_2 = w_o \times 4 = 4\ w_o \qquad \ldots\text{(i)}$$

$$\Sigma M_A = 0$$

Cross-section

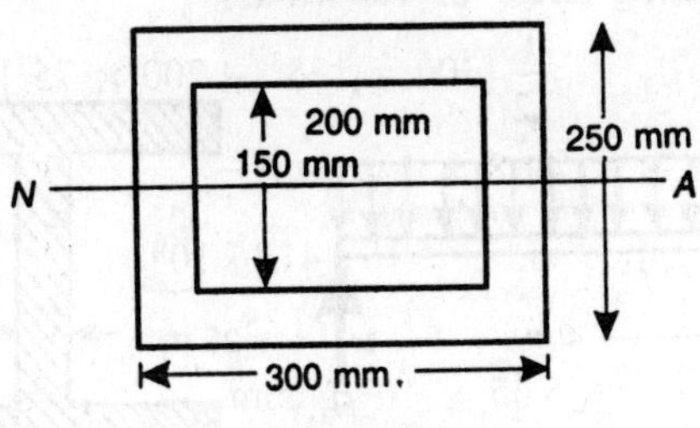

$I_{NA} = 41.8 \times 10^6\ mm^4$

Fig. P. 5.9 (b)

$$R_2 \times 3 = w_o \times 4 \times 2$$

$$\therefore \quad R_2 = \frac{8w_o}{3} = 2.67\ w_o$$

$$\therefore \quad R_1 = 1.33\ w_o$$

S.F.D is zero at $x = 1.33$ m and at $x = 3$ m,

$\therefore \quad M_{max}$ at $x = 1.33$ m

$$M_{max} = \frac{1}{2} \times 1.33\ w_o \times 1.33 = 0.884\ w_o$$

$$\text{or,} \quad M = \frac{\sigma . I}{y} = \frac{10 \times 41.8 \times 10^6}{125} = 3.344 \times 10^6 \text{ N.mm}$$

equating these from above moment eq.

$$0.884\ w_o \times 10^6 = 3.344 \times 10^6$$

$$\therefore \quad w_o = 3.782 \text{ kN/m}$$

Maximum shear fore (V_{man}) = 1.67 w_o

$$\tau_{max} = \frac{V}{Ib} \times A\bar{y} = \frac{1.67 w_o \times 10^3}{41.8 \times 10^6 \times 100} \times [125 \times 300 \times 62.5 - 200 \times 75 \times 37.5]$$

$$= \frac{1.67 w_o \times 10^3}{41.8 \times 10^6 \times 100} \times 17181250 = 0.71\ w_o$$

$$\therefore \quad 0.71\ w_o = 1$$

$$\therefore \quad w_o = 1.4 \text{ kN/m}$$

Take minimum value 1.4 kN/m

Problem 5.10: The distributed load shown in Fig p-5.10 (a) is supported by a wide-flange section of the given dimensions. Determine the maximum value of w_0 that will not exceed a flexural stress of 10 MPa or a shearing stress of 1.0 MPa.

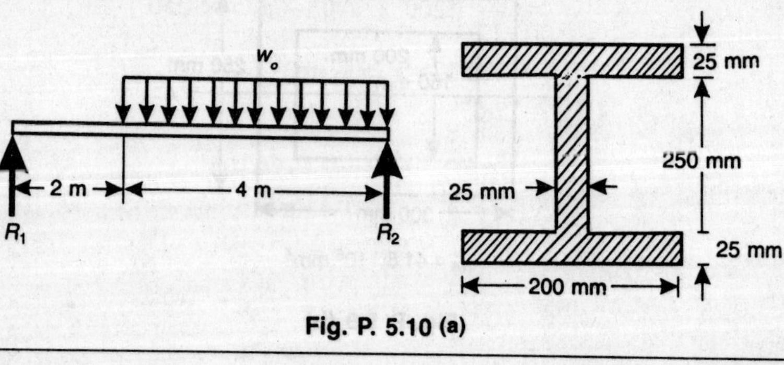

Fig. P. 5.10 (a)

Solution:

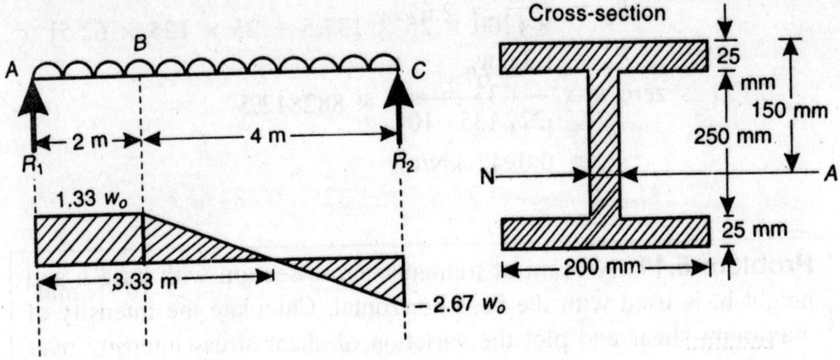

Fig. P. 5.10 (b)

$$\Sigma F_y = 0,$$

$$R_1 + R_2 = 4\, w_0$$

$$\Sigma M_A = 0,$$

$$R_2 \times 6 = 4\, w_0 \times 4$$

$$\therefore \quad R_2 = \frac{16}{6}\, w_0 = 2.67\, w_0$$

$$R_1 = 1.33\, w_0$$

$\therefore$ Shear fore zero at $x = 3.33$ m

$$M_{max} = 1.33\, w_0 \times 2 + \frac{1}{2} \times 1.33 \times 1.33\, w_0 = 3.544\, w_0$$

$$I_{NA} = \frac{1}{12}[BD^3 - bd^3]$$

$$= \frac{1}{12}\,[200 \times 300^3 - 175 \times 250^3]$$

$$= 222.135 \times 10^6 \text{ mm}^4$$

$$V_{max} = 2.67\, w_0$$

$$M = \frac{\sigma I}{y} = \frac{10 \times 222.135 \times 10^6}{150} = 14.809 \times 10^6$$

$$\therefore \; 3.544\, w_0 \times 10^6 = 14.809 \times 10^6$$

$$\therefore \quad w_0 = 4.1786 \text{ kN/m}$$

$$\tau_{max} = I = \frac{V}{Ib} \times A\,\bar{y} = \frac{2.67 w_o \times 10^3}{222.135 \times 10^6}$$

$$\times [200 \times 25 \times 137.5 + 25 \times 125 \times 62.5]$$

$$= \frac{2.67 w_o}{222.135 \times 10^3} \times 882812.5$$

$$\therefore \quad w_o = 0.0942 \text{ kN/m}$$

Problem 5.11: A beam of triangular cross-section with base b and height h, is used with the base horizontal. Calculate the intensity of maximum shear and plot the variation of shear stress intensity over the section.

Solution:

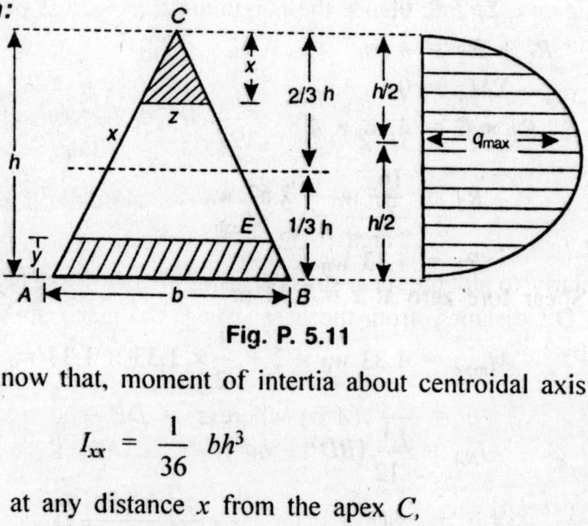

Fig. P. 5.11

We know that, moment of intertia about centroidal axis

$$I_{xx} = \frac{1}{36} bh^3$$

Now, at any distance x from the apex C,

$$q = \frac{F}{Iz}(A\,\bar{y})$$

where $$z = b.\frac{x}{h} \text{ and}$$

$$A\,\bar{y} = \left(\frac{1}{2}zx\right)\left(\frac{2}{3}h - \frac{2}{3}x\right) = \frac{1}{3}\,zx\,(h - x)$$

$$q = \frac{F}{Iz} \times \frac{1}{3}\,zx\,(h - x) = \frac{F}{3I}x\,(h - x)$$

The variation of shear stress is parabolic $q = 0$ at $x = 0$

At the neutral $\quad q_n = \dfrac{F}{3I}\left(\dfrac{2}{3}h\right)\left(h - \dfrac{2}{3}h\right) = \dfrac{2}{27}\dfrac{F}{I}h^2$

$$= \frac{2}{27}\frac{36F}{bh^3}h^2 = \frac{72}{27}\frac{F}{bh} = \frac{8}{3}\frac{F}{bh}$$

$$= \frac{4}{3}\frac{F}{\text{area}} = \frac{4}{3}q_{\text{mean}}$$

Hence the intensity of shear stress at the N.A. is 4/3 of the mean intensity.

For maximum intensity above the N.A., $\dfrac{dq}{dx} = 0 = h - 2x.$

This gives $x = h/2$. Hence the maximum shear stress occurs at the mid-height from base

$$q_{\max} = \frac{F}{3I}\frac{h}{2}\cdot\frac{h}{2} = \frac{F}{12I}\cdot h^2 = \frac{3F}{bh^3}\cdot h = \frac{3F}{bh}$$

$$= \frac{3F}{2\text{ area}} = \frac{3}{2}q_{\text{mean}}.$$

Similarly, to plot the shear stress distribution below the N.A consider a section DE distant y from the base, where the shear stress intensity is given by

$$q = \frac{F}{Iz'}(A\bar{y}) \text{ where } z' = DE = \frac{b}{h}(h - y)$$

and $\quad A\bar{y} = \dfrac{1}{2}(b + z')\,y\left[\dfrac{h}{3} - \dfrac{y}{3}\left(\dfrac{b+2z'}{b+z'}\right)\right]$

$$= \frac{y}{6}\left[b + \frac{b}{h}(h-y)\right]\left[h - y\left\{\frac{b + \dfrac{2b}{h}(h-y)}{b + \dfrac{b}{h}(h-y)}\right\}\right]$$

$$= \frac{by}{6h}(2h - y)\left[h - y\left(\frac{3h-2y}{2h-y}\right)\right]$$

$$= \frac{by}{3h}(h - y)^2$$

$$q = \frac{F}{I\frac{b}{h}(h-y)} \cdot \frac{by}{3h}(h-y)^2 = \frac{Fy}{3I}(h-y)$$

This variation is also parabolic, giving $q = 0$, at $y = 0$

Also, at

$$y = \frac{h}{3},\ q_n = \frac{F}{3I}\ \frac{h}{3} \times \frac{2h}{3} = \frac{2}{27}\ \frac{F}{I}\ h^2$$

$$= \frac{8F}{3bh},$$ which is the same as found earlier.

Problem 5.12: A cast iron bracket, subjected to bending has a cross-section of I-shape with unequal flanges, as shown. If the section is subjected to a shear force of 1600 kN, draw the shear stress distribution over the depth of the section, indicating the principal values.

Solution:

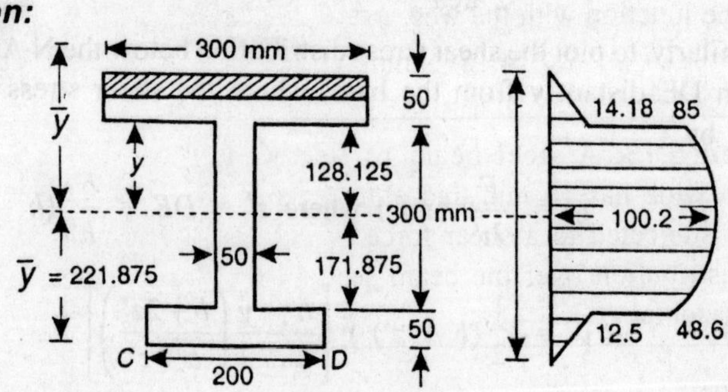

Fig. P. 5.12

Let $\bar{y}$ the distance of the N.A. form the top fibre of the I-Section

Total Area $= 300 \times 50 + 300 \times 50 + 200 \times 50 = 40000\ \text{mm}^2$

$$\bar{y} = \frac{1}{40000}[250 \times 50 \times 25 + 350 \times 50 \times 175$$
$$+ 200 \times 50 \times 375]$$
$$= 178.125\ \text{mm}$$

$$I_{AB} = \frac{1}{3} \times 250 \times 50^3 + \frac{1}{3} \times 50^3 \times 350^3$$

$$+ \left[\frac{1}{12} \times 200\,(50)^3 + 200 \times 50\,(375)^2\right]$$

$$= 213333 \times 10^4 \text{ mm}^4$$

$$\therefore \quad I_{xx} = 213333 \times 10^4 - 40000\,(178.125)^2 = 86419 \times 10^4 \text{ mm}^4$$

Now, at any distance y from the

$$q = \frac{E}{Iz}\,(A\bar{y})$$

where $z = 300$ mm and $A\bar{y} = 300\,(\bar{y} - y)\,(\bar{y} + y)\frac{1}{2}$

$$= 150\,(\bar{y}^2 - y^2) = 150\,(178.125^2 - y^2)$$

$$= 150\,(31729 - y^2)$$

$$\therefore \quad q = \frac{F}{Iz}\,(A\bar{y}) = \frac{1600 \times 10^3}{86419 \times 10^4 \times 300}$$

$$\times 150\,(31729 - y^2) = 9.257 \times 10^{-4}\,(31729 - y^2)$$

The variation of shear stress is parabolic, giving $q = 0$ at $y = 178.125$ mm

At the junction with the web, $q_f = 9.257 \times 10^{-4}\,(31729 - 128.125^2)$

$$= 14.18 \text{ N.mm}$$

Problem 5.13: A steel beam of I-section , 200 mm deep and 160 mm wide has 16 mm thick flanges and 10 mm thick web. The beam is subjected to a shear force of 200 kN. Determine the shear stress distribution over the beam section if the web of the beam is kept horizontal.

Solution:

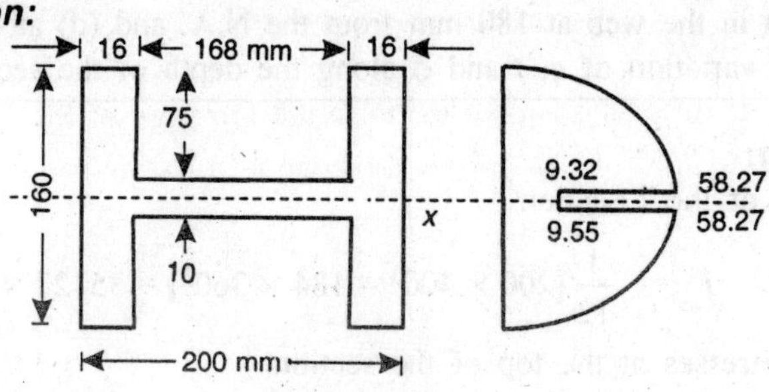

Fig. P. 5.13

moment of inertia about centroidal axis

$$I_{xx} = 2 \times \frac{1}{12} \times 16\,(160)^3 + \frac{1}{12} \times 168(10)^3$$

$$= 1094 \times 10^4 \text{ mm}^4$$

Shear stress in the flange, just above the web

$$q_f = \frac{F}{Iz}(A\,\bar{y}) \text{ where } z = 2 \times 16 = 32 \text{ mm}$$

$$A\bar{y} = 2 \times 16 \times 75\,(37.5 + 5) = 102000 \text{ mm}^3$$

$$\therefore \quad q_f = \frac{200 \times 10^3}{1094 \times 10^4 \times 32} \times 102000 = 58.27 \text{ N/mm}^2$$

Shear stress in the web, just at the junction

$$q_w = q_f \times \frac{32}{200} = 58.27 \times \frac{32}{200} = 9.32 \text{ N/mm}^2$$

Shear stress at neutral axis

$$q_n = \frac{F}{Iz}(A\,\bar{y}), \text{ where } z = 200 \text{ mm}$$

$$A\,\bar{y} = 2 \times 16 \times 75\,(37.5 + 5) + 200 \times 5 \times 2.5$$

$$= 104500 \text{ mm}^3$$

$$\therefore \quad q_n = \frac{200 \times 10^3}{1094 \times 10^4 \times 200} \times 104500 = 9.55 \text{ N/mm}^2$$

Problem 5.14: A 200 mm × 400 mm R.S.J. has flange 20 mm thick and web 16 mm thick. At a certain cross-section it has to resist a bending moment of 160 kN.m and a shear force of 300 kN. Find the principal stresses (a) at the top (b) in the flange at 180 mm from the N.A. (c) in the web at 180 mm from the N.A. and (d) at the N.A Plot the variation of q, f and σ along the depth of the section.

Solution:

M.I. of the I section

$$I = \frac{1}{12}[200 \times 400^3 - 184 \times 360^3] = 35127 \times 10^4 \text{ mm}^4$$

(a) Stresses at the top of the section

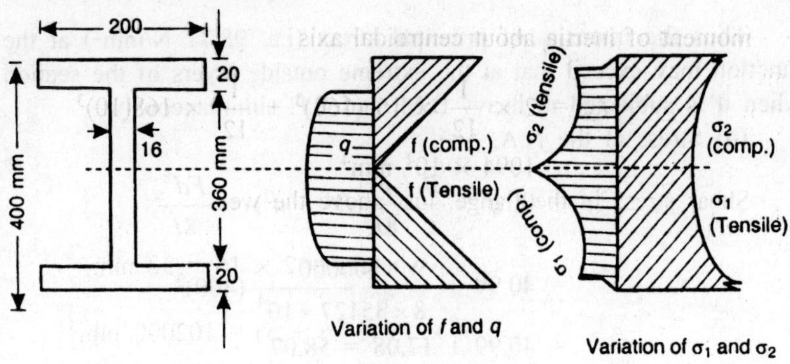

Fig. P. 5.14

$$f = \frac{M}{I}y = \frac{160 \times 10^6}{35127 \times 10^4} \times 200$$

$$= 91.1 \text{ N/mm}^2 \text{ (comp.)}$$

$$q = 0$$

$$\therefore \quad \sigma_1 = 91.1 \text{ N/mm}^2 \text{ (comp.)}$$

$$\sigma_2 = 0$$

(b) Stresses in the flange at 180 mm from the neutral

$$f = \frac{160 \times 10^6}{35127 \times 10^4} \times 180 = 81.99 \text{ N/mm}^2 \text{ (comp.)}$$

$$q = \frac{F}{8I}(D^2 - d^2) = \frac{300000}{8 \times 35127 \times 10^4}(400^2 - 360^2)$$

$$= 3.25 \text{ N/mm}^2$$

$$\therefore \quad \sigma = \frac{81.99}{2} \pm \sqrt{\left(\frac{81.99}{2}\right)^2 + (3.25)^2} = 40.99 \pm 41.12$$

$$\therefore \quad \sigma_1 = 82.11 \text{ N/mm}^2 \text{ (comp)}; \sigma_2 = 0.13 \text{ N/mm}^2 \text{ (tensile)}$$

(c) Stresses in the web at 180 mm from the neutral axis

$$f = 81.99 \text{ N/mm}^2 \text{ (comp.)};$$

$$q = 3.25 \times \frac{200}{16} = 40.63 \text{ N/mm}^2$$

$$\sigma = \frac{81.99}{2} \pm \sqrt{\left(\frac{81.99}{2}\right)^2 + (40.63)^2} = 40.99 \pm 57.72$$

$$\therefore \quad \sigma_1 = 98.71 \text{ N/mm}^2 \text{ (comp) and}$$

$$\sigma_1 = 16.73 \text{ N/mm}^2 \text{ (tensile)}$$

Thus the intensity of principal stress (i.e. 98.71 N/mm^2) at the junction may exceed that at the extreme outside layers of the section when it is subjected to heavy B.M. and S.F. simultaneously.

(d) Stress at the N.A.

$$f = 0;\ q = \frac{B}{t} \cdot \frac{F}{8I}(D^2 - d^2) + \frac{Fd^2}{8I}$$

$$= 40.99 + \frac{300000}{8 \times 35127 \times 10^4}(400)^2$$

$$= 40.99 + 17.08 = 58.07$$

$\therefore \quad \sigma_1 = 58.07$ N/mm^2 and $\sigma_1 = 58.07$ N/mm^2

Problem 5.15: A bar of hexagonal cross-section of side length b mm is used as cantilever, one diagonal being horizontal. Plot the shear stress distribution diagram if it is subjected to a shear force of 8 kN and if b = 25 mm.

Solution:

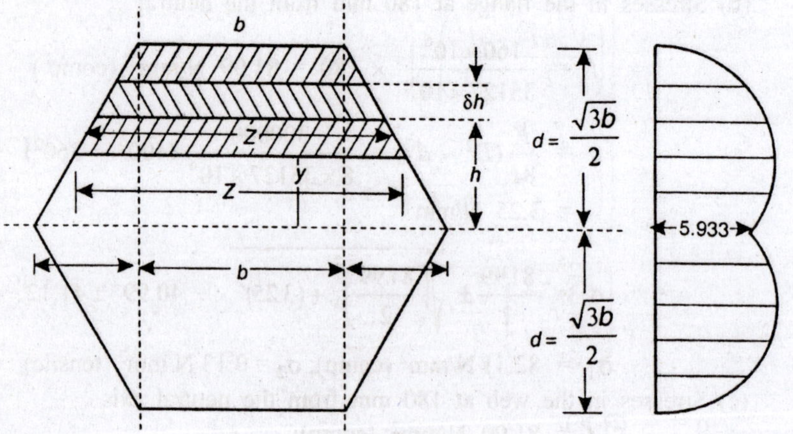

Fig. P. 5.15

Shear stress at any layer distant $\pm\ y$ from the N.A., $q = \frac{F}{Iz}(A\,\bar{y})$

where $\quad I = I_x$ of rectangle + I_x of triangles.

$$= 2\left[\frac{1}{3}b\,d^3 + \frac{1}{12}b\,d^3\right]$$

$$= \frac{5}{6} bd^3 = \frac{5}{6} b \frac{3\sqrt{3}}{8} b^3 = 0.54\ b^4$$

$$z = b + 2\left(\frac{b}{2}\right)\left(\frac{d-y}{d}\right) = b + \frac{2}{\sqrt{3}}\left(\frac{\sqrt{3}}{2}b - y\right)$$

$$= 2\,b - \frac{2}{\sqrt{3}} y = \frac{2}{\sqrt{3}}(\sqrt{3}\,b - y)$$

At $y = h,\ z' = \frac{2}{\sqrt{3}}(\sqrt{3}\,b - h)$

$A\,\bar{y}$ = moment of the shaded area about the neutral axis

$$= \sum_{h=y}^{h=d}\left[\frac{2}{\sqrt{3}}\left(\sqrt{3}\,b - h\right)\delta h\right] h = \frac{2}{\sqrt{3}}\left(\frac{\sqrt{3}bh^2}{2} - \frac{h^3}{3}\right)_y^d$$

$$= \frac{2}{\sqrt{3}}\left[\left(\frac{\sqrt{3}bd^2}{2} - \frac{d^3}{3}\right) - \left(\frac{\sqrt{3}by^2}{2} - \frac{y^3}{3}\right)\right]$$

$$= \frac{2}{\sqrt{3}}\left[\left(\frac{\sqrt{3}}{2}\frac{3}{4}b3 - \frac{1}{3}\frac{3\sqrt{3}}{8}b^3\right) - \left(\frac{\sqrt{3}by^2}{2} - \frac{y^3}{3}\right)\right]$$

$$= \frac{2}{\sqrt{3}}\left[\frac{\sqrt{3}}{4}b^3 - \left(\frac{\sqrt{3}}{2}by^2 - \frac{y^3}{3}\right)\right]$$

$$\therefore\ q = \frac{F}{Iz}(A\,\bar{y})$$

$$= \frac{F\sqrt{3}}{(0.54b^4)2(\sqrt{3}b - y)} \times \frac{2}{\sqrt{3}}\left[\frac{\sqrt{3}}{4}b^3 - \left(\frac{\sqrt{3}}{2}by^2 - \frac{y^3}{3}\right)\right]$$

or, $q = \dfrac{F}{0.54b^4(1.73b - y)}\ [0.433\ b^3 - y^2\ (0.865\ b - 0.333\ y)]$

This is the required equation for the shear stress.

when b = 25 mm and F = 8 kN = 8000 N

$$d = \frac{\sqrt{3}}{2} b = \frac{\sqrt{3} \times 25}{2} = 21.65 \text{ mm}$$

$$\therefore \quad q = \frac{8000}{0.54(25)^4 (1.73 \times 25 - 4)} \times$$

$$\{0.433\ (25)^3 - y^2\ (0.865 \times 25 - 0.333\ y)\}$$

$$= \frac{0.03793}{43.25 - y} [6765.6 - y^2 (21.625 - 0.333\ y)]$$

Since area of hexagon $= (b \times 2\ d) + 4\left[\frac{1}{2} \times \frac{b}{2} \cdot d\right]$

$$= 2\ bd + bd = 3\ bd = 3\ b \times \frac{\sqrt{3}}{2} b$$

$$= \frac{3\sqrt{3}}{2} b^2 = \frac{3\sqrt{3}}{2} (25)^2 = 1623.8 \text{ mm}^2$$

$$q_{mean} = \frac{8000}{1623.8} = 4.927 \text{ N/mm}^2$$

At $y = 0$, $q_{na} = \dfrac{0.03793}{43.25} \times 6765.6 = 5.933 \text{ N/mm}^2$

Problem 5.16: Find the ratio of maximum to mean intensity of vertical shear in the cross-section of a beam of hollow circular section, the outside diameter being twice the inside diameter.

Solution:

Take a strip of thickness dy at a distance y from the N.A we get

$$y = \frac{D}{2} \sin\theta = \frac{d}{2} \sin\varphi$$

$$dy = \frac{D}{2} \cos\theta\ d\theta = \frac{d}{2} \cos\varphi\ d\varphi$$

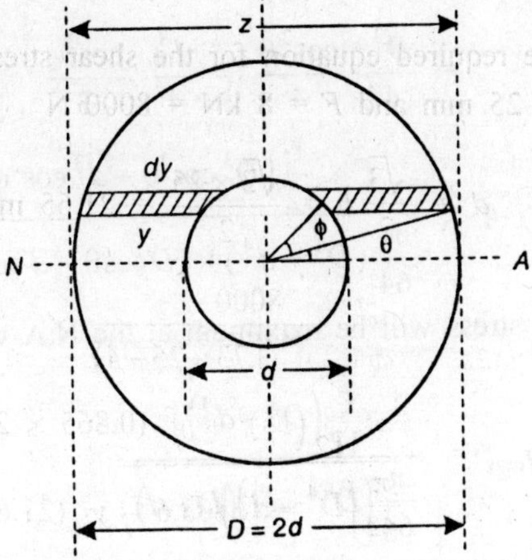

Fig. P. 5.16

$$q = \frac{F}{Iz}(A\,\bar{y})$$

where $$z = 2\left[\frac{D}{2}\cos\theta - \frac{d}{2}\cos\varphi\right] = D\cos\theta - d\cos\varphi$$

Area of stripof circular section

$$= (D\cos\theta - d\cos\varphi)\,dy$$

$$= D\cos\theta \,.\, \frac{D}{2}\cos\theta\, d\theta - d\cos\varphi \,.\, \frac{d}{2}\cos\varphi\, d\varphi$$

$$= \frac{D^2}{2}\cos^2\theta\, d\theta - \frac{d^2}{2}\cos^2\varphi\, d\varphi$$

$$A\,\bar{y} = \int_0^{\frac{\pi}{2}} \frac{D^2}{2}\cos^2\theta\, d\theta \,.\, y - \int_0^{\frac{\pi}{2}} \frac{d^2}{2} \times$$

$$\cos^2\varphi\, d\varphi \,.\, \frac{d}{2}\sin\varphi$$

$$= \frac{D^3}{4}\int_0^{\frac{\pi}{2}} \cos^2\theta\sin\theta\, d\theta - \frac{d^3}{4}\int_0^{\frac{\pi}{2}} \cos^2\varphi\sin\varphi\, d\varphi$$

$$= \frac{D^3}{4} \cdot \frac{\cos^3\theta}{4} - \frac{d^3}{4} \quad \frac{\cos^3\varphi}{3}$$

$$\therefore \quad q = \frac{F \times \frac{1}{12}[D^3\cos^3\theta - d^3\cos^3\varphi]}{\frac{\pi}{64}(D^4 - d^4) \times (D\cos\theta - d\cos\varphi)}$$

Now shear stress will be maximum at the N.A where $\theta = \varphi = 0$

$$\therefore \quad q_{\max} = \frac{\frac{F}{12}\left(D^3 - d^3\right)}{\frac{\pi}{64}\left(D^4 - d^4\right)(D - d)}$$

$$q_{av} = \frac{F}{\frac{\pi}{4}\left(D^2 - d^2\right)}$$

$$\therefore \quad \frac{q_{\max}}{q_{av}} = \frac{\frac{F}{12}\left(D^3 - d^3\right)}{\frac{\pi}{64}\left(D^4 - d^4\right)(D - d)} \times \frac{\frac{\pi}{4}\left(D^2 - d^2\right)}{F}$$

$$= \frac{4\left(D^3 - d^3\right)}{3\left(D^2 + d^2\right)(D - d)}$$

We know from problem

$$d = \frac{D}{2}$$

$$\therefore \quad \frac{q_{\max}}{q_{av}} = \frac{4\left[D^3 - \left(\frac{D}{2}\right)^3\right]}{3\left[D^2 + \left(\frac{D}{2}\right)^2\right]\left[D - \frac{D}{2}\right]}$$

$$= \frac{4 \times \frac{7}{8}D^3}{3 \times \frac{5}{4}D^2 \times \frac{1}{2}D} = \frac{28}{15}$$

Chapter 6

Combined Stresses

Problem 6.1: Two wooden joists 50 mm × 100 mm are glued together along the joint *AB* as shoiwn in Fig. P. 6.1 (a). Determine the normal stress and shearing stress in the glue if P = 200 kN.

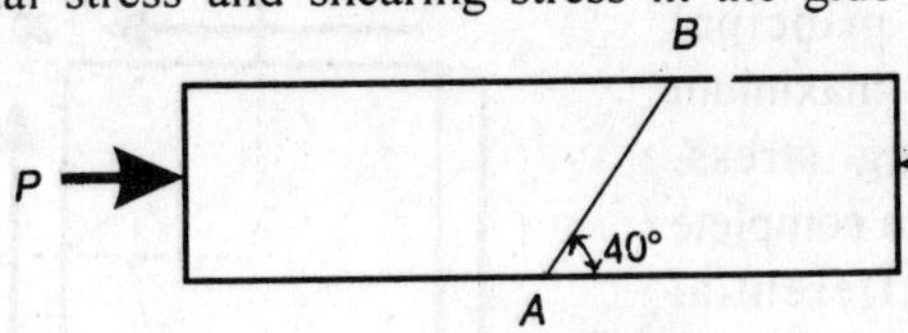

Fig. P. 6.1 (a)

Solution: Given:

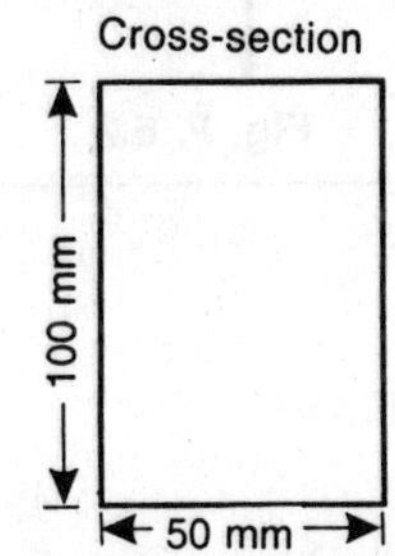

Fig. P. 6.1 (b)

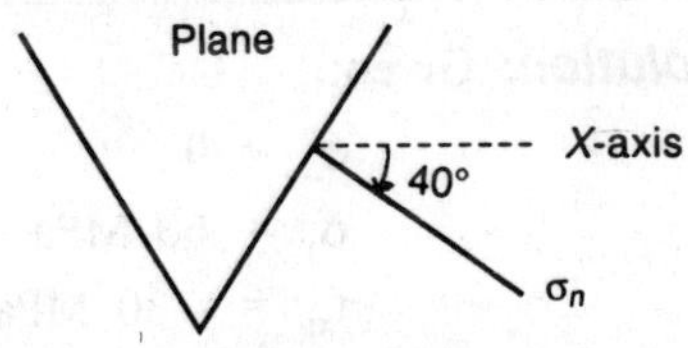

Fig. P. 6.1 (c)

$$P = 200 \text{ kN} = 200 \times 10^3 \text{ N} = 2 \times 10^5 \text{ N}$$

Normal stresses (σ_n)

$$= \frac{\sigma_n}{2} - \frac{\sigma_n}{2} \cos 2\theta = \frac{P}{2A} - \frac{P}{2A} \cos 2\theta$$

$$= \frac{2 \times 10^5}{2 \times 100 \times 50} - \frac{2 \times 10^5}{2 \times 100 \times 50} \cos 2 \times 40$$

$$= \frac{100}{5} - \frac{100}{5} \cos 80^\circ$$

$$= 20 - 20 \cos 80^\circ$$
$$= 16.5 \text{ N/mm}^2 = 16.5 \text{ MPa}$$

Shearing stress (τ)

$$= \frac{\sigma_n}{2} \sin 2\theta = \frac{P}{2A} \sin 2\theta$$

$$= \frac{2 \times 10^5}{2 \times 100 \times 50} \sin 80^\circ = 20 \sin 80^\circ$$

$$= 19.69 \text{ N/mm}^2 \approx 19.7 \text{ MPa}$$

Problem 6.2: For the state of stress shown in figure, determine the principal stresses and the maximum in-plane shearing stress. Show all results on complete sketches of differential elements.

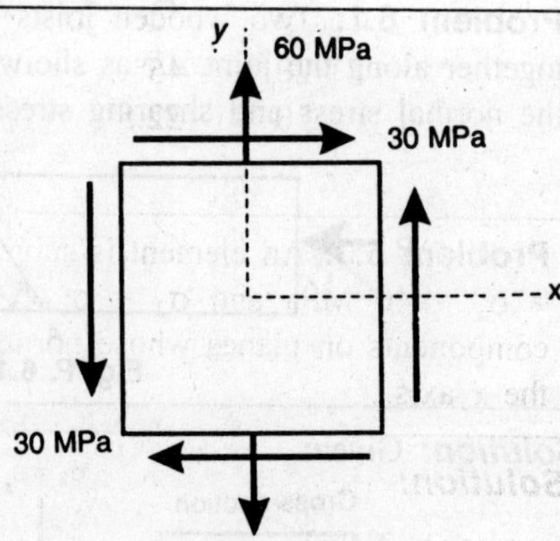

Fig. P. 6.2

Solution: Given:

$$\sigma_x = 0$$
$$\sigma_y = 60 \text{ MPa} = +60 \text{ N/mm}^2$$
$$\tau_{xy} = -30 \text{ MPa}$$
$$= -30 \text{ N/mm}^2 \text{ (anticlockwise in } x \text{ face)}$$

$\therefore$ Principal stress ($\sigma_{1,2}$)

$$= \frac{\sigma_x + \sigma_x}{2} \pm \sqrt{\left(\frac{\sigma_x - \sigma_y}{2}\right)^2 + \left(\tau_{xy}\right)^2}$$

$$= \frac{\sigma_y}{2} \pm \sqrt{\left(\frac{\sigma_y}{2}\right)^2 + \left(\tau_{xy}\right)^2} \quad [\because \sigma_x = 0]$$

$$= \frac{60}{2} \pm \sqrt{\left(\frac{60}{2}\right)^2 + (-30)^2}$$

$$= 30 \pm \sqrt{900+900}$$

$$= 30 \pm \sqrt{1800} = 30 \pm 42.42$$

$$\therefore \quad \sigma_1 = 72.4 \text{ MPa}$$

$$\sigma_2 = -12.4 \text{ MPa}$$

$$\tau_{max} = \frac{\sigma_1 - \sigma_2}{2} = \frac{72.4 - [-12.4]}{2}$$

$$= 42.4 \text{ MPa}$$

$$\tan 2\theta = -\frac{2\tau_{xy}}{\sigma_x - \sigma_y} = \frac{2 \times 30}{-60} = -1$$

$$2\theta = -45^\circ$$

$$\therefore \quad \theta = -22.5^\circ$$

> **Problem 6.3:** An element is subjected to the principal stresses $\sigma_1 = \sigma_x = 40$ MPa and $\sigma_2 = \sigma_y = -30$ MPa. Compute the stress components on planes whose normals are at + 30° and + 120° with the x axis.

Solution:

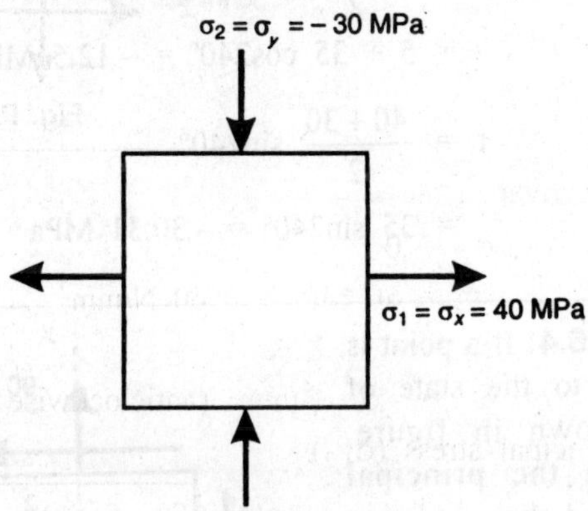

Fig. P. 6.3

$$\tau_{xy} = 0$$

$$\sigma_x = +40 \text{ MPa}$$

$$\sigma_y = -30 \text{ MPa}$$

Compute the stress components on planes whose normals are at + 30° and + 120° with the x-axis.

If $\quad \theta = 30^\circ$ then

$$\sigma_n = \frac{\sigma_x + \sigma_y}{2} + \frac{\sigma_x - \sigma_y}{2}\cos 2\theta - \tau_{xy}\sin 2\theta$$

$$= \frac{\sigma_x + \sigma_y}{2} + \frac{\sigma_x - \sigma_y}{2}\cos 2\theta$$

$$= \frac{40 - 30}{2} + \frac{40 + 30}{2}\cos 60^\circ$$

$$= 5 + 17.5 = 22.5 \text{ N/mm}^2$$

$$\tau = \frac{\sigma_x - \sigma_y}{2}\sin 2\theta + \tau_{ny}\cos 2\theta$$

$$\therefore \quad \tau = \frac{\sigma_x - \sigma_y}{2}\sin 2\theta \qquad [\because \tau_{xy} = 0]$$

$$= \frac{40 + 30}{2}\sin 60^\circ = 30.31 \text{ MPa}$$

If $\theta = 120^\circ$ then,

$$\sigma_n = \frac{40 - 30}{2} + \frac{40 + 30}{2}\cos 240^\circ$$

$$= 5 + 35\cos 240^\circ = -12.5 \text{ MPa}$$

$$\tau = \frac{40 + 30}{2}\sin 240^\circ$$

$$= 35\sin 240^\circ = -30.31 \text{ MPa}$$

Problem 6.4: If a point is subjected to the state of stress shown in figure, determine the principal stresses and the maximum in-plane shearing stress.

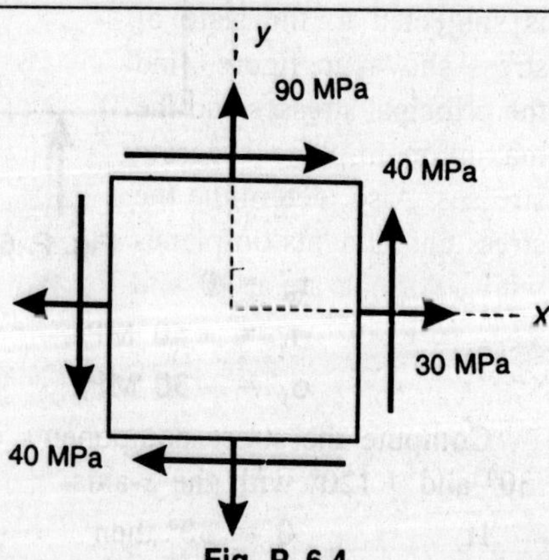

Fig. P. 6.4

Solution: Given

$$\sigma_x = +80 \text{ MPa}$$
$$\sigma_y = +90 \text{ MPa}$$
$$\tau_{zy} = -40 \text{ MPa}$$

Principal stresses (σ_1, σ_2)

$$= \frac{\sigma_x + \sigma_y}{2} \pm \sqrt{\left(\frac{\sigma_x - \sigma_y}{2}\right)^2 + (\tau_{xy})^2}$$

$$= \frac{30+90}{2} \pm \sqrt{\left(\frac{30-90}{2}\right)^2 + (-40)^2}$$

$$= 60 \pm \sqrt{900+1600} = 60 \pm 50$$

$$\therefore \quad \sigma_1 = 60 + 50 = 110 \text{ MPa}$$
$$\sigma_2 = 60 - 50 = 10 \text{ MPa}$$

$\therefore$ Major principal stress (σ_1) = 110 MPa,
Minor principal stress (σ_2) = 10 MPa
Maximum in plane shearing stress (τ_{max})

$$= \pm \sqrt{\left(\frac{\sigma_x - \sigma_y}{2}\right)^2 + (\tau_{xy})^2} \text{ or } \pm \frac{\sigma_1 - \sigma_2}{2}$$

$$\therefore \quad \tau_{max} = \frac{\sigma_1 - \sigma_2}{2} = \frac{110-10}{2} = \frac{100}{2} = 50 \text{ MPa}$$

Problem 6.5: If an element is subjected to the state of stress shown in figure, find the principal stresses and the maximum in-plane shearing stresses. Also, determine the stress components on planes whose normals are at 45° and 135° with the x-axis.

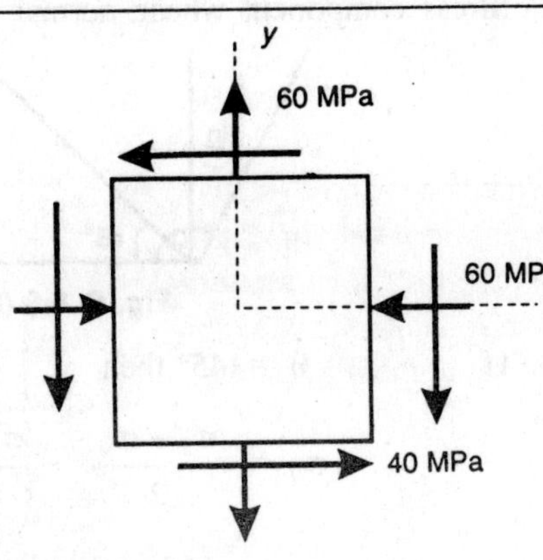

Fig. P. 6.5 (a)

Solution: Given

$$\sigma_x = -60 \text{ MPa}$$
$$\sigma_y = +60 \text{ MPa}$$
$$\tau_{xy} = 40 \text{ MPa}$$

Principal stresses (σ_1, σ_2)

$$= \frac{\sigma_x + \sigma_y}{2} \pm \sqrt{\left(\frac{\sigma_x - \sigma_y}{2}\right)^2 + (\tau_{xy})^2}$$

$$= \frac{-60+60}{2} \pm \sqrt{\left(\frac{-60-60}{2}\right)^2 + (40)^2}$$

$$= 0 \pm \sqrt{(-60)^2 + (40)^2}$$

$$= \pm\sqrt{3600+1600}$$

$$= \pm\sqrt{5200} = \pm 72.11 \text{ MPa}$$

$$\therefore \quad \sigma_1 = +72.11 \text{ MPa}$$
$$\sigma_2 = -72.11 \text{ MPa}$$

Maximum shearing stress (τ_{max})

$$= \frac{\sigma_1 - \sigma_2}{2} = \frac{72.11-(-72.11)}{2} = +72.11 \text{ MPa}$$

Stress component whose normal are at 45° and 135°

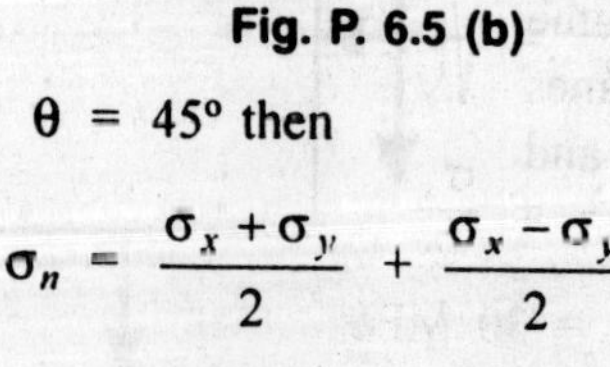

Fig. P. 6.5 (b)

If $\theta = 45°$ then

$$\sigma_n = \frac{\sigma_x + \sigma_y}{2} + \frac{\sigma_x - \sigma_y}{2}\cos 2\theta - \tau_{xy}\sin 2\theta$$

$$= \frac{-60+60}{2} + \frac{-60-60}{2}\cos 90° - 40\sin 90°$$

$$= 0 + 0 - 40 = -40 \text{ MPa}$$

$$\tau = \frac{\sigma_x - \sigma_y}{2} \sin 2\theta + \tau_{ny} \cos 2\theta$$

$$= 0 + 40 \cos 90^\circ = 0 + 0 = 0 \text{ MPa}$$

If $\theta = 135^\circ$ then

$$\sigma_x = \frac{-60+60}{2} + \frac{-60-60}{2} \cos 270 - \tau_{xy} \sin 270^\circ$$

$$= 0 + 0 = 40 \times -1 = 40 \text{ MPa}$$

$$\tau = \frac{\sigma_x - \sigma_y}{2} \sin 2\theta + \tau_{xy} \cos 2\theta$$

$$= 0 + 40 \cos 270^\circ = 0 + 0 = 0 \text{ MPa}$$

Problem 6.6: The state of stress at a point is the result of the three separate actions that produce the three states of stress shown in figure P. 6.6. Determine the principal stresses and principal planes caused by the superposition of these three stress states.

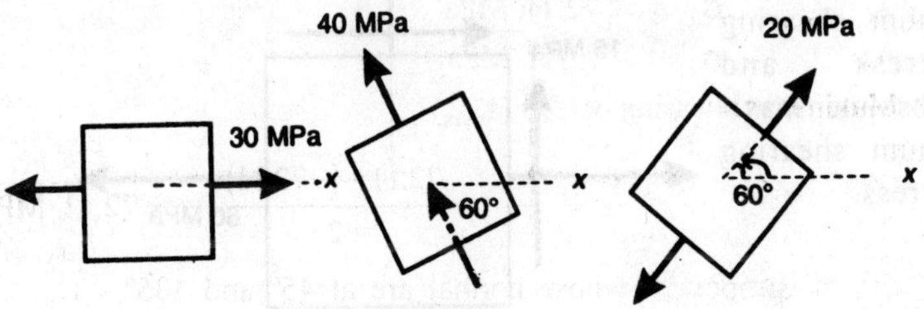

Fig. P. 6.6

Solution:

Case (a)

$$\text{Principal stress} = \frac{\sigma_x}{2} \pm \sqrt{\left(\frac{\sigma_x}{2}\right)^2}$$

$$= \frac{\sigma_x}{2} \pm \frac{\sigma_x}{2}$$

or $\sigma_1 = \sigma_x = 30 \text{ MPa}$

$$\tan 2\theta = \frac{-2\tau_{xy}}{\sigma_x - \sigma_y} = 0$$

$\therefore \quad \theta = 0$

Case (b) $\quad \sigma_x = -40$ MPa

$$\text{Principal stress} = \frac{\sigma_x}{2} - \frac{\sigma_x}{2}\cos 2\theta$$

$$= -\frac{40}{2} + \frac{40}{2}\cos 120^\circ$$

$$= -20 + (-10) = -30 \text{ MPa}$$

Case (c)

$$\text{Principal stress} = \frac{\sigma_x}{2} + \frac{\sigma_x}{2}\cos 2\theta$$

$$= -\frac{40}{2} + -\frac{40}{2}\cos 120$$

$$= -20 - 20\cos 120^\circ$$

$$= -20 + 10 = -10 \text{ MPa}$$

Problem 6.7: Find the maximum shearing stress and absolute maximum shearing stress.

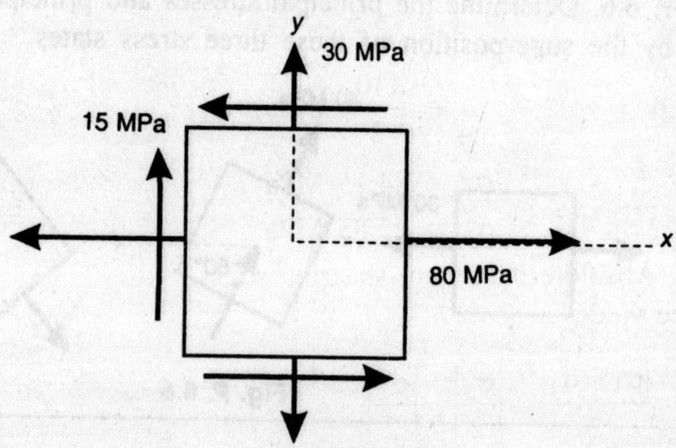

Fig. P. 6.7

Solution: Given

$$\sigma_x = 80 \text{ MPa}$$
$$\sigma_y = 30 \text{ MPa}$$
$$\tau_{xy} = 15 \text{ MPa}$$

(a) Maximum in plane shearing stress (τ_{max})

$$= \pm\sqrt{\left(\frac{\sigma_x - \sigma_y}{2}\right)^2 + (\tau_{xy})^2}$$

$$= \pm \sqrt{\left(\frac{80-30}{2}\right)^2 + (15)^2}$$

$$= \pm \sqrt{\left(\frac{50}{2}\right)^2 + 15^2}$$

$$= \pm \sqrt{(25)^2 + (15)^2} = \pm\ 29.15 \text{ MPa}$$

(b) Absolute maximum shearing stress

$$\sigma_1,\ \sigma_2 = \frac{\sigma_x + \sigma_y}{2} \pm \sqrt{\left(\frac{\sigma_x - \sigma_y}{2}\right)^2 + \left(\tau_{xy}\right)^2}$$

$$= \frac{80+30}{2} \quad \sqrt{\left(\frac{80-30}{2}\right)^2 + (15)^2}$$

$$= \frac{110}{2} \quad \sqrt{(25)^2 + (15)^2} = 55 \pm 29.15$$

$$\therefore \quad \sigma_1 = 84.15 \text{ MPa}$$

$$\sigma_2 = 25.85 \text{ MPa}$$

Absolute maximum shearing stress is the largest of the following three values

$$\frac{|\sigma_1 - \sigma_1|}{2},\ \left|\frac{\sigma_1}{2}\right|,\ \left|\frac{\sigma_2}{2}\right|$$

$$\frac{|\sigma_1 - \sigma_1|}{2} = \frac{|84.15 - 25.85|}{2} = 29.15 \text{ MPa}$$

$$\left|\frac{\sigma_1}{2}\right| = \frac{|84.151|}{2} = 42.075 \text{ MPa}$$

$$\left|\frac{\sigma_2}{2}\right| = \frac{|25.85|}{2} = 12.925 \text{ MPa}$$

$$\Rightarrow \quad 42.075 > 29.15 > 12.925$$

$\therefore$ Absolute maximum shearing stresses is 42.075 MPa

Problem 6.8: For a state of plane stress with $\sigma_1 = -40$ MPa and $\sigma_2 = -80$ MPa, determine (a) the maximum in-plane shearing stress, (b) the absolute maximum stress.

Solution: Given $\sigma_1 = -40$ MPa

$\sigma_2 = -80$ MPa

(a) The maximum in-plane shearing stress is

$$\tau_{max} = \pm \frac{\sigma_1 - \sigma_2}{2} = \pm \frac{-40-(-80)}{2}$$

$$= \pm \frac{-40+80}{2} = \pm \frac{40}{2} = \pm 20$$

(b) The absolute maximum shearing stress is the largest of the following three values?

$$\frac{|\sigma_1 - \sigma_2|}{2}, \frac{|\sigma_1|}{2}, \frac{|\sigma_2|}{2} \quad \therefore \quad \frac{|\sigma_1 - \sigma_2|}{2} = \frac{|-40+80|}{2} = \frac{|+40|}{2} = \frac{40}{2} = 20$$

$$\frac{|\sigma_1|}{2} = \frac{|-40|}{2} = \frac{40}{2} = 20$$

$$\frac{|\sigma_2|}{2} = \frac{|-80|}{2} = \frac{80}{2} = 40$$

$\therefore$ Absolute maximum shearing stress is 40 MPa

Problem 6.9: Compute the maximum in plane shearing stress and absolute maximum shearing stress.

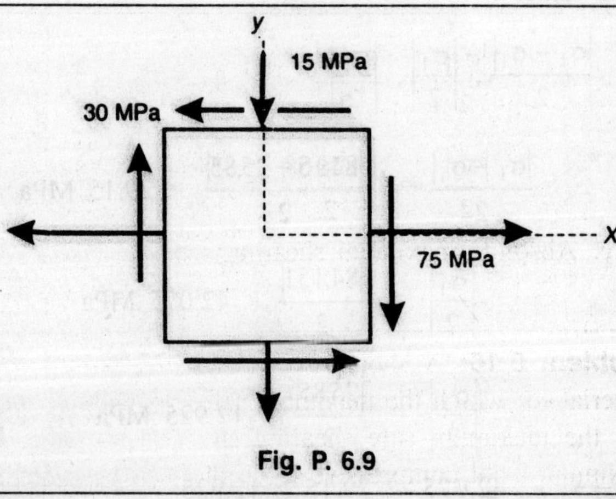

Fig. P. 6.9

Solution: Given

$$\sigma_x = 75 \text{ MPa}$$
$$\sigma_y = -15 \text{ MPa}$$
$$\tau_{xy} = +30 \text{ MPa}$$

$$\sigma_1, \sigma_2 = \frac{\sigma_x + \sigma_y}{2} \pm \sqrt{\left(\frac{\sigma_x - \sigma_y}{2}\right)^2 + (\tau_{xy})^2}$$

$$= \frac{75-15}{2} \pm \sqrt{\left(\frac{75+15}{2}\right)^2 + (30)^2}$$

$$= \frac{60}{2} \pm \sqrt{(45)^2 + (30)^2} = 30 \pm 54.08$$

$$\therefore \quad \sigma_1 = 84.08 \text{ MPa}$$
$$\sigma_2 = -24.08 \text{ MPa}$$

(a) The maximum in plane shearing stress

$$= \pm \frac{\sigma_1 - \sigma_2}{2} = \pm \frac{84.08 + 24.08}{2}$$

$$= \pm 54.08 \text{ MPa}$$

(b) Absolute maximum shearing stress is the largest of the following three values:

$$\frac{|\sigma_1 - \sigma_2|}{2} = \frac{|84.08 + 24.08|}{2} = 54.08 \text{ MPa}$$

$$\frac{|\sigma_1|}{2} = \frac{84.08}{2} = 42.04 \text{ MPa}$$

$$\frac{|\sigma_2|}{2} = \frac{|-24.08|}{2} = \frac{24.08}{2} = 12.04 \text{ MPa}$$

$\therefore$ Absolute maximum shearing stress is 54.08 MPa

Problem 6.10: A short 60 mm diameter circular bar is made of material for which the maximum safe compressive stress is 80 MPa and the maximum safe shearing stress is 30 MPa. Determine the maximum axial compressive force that can be safely applied to the bar.

Solution: Given

Diameter of circular bar = 60 mm

Maximum safe compressive stress = 30 MPa

$$\text{Area of the bar} = \frac{\pi}{4} d^2$$

$$= \frac{\pi}{4} \times (60)^2 = 2827.43 \text{ mm}^2$$

$$\therefore \quad 30 = \frac{F}{2827.43}$$

$$\therefore \quad F = 30 \times 2827.43 = 84.82 \text{ kN}$$

> **Problem 6.11:** A solid shaft 100 mm in diameter is subjected simultaneously to an axial compressive force of 600 kN and to a torque that twists the shaft through an angle of 1.5° in a length of 8 m. If G = 80 GPa, compute the maximum normal and shearing stresses in the shaft.

Solution: Given:

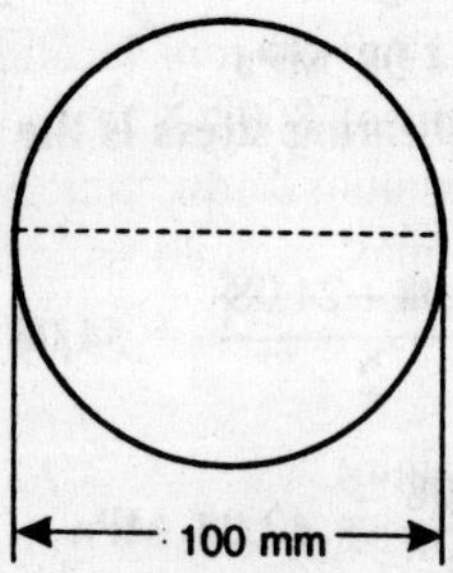

Fig. P. 6.11

Axial compressive force = 600 kN

$$\text{Angle} = 1.5° = 1.5 \times \frac{\pi}{180} \text{ radian} = 0.02617 \text{ radian}$$

$$\text{Area} = \frac{\pi}{4} 100^2$$

$$= 7853.98 \text{ mm}^2$$

$$G = 80 \text{ GPa} = 80 \times 10^3 \text{ N/mm}^2$$

$$\text{Length} = 8 \text{ m} = 8 \times 10^3 \text{ mm}$$

Maximum normal stresses

$$= \frac{600 \times 100}{7853.98} = 76.39 \text{ N/mm}^2$$

Maximum shearing stress (q_{max})

$$= \frac{N\theta}{l} \times r = \frac{80 \times 10^3 \times 0.02614}{8 \times 10^3} \times 50 = 13.085$$

$$\text{Torque} = \frac{N\theta}{l} \times J$$

$$= \frac{80 \times 10^3 \times 0.02617}{8 \times 10^3} \times \frac{\pi}{32}(100)^4$$

$$= 2.56 \text{ kN.m}$$

$$q_{max} = \frac{T}{J} \times r_{max} = 13.03 \text{ MPa}$$

Problem 6.12: A solid shaft 80 mm in diameter is subjected to a maximum bending moment of 800π N.m and and axial tensile load of 40π kN. FInd the maximum torque that can be safely applied if max. $\sigma \leq 100$ MPa and max. $\tau \leq 80$ MPa.

Solution: Given

Diameter = 80 mm

Maximum B.M.= 800π N.m

Axial tensile load = 40π kN

Find the maximum torque that can be safely applied if

$$\sigma_{max} \leq 100 \text{ MPa}$$

and $$\tau_{max} \leq 80 \text{ MPa}$$

We know that,

$$\tau_{max} = \sqrt{\left(\frac{2M}{\pi r^3}\right)^3 + \left(\frac{2T}{\pi r^3}\right)^2}$$

or $$80 = \sqrt{\left(\frac{2 \times 800\pi \times 10^3}{\pi \times (40)^3}\right)^2 + \left(\frac{2T}{\pi \times (40)^3}\right)^2}$$

$$= \frac{1}{\pi \times 40^3} \sqrt{\left(2 \times 800\pi \times 10^3\right) + (2T)^2}$$

$$(2T)^2 = 2.33 \times 10^{14}$$

$$\therefore \quad T = 3.81 \text{ kN.m}$$

Problem 6.13: A shaft 50 mm in diameter carries a maximum bending moment of 1.2 kN.m. What torque can also be applied without exceedig a maximum shearing stress of 60 MPa or a maximum normal stress of 120 MPa?

Solution: Given

Diameter of shaft = 50 mm

Max. B.M. = 1.2 kN.m

$$\tau_{max} = 60 \text{ MPa},$$

$$\sigma_{max} = 120 \text{ MPa}$$

$$\tau_{max.} = \frac{2}{\pi r^3} \sqrt{m^2 + T^2}$$

$$\therefore \quad 60 = \frac{2}{\pi \times (25)^3} \sqrt{\left(1.2 \times 10^6\right)^2 + T^2}$$

$$\therefore \quad T = 1.47 \text{ kN.m}$$

Problem 6.14: A solid circular shaft is used to transmit simultaneously a torque of 3000 N.m and a maximum bending moment of 2400 N.m. Determine the radius of the smallest shaft that can be used if max. $\sigma \leq 120$ MPa and max $\tau \leq 60$ MPa.

Solution: Given :

Torque (T) = 3000 N.m

Max B.M. = 2400 N.m

Determine the radius of the smallest shaft that can be used if,

$$\sigma_{max.} \leq 120 \text{ MPa and}$$

$$\tau_{max.} \leq 60 \text{ MPa}$$

Equivalent torque

$$(\tau_e) = \sqrt{M^2 + T^2}$$

$$= \sqrt{(2400 \times 100)^2 + (3000 \times 1000)^2} = 3.84 \text{ kN.m}$$

Equivalent moment (M_e)

$$= \frac{1}{2} [M + T_e]$$

$$= \frac{1}{2} [2.4 + 3.84] = 3.12 \text{ kN.m}$$

The shaft radius required so that the maximum shearing stress will not be exceeded is found from,

$$\tau_{max} = \frac{2Te}{\pi r^3}$$

$$\therefore \quad r^3 = \frac{2Te}{\pi \, \tau_{max}} = \frac{2 \times 3.84 \times 10^6}{\pi \times 60}$$

or $\quad r = 34.41$ mm

The radius that will avoid exceeding the maximum normal stress is, from

$$\sigma = \frac{4\,Me}{\pi r^3}$$

$$\therefore \quad r^3 = \frac{4\,Me}{\pi \, \sigma_{max}} = \frac{4 \times 3.12 \times 10^6}{\pi \times 120}$$

or $\quad r = 32.10$ mm

The larger of these two radii will satisfy both stress conditions; hence the proper diameter is,

$$d = 2 \times 34.41 = 68.82 \text{ mm}$$

Problem 6.15: A line shaft 50 mm in diameter is subjected to the loads shown in Fig. 6.15. The belt pulls on pulley *A* are horizontal, and those on pulley *B* are vertical. Calculate the maximum normal and shearing stresses developed in the shaft.

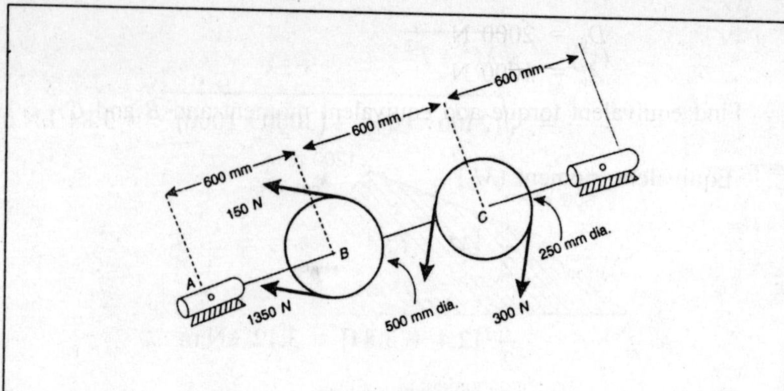

Fig. P. 6.15 (a)

Solution: Given

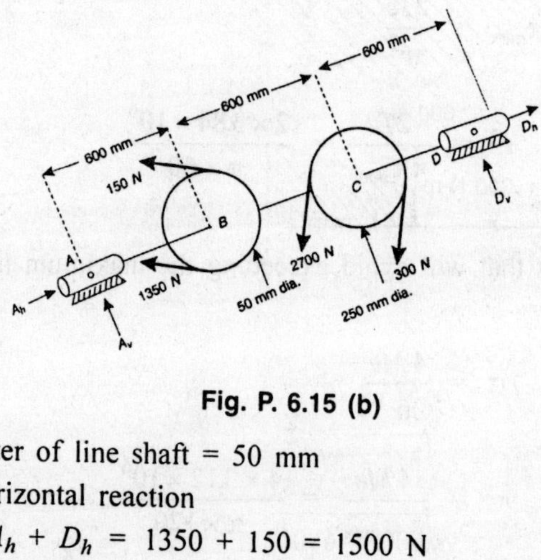

Fig. P. 6.15 (b)

Diameter of line shaft = 50 mm

For horizontal reaction

$$A_h + D_h = 1350 + 150 = 1500 \text{ N}$$

$$\Sigma M_A = 0,$$

$$D_h \times 1.8 + 1500 \times 0.6 = 0$$

$$\therefore \quad D_h = -500 \text{ N}$$

$$\therefore \quad D_h = -500 \text{ N}$$

$$\therefore \quad A_h = 1000 \text{ N}$$

For vertical rectifier

$$A_v + D_v = 3000$$

$$\Sigma M_A = 0,$$

$$3000 \times 1.2 = D_v \times 1.8$$

$$\therefore \quad D_v = 2000 \text{ N}$$

$$\therefore \quad A_v = 1000 \text{ N}$$

Find equivalent torque and equivalent moment and B and C,

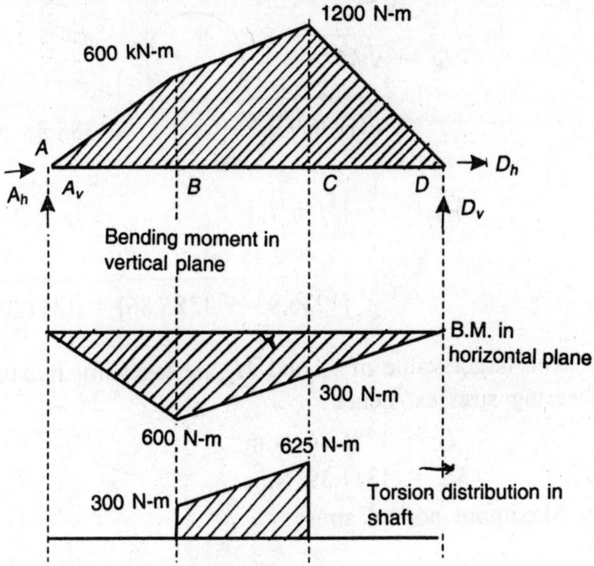

Fig. P. 6.15 (c)

At B
$$T_e = \sqrt{M^2 + T^2}$$

$$M = \sqrt{M_h^2 + M_v^2}$$

$$= \sqrt{600^2 + 600^2}$$

$$= 848.52 \text{ N.m}$$

$$\therefore \quad T_e = \sqrt{(848.52)^2 + (300)^2}$$

$$= 900 \text{ N.m}$$

$$\therefore \quad M_e = \frac{1}{2}[M + T_e]$$

$$= \frac{1}{2}[848.52 + 900] = 874.25 \text{ N.m}$$

At C

$$M = \sqrt{M_h^2 + M_v^2}$$

$$= \sqrt{(300)^2 + (1200)^2} = 1236.93 \text{ N.m}$$

$$\therefore \quad T_e = \sqrt{M^2 + T^2}$$

$$= \sqrt{(1236.93)^2 + (625)^2} = 1385.86 \text{ N.m}$$

$$M_e = \frac{1}{2}\,[M + T_e]$$

$$= \frac{1}{2}\,[1236.93 + 1385.86] = 1311.39 \text{ N.m}$$

$\therefore$ The larger value of T_e and M_e are taken, for maximum normal and sheering stresses, hence

$$T_e = 1385.86 \text{ N.m}$$

$$M_e = 1311.39 \text{ N.m}$$

$\therefore$ Maximum normal stress

$$(\sigma_{max}) = \frac{4\,M_e}{\pi r^3} = \frac{4 \times 1311.39 \times 10^3}{\pi \times (25)^3} = 106.86 \text{ N/mm}^2$$

Maximum shearing stress

$$(\tau_{max.}) = \frac{2T_e}{\pi r^3}$$

$$= \frac{2 \times 1385.86 \times 10^3}{\pi \times (25)^3} = 56.46 \text{ N/mm}^2$$

Problem 6.16: A masonry pier of 2 m × 3 m supports a vertical load of 50 kN as shown in Fig. 6.16. Find (a) stresses developed at each corner of the pier (b) what additional load should be placed at the centre of the pier, so that there is not tension anywhere in the pier section? (c) what are the stresses at the corners with the additional load in the centre?

Solution:

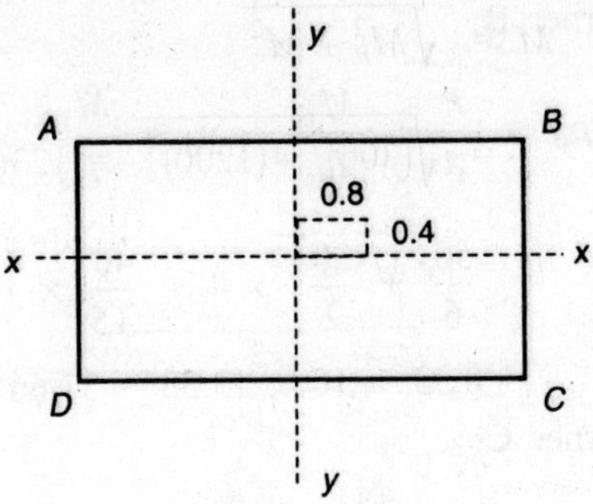

Fig. P. 6.16

We know from question

$$B = 3 \text{ m}$$

$$D = 2 \text{ m}$$

$$A = B \times D = 3 \times 2 = 6 \text{ m}^2$$

$$I_{xx} = \frac{1}{12} \times 3(2)^3 = 2 \text{ m}^4$$

$$I_{yy} = \frac{1}{12} \times 2\,(3)^3 = 4.5 \text{ m}^4$$

Eccentricity about $x - x$ axis

$$= e_x = 0.4 \text{ m}$$

Eccentricity about $y - y$ axis

$$= e_y = 0.8 \text{ m}$$

Moment about x-axis

$$M_{xx} = P \times e_x = 50 \times 0.4 = 20 \text{ kN.m}$$

Moment about y-axis

$$M_{yy} = P \times e_y = 50 \times 0.8 = 40 \text{ kN.m}$$

Distance of A and B from y-axis

$$= x_A = x_B = 1.5 \text{ m}$$

Distance of A and D from x-axis

$$= y_A = y_D = 1 \text{ m}$$

Stress at corner A:

$$p_A = \frac{P}{A} + \frac{M_{xx}}{I_{xx}} \times y_A - \frac{M_{yy}}{I_{yy}} \times x_A$$

$$= \frac{50}{6} + \frac{20}{2} \times 1 - \frac{40}{4.5} \times 1.5$$

$$= 8.33 + 10 - 13.35 = 5 \text{ kN/m}^2$$

Stress at corner B:

$$p_B = \frac{P}{A} + \frac{M_{xx}}{I_{xx}} \times y_B + \frac{M_{yy}}{I_{yy}} \times x_B$$

$$= \frac{50}{6} + \frac{20}{2} \times 1 + \frac{40}{4.5} \times 1.5$$

$$= 8.33 + 10 + 13.33 = 31.66 \text{ kN/m}^2$$

Stress at corner C:

$$p_C = \frac{P}{A} - \frac{M_{xx}}{I_{xx}} \times y_C + \frac{M_{yy}}{I_{yy}} \times x_C$$

$$= \frac{50}{6} - \frac{20}{2} \times 1 + \frac{40}{4.5} \times 1.5$$

$$= 8.33 - 10 + 13.33 = 11.66 \text{ kN/m}^2$$

Stress at corner D:

$$p_D = \frac{P}{A} - \frac{M_{xx}}{I_{xx}} \times y_D - \frac{M_{yy}}{I_{yy}} \times x_D$$

$$= \frac{50}{6} - \frac{20}{2} \times 1 - \frac{40}{4.5} \times 1.5$$

$$= 8.33 - 10 - 13.33 = -15 \text{ kN/m}^2 \text{ (i.e., tension)}$$

Additional load at the centre for no tension any where in the section

Let W (kN) be the additional central load.

$\therefore$ Additional compressive stress

$$p_0 = \frac{W}{A} = \frac{W}{6} \text{ kN/m}^2$$

For no tension to develop at D, p_0 should be equal to P_D, i.e. $p_0 = p_D$

$$\therefore \quad \frac{W}{6} = 15, \text{ from which } W = 90 \text{ kN}$$

Final stresses at corners due to the additional load

$$W = 90 \text{ kN};$$

$$\therefore \quad p_0 = 90/6 = 15 \text{ kN/m}^2$$

$$p_{AF} = p_A + p_0 = 5 + 15 = 20 \text{ kN/m}^2 \text{ (comp.)}$$
$$p_{BF} = p_B + p_0 = 31.66 + 15 = 46.66 \text{ kN/m}^2 \text{ (comp.)}$$
$$p_{CF} = p_C + p_0 = 11.66 + 15 = 26.66 \text{ kN/mm}^2 \text{ (comp.)}$$
$$p_{DF} = p_D + p_0 = -15 + 15 = \text{zero}$$

Problem 6.17: A short hollow cylindrical cast iron column having outside diameter 400 mm and inside diameter 300 mm was cast in a factory. On inspection, it was found that the base is eccentric in such a way that the thickness varies from 30 mm at one end to 70 mm at the other. Calculate the extreme intensities of stress induced in the section if the column carries a load of 2000 kN along the axis of the bore.

Solution:

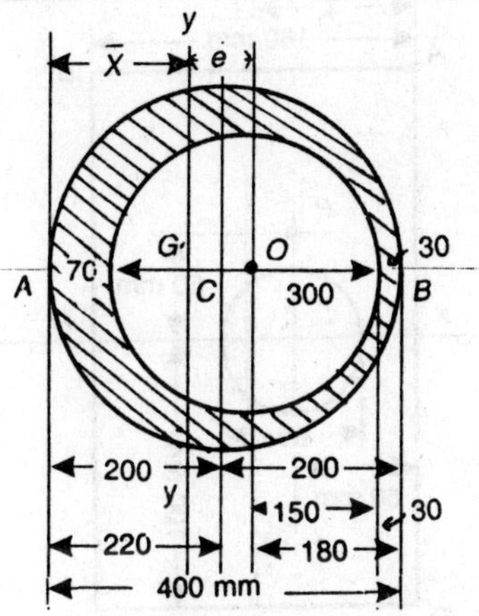

Fig. P. 6.17

Net area of cross-section of the section

$$A = \frac{\pi}{4}(400^2 - 300^2) = 54978 \text{ mm}^2$$

Let $\bar{x}$ be the distance of centre of gravity from end A.

$$\therefore \quad 54978\,\bar{x} = \frac{\pi}{4}(400)^2 \times 200 - \frac{\pi}{4}(300)^2 \times 220$$
$$= 25132741 - 15550884 = 9581857$$

From which $\bar{x} = 174.29$ mm

$\therefore$ Eccentricity $e = AO - AG = 220 - 174.29$

$= 45.71$ mm

$$I_{yy} = \left[\frac{\pi}{64}\times 400^4 + \frac{\pi}{4}(400)^2(200-174.29)^2\right] - \left[\frac{\pi}{64}\times 300^4 + \frac{\pi}{4}(300)^2(220-174.29)^2\right]$$

$$= 794.40 \times 10^6 \text{ mm}^4$$

Problem 6.18: A short C.I. column has a rectangular section 160 mm × 200 mm with a circular hole of 80 mm diameter as shown in Fig. 6.18. It carries an eccentric load of 100 kN, located as shown in the Fig. Determine the values of thre stresses at the four corners of the section.

Solution:

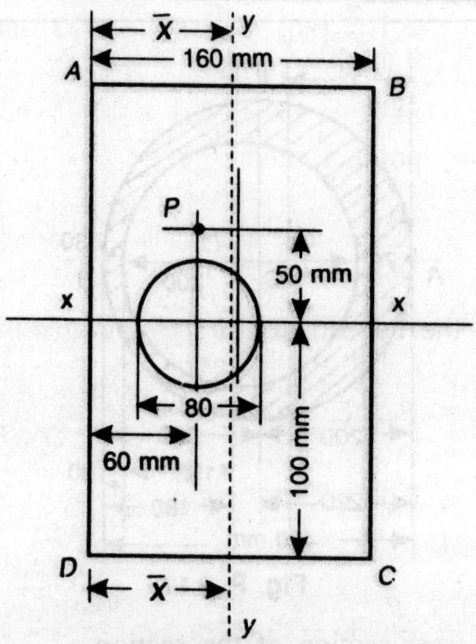

Fig. P. 6.18

The net section is symmetrical about *X-X* axis. Let the *Y-Y* axis be situated at a distance $\bar{x}$ from the face *AD*.

Let a_1 = area of solid rectangle = 160 × 200 = 32000 mm²

$\bar{x}_1$ = distance of centre of gravity of rectangle from *AD* = 80 mm

Let a_2 = area of the circular hole = $\frac{\pi}{4}(80)^2$ = 5026 mm²

$\bar{x}_2$ = distance of c.g. of hole from *AD* = 60 mm

Hence $\bar{x} = \dfrac{a_1\bar{x}_1 - a_2\bar{x}_2}{a_1 - a_2}$

$$= \frac{32000 \times 80 - 5026 \times 60}{32000 - 5026}$$

$$= \frac{2258440}{26974} = 83.73 \text{ mm}$$

$$I_{xx} = \frac{1}{12} \times 160\ (200)^3 - \frac{\pi}{64}\ (80)^4$$

$$= 10466 \times 10^4 \text{ mm}^4$$

$$I_{yy} = [\frac{1}{12} \times 200\ (160)^3 + 200 \times 160\ (83.73 - 80)^2]$$

$$- \left[\frac{\pi}{64}(80)^4 + \frac{\pi}{4}(80)^2(83.73 - 60)^2\right]$$

$$= 6871 \times 10^4 - 484 \times 10^4 = 6387 \times 10^4 \text{ mm}^4$$

The stress at any point, is given by

$$p = \frac{P}{A} + \frac{M_{xx}}{I_{xx}} \times y + \frac{M_{yy}}{I_{yy}} \times x$$

where x and y are the coordinates of any point with respect to the coordinate axes,

$$M_{xx} = P \times e_x = 100 \times 10^3 \times 50 = 5 \times 10^6 \text{ N.mm};$$

$$M_{yy} = P \times e_x = 100 \times 10^3 \times (83.73 - 60)$$

$$= 2.373 \times 10^6 \text{ N.mm};$$

Stress at point A:

$$P_A = \frac{P}{A} + \frac{M_{xx}}{I_{xx}} \times y_A - \frac{M_{yy}}{I_{yy}} \times x_A$$

where $y_A = 100$ mm and $x_A = 83.73$ mm

$$\therefore \quad p_A = \frac{100 \times 10^3}{26974} + \frac{5 \times 10^6}{10466 \times 10^4}$$

$$\times 100 + \frac{2.373 \times 10^6}{6387 \times 10^4} \times 83.73 = 11.6 \text{ N/mm}^2$$

Stress at point B:

$$p_B = \frac{P}{A} + \frac{M_{xx}}{I_{xx}} \times y_B - \frac{M_{yy}}{I_{yy}} \times x_B$$

where $y_B = 100$ mm and x_B

$= 160 - 83.73 = 76.27$ mm

$$\therefore \quad p_B = \frac{100\times 10^3}{26974} + \frac{5\times 10^6}{10466\times 10^4} \times 100 - \frac{2.373\times 10^6}{6387\times 10^4} \times 76.27$$

$$= 5.65 \text{ N/mm}^2$$

Stress at point C

$$p_C = \frac{P}{A} - \frac{M_{xx}}{I_{xx}} \times y_C - \frac{M_{yy}}{I_{yy}} \times x_C$$

where $y_C = 100$ mm and $x_C = 160 - 83.73 = 76.27$ mm

$$\therefore \quad p_C = \frac{100\times 10^3}{26974} - \frac{5\times 10^6}{10466\times 10^4} \times 100 - \frac{2.373\times 10^6}{6387\times 10^4} \times 76.27$$

$$= -3.90 \text{ N/mm}^2 = 3.90 \text{ N/mm}^2 \text{ (tension)}$$

Stress at point D

$$p_D = \frac{P}{A} - \frac{M_{xx}}{I_{xx}} \times y_D + \frac{M_{yy}}{I_{yy}} \times x_D$$

where $y_D = 100$ mm and $x_D = \bar{x} = 83.23$

$$\therefore \quad p_D = \frac{100\times 10^3}{26974} + \frac{5\times 10^6}{10466\times 10^4} \times 100 + \frac{2.373\times 10^6}{6387\times 10^4} \times 83.23 = 2.02 \text{ N/mm}^2$$

Problem 6.19: A piece of material is subjected to tensile stresses of p_1 and p_2 at right angles to each other ($p_1 > p_2$). Find the plane across which the resultant stress is most inclined to the normal. Find the value of this inclination and the resultant stress when $p_1 = 60$ N/mm² and $p_2 = 40$ N/mm² (both tensile).

Solution:

Let φ be the inclination of the resultant with the normal

then $$\tan\varphi = \frac{p_t}{p_n} = \frac{(p_1 - p_2)\sin\theta\cos\theta}{p_1\cos^2\theta + p_2\sin^2\theta} \quad \text{...(i)}$$

For φ to be maximum

$$\frac{d(\tan\varphi)}{d\theta} = 0$$

Hence, differentiating the dividing out common factors,

$(p_1 \cos^2\theta + p_2 \sin^2\theta) \cos 2\theta + (p_1 - p_2) \sin\theta \cos\theta \times \sin 2\theta = 0$

or $p_n \cos 2\theta + p_t \sin 2\theta = 0$

From this we get

$$\tan 2\theta = -\frac{p_n}{p_t} = -\cot\varphi = \tan\left(\frac{\pi}{2} + \varphi\right)$$

or $$2\theta = \frac{\pi}{2} + \varphi$$

Substituting the value of θ in (i), we get

$$\tan\varphi = \frac{(p_1 - p_2)\cos\varphi}{p_1(1 - \sin\varphi) + p_2(1 + \sin\varphi)}$$

or $$\frac{p_2}{p_1} = \frac{1 - \sin\varphi}{1 + \cos\varphi}$$

or $$\sin\varphi = \frac{p_1 - p_2}{p_1 + p_2} \quad \text{...(iii)}$$

Hence from (ii)

$$\theta = \frac{1}{2}\left(\frac{\pi}{2} + \sin^{-1}\frac{p_1 - p_2}{p_1 + p_2}\right)$$

which gives the position of the required plane.

Substituting the values of p_1 and p_2, we get required value of θ

$$\theta = \frac{1}{2}\left(90^\circ + \sin^{-1}\frac{60^\circ - 40^\circ}{60^\circ + 40^\circ}\right)$$

$$= \frac{1}{2}(90° + 11°32') = 50°46'$$

$$\varphi = \sin^{-1}\frac{p_1 - p_2}{p_1 + p_2} = \sin^{-1}\frac{60° - 40°}{60° + 40°} = 11°32'$$

$$p_r = \sqrt{(60\cos 50°46')^2 + (40\sin 50°46')^2}$$

$$= 48.9 \text{ N/mm}^2$$

Problem 6.20: A plane stress condition exists at a point in a loaded structure. The stresses have the magnitude and directions shown on the stress element of Fig. 6.20 (a). Calculate the stresses acting on the planes obtained by rotating the element clockwise through an angle of 16°.

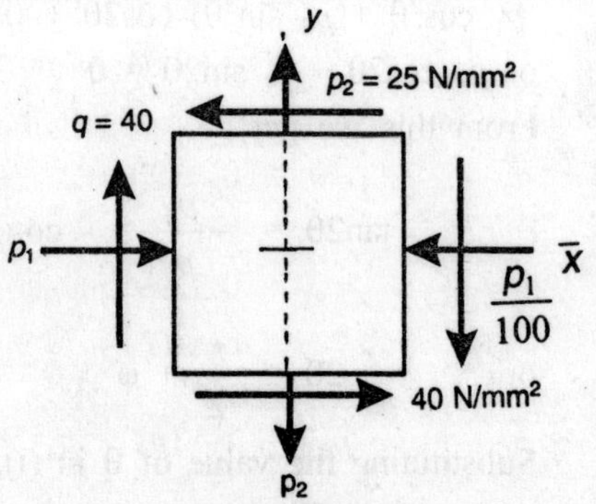

Fig. P. 6.20 (a)

Solution:

According to question we have

$$p_1 = -100 \text{ N/mm}^2$$

$$p_2 = 25 \text{ N/mm}^2$$

and $$q = -40 \text{ N/mm}^2$$

$$\frac{p_1 + p_2}{2} = \frac{-100 + 25}{2}$$

$$= -37.5 \text{ N/mm}^2$$

$$\frac{p_1 - p_2}{2} = \frac{-100 - 25}{2}$$

$$= -62.5 \text{ N/mm}^2$$

$$q = -40 \text{ N/mm}^2$$

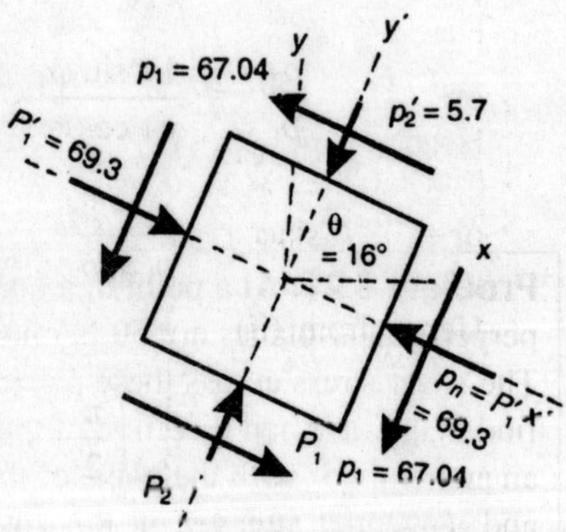

Fig. P. 6.20 (b)

For clockwise rotation,

$$\theta = -16^\circ \text{ (rotating angle)}$$

$$\therefore \quad \sin 2\theta = \sin(-32^\circ)$$

$$= -0.5299 \text{ and } \cos 2\theta$$

$$= \cos(-32^\circ) = 0.8480$$

$$= -37.5 - 62.5\,(0.8480) - 40\,(-0.5299)$$

$$= -69.3 \text{ N/mm}^2 \text{ (i.e. compressive)}$$

$$\therefore \quad p_t' = -\frac{p_1 - p_2}{2} \sin 2\theta + q \cos 2\theta$$

$$= +\,62.5\,(-0.5299) + (-40)\,(0.8480)$$

$$= -67.04 \text{ N/mm}^2$$

For finding stresses on the y' face, substitute

$$\theta = \theta + 90^\circ = -16^\circ + 90^\circ = 74^\circ$$

Hence $\quad \sin 2\theta = \sin(2 \times 74^\circ)$

$$= 0.5299 \text{ and } \cos 2\theta$$

$$= \cos(2 \times 74^\circ) = -0.8480$$

$$\therefore \quad p_2' = \frac{1}{2}(p_1 + p_2) + \frac{1}{2}(p_1 - p_2)\cos 2\theta + q \sin 2\theta$$

$$= -37.5 - 62.5\,(-0.8480) - 40\,(0.5299)$$

$$= -5.7 \text{ N/mm}^2 \text{ (i.e. compression)}$$

Here $\quad p_1' + p_2' = -69.3 - 5.7 = -75$

and $\quad p_1 + p_2 = -100 + 25 = -75$

Hence $\quad p_1' + p_2' = p_1 + p_2$

Problem 6.21: At a point in a material, the stresses on two mutually perpendicular planes are 50 N/mm^2 (tensile) and 30 N/mm^2 (tensile). The shear stress across these planes is 12 N/mm^2. Using Mohr circle, find magnitude and direction of the resultant stress on a plane making an angle of 35° with the plane of the first stress. FInd also, the normal and tangential stresses on this plane.

Solution:

To draw Mohr's circle, make OA = 50 N/mm^2 and OB = 30 N/mm^2, to the same side of origin. Erect perpendicular AG = 12 N/mm^2

in the downward direction (since q is positive on x-face) and erect BF = 12 N/mm² (since q is negative on y-face). Join F and G to get point C, the centre of the stress circle. With C as the centre and CG (or CF) as radius at Mohr's circle, draw Mohr's circle. Draw CD at angle $2\theta = 70°$ in the anti-clockwise direction with CG. The coordinates of point D give the values of p_n and p_r. Thus, get

$$p_n = OE = 54.5 \text{ N/mm}^2$$
$$p_t = ED = -5.5 \text{ N/mm}^2$$
$$p_r = OD = 54.5 \text{ N/mm}^2$$
$$\varphi = -5.5°$$

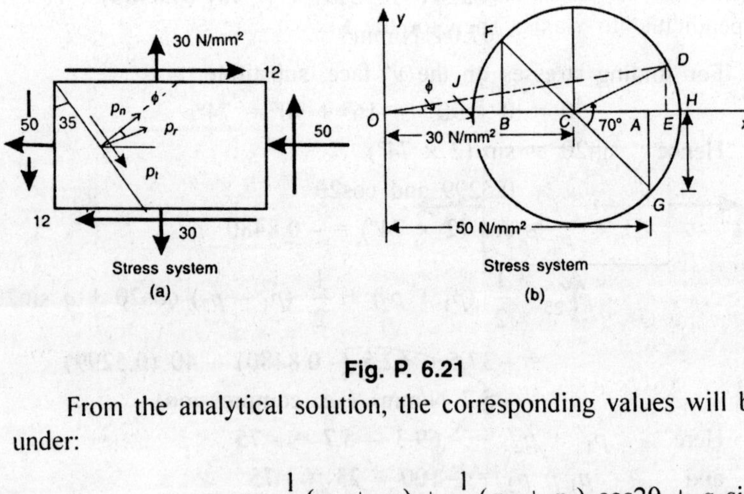

Fig. P. 6.21

From the analytical solution, the corresponding values will be as under:

$$p_n = \frac{1}{2}(p_1 + p_2) + \frac{1}{2}(p_1 + p_2)\cos 2\theta + q \sin 2\theta$$

$$= \frac{1}{2}(50 + 30) + \frac{1}{2}(50 - 30)\cos 70° + 12 \sin 70°$$

$$= 54.7 \text{ N/mm}^2$$

$$p_t = -\frac{p_1 - p_2}{2}\sin 2\theta + q\cos 2\theta$$

$$= -\frac{50 - 30}{2}\sin 70° + 12\cos 70° = -5.29 \text{ N/mm}^2$$

$$p_r = \sqrt{(54.7)^2 + (-5.29)^2} = 54.96 \text{ N/mm}^2$$

$$\varphi = \tan^{-1}\left(\frac{p_t}{p_n}\right) = \tan^{-1}\left(\frac{-5.29}{54.7}\right) = -5.52°$$

Problem 6.22: Draw Mohr's stress circle for direct stresses of 45 N/mm^2 (tensile) and 25 N/mm^2 (comp.) and find the magnitude and direction of resultant stresses on planes making angles of 30° and 60° with the plane of first principle stress. Also, find the normal and tangential stresses on these planes.

Solution:

To draw Mohr's circle, plot OA = 45 N/mm^2 to the right and OB = 25 N/mm^2 to the left of origin O. Bisect AB to get C. WIth C as centre and CA (or CB) as radius, draw the Mohr's stress circle.

(a) Draw CD_1 at an angle $2\theta_1 = 2 \times 30° = 60°$ with CA. Draw D_1E_1 perpendicular to x-axis. Then for the first plane, we get

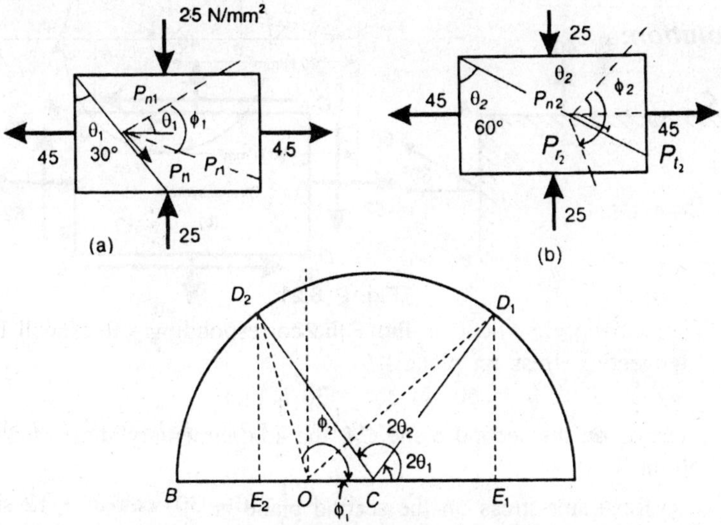

Fig. P. 6.22

$$p_{n1} = OE_1 = 27.5 \text{ N/mm}^2 \text{ (tensile)}$$

$p_{t1} = E_1D_1 = -30$ N/mm^2 (i.e. clockwise rotation as marked in Fig. 6.22 (a)

$$p_{r1} = OD_1 = 41 \text{ N/mm}^2$$

and $\quad \varphi_1 = -48°$

(b) Draw CD_2 at an angle $2\theta_2$ at an angle $2\theta_2 = 2 \times 60° = 120°$ with CA. Draw D_2E_2 perpendicular to x-axis. From the second plane we get,

$$p_{n2} = OE_2 = -7.5 \text{ N/mm}^2 \text{ (i.e. comp.)}$$
$$p_{t2} = E_2D_2 = -30 \text{ N/mm}^2 \text{ (i.e. clockwise rotation)}$$
$$p_{r2} = OD_2 = 31 \text{ N/mm}^2$$
and
$$\varphi_2 = -104^\circ$$

Problem 6.23: At a point in a material under stress, the intensity of resultant stress on a certain plane is 60 N/mm² (tensile) inclined at 30° to the normal of that plane. The stress on a plane at right angles to this has a normal tensile component of intensity 40 N/mm². Find fully (i) the resultant stress on the second plane (ii) the principal planes and stresses, and (iii) the plane of maximum shear and its location (Based on U.L.).

Solution:

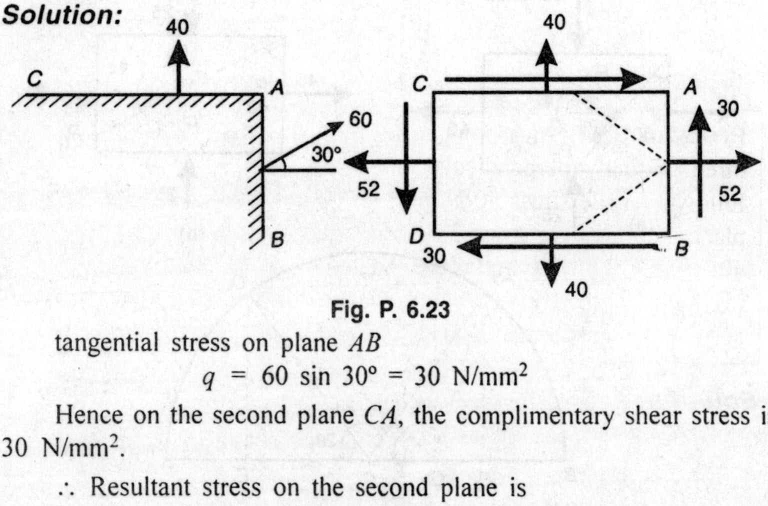

Fig. P. 6.23

tangential stress on plane *AB*

$$q = 60 \sin 30^\circ = 30 \text{ N/mm}^2$$

Hence on the second plane *CA*, the complimentary shear stress is 30 N/mm².

∴ Resultant stress on the second plane is

$$p_r = \sqrt{(40)^2 + (30)^2} = 50 \text{ N/mm}^2$$

The intensity of stress normal to the first plane (*AB*) is

$$p_n = 60 \cos 30^\circ \approx 52 \text{ N/mm}^2$$

The final stress system has been represented in figure (b) from which the principal stresses are

$$\sigma = \frac{p_1 + p_2}{2} \pm \sqrt{\left(\frac{p_1 - p_2}{2}\right)^2 + q^2}$$

$$= \frac{52+40}{2} \pm \sqrt{\left(\frac{52-40}{2}\right)^2 + (30)^2} = 46 \pm 30.6$$

$\therefore \quad \sigma_1 = 76.6 \text{ N/mm}^2$ (tension)

and $\quad \sigma_2 = 15.4 \text{ N/mm}^2$ (tension)

$$\tan 2\theta_p = \frac{2q}{p_1 - p_2} = \frac{2 \times 30}{52 - 40} = \frac{60}{12}$$

$$= 5, \text{ from which } 2\theta_p = 78°42'$$

$\therefore \quad \theta_{p1} = 39°21'$

and $\quad \theta_{p2} = 129°21'$

$$p_{t,\max} = \frac{\sigma_1 - \sigma_2}{2} = \frac{76.6 - 15.4}{2} = 30.6 \text{ N/mm}^2$$

$$\theta_{s1} = 45° + \theta_{p1} = 45° + 39°21' = 84°21'$$

Problem 6.24: At a point in a piece of elastic material, there are three mutually perpendicular planes on which the stresses are as follows: tensile stress 50 N/mm² and shear stress 40 N/mm² on one plane, compressive stress 35 N/mm² and complimentary shear stress 40 N/mm² on the second plane, and no stress on the third pl ıne. Find (a) the principal stresses and the positions of the planes on which they act (b) the position of planes on which there is no normal stress.

Solution:

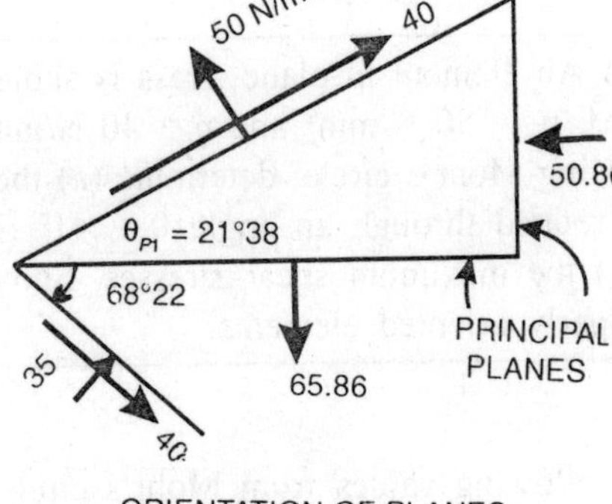

Fig. P. 6.24

$$\sigma = \frac{50-35}{2} \pm \sqrt{\left(\frac{50+35}{2}\right)^2 + (40)^2}$$

$$= 7.5 \pm 58.36$$

$\therefore$ $\sigma_1 = 65.86$ N/mm² (tensile)

and $\sigma_2 = -50.86$ N/mm² (i.e. compressive)

$$2\theta_p = \tan^{-1}\frac{2\times 40}{50+35} = 43°16'$$

$\therefore$ $\theta_{p1} = 21°38'$

If there is no normal stress on a plane, then for that plane θ and A will coincide and therefore the radius vector for that plane will be CG' (Fig. 6.24 a)

Then $$2\theta' = 180° - \cos^{-1}\frac{OC}{CG'},$$

where $OC = OA - CA = 50 - \frac{1}{2}(50 + 35) = 7.5$

and $CG' = R = CG = \sqrt{(40)^2 + (42.5)^2} = 58.36$

$\therefore$ $$2\theta' = 180° - \cos^{-1}\frac{7.5}{58.36} = 180° - 82°37' = 97°23'$$

$\therefore$ $\theta' = 48°41.5'$ to the principal plane

Problem 6.25: An element in plane stress is subjected to stresses $p_1 = 150$ N/mm², $p_2 = 50$ N/mm² and $q = 40$ N/mm², as shown in Fig. 6.25 (a). Using Mohr's circle, determine (a) the stresses acting on an element rotated through an angle $\theta = 41°$ (b) the principal stresses, and (c) the maximum shear stresses. Show all results on sketches of properly oriented elements.

Solution:

We get the following values from Mohr's circle

$(p_n)_D = OE = +147$ N/mm²;

$(p_t) = ED = -44$ N/mm²;

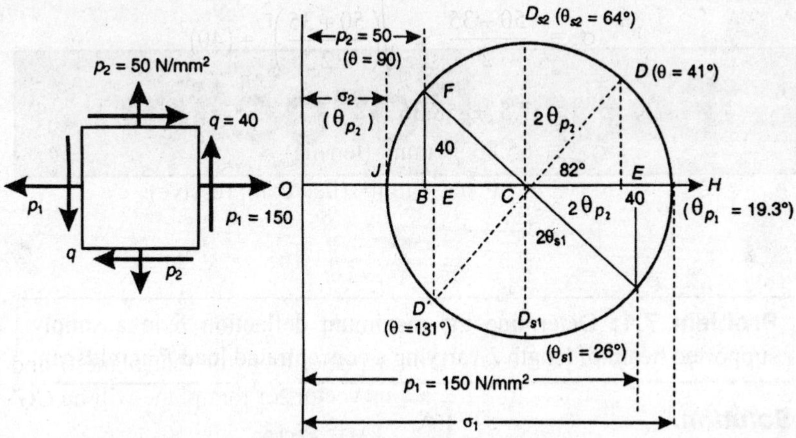

Fig. P. 6.25

$$\left(P_n\right)_D = OE' = 54 \text{ N/mm}^2$$

$$\sigma_1 = OH = +164 \text{ N/mm}^2$$

$$\sigma_2 = OJ = 36 \text{ N/mm}^2$$

$$2\theta_{p1} = +38.6^\circ$$

$$\theta_{p1} = 19.3^\circ$$

$$(p_t)_{max} = CD_1 = +64 \text{ N/mm}^2$$

$$(p_t)_{min} = CD_2 = -64 \text{ N/mm}^2$$

$$2\theta_{s1} = -52^\circ;\ \theta_{s1} = -26^\circ;$$

$$2\theta_{s2} = +128^\circ;\ \theta_{s2} = +64^\circ$$

$$\left(P_n\right)_{D_1} = p_{av} = 100 \text{ N/mm}^2 = \left(P_n\right)_{D_2}$$

Chapter 7

Beam Deflection

Problem 7.1: Determine the maximum deflection δ in a simply supported beam of length L carrying a concentrated load P at midspan.

Solution:

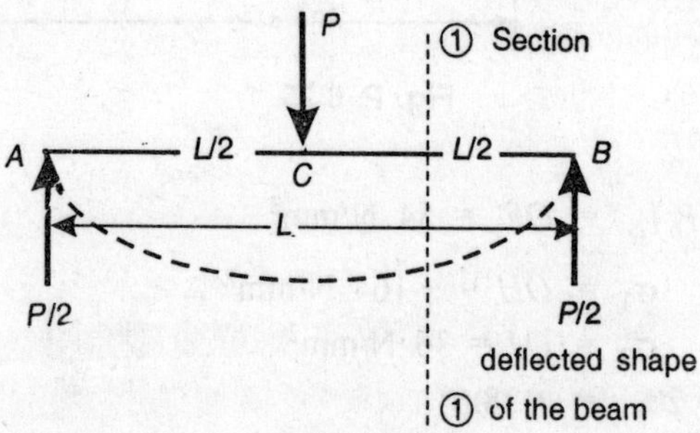

Fig. P. 7.1

First write general equation,

$$EI\frac{d^2y}{dx^2} = \frac{P}{2}x - P\left\langle x-\frac{L}{2}\right\rangle$$

$$\therefore \quad EI\frac{dy}{dx} = \frac{P}{2}\cdot\frac{x^2}{2} - \frac{P}{2}\left\langle x-\frac{L}{2}\right\rangle^2 + c_1 \quad \text{...(i)}$$

$$EIy = \frac{P}{2}\cdot\frac{x^3}{2\times 3} - \frac{P}{2}\frac{\left\langle x-\frac{L}{2}\right\rangle^3}{3} + c_1x + c_2$$

$$= \frac{px^3}{12} - \frac{P}{6}\left\langle x-\frac{L}{2}\right\rangle^3 + c_1x + c_2 \quad \text{...(ii)}$$

From condition equation we know that,

$$\left.\begin{aligned} \text{At} \quad & x = 0, y = 0 \\ \text{At} \quad & x = \frac{L}{2}, \frac{dy}{dx} = 0 \end{aligned}\right\} \quad \ldots\text{(iii)}$$

Putting these value in equation (i) and (ii) then we get

$$0 = \frac{P}{2} \times \frac{1}{2} \times \frac{L^2}{4} - 0 + c_1$$

$$\therefore \quad C_1 = -\frac{PL^2}{16}$$

From equation (2) $C_2 = 0$

$\therefore$ Slope equation

$$EI\frac{dy}{dx} = \frac{px^2}{4} - \frac{p}{2}\left\langle x - \frac{L}{2}\right\rangle^2 - \frac{PL^2}{16} \quad \ldots(4)$$

$$EIy = \frac{Px^3}{12} - \frac{p}{6}\left\langle x - \frac{L}{2}\right\rangle^3 - \frac{PL^2}{16}x \quad \ldots(5)$$

From loading condition we observed that diflection is maximum at mid span of the beam.

$$\therefore \quad \text{At } x = \frac{L}{2},$$

$$EIy_{\max} = \frac{P}{12} \times \left(\frac{L}{2}\right)^3 - 0 - \frac{PL^2}{16} \times \frac{L}{2}$$

$$= \frac{PL^3}{12 \times 8} - \frac{PL^3}{32}$$

$$= \frac{PL^3}{96} - \frac{PL^3}{32} = \frac{PL^3 - 3PL^3}{96}$$

$$= -\frac{2PL^3}{96} = -\frac{PL^3}{48}$$

$$\therefore \quad y_{\max} = -\frac{PL^3}{48\,EI}$$

Problem 7.2: Determine the maximum deflection δ in a simply supported beam of length L carrying a uniformly distributed load intensity w_0 applied over its entire length.

Solution:

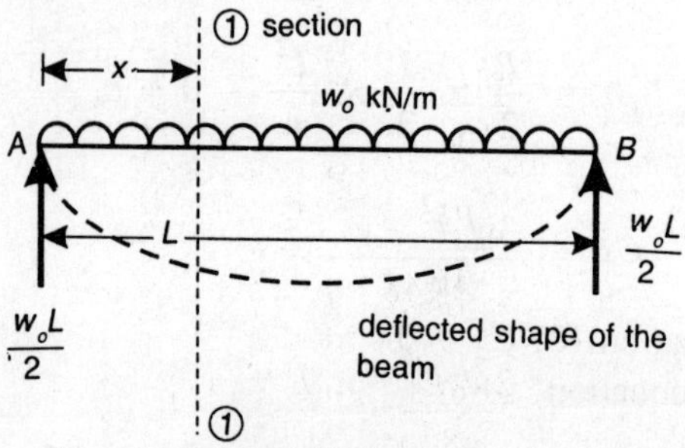

Fig. P. 7.2

First take any section (i)-(i) at a distance x from the end A of the beam, then general expression of the beam is,

$$EI\frac{d^2y}{dx^2} = \frac{w_0 L}{2}x - w_0 \cdot \frac{x^2}{2} \quad \text{...(i)}$$

$$EI\frac{dy}{dx} = \frac{w_0 L}{2} \cdot \frac{x^2}{2} - \frac{w_0}{2} \cdot \frac{x^3}{3} \cdot + C_1$$

$$EI\frac{dy}{dx} = \frac{w_0 L}{4}x^2 - \frac{w_0}{6}x^3 + C_1 \quad \text{...(2)}$$

$$EIy = \frac{w_0 L}{4} \cdot \frac{x^3}{3} - \frac{w_0}{6} \cdot \frac{x^4}{4} + C_1x + C_2$$

$$EIy = \frac{w_0 L}{12}x^3 - \frac{w_0}{24} \cdot x^4 + C_1x + C_2 \quad \text{...(3)}$$

From boundary condition we know that,

$$\left.\begin{array}{ll} \text{At} & x = 0,\ y = 0 \\ \text{At} & x = \dfrac{L}{2},\ \dfrac{dy}{dx} = 0 \end{array}\right\} \quad \text{...(4)}$$

From help of equation 4 we find the value of integration constants,

$\therefore$ From equation (3) and (4)

$$C_2 = 0$$

From equation (2) and (4)

$$0 = \frac{w_0 L}{4} \cdot \left(\frac{L}{2}\right)^2 - \frac{w_0}{6} \times \left(\frac{L}{2}\right)^3 + C_1$$

$$\therefore \quad C_1 = \frac{w_0 L^3}{16} - \frac{w_0 L^3}{48}$$

$$= \frac{3w_0 L^3 - w_0 L^3}{48}$$

$$= \frac{2w_0 L^3}{48} = \frac{w_0 L}{24}$$

$\therefore$ putting this value in equation (3) then we get

$$EIy = \frac{w_0 L}{12} x^3 - \frac{w_0}{24} x^4 - \frac{w_0 L^3}{24} x$$

$\therefore$ Maximum diflection at $x = \frac{L}{2}$,

$$\therefore \quad E\,I\,y_{\max} = \frac{w_0 L}{12} \times \left(\frac{L}{2}\right)^3 - \frac{w_0}{24}\left(\frac{L}{2}\right)^4 - \frac{w_0 L^3}{24} \times \frac{L}{2}$$

$$= \frac{w_0 L}{12} \times \frac{L^3}{8} - \frac{w_0}{24} \times \frac{L^4}{16} - \frac{w_0 L^3}{24} \times \frac{L}{2}$$

$$= \frac{w_0 L^4}{96} - \frac{w_0 L^4}{384} - \frac{w_0 L^4}{48}$$

$$= \frac{12w_0 L^4 - 3w_0 L^4 - 24w_0 L^4}{1152}$$

$$= \frac{-15w_0 L^4}{1152}$$

$$= -\frac{5\,w_0 L^4}{384\,EI}$$

Problem 7.3: Determine the maximum value of EIy for the cantilever beam loaded as shown in Fig. 7.3 (a). Take the origin at the wall.

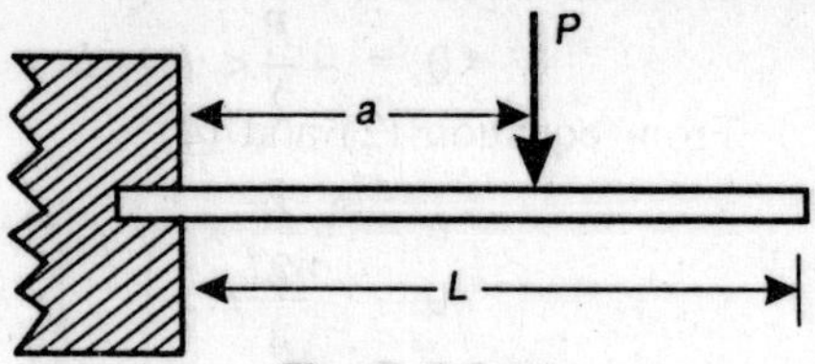

Fig. P. 7.3 (a)

Solution:

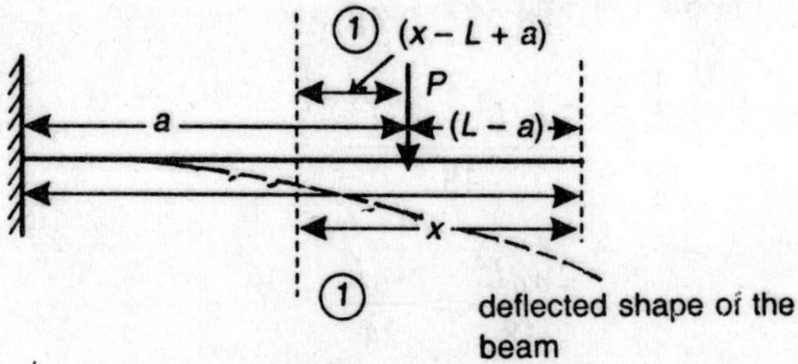

Fig. P. 7.3 (b)

Hence general expression of the beam is

$$EI\frac{d^2y}{dx^2} = -p < x - L + a >$$

$$EI\frac{dy}{dx} = -\frac{P}{2} < x - L + a >^2 + C_1 \quad \text{...(i)}$$

$$EIy = -\frac{P}{2}\frac{\langle x - L + a\rangle^3}{3} + C_1 x + C_2$$

$$= -\frac{P}{6} < x - L + a >^3 + C_1 x + C_2 \quad \text{... (2)}$$

From condition we know that,

$$\left.\begin{array}{ll} \text{At} & x = L, y = 0 \\ \text{At} & x = L, \dfrac{dy}{dx} = 0 \end{array}\right\} \quad \text{...(3)}$$

∴ From (i) and (3)

$$0 = -\frac{P}{2} < x - L + a >^2 + C_1$$

$$\therefore \quad 0 = -\frac{P}{2} < L - L + a >^2 + C_1$$

$$0 = -\frac{P}{2} \times a^2 + C_1$$

$$\therefore \quad C_1 = \frac{P}{2} a^2$$

From equation (2) and (3)

$$0 = -\frac{P}{6} < L - L + a >^3 + \frac{P}{2} a^2 \times L + C_2$$

$$= -\frac{P}{6} < a >^3 + \frac{Pa^2 L}{2} = -\frac{P}{6} a^3 + \frac{Pa^2 L}{2} + C_2$$

$$C_2 = \frac{P}{6} a^3 - \frac{Pa^2 L}{2}$$

Putine these value in equation (2), we get

$$EIy = \frac{P}{6} < x - L + a >^3 + \frac{Pa^2}{2} x + \frac{P}{6} a^3 - \frac{Pa^2 L}{2} \quad \ldots(4)$$

We from elastic deformation of the beam, y_{max} . at $x = 0$

$$\therefore \quad EIy_{max} = -\frac{p}{2} < 0 - L + a >^3 + 0 + \frac{P}{6} a^3 - \frac{Pa^2 L}{2}$$

$$= \frac{p}{6} a^3 - \frac{Pa^2 L}{2} = \frac{Pa^2}{2} \left[\frac{a}{3} - L \right]$$

Problem 7.4: Find the equation of the elastic curve for the cantilever beam shown in Fig. P 7.4 (a) it carries a load that varies from zero at the wall to w_0 the free end. Take the origin at the wall.

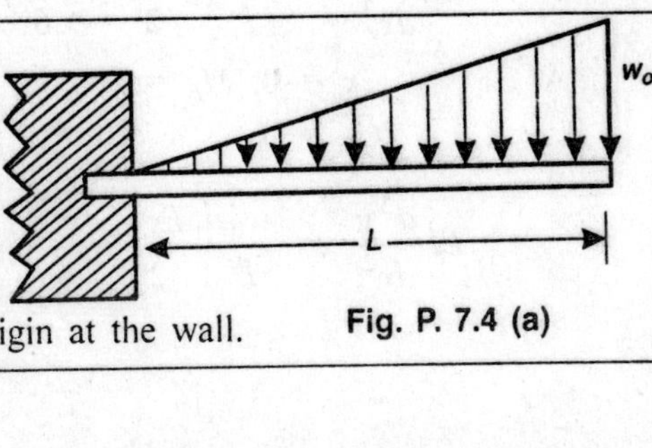

Fig. P. 7.4 (a)

Solution:

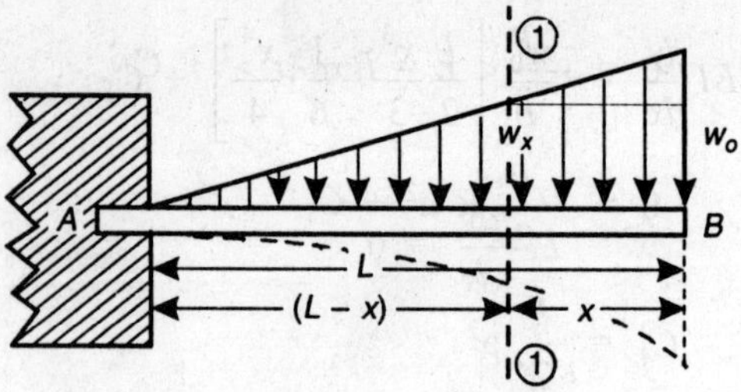

Fig. P. 7.4 (b)

From similar triangle we know that,

$$\frac{w_0}{L} = \frac{w_x}{(L-x)}$$

$$\therefore \quad w_x = \frac{w_0(L-x)}{L}$$

At any distance x from the free end

$$EI\frac{d^4y}{dx^4} = \text{rate of loading} = w_x = \frac{w_0}{L}(L-x)$$

$$\therefore \quad EI\frac{d^3y}{dx^3} = +\frac{w_0}{L}(Lx - \frac{x^2}{2}) + C_1 = +Fx$$

At $\quad x = 0$, force $(f_B) = 0$

$$\therefore \quad C_1 = 0$$

$$\therefore \quad EI\frac{d^3y}{dx^3} = \frac{w_0}{L}(Lx - \frac{x^2}{2})$$

$$\therefore \quad EI\frac{d^2y}{dx^2} = \frac{w_0}{L}(L\frac{x^2}{2} - \frac{x^3}{6}) + C_2$$

At $\quad x = 0, M_B = 0$

$$\therefore \quad C_2 = 0$$

$$\therefore \quad EI\frac{d^2y}{dx^2} = \frac{w_0}{L}\left(\frac{Lx^2}{2} - \frac{x^3}{6}\right) \quad \ldots(i)$$

$$\therefore \quad EI\frac{dy}{dx} = \frac{w_0}{L}\left[\frac{L}{2}\frac{x^3}{3} - \frac{1}{6}\cdot\frac{x^4}{4}\right] + C_1$$

At $\quad x = L, \dfrac{dy}{dx} = 0$

$$\therefore \quad 0 = \frac{w_0}{L}\left[\frac{L}{2}\times\frac{L^3}{3} - \frac{1}{6}\cdot\frac{L^4}{4}\right] + C_1$$

$$0 = \frac{w_0}{L}\left[\frac{L^4}{6} - \frac{L^4}{24}\right] + C_1$$

$$\therefore \quad C_1 = -\frac{w_0}{L}\left[\frac{4L^4 - L^4}{24}\right] = -\frac{w_o}{L}\times\frac{3L^4}{24}$$

$$= -\frac{w_0}{L}\times\frac{L^4}{8} = -\frac{w_0 L^4}{8L}$$

$$\therefore \quad C_1 = -\frac{w_0 L^4}{8L}$$

$$EIy = \frac{w_0}{L}\left[\frac{L}{2}\cdot\frac{1}{3}\cdot\frac{x^4}{4} - \frac{1}{6}\cdot\frac{1}{4}\cdot\frac{x^5}{5}\right] + C_1\,x + C_2$$

$$= \frac{w_0}{L}\left[\frac{Lx^4}{24} - \frac{x^5}{120}\right] - \frac{w_0 L^4}{8L}\times x + C_2$$

At $\quad x = L, y = 0$, from boundary condition

$$0 = \frac{w_0}{L}\left[\frac{L\times L^4}{24} - \frac{L^5}{120}\right] - \frac{w_0 L^4}{8L} + C_2$$

$$0 = w_0\left[\frac{L^4}{24} - \frac{L^4}{120}\right] - \frac{w_0 L^4}{8L} + C_2$$

$$0 = w_0\left[\frac{5L^4 - L^4}{120}\right] - \frac{w_0 L^4}{8L} + C_2$$

$$0 = \frac{4w_0L^4}{120} - \frac{w_0L^4}{8L} + C_2$$

$$\therefore \quad C_2 = \frac{w_0L^4}{8L} - \frac{w_0L^4}{30} = \frac{15w_0L^4 - 4w_0L^4}{120} = \frac{11}{120}w_0L^4$$

∴ Final equation of deflection is,

$$EIy = \frac{w_0}{L}\left[\frac{Lx^4}{24} - \frac{x^5}{120}\right] - \frac{w_0L^4}{8L}x + \frac{11}{120}w_0L^4$$

We know from elastic curve, deflection is maximum at $x = 0$, so

$$E\,I\,y_{max} = \frac{w_0}{L}[0 - 0] - 0 + \frac{11}{120}w_0L^4 = \frac{11}{120}w_0L^4$$

$$\therefore \quad y_{max} = \frac{11}{120EI}\cdot w_oL^4$$

Problem 7.5: As shown in Fig 7.5 (a) a simply supported beam carries two symmetrically placed concentrated loads. Compute the maximum deflection δ.

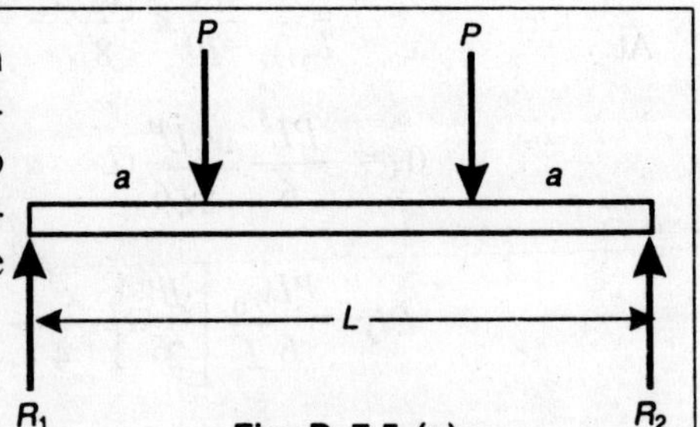

Fig. P. 7.5 (a)

Solution:

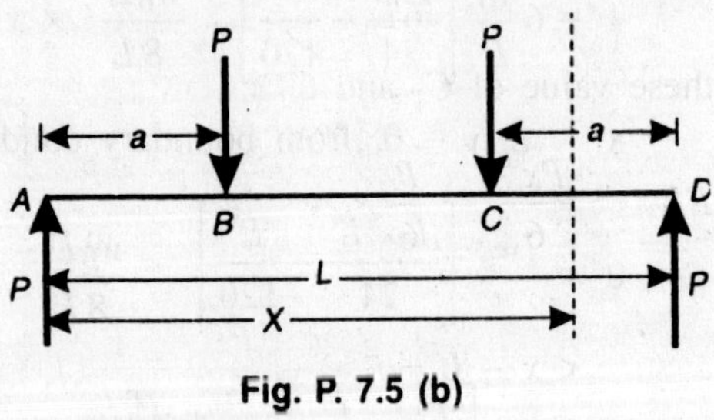

Fig. P. 7.5 (b)

General expression for the beam,

$$EI\frac{d^2y}{dx^2} = Px - P < x - a > - P < x - L + a >$$

$$EI\frac{dy}{dx} = \frac{Px^2}{2} - \frac{P\langle x-a\rangle^2}{2} - \frac{P\langle x-L+a\rangle^2}{2} + C_1 \ldots\text{(i)}$$

$$EIy = \frac{P}{2}\frac{x^3}{3} - \frac{P}{2}\frac{\langle x-a\rangle^3}{3} - \frac{P}{2}$$

$$\frac{\langle x-L+a\rangle^3}{3} + C_1x + C_2$$

$$\therefore \quad EIy = \frac{Px^3}{6} - \frac{P}{6} <x - a>^3 - \frac{P}{6} < x - L + a >^3 +$$

$$C_1x + C_2$$

At $x = 0, y = 0$

From equation (ii)

$$0 = 0 - 0 - 0 + 0 + C_2$$

$$\therefore \quad C_2 = 0$$

At $x = L, y = 0$

$$0 = \frac{PL^3}{6} - \frac{P}{6}(L - a)^3 - \frac{P}{6}\text{ a}^3 + C_1L + 0$$

$$= \frac{PL^3}{6} - \frac{P}{6}(L - a)^3 - \frac{Pa^3}{6L} + C_1L$$

$$C_1 = \frac{PL^2}{6} - \frac{P}{6L}(L - a)^3 - \frac{Pa^3}{6L}$$

$\therefore$ Putting these value of C_1 and C_2 in

$$EIy = \frac{Px^3}{6} - \frac{P}{6} < x - a >^3 - \frac{P}{6}$$

$$< x - L + a >^3 + \left[\frac{PL^2}{6} - \frac{P}{6L}(L-a)^3 - \frac{Pa^3}{6L}\right]x$$

And, $$EI\frac{dy}{dx} = \frac{Px^2}{2} - \frac{P\langle x-a\rangle^2}{2} - \frac{P\langle x-L+a\rangle^2}{2}$$

$$+\ \frac{PL^2}{6} - \frac{P}{6L}(L-a)^3 - \frac{Pa^3}{6L}$$

For maximum value of deflection, $\frac{dy}{dx} = 0$ and maximum value of deflection exist between B and C, so equate slope equation is equal to 0 at c.

$$0 = \frac{Px^2}{2} - \frac{P(x-a)^2}{2} + \frac{PL^2}{6} - \frac{P}{6L} \times$$

$$(L-a)^3 - \frac{Pa^3}{6L}$$

$$0 = \frac{Px^2}{2} - \frac{P}{2}(x^2 - 2\,ax + a^2) + \frac{PL^2}{6}$$

$$- \frac{P}{6L}(L-a)^3 - \frac{Pa^3}{6L}$$

$$0 = \frac{Px^2}{2} - \frac{Px^2}{2} + Pax - \frac{Pa^2}{2} + \frac{PL^2}{6}$$

$$- \frac{P}{6L}(L-a)^3 - \frac{Pa^3}{6L}$$

$$\therefore \quad Pax = +\frac{Pa^2}{2} + \frac{P}{6L}(L-a)^3 + \frac{Pa^3}{6L} - \frac{PL^2}{6}$$

$$\therefore \quad x = \frac{Pa^2}{2Pa} + \frac{P}{6Pa}(L-a)^3 + \frac{Pa^3}{6PaL} - \frac{PL^2}{6Pa}$$

$$= \frac{a}{2} + \frac{1}{6a}(L-a)^3 + \frac{a^2}{6L} - \frac{L^2}{6a}$$

At this value of x, y is maximum

$$\therefore \quad EIy_{max} = \frac{Pa}{24\,E\,I}(3\,L^2 - 4\,a^2)$$

Problem 7.6: Compute the value of $EI\delta$ at midspan for the beam loaded as shown in Fig 7.6. If E = 10 GPa, what value of I is required to limit the midspan deflection to 1/360 of the span ?

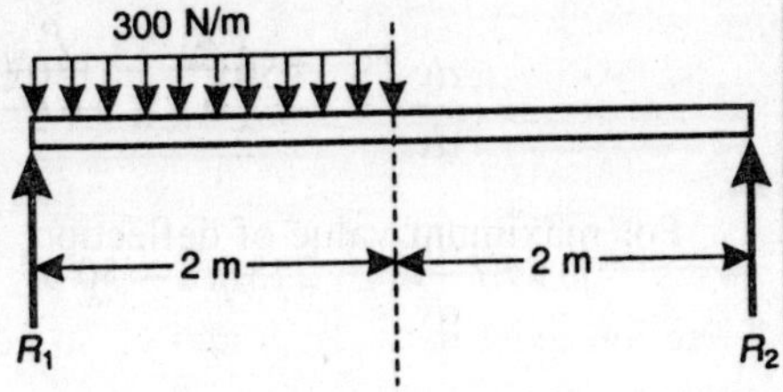

Fig. P. 7.6 (a)

Solution:

Given $E = 10 \text{ GPa} = 10 \times 10^3 \text{ MPa} = 10^4 \text{ N/mm}^2$

What value of I is required to limit the mid span deflection to $\frac{1}{360}$ of the span, i.e. $\frac{L}{360}$

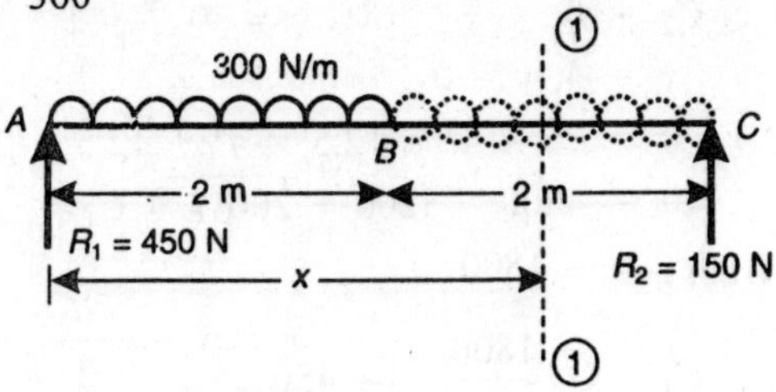

Fig. P. 7.6 (b)

$$\Sigma F_y = 0,$$

$$R_1 + R_2 = 300 \times 2 = 600$$

$$\Sigma M_A = 0,$$

$$R_2 \times 4 = 300 \times 2 \times 1$$

$$R_2 = \frac{600}{4} = 150$$

$$R_1 = 450 \text{ N}$$

$\therefore$ General expression for the beam,

$$EI\frac{d^2y}{dx^2} = 450\,x - \frac{300}{2}x^2 + \frac{300}{2} < x - 2 >^2$$

$$= 450\,x - 150\,x^2 + 150 < x - 2 >^2 \qquad \ldots(1)$$

$$EI\frac{dy}{dx} = \frac{450x^2}{2} - \frac{150x^3}{3} + \frac{150}{3} < x - 2 >^3 + C_1$$

$$\therefore \quad EI\frac{dy}{dx} = 225\,x^2 - 50\,x^3 + 50 < x - 2 >^3 + C_1 \quad \ldots(2)$$

$$EIy = 225\,\frac{x^3}{3} - \frac{50}{4}\,x^4 + \frac{50}{4} < x - 2 >^4$$

$$+ C_1\,x + C_2$$

$$EIy = 75\,x^3 - 12.5\,x^4 + 12.5 < x - 2 >^4$$

$$+ C_1\,x + C_2 \quad \ldots(3)$$

At $x = 0,\ y = 0$

$$0 = 0 - 0 + 0 + 0 + C_2$$

$$\therefore \quad C_2 = 0$$

At $x = 4,\ y = 0$

$$0 = 75 \times 4^3 - 12.5 \times 4^4 + 12.5 \times 2^4 + 4\,C_1 + 0$$

$$0 = 4800 - 3200 + 200 + 4\,C_1$$

$$4c_1 = -\,1800$$

$$\therefore \quad C_1 = -\,\frac{1800}{4} = -\,450$$

$$\therefore \quad EI\frac{dy}{dx} = 225\,x^2 - 50\,x^3 + 50 < x - 2 >^3 - 450 \quad \ldots(4)$$

$$EIy = 75\,x^3 - 12.5\,x^4 + 12.5 < x - 2 >^4 - 450 \quad \ldots(5)$$

For maximum deflection, $\frac{dy}{dx} = 0$ maximum deflection will evidently occur in range *AB*. putting $\frac{dy}{dx} = 0$ in eq. (4) and including all terms upto *B*.

$$0 = 225\,x^2 - 50\,x^3 \quad 450$$

or $225\,x^2 - 50\,x^3 - 450 = 0$

$$4.5\,x^2 - x^3 - 9 = 0$$

From trial and error method,

$$x = 1.8 \text{ m}$$

$$\therefore \quad EIy_{max} = 75 \times (1.8)^3 - 12.5 \times (1.8)^4 +$$

$$12.5 < 1.8 - 2 >^4 - 450 \times 1.8$$
$$= 437.4 - 131.22 - 810$$
$$EIy_{max} = -503.82 \text{ N.m}^3$$
$$EIy_{\text{at } x = 2} = 75 \times 2^3 - 12.5 \times 2^4 - 450 \times 2$$
$$= 600 - 200 - 900 = -500 \text{ Nm}^3$$
$$EIy_{\text{ at } x = 2} = -500 \times 10^6 \text{ mm}^3$$
$$y \text{ at } x = 2 \text{ m} = -\frac{500 \times 10^6}{E\,I} = \frac{4}{360}$$
$$\therefore \quad I = \frac{500 \times 360}{4 \times 10^4} \times 10^6 = 4.5 \times 10^6 \text{ mm}^4$$

Problem 7.7: Compute the midspan value of $EI\delta$ for the beam loaded as shown in Fig 7.7

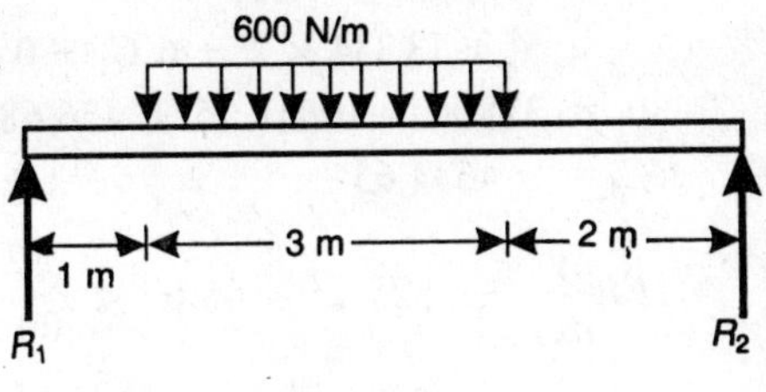

Fig. P. 7.7

Solution:

$$\Sigma F_y = 0$$
$$R_1 + R_2 = 600 \times 3 = 1800$$
$$\Sigma M_A = 0$$
$$R_2 \times 6 = 600 \times 3 \times 2.5 = 4500$$
$$\therefore \quad R_2 = 750 \text{ N}$$

General expression for the beam,

$$EI\frac{d^2y}{dx^2} = 1050\,x - \frac{600}{3} < x - 1 >^2 + \frac{600}{3} < x - 4 >^2$$
$$= 1050\,x - 200 < x - 1 >^2 + 200 < x - 4 >^2$$
$$EI\frac{dy}{dx} = 1050\,\frac{x^2}{2} - 200\,\frac{\langle x-2\rangle^3}{3} + 200\,\frac{\langle x-4\rangle^3}{3} + C_1$$
$$= 525\,x^2 - 66.67 < x - 2 >^3 + 66.67 < x - 4 >^3 + C_1$$
$$EI\frac{dy}{dx} = 525\,x^2 - 66.67 < x - 1 >^3 + 66.67 \times$$
$$< x - 4 >^4 + C_1 \qquad \ldots(1)$$

$$EIy = 525\ \frac{x^3}{3} - \frac{66.67}{4} < x - 1 >^4$$

$$+ \frac{66.67}{5} < x - 1 >^5 + C_1\ x + C_2$$

$$EIy = 175\ x^3 - 16.67 < x - 1 >^4 + 13.334$$

$$< x - 1 >^5 + C_1 x + C_2 \qquad ...(2)$$

From boundary condition,

At $x = 0$, $y = 0$

At $x = 6$ m, $y = 0$

$$0 = 0 + 0 + 0 + 0 + C_2$$

$$\therefore C_2 = 0$$

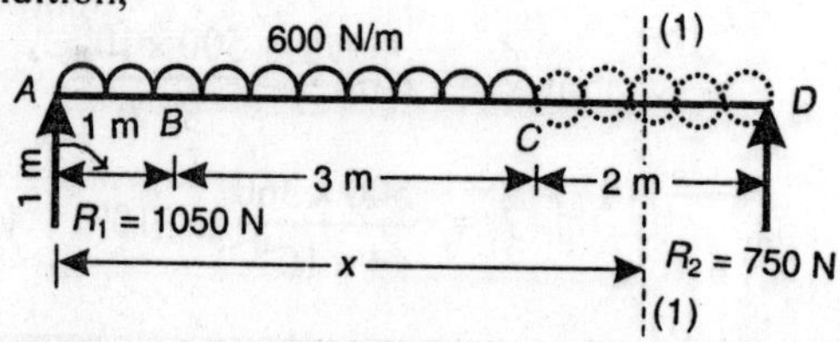

Fig. P. 7.7 (a)

$$0 = 175 \times 6^3 - 16.67 \times 5^4 + 13.334 \times 2^5 + 6\ C_1 + 0$$

$$0 = 37800 - 10418.75 + 426.688 + 6\ C_1$$

$$\therefore C_1 = -\ 4634.65$$

$$\therefore EI\frac{dy}{dx} = 525\ x^2 - 66.67 < x - 1 >^3 + 66.67 < x - 4 >^4 - 4634.65\ x$$

$$EIy = 175\ x^3 - 16.67 < x - 1 >^4 + 13.334 < x - 4 >^5 - 4634.65$$

$E\ I\ y$ at midspan i.e. at $x = 3$ m

$$EIy = 175 \times 3^3 - 16.67 < 3 - 1 >^4 + 13.334 < 3 - 4 >^5 - 4634.65 \times 3$$

$$= 4725 - 266.72 + 0 - 13903.95 = -\ 9445.67\ \text{N.m}^3$$

Problem 7.8: A cantilever *AB* of length *L* is subjected to a concentrated load *W* and a couple μ acting at the free end, as shown in Fig. 7.8 (a). Determine the slope and deflection at the free end.

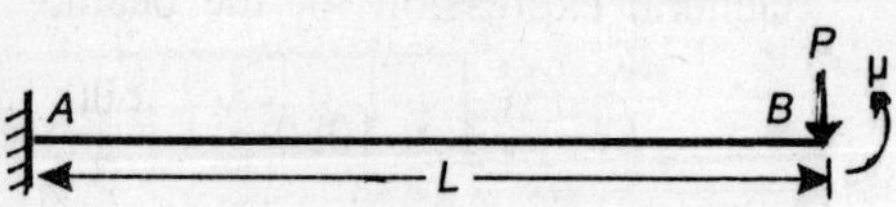

Fig. P. 7.8 (a)

Solution:

The component B.M. diagrams are shown in Fig. 7.8 (b) and (c) respectively.

Now $\theta_B^A = \sum_B^A A_{M/EI} = -\left[\mu L - \frac{1}{2} PL.L\right] \frac{1}{E\ I}$

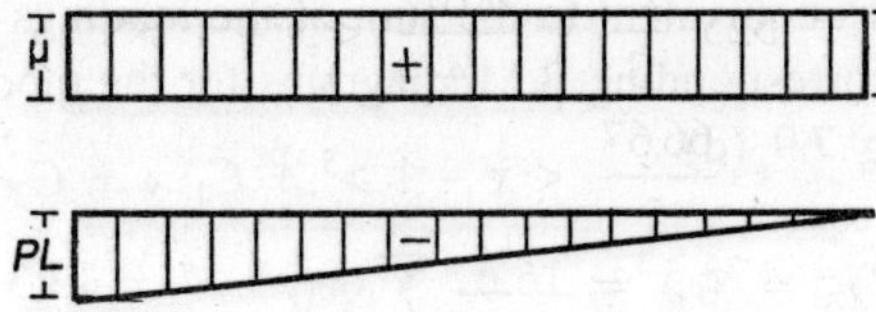

Fig. P. 7.8 (b)

$$= \frac{1}{2\,EI}[PL - 2\,\mu]$$

and $\qquad y_B^A = -\sum_B^A A_{M/EI}.\bar{x} = -\frac{1}{EI}\left[\mu L.\frac{L}{2} - \frac{1}{2}.PL.L.\frac{2L}{3}\right]$

or $\qquad y_B^A = \frac{PL^3}{3\,EI} - \frac{\mu L^2}{2\,EI} = \frac{L^2}{6\,EI}[2\,PL - 3\,\mu]$

Problem 7.9: Compute the maximum slope and deflection for a cantilever beam, shown Fig. 7.9 (a), that carries varying load from zero at the wall to w_0 at the free end.

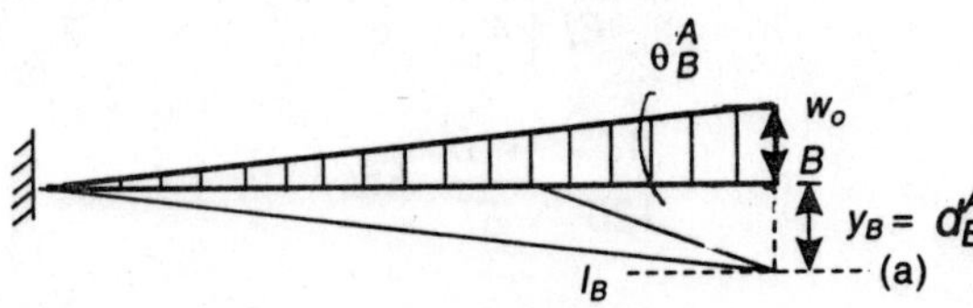

Fig. P. 7.9 (a)

Solution:

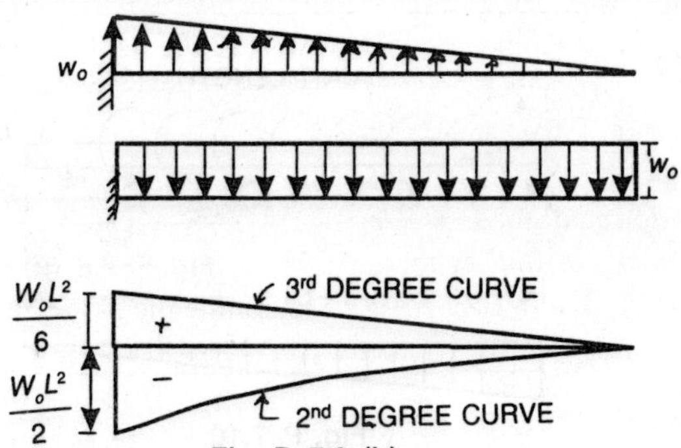

Fig. P. 7.9 (b)

The elastic curve in Fig. 7.9 (a) shows that the maximum slope end maximum deflection occurs at the free end B. Also, the given loading shown in Fig (a) is equivalent to the sum of the loadings shown in Fig. (b) and (c). The corresponding B.M. diagrams for the modified loadings are shown in Fig 7.9 (d)

Now $$i_B = \theta_B^A = -\frac{1}{EI}\sum_B^A A_M$$

$$= -\frac{1}{EI}\left[\frac{1}{4}\times\frac{w_0 L^2}{6}\times L - \frac{1}{3}\frac{w_0 L^2}{2}\times L\right]$$

$$= \frac{w_0 L^2}{8\,EI} \quad \textbf{Ans}$$

Also $$Y_B = d_B^A = -\frac{1}{EI}\sum_B^A A_M\,.\,\bar{x}$$

$$= -\frac{1}{EI}\left[\frac{1}{4}\times\frac{w_o L^2}{6}\times L\times\frac{4}{5}L - \frac{1}{3}\frac{w_o L^2}{2}\cdot L\times\frac{3}{4}L\right]$$

$$= \frac{11}{120}\;\frac{w_0 L^4}{EI} \quad \textbf{Ans}$$

Problem 7.10: A cantilever is acted on by a uniformly distributed moment of intesity *m* per unit distance along the axis of the beam. Find the slope and deflection at the free end.

Solution:

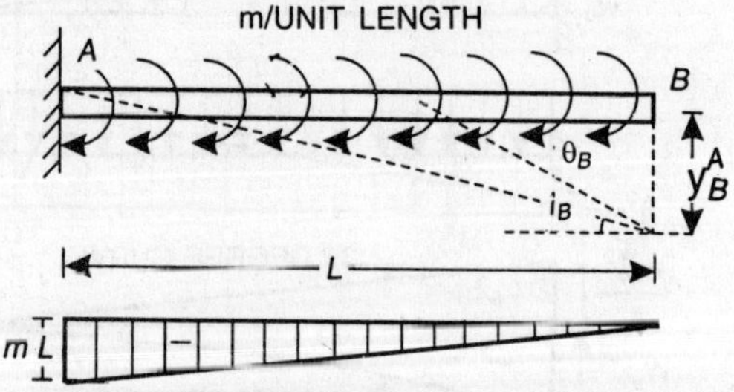

Fig. P. 7.10

$$M_B = 0;\; M_A = mL$$

The B.M. diagram is shown in Fig. 7.10 (b), the variation being linear.

$$\text{Now} \quad i_B = \theta_B^A = -\sum_B^A \frac{A_M}{EI} = \frac{1}{EI}\left[-\frac{1}{2}\times mL\times L\right] = \frac{mL^2}{2\,EI} \quad \textbf{Ans}$$

$$\text{Also,} \quad y_B = d_B^A = -\sum_B^A \frac{A_M}{EI}.\bar{x} = -\frac{1}{EI}\left[-\frac{1}{2}\times mL\times L\times\frac{2}{3}L\right] = \frac{mL^3}{3EI}$$

Problem 7.11: A beam *ACB*, simply supported at the ends, has moment of inertia 4*I* for the length *AC* and *I* for the length *CB*, and is loaded with point load *W* at *C*. Determine (i) Slope at end *A*, (ii) Deflection at mid-span, and (iii) Maximum deflection.

Solution:

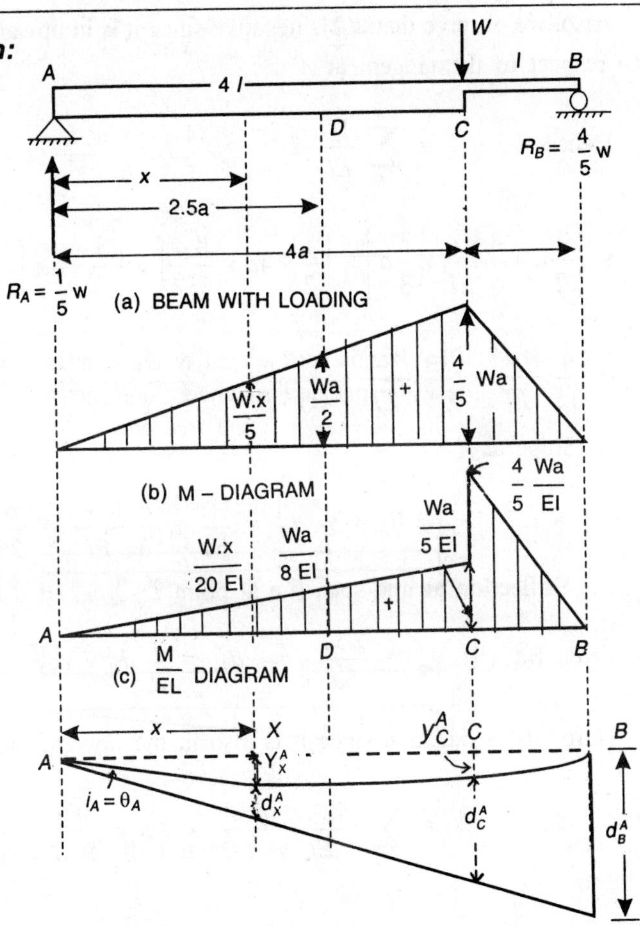

Fig. P. 7.11

Fig. 7.11 (b) shows the M diagram, while Fig. 7.11 (c) shows $\frac{M}{EI}$ diagram

Fig. 7.11 (d) shows the geometry of the deflection curve from which we find that

$$i_A = \theta_A = \frac{1}{L} d_B^A \qquad \text{...(i)}$$

Also, for any point X, distant x from A, we have

$$y_X^A = \frac{x}{L} d_B^A - d_X^A \qquad \text{...(ii)}$$

Also, we observe that d_B^A is negative since it is in upward direction with respect to the tangent at A.

Hence $$d_B^A = \sum_B^A \frac{A_M}{EI} \cdot \bar{x}$$

$$= \left[\frac{1}{2} a \times \frac{4}{5} \frac{Wa}{EI} \times \frac{2}{3} a\right] + \left[\frac{1}{2} \times 4a \times \frac{Wa}{5EI}\left(a + \frac{1}{3} \times 4a\right)\right]$$

$$= \frac{4}{15} \frac{Wa^3}{EI} + \frac{14}{15} \frac{Wa^3}{EI} = \frac{18\, w_o^3}{15\, EI} = \frac{6}{5} \frac{Wa^3}{EI} \qquad \text{...(iii)}$$

(i) Slope at A :

$$i_A = \theta_A = \frac{1}{L} d_B^A = \frac{1}{5a} \times \frac{6}{5} \frac{Wa^3}{EI} = \frac{6}{25} \frac{Wa^2}{EI} \quad \textbf{Ans}$$

(ii) Deflection at mid-span (i.e at point D)

From Eq. (ii) $$y_D^A = \frac{2.5a}{5a} d_B^A - d_D^A = \frac{1}{2} d_B^A - d_D^A$$

Here, d_D^A is negative since it is also in the upward direction.

$$d_D^A = \sum_D^A \frac{A_M \cdot \bar{x}}{EI} = \frac{1}{2} \times 2.5\text{ a } \frac{Wa}{8\, EI} \left(\frac{1}{3} \times 2.5a\right)$$

$$= \frac{25}{192}\frac{Wa^3}{EI}$$

Substituting the values of d_B^A and d_D^A in (iv), we get

$$y_D^A = \frac{1}{2}\left(\frac{6}{5}\frac{Wa^3}{EI}\right) - \frac{25}{192}\frac{Wa^3}{EI}$$

$$= \frac{902}{1920}\frac{Wa^3}{EI} = 0.4698\ \frac{Wa^3}{EI}$$

(iii) Maximum deflection

Maximum deflection occurs at point X, distance x from A. Since its location quite near to the midspan, it is definitely to the left of point C. Hence only the triangle of the $\frac{M}{EI}$ diagram Fig.7.11 (c) will be considered. The definitely y_X^A is Eq. (ii) in which the deviations d_X^A is negative since it is in the upward direction

$$d_X^A = \sum_X^A \frac{A_M.\bar{x}}{EI} = \frac{1}{2}\,x.\frac{Wx}{20EI}\left(\frac{x}{3}\right) = \frac{Wx^3}{120EI}$$

Hence $y_X^A = \frac{x}{L}\,d_B^A - d_X^A$

$$= \frac{x}{5a}\times\frac{6}{5}\frac{Wa^3}{EI} - \frac{Wx^3}{120EI} = \frac{6}{25}\frac{Wxa^2}{EI} - \frac{Wx^3}{120EI}$$

For maxima,

$$\frac{dy_X^A}{dx} = 0 = \frac{6}{25}\frac{Wa^2}{EI} - \frac{3Wx^2}{120EI}$$

From which, we get

$$x = 3.0984\ a$$

Hence $\quad y_{max} = y_{X=3.0984a}^A = \frac{6}{25}\frac{Wa^2}{EI}(3.0984a)$

$$= \frac{W}{120EI}(3.0984\ a)^3$$

or $$y_{max} = 0.4957\ \frac{Wa^3}{EI}$$

Problem 7.12: Calculate the slope at the ends and deflection at the centre of a beam loaded with a couple μ at the centre.

Solution:

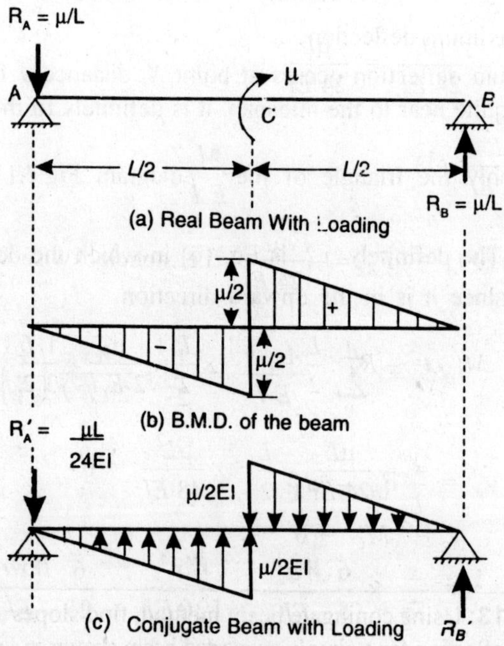

Fig. P. 7.12

Fig 7.12 (a) Show the real beam with the loading, while Fig.7.12 (b) shows its B.M.D.

Evidently $$R_A = \frac{\mu}{L}\ (\downarrow)$$

and $$R_B = \frac{\mu}{L}\ (\uparrow)$$

$$\therefore \quad M_A = 0; M_C \text{ (left)} = -\frac{\mu}{L} \times \frac{L}{2} = -\frac{\mu}{2}$$

$$M_C \text{ (right)} = -\frac{\mu}{L} \times \frac{L}{2} + \mu = +\frac{\mu}{2}$$

$$M_B = 0$$

Fig. 7.12 (c) show the conjugate beam with M/*E I* loading.

For the conjugate beam (Fig. 7.12 c)

$$R_A' = \frac{1}{L}\left[\frac{1}{2} \times \frac{L}{2} \times \frac{\mu}{2\,E\,I}\left(\frac{L}{2} + \frac{1}{3}\frac{L}{2}\right) - \frac{1}{2} \times \frac{L}{2} \times \frac{\mu}{2\,E\,I}\left(\frac{2}{3} \cdot \frac{L}{2}\right)\right]$$

$$= \frac{\mu L}{12\,EI} - \frac{\mu L}{24\,EI} = \frac{\mu L}{24\,EI} \ (\downarrow)$$

$$R_B' = \frac{\mu L^2}{48\,EI} \ (\uparrow)$$

For $\qquad i_A = F_A' = -\frac{\mu L}{24\,EI}$ (i.e. anticlockwise wise)

Also, $\qquad M_C' = -R_A' \cdot \frac{L}{2} + \left(\frac{1}{2} \times \frac{L}{2} \times \frac{\mu}{2\,E\,I} \times \frac{1}{3}\frac{L}{2}\right)$

$$= -\frac{\mu L}{24\,EI} \cdot \frac{L}{2} + \frac{\mu L^2}{48\,EI} = 0$$

$$\therefore \quad y_c = M_C' = 0$$

Problem 7.13: Using conjugate beam method, find slopes at the ends and central deflection for a simply supported beam shown in Fig. 7.13 (a)

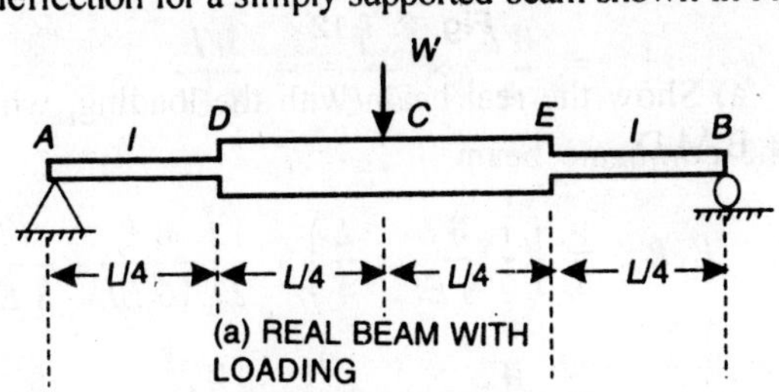

Fig. P. 7.13 (a)

Solution:

For the real beam, $M_C = \dfrac{WL}{4}$

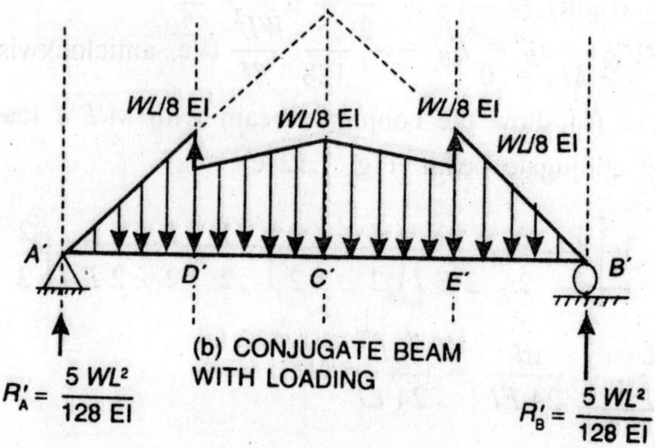

Fig. P. 7.13 (b)

$$M_D = \frac{W}{2} \cdot \frac{L}{4} = \frac{WL}{8} = M_E.$$

Also, $M_A = M_B = 0$

Fig. 7.13 (b) shows the conjugate beam with its loading, $\dfrac{M}{EI}$

Ordinate of loading at D (left)

$$= \frac{WL}{8} \cdot \frac{1}{EI} = \frac{WL}{8\,EI}$$

Ordinate of loading at D (right)

$$= \frac{WL}{8} \cdot \frac{1}{2\,EI} = \frac{WL}{16\,EI}$$

Ordinate of loading at C

$$= \frac{WL}{4} \times \frac{1}{2\,EI} = \frac{WL}{8\,EI}$$

For the conjugate beam,

$$R_A' = \left(\frac{1}{2} \frac{WL}{8\,EI} \times \frac{L}{4}\right) + \frac{1}{2}\left(\frac{WL}{16\,EI} + \frac{WL}{8\,EI}\right)\frac{L}{4}$$

$$= \frac{WL^2}{64\,EI} + R_A = \frac{3\,WL^2}{128\,EI} = \frac{5}{128}\frac{WL^2}{EI}$$

Hence $\quad i_A = F_A' = \dfrac{5}{128}\dfrac{WL^2}{EI}$ (clockwise)

By symmetry, $\quad i_B = F_B' = -\dfrac{5}{128}\dfrac{WL^2}{EI}$ (i.e. anticlockwise)

Also, $\quad y_c = M_C' = R_A'\left(\dfrac{L}{2}\right) - \dfrac{WL^2}{64\,EI}\left(\dfrac{L}{4}+\dfrac{1}{3}\dfrac{L}{4}\right)$

$$-\frac{3}{128}\frac{WL^2}{EI}\left[\frac{1}{3}\frac{L}{4}\frac{\frac{2\,WL}{16\,EI}+\frac{WL}{8\,EI}}{\frac{WL}{16\,EI}+\frac{WL}{8\,EI}}\right]$$

$$= \frac{5}{28}\frac{WL^2}{EI}\cdot\frac{L}{2} - \frac{WL^3}{192\,EI} - \frac{1}{128\times 3}\frac{WL^3}{EI} = \frac{3}{256}\frac{WL^3}{EI} \text{ Ans}$$

Problem 7.14: A simple beam *AB* is acted upon by couples μ and 2 μ at the ends as shown in Fig. 7.14 (a). Determine the slope at *A* and *B*, and (ii) the deflection at the mid-span.

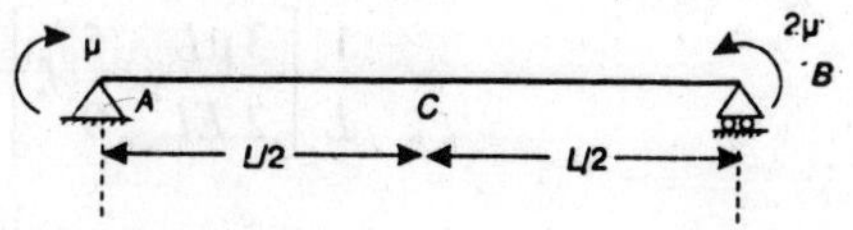

Fig. P. 7.14 (a)

Solution:

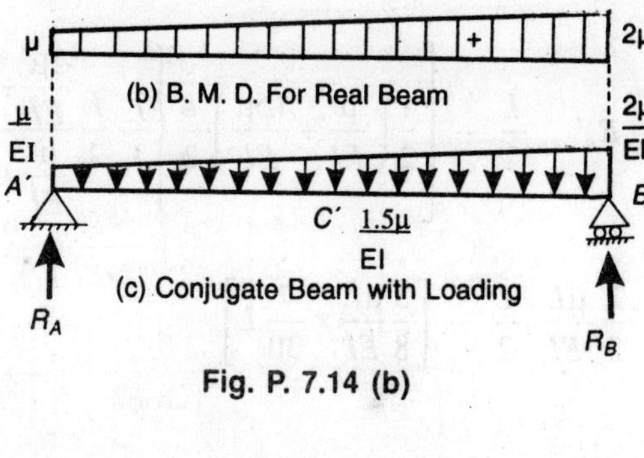

Fig. P. 7.14 (b)

Fig 7.14 (a) shows the real beam while Fig 7.14 (c) shows the conjugate beam with the loading M/EI. Load at A′ = μ/EI while the load at B′

= 2 μ/EI

For the conjugate beam,

$$\text{Reaction}\quad R_A' = \frac{1}{L}\left[\frac{1}{2}\left(\frac{\mu}{EI}+\frac{2\mu}{EI}\right)\times L\left\{\frac{L}{3}\,\frac{\frac{2\mu}{EI}+\frac{2\mu}{EI}}{\frac{\mu}{EI}+\frac{2\mu}{EI}}\right\}\right]$$

$$= \frac{1}{L}\left[\frac{3}{2}\frac{\mu L}{EI}\times\frac{4}{9}L\right] = \frac{2}{3}\frac{\mu L}{EI}$$

$$R_B' = \frac{1}{L}\left[\frac{1}{2}\left(\frac{\mu}{EI}+\frac{2\mu}{EI}\right)L\left\{\frac{L}{3}\,\frac{\frac{2\times 2\mu}{EI}+\frac{\mu}{EI}}{\frac{2\mu}{EI}+\frac{\mu}{EI}}\right\}\right]$$

$$= \frac{1}{L}\left[\frac{3}{2}\frac{\mu L}{EI}\times\frac{5}{9}L\right] = \frac{5}{6}\frac{\mu L}{EI}$$

$$\text{Now } i_A = F_A' = +R_A' = +\frac{2}{3}\frac{\mu L}{EI}$$

$$i_B = F_B' = -R_B' = -\frac{5}{6}\frac{\mu L}{EI}$$

$$y_c = M_C'$$

$$= R_A'\cdot\frac{L}{2} - \left[\frac{1}{2}\left(\frac{\mu}{EI}+\frac{1.5\mu}{EI}\right)\frac{L}{2}\left\{\frac{1}{3}\cdot\frac{L}{2}\,\frac{\frac{2\mu}{EI}+\frac{1.5\mu}{EI}}{\frac{\mu}{EI}+\frac{1.5\mu}{EI}}\right\}\right]$$

$$= \frac{2}{3}\frac{\mu L}{EI}\cdot\frac{L}{2} - \left[\frac{5}{8}\frac{\mu L}{EI}\times\frac{7}{30}L\right]$$

$$= \frac{\mu L^2}{3\,EI} - \frac{7}{48}\frac{\mu L^2}{3\,EI} = \frac{3}{16}\frac{\mu L^2}{EI}$$

Problem 7.15: A beam *ABC* is fixed at end A, and has a roller support at end C. It is also provided with an internal hinge at B. Determine the slope and deflection at the hinge B, when loaded with a point load W as shown in Fig. P. 7.15 (a).

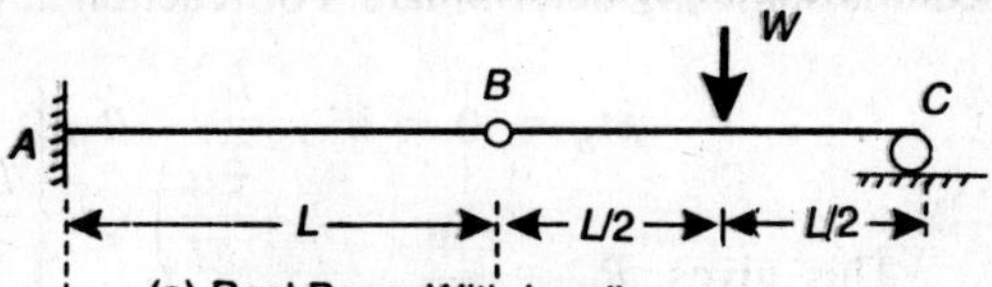

(a) Real Beam With Loading

Fig. P. 7.15 (a)

Solution:

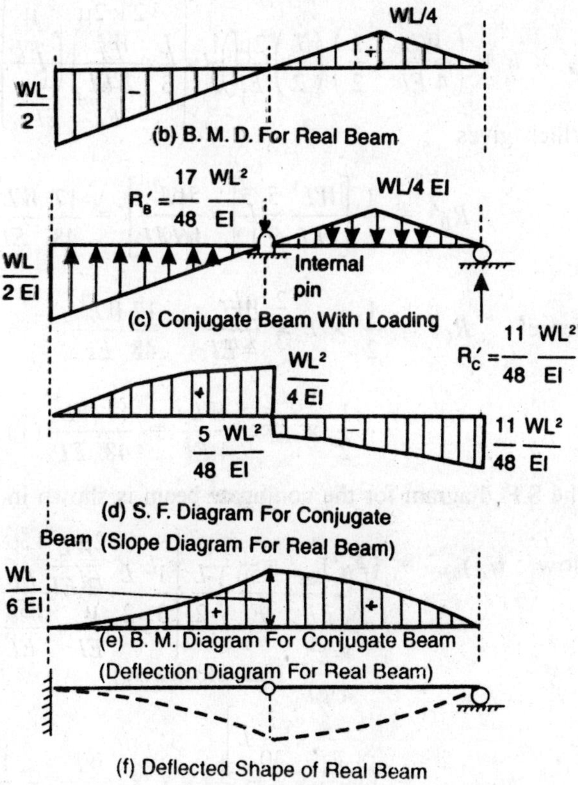

Fig. P. 7.15 (b)

Fig 7.15 (a) shows the real beam ABC. There are three reaction components: R_A, M_A and R_C. Against this, there are three condition equation available from statical equilibrium

$\Sigma V = 0$, $M_C = 0$ and $M_B = 0$ (because of internal hinge at B) Hence the beam is statically determinate. For reaction at C, take moments about B.

$$\therefore \quad M_B = 0 = W \cdot \frac{L}{2} - R_A L$$

This gives $R_A = \dfrac{W}{2}$

The B.M.D. for the real beam is shown in Fig 7.15 (b). The conjugate beam $A'\ B'\ C'$, will have free end at A', roller support at C' and will have as interior pin support at B'. The conjugate beam with the elastic loading equal to M/E I is shown in Fig 7.15 (c). For reaction R_B' at B' of the conjugate beam, take moment about C'.

$$R_B' \times L + \left(\frac{WL}{4\,EI} \times \frac{L}{2}\right)\left(\frac{L}{2}\right) - \left(\frac{1}{2} \times L \times \frac{WL}{2\,EI}\right)\left(L + \frac{2}{3}L\right) = 0$$

Which gives

$$R_B' = \frac{1}{L}\left[\frac{WL^3}{EI} \times \frac{5}{12} - \frac{WL^2}{16\,EI}\right] = \frac{17}{48}\frac{WL^2}{EI}\ (\downarrow)$$

Hence $\quad R_C' = \dfrac{1}{2} \times L \times \dfrac{WL}{4\,EI} + \dfrac{17}{48}\dfrac{WL^2}{EI}$

$$- \frac{1}{2} \times L \times \frac{WL}{4\,EI} = \frac{11}{48}\frac{WL^2}{EI}\ (\uparrow)$$

The S.F. diagram for the conjugate beam is shown in Fig. 7.15(d).

Now $\quad (i_B)_{left} = (F_B')_{left} = + \dfrac{1}{2} \times L \times \dfrac{WL}{2\,EI}$

$$= + \frac{WL^2}{4\,EI} \quad \text{(i.e. clockwise)}$$

$$(i_B)_{right} = (F_B')_{right} = \frac{1}{2} \times L \times \frac{WL}{2\,EI} - \frac{17}{48}\frac{WL^2}{EI}$$

$$= -\frac{5}{48}\frac{WL^2}{EI} \text{ (i.e. anticlockwise)}$$

Also, $$y_B = M_B' = +\frac{1}{2} \times L \times \frac{WL}{2\,EI}\left(\frac{2}{3}L\right) = +\frac{Wl^3}{6EI} \text{ (i.e. } \downarrow)$$

Also, $y_A = M_A' = 0;\ y_C = M_C' = 0$

The B.M.D. for the conjugate beam is shown in Fig. 7.15 (e). Fig. 7.15 (f) shows the deflected shape of the real beam.

Problem 7.16: A cantilever of uniform section has a length $AB = L$. End B is free end and carries a point load W, while end A is fixed end. Find the slope and deflection at a point C, distant $L/4$ from the free end A.

Solution:

Take section X-X from distance x from end A.

$$EI\frac{d^2y}{dx^2} = -M = -[-W(L-x)] = W(L-x)$$

Integrating it we get

$$EI\frac{dy}{dx} = W\left(Lx - \frac{x^2}{2}\right) + C_1$$

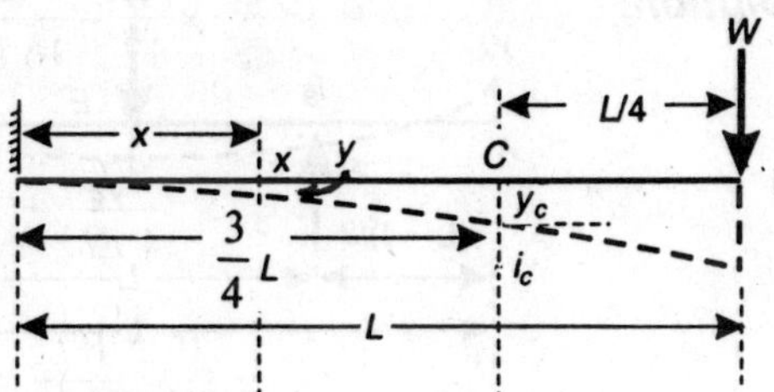

Fig. P. 7.16

At A, $x = 0$ and $\frac{dy}{dx} = 0$.

Hence $C_1 = 0$, where C_1 is constant of integration.

$$EI\frac{dy}{dx} = W\left(Lx - \frac{x^2}{2}\right) \qquad ...(1)$$

Integrating again,

$$EIy = W\left(\frac{Lx^2}{2} - \frac{x^3}{6}\right) + C_2$$

At A, $x = 0$ and $y = 0$. Hence $C_2 = 0$

Hence $EIy = W\left(\frac{Lx^2}{2} - \frac{x^3}{6}\right)$ ...(2)

which is the equation for deflection at any point of the beam

At C, $x = \frac{3}{4}L$

Hence, $i_c = \frac{1}{EI} \cdot W\left[L \times \frac{3}{4}L - \frac{1}{2}\left(\frac{3}{4}L\right)^2\right] = \frac{15}{32}\frac{WL^2}{EI}$

and $y_c = \frac{1}{EI} \cdot W\left[\frac{L}{2}\left(\frac{3}{4}L\right)^2 - \frac{1}{6}\left(\frac{3}{4}\right)^3\right] = \frac{27}{128}\frac{WL^3}{EI}$

Problem 7.17: A uniform beam of length L is simply and symmetrically supported over of span L. Find the ratio L/l so that the upward deflection at each end equals the downward deflection at the mid-span due to a central point load.

Solution:

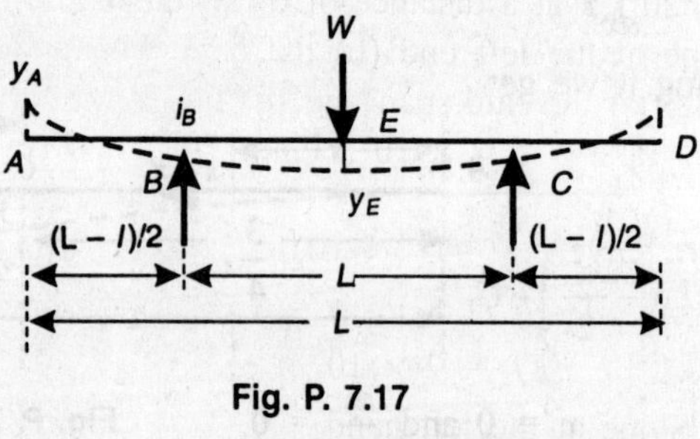

Fig. P. 7.17

The clotted line (---) represents the deflected shape of the beam. Due to a central concentrated load W, the slope at the ends of a simply supported beam are each $\frac{Wl^2}{16\,EI}$ and the deflection at centre of the span is $\frac{Wl^2}{48\,EI}$.

$$i_B = \frac{Wl^2}{16\,EI}$$

Since this angle is very small,

$$y_A = i_B\,(AB) = \frac{W\,l^2}{16\,E\,I}\cdot\left(\frac{L-l}{2}\right)(\uparrow) \qquad \text{...(1)}$$

Also, $$y_E = \frac{wL^3}{48\,EI}(\downarrow) \qquad \text{...(2)}$$

Equating the two, we get

$$\frac{Wl^2}{16\,EI}\left(\frac{L-1}{2}\right) = \frac{Wl^2}{48\,EI}$$

$$L - l = \frac{2}{3}l$$

or $$\frac{L}{l} = \frac{2}{3}+1 = \frac{5}{3}$$

Problem 7.18: A simply supported beam of span 8 m carries a point load of 20 kN at a distance of 6 m from the left end. Compute (a) the slope at the left end (b) the deflection under the load (c) the deflection at the mid-span and (d) the maximum deflection and its location Take $E = 2 \times 10^5$ N/mm^2 and $I = 6 \times 10^8$ mm^4

Solution: Given :

$$L = 8 \text{ m};\ a = 6 \text{ m};\ b = 2 \text{ m};\ W = 20 \text{ kN}$$

$$EI = 2 \times 10^5 \times 6 \times 10^8 = 12 \times 10^{13} \text{ N.mm}^2 \text{ units}$$

(a) Slope at the left hand

$$i_A = \frac{Wab}{6\,EIL}(L+b)$$

$$= \frac{20000(6\times 2)}{6\times 12\times 10^3\times 8}\quad (8+2)\times(1000)^2$$

$$= 4.167 \times 10^{-4} \text{ radian} = 0.0004167 \text{ radian}$$

(b) Deflection under the load

$$y_c = \frac{Wa^2b^2}{3\,EIL}$$

$$= \frac{20000}{3 \times 12 \times 10^3 \times 8} (6)^2 \ (2)^2 \ (1000)^3$$

$$= 1 \text{ mm}$$

(c) Deflection at mid-span

$$[y]_{x=L/2}^{a>b} = \frac{Wb}{48\ EI}(3L^2 - 4b^2)$$

$$= \frac{2000 \times 2}{48 \times 12 \times 10^{13}} [3(8)^2 - 4(2)^2]\ (1000)^3$$

$$= 1.222 \text{ mm}$$

(d) Maximum deflection

Maximum deflection occurs at $x = \sqrt{\dfrac{L^2 - b^2}{3}}$

$$= \sqrt{\frac{(8)^2 - (2)^2}{3}} = 4.472 \text{ m}$$

$$y_{max} = \frac{Wb\,(L^2 - b^2)^{3/2}}{9\sqrt{3}\ EIL}$$

$$= \frac{2000 \times 2}{9\sqrt{3}\left(12 \times 10^{13}\right)8}$$

$$[(8)^2 - (2)^2]^{3/2} \times (1000)^3$$

$$= 1.242 \text{ mm}$$

Problem 7.18: A uniform circular bar of length L extends by an amount under a tensile pull P. Show that if the bar is used as beam simply supported at its ends and carrying a central load W, the maximum deflection is given by $y = \dfrac{W\delta L^2}{3\ Pd^2}$ where d is the diameter of the beam. If $L = 80\ d$, and maximum bending stress due to W

is equal to 0.9 times the tensile stress due to pull P, find the ratio y/δ.

Solution:

Due to tensile load P, the extension δ of the circular bar from hook'slaw is given by

$$\delta = \frac{PL}{AE} = \frac{PL}{\frac{\pi}{4}d^2 E} = \frac{4\,PL}{\pi d^2 E}$$

From which $E = \dfrac{4\,PL}{\pi d^2 \delta}$ ...(1)

Due to central point load, the cental deflection is given by

$$y = y_{\max} = \frac{WL^3}{48\,EI} = \frac{WL^3}{48\,E\frac{\pi}{64}d^4} = \frac{4WL^3}{3\,E\,\pi d^4} \quad ...(2)$$

Substituting the value of E in (2), we get

$$y = \frac{4\,WL^3}{3\pi d^4} \times \frac{\pi d^2 \delta}{4\,PL} = \frac{W\delta L^2}{3\,Pd^2} \quad ...(3)$$

Now, maximum bending stress $= f_{\max} = \dfrac{M}{Z} = \dfrac{WL}{4} \times \dfrac{32}{\pi d^3}$

or $\quad f_{\max} = \dfrac{8W}{\pi} \times \dfrac{80d}{d^3} = \dfrac{640\,W}{\pi d^3}$ (since L =80 d) ...(4)

Max. tensile stress

$$p = \frac{P}{\frac{\pi}{4}d^2} = \frac{4\,P}{\pi d^2} \quad ...(5)$$

Since $\quad f = 0.9\ p$, we have, from (4) and (5)

$$\frac{640\,W}{\pi\,d^2} = 0.9 \times \frac{4\,P}{\pi\,d^2}$$

or, $\quad \dfrac{P}{W} = \dfrac{640}{3.6}$ ...(6)

But from (3),

$$\frac{P}{W} = \frac{\delta L^2}{3yd^2} = \frac{\delta(80d)^2}{3yd^2} = \frac{6400\,\delta}{3\,y} \qquad \ldots(7)$$

Equating (6) and (7), we get $\dfrac{6400\,\delta}{3\,y} = \dfrac{640}{3.6}$

or, $$\frac{y}{\delta} = \frac{6400}{3} \times \frac{3.6}{640} = \frac{36}{3} = 12$$

$\therefore$ $$y = 12\,\delta$$

Problem 7.19: An over hanging beam *ABC* is loaded as shown in Fig. 7.19 (a) Determine the slope and deflection of the beam at point *C*, in terms of *E*, *I*, *W* and *L*.

Solution:

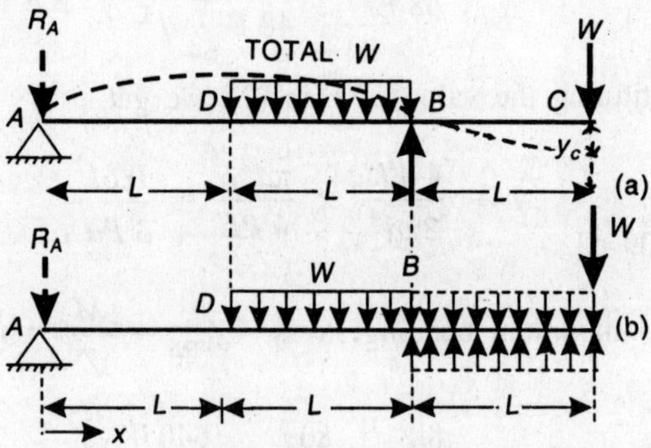

Fig. P. 7.19

For reaction at end *A*, take moment about *B*,

$\therefore$ $$R_A = \frac{1}{2\,I}\left[\frac{WL}{2} - WL\right] = -\frac{W}{4} \quad \left(\text{i.e. } \frac{W}{4}\downarrow\right)$$

Hence $$R_B = W + \frac{W}{4} + W = \frac{9}{4}W\ (\uparrow)$$

Total U.D.L $= W$

$$w = \frac{W}{L}$$

In order to effectively apply Macaulay's method, extend U.D.L from *B* to *C*, and apply upward load from *B* to *C*, as shown in Fig. 7.19 (b) Measuring *x* from *A*, we have

$$EI\frac{d^2y}{dx^2} = \frac{W}{4}x\left\langle +\frac{W}{L}\frac{(x-L)^2}{2}\right\rangle$$

$$-\frac{9W}{4}(x-2L) - \frac{W}{L}\frac{(x-2L)^2}{2}$$

Hence $$EI\frac{dy}{dx} = \frac{Wx^2}{8} + C_1\left\langle +\frac{W}{6L}(x-L)^3\right\rangle$$

$$-\frac{9}{8}W(x-2L)^2 - \frac{W}{6L}(x-2L)^3 \quad \ldots(1)$$

and $$EIy = \frac{Wx^3}{24} + C_1x + C_2\left\langle +\frac{W}{24L}(x-L)^4\right\rangle$$

$$-\frac{9}{24}W(x-2L)^3 - \frac{W}{24L}(x-2L)^4 \quad \ldots(2)$$

and At $x = 0$, $y = 0$. Hence $C_2 = 0$

Also, at $$x = 2L,\ y = 0 = \frac{W}{24}(2L)^3 + C_1(2L) +$$

$$\frac{W}{24L}(2L-L)^4$$

From which $C_1 = -\frac{3}{16}WL^2$

Substituting the value of C_1 in eqs. (1) and (2)

$$EI\frac{dy}{dx} = \frac{Wx^2}{8} - \frac{3}{16}WL^2\left\langle +\frac{W}{6L}(x-L)^3\right\rangle$$

$$-\frac{9}{8}W(x-2L)^2 - \frac{W}{6L}(x-2L)^3 \quad \ldots(I)$$

and $$EIy = \frac{Wx^3}{24} - \frac{3}{16}WL^2x\left\langle +\frac{W}{24L}(x-L)^4\right\rangle - \frac{9}{24}$$

$$\times (x - 2L)^3 - \frac{W}{24L}(x - 2L)^4 \qquad \ldots\text{(II)}$$

At C, $x = 3L$. hence from (I),

$$EIy_c = \frac{W(3L)^2}{8} - \frac{3}{16}WL^2 + \frac{W}{6L}(3L - 2L)^2 - \frac{W}{6L}$$

$$(3L - 2L)^3$$

From which $i_c = \frac{47}{48}\frac{WL^2}{EI}$ **Ans**

Also, at C, $x = 3L$. Hence from (II),

$$EIy_c = \frac{W}{24}(3L)^3 - \frac{3}{16}WL^2(3L) + \frac{W}{24L}$$

$$(3L - L)^4 - \frac{9}{24}W(3L - 2L)^3 - \frac{W}{24L}(3L - 2L)^4$$

From which $y_c = \frac{13}{16}\frac{WL^3}{EI}$ (Ans.)

Problem 7.20: A beam ABC is loaded as shown Fig. 7.20 Find (i) slopes at A and B and C (ii) deflection at C and (iii) maximum deflection between the supports

Solution:

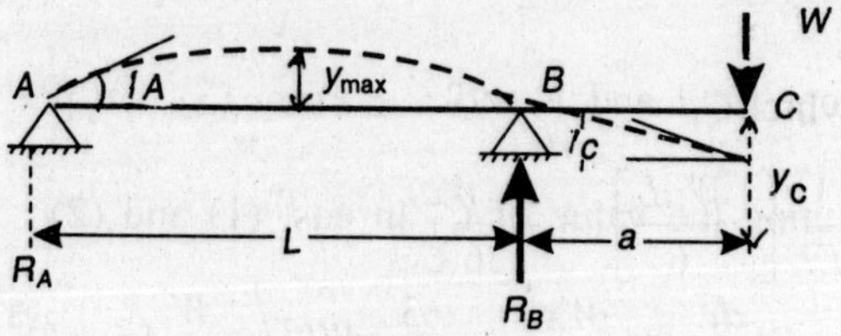

Fig. P. 7.20

$$R_A = \frac{1}{L}(W_a) = \frac{Wa}{L} \quad (\downarrow)$$

$$R_B = \frac{1}{L}W(a + L) = \frac{W(a + L)}{L} \quad (\uparrow)$$

$$EI\frac{d^2y}{dx^2} = -M_x = \frac{Wa}{L}x \left\langle -\frac{W(a+L)}{L}(x-L) \right\rangle$$

$$\therefore \quad EI\frac{dy}{dx} = \frac{Wax^2}{2L} + C_1 \left\langle -\frac{W(a+L)}{L}\frac{(x-L)^2}{2} \right\rangle \quad \ldots(1)$$

and $$EIy = \frac{Wax^3}{6L} + C_1 x + C_2 \left\langle -\frac{W(a+L)}{L}\frac{(x-L)^3}{6} \right\rangle \quad \ldots(2)$$

At $x = 0$, $y = 0$, Hence $C_2 = 0$

At $x = L$, $y = 0$.

Hence from (2), including appropriate terms

$$0 = \frac{Wa(L)^3}{6L} + C_1 L, \text{ from which } C_1 = -\frac{WaL}{6}$$

Hence the slope and deflection equations are as follows :

$$EI\frac{dy}{dx} = \frac{Wax^2}{2L} - \frac{WaL}{6} \left\langle -\frac{W(a+L)}{L}\frac{(x-L)^2}{2} \right\rangle \quad \ldots(3)$$

and $$EIy = \frac{Wax^3}{6L} - \frac{WaL}{6}x \left\langle -\frac{W(a+L)}{L}\frac{(x-L)^3}{6} \right\rangle \quad \ldots(4)$$

and slope at A and B and C : From (3), we get

$$i_A = \frac{1}{EI}\left[-\frac{WaL}{6}\right] = -\frac{WaL}{6EI}$$

$$i_B = \frac{1}{EI}\left[\frac{WaL^2}{2L} - \frac{WaL}{6}\right] = \frac{WaL}{3EI}$$

$$i_C = \frac{1}{EI}\left[\frac{Wa}{2L}(a+L)^2 - \frac{WaL}{6} - \frac{W(a+L)}{L}\frac{(a+L-L)^2}{2}\right]$$

$$= \frac{Wa}{6\,EI}(2\ L + 3a)$$

Deflection at C : From (4), at $x = a + L$, we get

$$y_C = \frac{1}{EI}\left[\frac{Wa}{6\,L}(a+L)^3 - \frac{WaL}{6}(a+L) - \frac{W(a+L)}{6\,L}(a+L-L)^3\right]$$

or $$y_C = \frac{1}{EI}\frac{Wa}{6\,L}\ [a^3 + L^3 + 3\,a^3\,L + 3\,L^2\,a - L^2.\,a - L^3 - a^3 - a^2\,L]$$

$$= \frac{Wa^2}{3\,EI}(a\ +L)$$

Maximum deflection betwen A and B: putting $\frac{dy}{dx} = 0$ in (3), we get

$$0 = \frac{Wax^2}{2\,L} - \frac{WaL}{6},$$ from which we get

$$x = \frac{L}{\sqrt{3}}$$

Hence $$EIy_{max} = \frac{Wa}{6\,L}\left(\frac{L}{\sqrt{3}}\right)^3 - \frac{WaL}{6}\left(\frac{1}{\sqrt{3}}\right) = -\frac{WaL^2}{9\sqrt{3}}$$

or $$y_{max} = -\frac{WaL^2}{9\sqrt{3}\ EI}$$

Problem 7.21: A beam *ABCD* has equal over hangs to both the sides and is loaded as shown in Fig.7.21 Find (a) slopes at the free ends and at the supports (b) deflections at the free ends and at the midspan.

Solution:

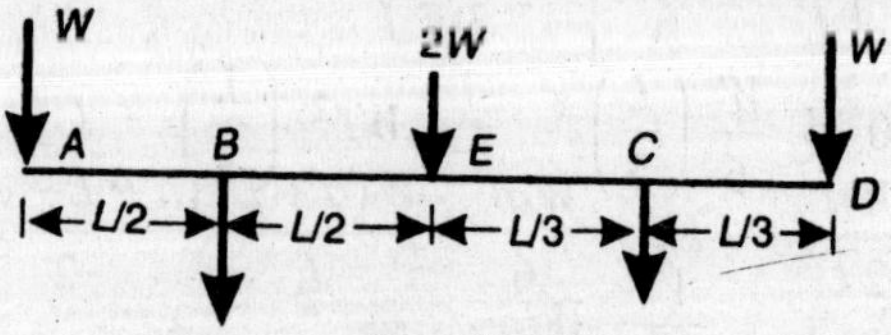

Fig. P. 7.21

$$R_B = R_C = \frac{4W}{2} = 2\,W$$

Taking x from end A

$$EI\frac{d^2y}{dx^2} = -M_x$$

$$= W.x \left\langle -2W\left(x-\frac{L}{3}\right), \right\rangle \text{ for portion upto point } E.$$

$$\therefore \quad EI\frac{dy}{dx} = \frac{Wx^2}{2} + C_1 \left\langle -\frac{2W\left(x-\frac{L}{3}\right)^2}{2} \right\rangle \quad \ldots(1)$$

$$\text{and} \quad EIy = \frac{Wx^3}{6} + C_1\,x + C_2 \left\langle -\frac{2W\left(x-\frac{L}{3}\right)^2}{6} \right\rangle \quad \ldots(2)$$

Due to symmetry, at

$$x = \left(\frac{L}{3}+\frac{L}{2}\right) = \frac{5}{6}L, \; \frac{dy}{dx} = 0$$

$$\therefore \quad 0 = \frac{W}{2}\left(\frac{5}{6}L\right)^2 + C_1 - W\left(\frac{5}{6}L-\frac{L}{3}\right)^2,$$

Which gives $C_1 = -\dfrac{7}{72}\,WL^2$

Also, at $x = \dfrac{L}{3},\; y = 0$

Hence from eq. (2)

$$0 = \frac{W}{6}\left(\frac{L}{3}\right)^3 - \frac{7}{72}\,WL^2\left(\frac{L}{3}\right) + C_2,$$

From which $C_2 = +\dfrac{17}{648}\,WL^3$

Hence the slope and deflection equation are

$$EI\frac{dy}{dx} = \frac{Wx^2}{2} - \frac{7}{72}WL^2 \left\langle -W\left(x-\frac{L}{3}\right)^2 \right\rangle \quad ...(3)$$

and $$EIy = \frac{Wx^3}{6} - \frac{7}{72}WL^2 x + \frac{17}{648}WL^3 \left\langle -\frac{W}{3}\left(x-\frac{L}{3}\right)^3 \right\rangle \quad ...(4)$$

and slope at free end : Substituting $x = 0$, in (3), we get

$$i_A = \frac{1}{EI}\left(-\frac{7}{72}WL^2\right) = -\frac{7WL^2}{72EI}$$

or $$P\left[1+\frac{6EI\lambda}{L^3}\right] = \frac{11WL^3}{96EI} \times \frac{6EI}{L^3} = \frac{11}{16}W$$

From which $P = \dfrac{11W}{16\left(1+\dfrac{6EI\lambda}{L^3}\right)}$. Hence proved.

Problem 7.22: Shows two cantilevers, the end of one being verticaly above the other, and its connected to it by a spring *AB*. Initially, the system is unstrained. A weight *W* placed at *A* causes a vertical deflection at *A* of δ_1 and vertical deflection at *B* of δ_2. when the spring is removed, the weight *W* at *A* causes a deflection at *A* of δ_3. Find the extension of the spring, when it is replaced and the weight *W* is transferred to *B*.

Solution:

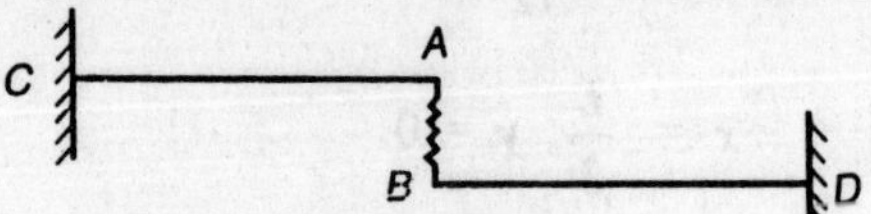

Fig. P. 7.22

Let L_1 = Length of cantilever *CA*; L_2 = Length of cantilever *DB*,

δ_4 = deflection of *A* with load at *B*; δ_5 = deflection of *B* with load at *B*.

Given δ_1 = deflection of *A* with load at *A*, with spring in position.

δ_2 = defletion of B with load at A, with spring in position

δ_3 = deflection of A with load at A, without spring

P_{SA} force in spring when load at A

P_{SB} force in spring when load at B

$$\delta_1 = \frac{(W - P_{SA})\, L_1^3}{3\, EI} \qquad \text{...(i)}$$

and $$\delta_2 = \frac{P_{SA} L_2^3}{3\, EI}$$

or $$\frac{\delta_2}{P_{SA}} = \frac{L_2^3}{3\, EI} \qquad \text{...(ii)}$$

Compression of spring $AB = \delta_1 - \delta_2$

$$\therefore \text{ Stiffness of spring} = s = \frac{\text{force}}{\text{compression}} = \frac{P_{SA}}{\delta_1 - \delta_2} \qquad \text{...(iii)}$$

Again, when the spring is removed, and load W is placed at A,

$$\text{Deflection of } A = \delta_3 = \frac{WL_1^3}{3\, EI} \text{ or } \frac{\delta_3}{W} = \frac{L_1^3}{3\, EI} \qquad \text{...(iv)}$$

Dividing (i) and (iv), we get $\dfrac{\delta_1}{\delta_3} = \dfrac{W - P_{SA}}{W} = 1 - \dfrac{P_{SA}}{W}$

or $$\frac{P_{SA}}{W} = 1 - \frac{\delta_1}{\delta_3} = \frac{\delta_3 - \delta_1}{\delta_3}$$

or $$P_{SA} = \frac{W(\delta_3 - \delta_1)}{\delta_3} \qquad \text{...(v)}$$

Substituting this value of P_{SA} in Eq. (iii), we get

$$s = \frac{W(\delta_3 - \delta_1)}{\delta_3(\delta_1 - \delta_2)} \qquad \text{...(vi)}$$

when the cantilever BD is loaded with W at B, the deflection of B is given by

$$\delta_5 = \frac{(W - P_{SB})L_2^3}{3EI}$$

and deflection of A is given by $\delta_4 = \dfrac{P_{SB}L_1^3}{3EI} = \dfrac{P_{SB}\delta_3}{W}$ (from eq. iv)

Substituting the values of $\dfrac{L_2^3}{3EI}$ (from eq. ii), and P_{SA} (from v), we get

$$\delta_5 = \frac{(W - P_{SB})\delta_2}{P_{SA}} = \frac{(W - P_{SB})\delta_2\delta_3}{W(\delta_3 - \delta_1)} \quad \text{...(vii)}$$

$$\text{Now stiffness} = \frac{\text{force}}{\text{extension}} = \frac{P_{SB}}{\delta_5 - \delta_4}$$

$$\delta_5 - \delta_4 = \frac{P_{SB}}{S} = \frac{P_{SB}\,\delta_3(\delta_1 - \delta_2)}{W(\delta_3 - \delta_1)} \text{ (from vi)} \quad \text{...(I)}$$

$$\therefore \quad \delta_5 - \delta_4 = \frac{(W - P_{SB})\delta_2\delta_3}{W(\delta_3 - \delta_1)} - \frac{P_{SB}\delta_3}{W} =$$

$$\frac{\delta_3}{W}\left[\frac{(W - P_{SB})\delta_2}{\delta_3 - \delta_1} - P_{SB}\right] \quad \text{...(II)}$$

Equating (I) and (II), we get

$$\frac{P_{SB}.\delta_3(\delta_1 - \delta_2)}{W(\delta_3 - \delta_1)} = \frac{\delta_3}{W}\left[\frac{(W - P_{SB})\delta_2}{\delta_3 - \delta_1} - P_{SB}\right]$$

or
$$\frac{P_{SB}(\delta_1 - \delta_2)}{\delta_3 - \delta_1} = \frac{(W - P_{SB})\delta_2 - P_{SB}(\delta_3 - \delta_1)}{\delta_3 - \delta_1}$$

or
$$P_{SB}\,\delta_1 - P_{SB}\,\delta_2 = W\,\delta_2 - P_{SB}.\delta_2 - P_{SB}\,\delta_3 + P_{SB}\,\delta_1$$

or, $w\delta_2 = P_{SB}.\delta_3$ from which $P_{SB} = \dfrac{W\delta_2}{\delta_3}$. Substituting the value of P_{SB} in eq. (I), We get

$$\delta_5 - \delta_4 = \frac{\frac{W\delta_2}{\delta_3}\cdot\delta_3(\delta_1-\delta_2)}{W(\delta_3-\delta_1)} = \frac{\delta_2(\delta_1-\delta_2)}{\delta_3-\delta_1} \quad \textbf{Ans}$$

> **Problem 7.23:** A freely supported beam of span L carries a central load W. The sectional area of the beam is so designed that the moment of inertia of section increases uniformly from 1 at the ends to 1.51 at the middle. Calculate the central deflection.

Solution:

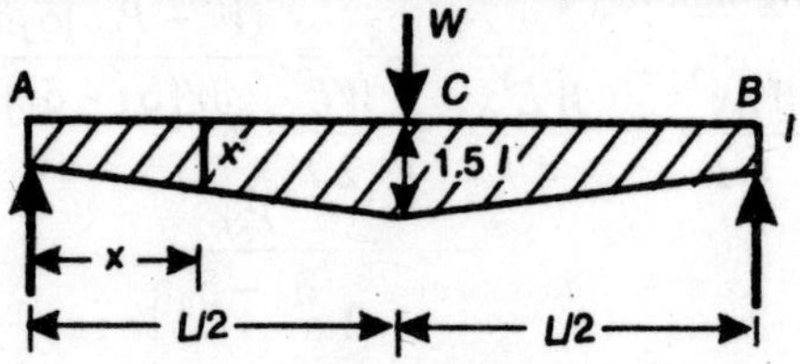

Fig. P. 7.23

At any section x from end A at the beam AB,

$$I_x = I + 0.5\,I\,\frac{2x}{L} = \text{I}\left(1+\frac{x}{L}\right)$$

$$\text{Hence } E\frac{d^2y}{dx^2} = -\frac{M_x}{I_x} = -\frac{\frac{W}{2}x}{I\left(1+\frac{x}{L}\right)} = -\frac{W}{2\,I}\left[\frac{L\left(1+\frac{x}{L}\right)-L}{1+\frac{x}{L}}\right]$$

$$\text{or} \quad E\frac{d^2y}{dx^2} = -\frac{WL}{2\,I} + \frac{WL}{2\,I\left(1+\frac{x}{L}\right)}$$

Integrating,

$$E\frac{dy}{dx} = -\frac{WLx}{2\,I} + \frac{WL^2}{2\,I}\log_e\left(1+\frac{x}{L}\right) + C_1$$

Where C_1 constant of integration determine from boundary condition at the beam support

$$\text{At} \quad x = \frac{L}{2},\ \frac{dy}{dx} = 0 = -\frac{WL}{2\,I}\left(\frac{L}{2}\right)$$

$$+ \quad \frac{WL^2}{2\,I} \log_e \left(1+\frac{x}{L}\right) + C_1$$

From which $C_1 = \dfrac{WL^2}{4\,I} - \dfrac{wL^2}{2\,I}\log_e 1.5$

Hence $E\dfrac{dy}{dx} = -\dfrac{WLx}{2\,I} + \dfrac{WL^2}{4\,I} + \dfrac{WL^2}{2\,I}\log_e \dfrac{2}{3}\left(1+\dfrac{x}{L}\right)$

Integrating again,

$$Ey = -\frac{WLx^2}{4\,I} + \frac{WL^2 x}{4\,I} + \frac{WL^2}{2\,I} \times$$

$$\left[x\log_e \frac{2}{3}\left(1+\frac{x}{L}\right) - \int \frac{\frac{2}{3}\frac{x}{L}}{\frac{2}{3}\left(1+\frac{x}{L}\right)}dx \right] + C_2$$

or $$Ey = -\frac{WLx^2}{4\,I} + \frac{WL^2 x}{4\,I} + \frac{WL^2}{2\,I} \times$$

$$\left[x\log_e \frac{2}{3}\left(1+\frac{x}{L}\right) - \int \left\{1 - \frac{1}{1+\frac{x}{L}}\right\}dx \right] + C_2$$

or $$Ey = -\frac{WLx^2}{4\,I} + \frac{WL^2 x}{4\,I} + \frac{WL^2}{2\,I} \times$$

$$\left[x\log_e \frac{2}{3}\left(1+\frac{x}{L}\right) - x + L\log_e\left(1+\frac{x}{L}\right) \right] + C_2$$

At $x = 0$, $y = 0$ Hence $C_2 = 0$

Hence

$$Ey = -\frac{WLx^2}{4\,I} + \frac{WL^2 x}{4\,I} + \frac{WL^2}{2\,I} \times$$

$$\left[x \log_e \frac{2}{3}\left(1+\frac{x}{L}\right) - x + L \log_e\left(1+\frac{x}{L}\right)\right]$$

Hence the central deflection at $x = L/2$ is given by

$$Ey_c = -\frac{WL}{4I}\left(\frac{L}{2}\right)^2 + \frac{WL^2}{2I}\left(\frac{L}{2}\right) + \frac{WL^2}{2I} \times$$

$$\left[\frac{L}{2}\log_e \frac{2}{3}\left(1+\frac{1}{2}\right) - \frac{L}{2} + L\log_e\left(1+\frac{1}{2}\right)\right]$$

$$= -\frac{WL^3}{16I} + \frac{WL^3}{8I} - \frac{WL^3}{4I}\ \frac{WL^3}{2I}\log\frac{3}{2}$$

$$= -\frac{3}{16}\frac{WL^3}{I} + 0.2075\frac{WL^3}{I}$$

Hence $\quad y_c = 0.02\dfrac{WL^3}{EI}$ Ans

Problem 7.24: A cantilever of length L carries a point load W at its free end. The member is circular in section, having diameter D for a distance $L/2$ from the fixed end and a diameter $D/2$ for the remaining length. Find the deflection at the free end.

Solution:

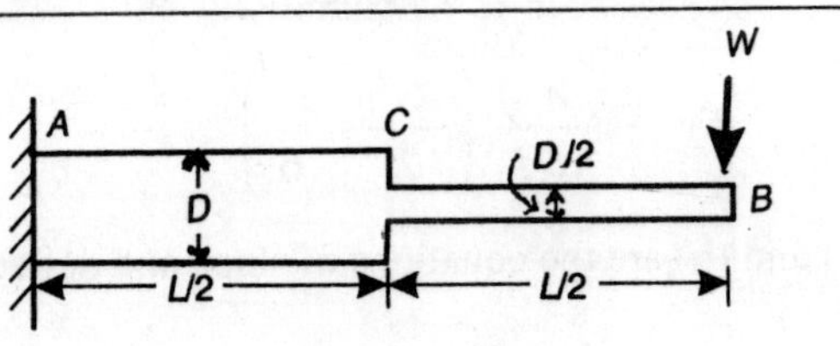

Fig. P. 7.24

For the portion AC

moment of inertia of portion AC

$$I_{AC} = \frac{\pi D^4}{64}$$

For the portion CB,

Moment of intertia of portion CB

$$I_{CB} = \frac{\pi}{64}\left(\frac{D}{2}\right)^4 = \frac{\pi D^4}{64 \times 16}$$

Hence if $I_{CB} = I$, $I_{AC} = 16\ I$

For portion AC : Measuring x from fixed end A, we get

$$E\frac{d^2y}{dx^2} = -\frac{M_x}{I_{AC}} = \frac{W(L-x)}{16\ I}$$

Integrating, $$E\frac{dy}{dx} = \frac{W}{16\ I}\left(Lx - \frac{x^2}{2}\right) + C_1$$

At $$x = 0,$$

$$\frac{dy}{dx} = 0. \text{ Hence } C_1 = 0$$

$$\therefore \quad E\frac{dy}{dx} = \frac{W}{16\ I}\left(Lx - \frac{x^2}{2}\right) \quad \text{...(i)}$$

Integrating further,

$$Ey = \frac{W}{16\ I}\left(\frac{Lx^2}{2} - \frac{x^3}{6}\right) + C_2$$

At $x = 0$, $y = 0$ Hence $C_2 = 0$

$$\therefore \quad Ey = \frac{W}{16\ I}\left(\frac{Lx^2}{2} - \frac{x^3}{6}\right) \quad \text{...(ii)}$$

Eqs. (i) and (ii) are the equations of slope and deflection for portion AC. At $x = \dfrac{L}{2}$, the slope and deflection at C are as under

$$i_C = \frac{W}{16\ EI}\left[L \cdot \frac{L}{2} - \frac{1}{2}\left(\frac{L}{2}\right)^2\right] = \frac{3\ WL^2}{128\ EI} \quad \text{...(a)}$$

and $$y_C = \frac{W}{16\ EI}\left[\frac{L}{2}\left(\frac{L}{2}\right)^2 - \frac{1}{6}\left(\frac{L}{2}\right)^3\right] = \frac{5\ WL^2}{16 \times 48\ EI} \quad \text{...(b)}$$

For portion CB : Measuring x from C, we have

$$E\frac{d^2y}{dx^2} = -\frac{M_x}{I_{CB}} = \frac{W\left(\frac{L}{2}-x\right)}{I}$$

Integrating

$$E\frac{dy}{dx} = \frac{W}{I}\left(\frac{L}{2}x-\frac{x^2}{2}\right)+C_3 \qquad \text{...(iii)}$$

At $\quad x = 0, \dfrac{dy}{dx} = i_c = \dfrac{3\,WL^2}{128\,EI}$

Hence from (iii), we get $C_3 = \dfrac{3\,WL^2}{128\,I}$

$$E\frac{dy}{dx} = \frac{W}{I}\left(\frac{L}{2}x-\frac{x^2}{2}\right)+\frac{3\,WL^2}{128\,I} \qquad \text{...(iv)}$$

Integrating further,

$$Ey = \frac{W}{I}\left(\frac{Lx^4}{4}-\frac{x^3}{6}\right)+\frac{3\,WL^2}{128\,I}x+C_4 \qquad \text{...(v)}$$

At $\quad x = 0, y = y_c = \dfrac{5\,WL^2}{16\times 48\,E\,I}$

Hence from (v), we get $C_4 = \dfrac{5\,WL^2}{16\times 48\,I}$

$$\therefore \quad Ey = \frac{W}{I}\left(\frac{Lx^2}{4}-\frac{x^3}{6}\right)+\frac{3\,WL^2}{128\,I}x+\frac{5\,WL^3}{16\times 48\,I} \qquad \text{...(vi)}$$

Eqs. (iv) and (vi) are the equations for slope and deflection for the portion *CB*

At the free end,

$x = \dfrac{L}{2}\cdot$ Hence from (vi), we get

$$Ey_B = \frac{W}{I}\left[\frac{L}{4}\left(\frac{L}{2}\right)^2 - \frac{1}{6}\left(\frac{L}{2}\right)^3\right] + \frac{3WL^2}{128I}\left(\frac{L}{2}\right) + \frac{5WL^3}{16 \times 48I}$$

From which we get

$$y_B = \frac{23}{384}\frac{WL^3}{EI}$$

Problem 7.25: An experimental beam of steel of uniform section $I = 5000\ \text{cm}^4$ is simply supported in a horizontal position by rigid supports 4 m apart. An elastic prop is placed under the beam at the mid-span with initial clearance between the top of the prop and the under side of the beam. The beam is centrally loaded with a series of increasing loads. The observed central deflection is 8 mm for a concentrated load of 100 kN, and when the results are plotted, the slope of the load deflection line is 27.3 kN per mm after the contact with the prop has occurred. Determine the stiffness of the prop and the initial clearance. Take E = 2.04×10^5 N/mm^2

Solution:

Let σ_b = stiffness of the beam

σ_p = stiffness of the prop

Hence

$\sigma_b + \sigma_p = 27.3$ (kN/mm)

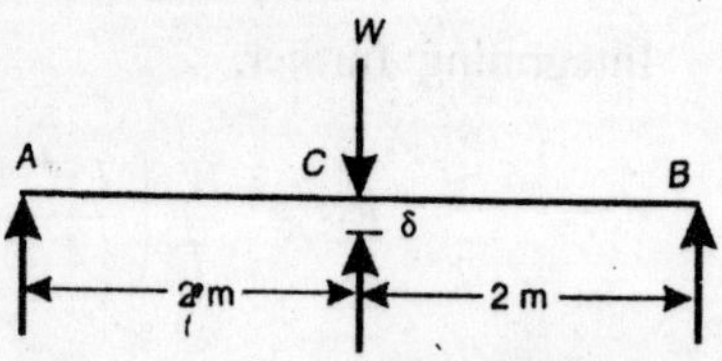

Fig. P. 7.25

But $$\sigma_B = \frac{\text{load}}{\text{deflection}} = \frac{W}{y} = \frac{W}{\dfrac{WL^3}{48EI}} = \frac{48EI}{L^3}$$

$$= \frac{48 \times 2.04 \times 10^5 \times 50 \times 10^6}{(4)^3(1000)^3}$$

$$= 7650 \text{ N/mm} = 7.65 \text{ kN/mm}$$

$\therefore \quad \sigma_p = 27.3 - 7.65 = 19.65$ kN/mm

Let δ = initial clearance, in mm

$\therefore$ Deflection of prop = $8 - \delta$

Force on prop = (stiffness of prop) × (deflection of prop)

$$P = \sigma_P\ (8 - \delta) = 19.65\ (8 - \delta)\ \text{kN}$$

Also, deflection of beam

$$= \frac{\text{Net load on the beam}}{\text{Stiffness of the beam}} = \frac{100 - p}{\sigma_h}$$

$$\therefore \quad 8 = \frac{100 - p}{\sigma_h} = \frac{100 - P}{7.65} = \frac{100 - 19.65(8 - \delta)}{7.65}$$

From above relation, we get

$$19.65\ \delta = 61.2 - 100 + 157.2 = 117.4$$

$$\therefore \text{Deflection}(\delta) = \frac{117.4}{19.65} = 5.97\ \text{mm}$$

Chapter 8

Fixed Beams or Restrained Beams

Problem 8.1: For the propped beam shown in figure P 8.1 (a), find R and draw the shear and moment diagrams.

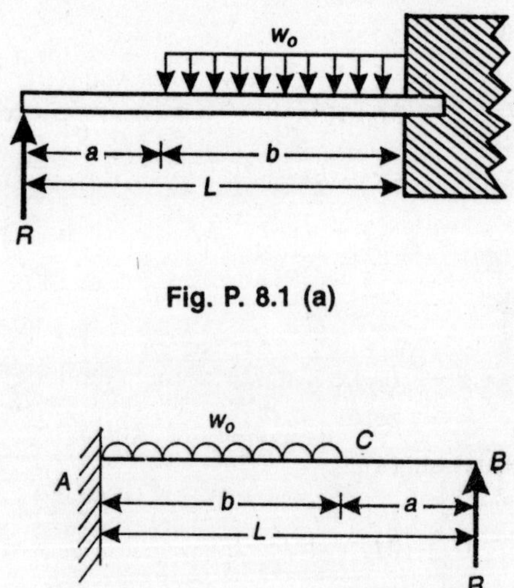

Fig. P. 8.1 (a)

Solution:

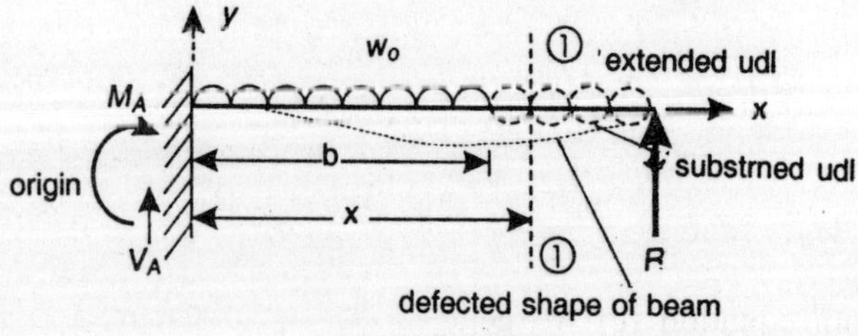

Fig. P. 8.1 (b)

General expression for bending moment,

$$EI\frac{d^2y}{dx^2} = M_A + V_A \times x - \frac{w_0}{2}$$

$$<x - 0>^2 + \frac{w_o}{2} <x - b>^2 \quad \text{...(a)}$$

$$\therefore \quad EI\frac{dy}{dx} = M_A \times x + \frac{V_A \times x^2}{2} - \frac{w_0}{6} <x - 0>^3 + \frac{w_0}{6}$$

$$<x - b>^3 + C_1$$

$$EIy = \frac{M_A\, x^2}{2} + \frac{V_A \times x^3}{6} - \frac{w_0}{24}$$

$$<x - 0>^4 + \frac{w_0}{24} <x - b>^4 + C_1 x + C_2$$

C_1 abd C_2 are both zero because origin is taken at fixed end, so putting C_1 and $C_2 = 0$.

$$EI\frac{dy}{dx} = M_A \times x + \frac{V_A}{2} x^2 - \frac{w_0}{6} <x - 0>^3 + \frac{w_0}{6}$$

$$<x - b>^3 \quad \text{...(b)}$$

$$EIy = \frac{M_A}{2} \times x^2 + \frac{V_A}{6} x^3 - \frac{w_0}{24} <x - 0>^4 + \frac{w_0}{24}$$

$$<x - b>^4 \quad \text{...(c)}$$

At $x = L$, i.e., at B, moment and deflection are both zero.

$\therefore$ From equation (a),

$$0 = M_A + V_A \times L - \frac{w_0}{2} <L - 0>^2 + \frac{w_0}{2} <L \;\; b>^2$$

$$= M_A + V_A L - \frac{w_0 L^2}{2} + \frac{w_0}{2} (L - b)^2$$

$$\therefore \; M_A + V_A L - \frac{w_0 L^2}{2} + \frac{w_0}{2} (L - b)^2 = 0 \quad \text{...(d)}$$

From equation (c) we get,

$$0 = \frac{M_A L^2}{2} + \frac{V_A L^3}{2} - \frac{w_0}{24} \times L^4 + \frac{w_0}{24}(L-b)^4$$

$$= M_A L^2 + \frac{V_A L^3}{3} - \frac{w_0}{12} L^4 + \frac{w_0}{12}(L-b)^4$$

$$\therefore\ M_A L^2 + \frac{V_A L^3}{3} - \frac{w_0}{12} L^4 + \frac{w_0}{12}(L-b)^4 = 0 \qquad \text{...(e)}$$

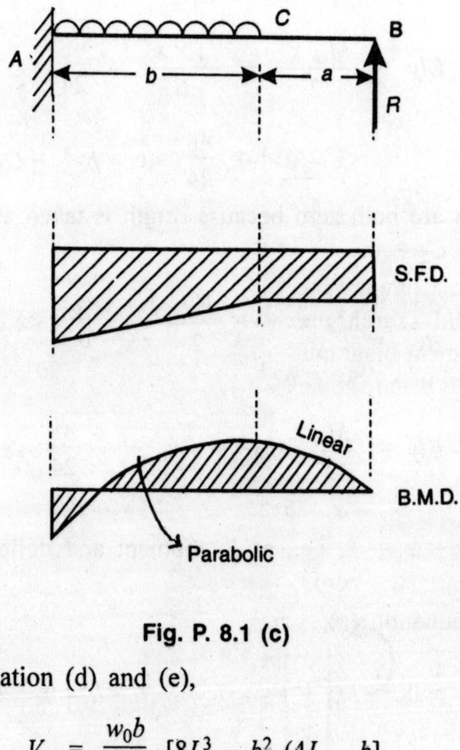

Fig. P. 8.1 (c)

From equation (d) and (e),

$$V_A = \frac{w_0 b}{8L^3}\,[8L^3 - b^2 (4L - b]$$

$\therefore$ Using statics equation

$$\Sigma F_y = 0,$$

$$V_A + R = w_0 b$$

$$\therefore\ \frac{w_0 b}{8L^3}\,[8L^3 - b^2 (4L - b] + R = w_0 b$$

$$\therefore \qquad R = w_0 b - \frac{w_0 b}{8L^3} \; [8L^3 - b^2 (4L - b]$$

$$= w_0 b \left[1 - \frac{1}{8}\left[8L^3 - b^2(4L - b)\right]\right]$$

$$= w_0 b \left[\frac{8 - 8L^3 - b^2(4L - b)}{8}\right]$$

$$= \frac{8w_0 b - 8w_0 bL^3 - w_0 b^3(4\ L - b)}{8}$$

$$= \frac{8w_0 b - 8w_0 bL^3 - 4w_0 b^3 L + q_0 b^4}{8}$$

$$= \left(\frac{w_0 b^3}{8L^3}\right) \times (4L - b)$$

Problem 8.2: Compute the reaction R and sketch the shear and moment diagrams for the propped beam shown in Fig. 8.2 (a).

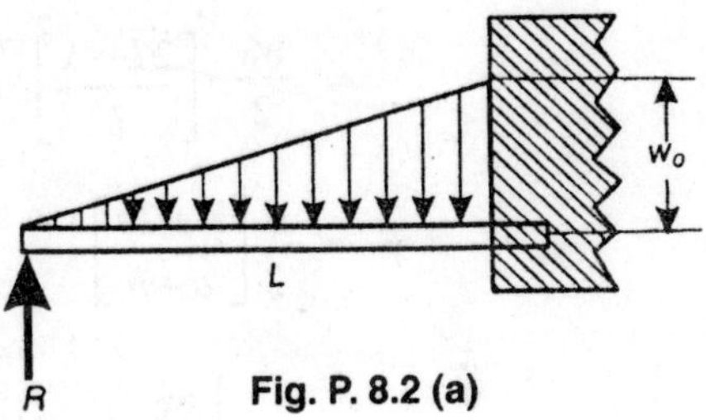

Fig. P. 8.2 (a)

Solution:

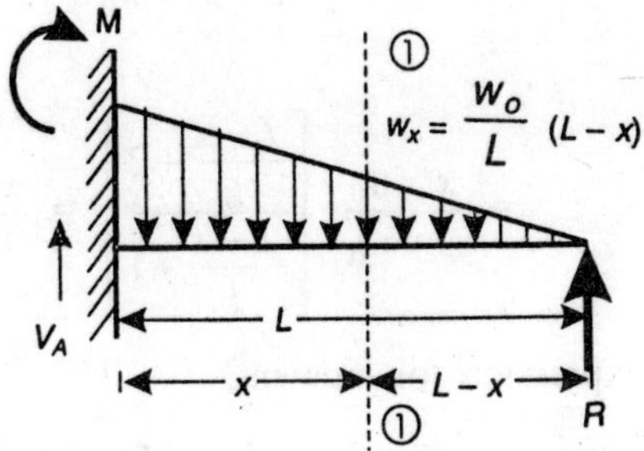

Fig. P. 8.2 (b)

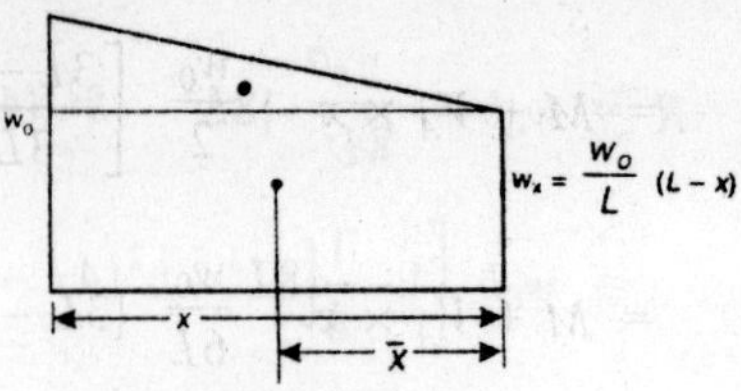

Fig. P. 8.2 (c)

$$\text{Area} = \left(\frac{a+b}{2}\right) \times d$$

$$= \frac{\frac{w_0}{L}(L-x)+w_0}{2} \times x$$

$$= \frac{w_0}{2}\left[\frac{L-x}{2}+1\right] x$$

$$= \frac{w_0}{2}\left[\frac{L-x+L}{L}\right] x$$

$$= \frac{w_0}{2}\left[\frac{2L-x}{L}\right] x$$

$$\bar{x} = \frac{d}{3}\left[\frac{a+2b}{a+b}\right]$$

$$\bar{x} = \frac{x}{3}\frac{\left[\frac{w_0}{L}(L-x)+2w_0\right]}{\frac{w_0}{L}(L-x)+w_0}$$

$$= \frac{x}{3}\frac{w_0}{w_0}\left[\frac{\frac{L-x}{L}+2}{\frac{L-x}{L}+1}\right] = \frac{x}{3}\left[\frac{3L-x}{2L-x}\right]$$

General expression for moment,

$$EI\frac{d^2y}{dx^2} = M+V_A\times x-\frac{w_0}{2}\left[\frac{2L-x}{L}\right]\times x\times\frac{x}{3}\left[\frac{3L-x}{2L-x}\right]$$

$$= M + V_A \times x - \frac{w_0}{2}\left[\frac{3L - x}{3L}\right] \times x^2$$

$$= M + V_A \times x - \frac{w_0}{6L}[3L - x]\, x^2$$

$$EI\frac{d^2y}{dx^2} = M + V_A x - \frac{w_0}{6L}[3L - x]\, x^2$$

$$EI\frac{d^2y}{dx^2} = M + V_A x - \frac{w_0}{6L}[3Lx^2 - x^3] \quad \ldots(1)$$

$$EI\frac{dy}{dx} = Mx + \frac{V_A}{2}x^2 - \frac{w_0}{6L}\left[3L\frac{x^3}{3} - \frac{x^4}{4}\right] + C_1$$

At $x = 0 \quad \frac{dy}{dx} = 0$

$\therefore \quad C_1 = 0$

$$EI\frac{dy}{dx} = Mx + \frac{V_A}{2}x^2 - \frac{w_0}{6L}\left[Lx^3 - \frac{x^4}{4}\right] \quad \ldots(2)$$

$$EIy = M\frac{x^2}{2} + \frac{V_A}{2}\frac{x^3}{3} - \frac{w_0}{6L}\left[\frac{Lx^4}{4} - \frac{x^5}{5}\right] \quad \ldots(3)$$

At $x = L$ moment and deflection are both zero,
From eq. (1)

$$\therefore \quad 0 = M + V_A \times L - \frac{w_0}{6L}[3L \times L^2 - L^3]$$

$$= M + V_A \times L - \frac{w_0}{6L}[3L^3 - L^3]$$

$$= M + V_A L - \frac{w_0}{6L}L^2$$

$$\therefore M + V_A L - \frac{w_0 L^2}{3} = 0 \quad \ldots(4)$$

From equation (3)

$$0 = \frac{M}{2}L^2 + \frac{V_A}{6}L^3 - \frac{w_0}{6L}\left[\frac{L}{4} \times L^4 - \frac{L^5}{5}\right]$$

$$= \frac{M}{2}L^2 + \frac{V_A}{6}L^3 - \frac{w_0}{6L} \times \left[\frac{5L^5 - 4L^5}{20}\right]$$

$$0 = \frac{M}{2}L^2 + \frac{V_A}{6}L^3 - \frac{w_0}{6L} \times \frac{L^5}{20}$$

$$0 = \frac{M}{2}L^2 + \frac{V_A}{6}L^3 - \frac{w_0}{120\,L}L^4$$

$$\frac{L^2}{2}\left[M + \frac{V_A L}{3} - \frac{w_0 L^2}{60}\right] = 0$$

$$\therefore\ M + \frac{V_A L}{3} - \frac{w_0 L^2}{60} = 0$$

$$M + 0.34\ V_A L - 0.0167\ w_0 L^2 = 0 \qquad \ldots(5)$$

From equation (4) and equation (5)

$$M + V_A L - 0.34\ w_0 L^2 = 0$$

$$M + 0.34\ V_A L - 0.0167\ w_0 L^2 = 0$$

$$-\qquad\qquad -\ +$$

$$0.66\ V_A L - 0.3233\ w_0 L^2 = 0$$

$$\therefore\ 0.66\ V_A L = 0.3233\ w_0 L^2$$

$$\therefore\quad V_A = \frac{0.3233}{0.66}\ w_0 L$$

$$= 0.4898\ w_0 L$$

$$\therefore\quad R_1 + V_A = \frac{1}{2} \times L \times w_0 = 0.5\ w_0 L$$

$$\therefore\quad R = 0.5\ w_0 L - V_A$$

$$= 0.5\ w_0 L - 0.4898\ w_0 L$$

$$= 0.5\ w_0 L - 0.4\ w_0 L - = 0.1\ w_0 L \text{ or } \frac{w_0 L}{10}$$

Problem 8.3: Find the reaction at A and B.

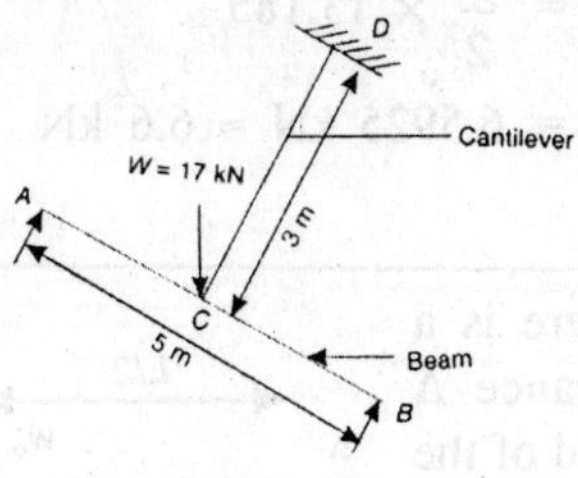

Fig. P. 8.3

Solution:

Load shared by beam = $W - P$

For the cantilever DC,

$$y_C = \frac{PL^3}{3EI} = \frac{P3^3}{3EI} = \frac{9P}{EI} \quad \ldots(1)$$

For the beam AB,

$$y_C = \frac{(W-P)\,l^3}{48\,EI} = \frac{(W-P)}{48\,EI} \times 5^3$$

$$= \frac{125}{48}\,\frac{(W-P)}{EI} \quad \ldots(2)$$

Equating the two, we have

$$\frac{9P}{EI} = \frac{125}{48}\,\frac{(W-P)}{EI}$$

From which $P = \dfrac{125}{557}$

$$W = \frac{125}{557} \times 17 = 3.815 \text{ kN}$$

Load shared by the beam

$= W - P$

$17 - 3.815 = 13.185$ kN

Hence reactions,

$$R_A = R_B$$

$$= \frac{1}{2} \times 13.185$$

$$= 6.5925 \text{ kN} \approx 6.6 \text{ kN}$$

Problem 8.4: There is a small initial clearance Δ between the left end of the beam shown in Fig. 8.4 (a) and the roller support. Determine the reaction at the roller support after the uniformly distributed load is applied.

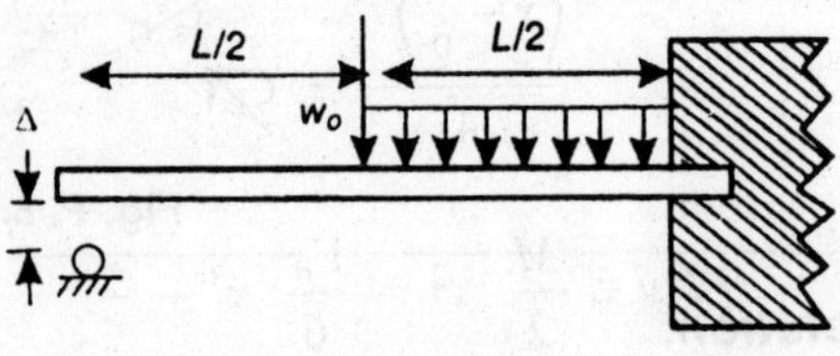

Fig. P. 8.4 (a)

Solution:

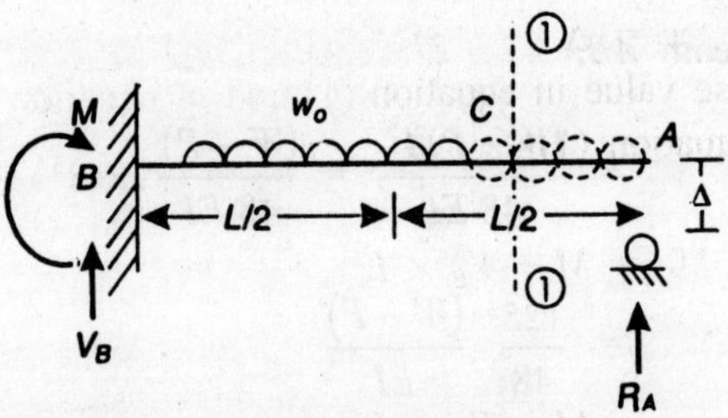

Fig. P. 8.4 (b)

General expression for moment,

$$EI\frac{d^2y}{dx^2} = M + V_B x - \frac{w_0}{2} <x - 0>^2 + \frac{w_0}{2}\left\langle x - \frac{L}{2}\right\rangle^2 \quad \ldots(1)$$

$$EI\frac{dy}{dx} = M \times x + V_B\frac{x^2}{2} - \frac{w_0}{2}\frac{\langle x - 0\rangle^3}{3} + \frac{w_0}{2} \times \frac{\left\langle x - \frac{L}{2}\right\rangle^3}{3}$$

$$EI\frac{dy}{dx} = M \times x + \frac{V_B}{2} \times x^2 - \frac{w_0}{6} <x - 0>^3 + \frac{w_0}{6}$$

$$\left\langle x-\frac{L}{2}\right\rangle^3 + C_1 = 0 \qquad \ldots(2)$$

$$EIy = M \times \frac{x^2}{2} + \frac{V_B}{2} \times \frac{x^3}{3} - \frac{w_0}{6}\,\frac{\langle x-0\rangle^4}{4} + \frac{w_0}{6}$$

$$\frac{\left\langle x-\frac{L}{2}\right\rangle^4}{4} + C_2 = 0$$

$$EI\;y = \frac{M}{2}\,x^2 + \frac{V_B}{6}\,x^3 - \frac{w_0}{24}\,\langle x-0\rangle^4 + \frac{w_0}{24}\left\langle x-\frac{L}{2}\right\rangle^4 \qquad \ldots(3)$$

Boundary condition

at $x = L$,

B.M. $= 0$

at $x = L$,

$y = \Delta$

Putting these value in equation (1) and in equation (3)

∴ From equation (1)

$$0 = M + V_B \times L - \frac{w_0}{2} \times L^2 + \frac{w_0}{2}\left(\frac{L}{2}\right)^2$$

$$= M + V_B \times L - \frac{w_0 L^2}{2} + \frac{w_0}{2} \times \frac{L^2}{4}$$

$$\therefore \qquad M + V_B \times L - \frac{w_0 L^2}{2} + \frac{w_0 L^2}{8} = 0$$

or

$$M + V_B \times L - \frac{3w_0 L^2}{8} = 0 \qquad \ldots(4)$$

From equation (3),

$$EI \times \Delta = \frac{M}{2}L^2 + \frac{V_B}{6} \times L^3 - \frac{w_0}{24} \times L^4 - \frac{w_0}{24} \times \left(\frac{L}{2}\right)^4$$

$$= \frac{M}{2}L^2 + \frac{V_B}{6} \times L^3 - \frac{w_0}{24} \times L^4 - \frac{w_0}{384}\,L^4$$

But we know,

∴ Putting the value of Δ in above eq. then we get,

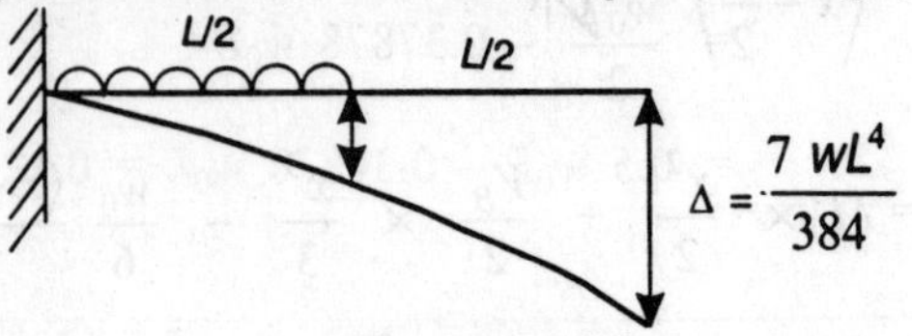

Fig. P. 8.4 (c)

$$EI \times \frac{7\ w_0 L^2}{384\ EI} = \frac{M}{2} L^2 + \frac{V_B}{6} L^3 - \frac{w_0}{24} \times L^4 - \frac{w_0}{384}\ L^4$$

$$\therefore \quad \frac{M}{2} L^2 + \frac{V_B}{6} L^3 = \frac{7\ w_0 L^4}{384} + \frac{w_0 L^4}{24} + \frac{w_0 L^4}{384}$$

$$= \frac{168\ w_0 L^4 + 384\ w_0 L^4 + 24\ w_0 l^4}{9216}$$

$$= \frac{576\ w_0 L^4}{9216} = \frac{w_0 L^4}{16}$$

$$\frac{M}{2}\ L^2 + \frac{V_B}{6}\ L^3 = \frac{w_0 L^4}{16}$$

$$\frac{L^2}{2}\left[M + \frac{V_B L}{3}\right] = \frac{w_0 L^4}{16}$$

$$\therefore\ \text{M} + \frac{V_B L}{3} = \frac{w_0 L^2}{8}$$

or $M + 0.34\ V_B L - 0.125\ w_0 L^2 = 0$ ...(5)

∴ From equation (4) and (5) we get

$M + V_B L - 0.375\ w_0 L^2 = 0$

$M + 0.34\ V_B L - 0.125\ w_0 L^2 = 0$

$- \quad - \qquad +$

$0.66\ V_B \times L - 0.25\ w_0 L^2 = 0$

$\therefore\ 0.66\ V_B \times L = 0.25\ w_0 L^2$

$$V_B = \frac{0.25\ w_0 L}{0.66} = 0.37878\ w_0 L$$

∴ Reaction of roller supports

$$= \frac{w_0 L}{2} - 0.37878\ w_0 L$$

$$= 0.5\ w_0 L - 0.37878\ w_0 L = 0.1212\ w_0 L$$

Problem 8.5: Determine the end moment and midspan value of $EI\delta$ for the restrained beam shown in Fig. 8.5 (a).

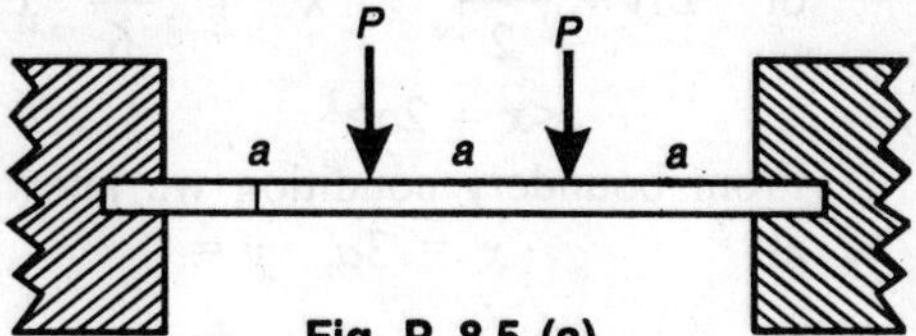

Fig. P. 8.5 (a)

Solution:

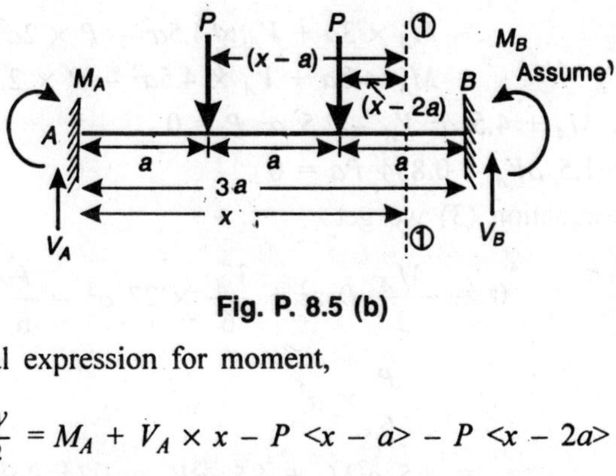

Fig. P. 8.5 (b)

General expression for moment,

$$EI\frac{d^2y}{dx^2} = M_A + V_A \times x - P <x - a> - P <x - 2a>$$

$$EI\frac{d^2y}{dx^2} = M_A + V_A \times x - P <x - a> - P <x - 2a> \qquad ...(1)$$

$$EI\frac{dy}{dx} = M_A \times x + V_A \times \frac{x^2}{2}$$

$$- \frac{P\langle x-a\rangle^2}{2} - \frac{P}{2} <x - 2a>^2 + C_1 \quad (= 0)$$

$$EI\frac{dy}{dx} = M_A x + \frac{V_A x^2}{2} - \frac{P}{2} <x - a>^2 - \frac{P}{2} <x - 2a>^2 \qquad ...(2)$$

$$EIy = M_A \times \frac{x^2}{2} + \frac{V_A}{2}\frac{x^3}{3}$$

$$-\frac{P}{2}\frac{\langle x-a\rangle^3}{3} - \frac{P}{2}\frac{\langle x-2a\rangle^3}{3} + C_2 \nearrow = 0$$

or $EIy = \frac{M_A}{2} \times x^2 + \frac{V_A}{6} x^3 - \frac{P}{6} <x - a>^3 - \frac{P}{6}$

$<x - 2a>^3$ ...(3)

From boundery condition we know that

at $x = 3a, \quad y = 0$

$$x = 3a \quad \frac{dy}{dx} = 0$$

From equation (2), we get,

$$0 = M_A \times 3a + \frac{V_A}{2}(3a)^2 - \frac{P}{2} \times (2a)^2 - \frac{P}{2} \times a^2$$

$$= M_A \times 3a + V_A \times 4.5a^2 - P \times 2a^2 - P \times 0.5a^2$$

$$= M_A \times 3a + V_A \times 4.5a^2 - P \times 2.5a^2$$

$\therefore 3a\, M_A + 4.5\, a^2\, V_A - 2.5\, a^2\, P = 0$

$M_A + 1.5\, aV_A - 0.834\, Pa = 0$ ...(4)

From equation (3) we get

$$0 = \frac{M_A}{2} 9\,a^2 + \frac{V_A}{6} \times 27\, a^3 - \frac{P}{6} \times 8\, a^3$$

$$-\frac{P}{6} \times a^3$$

$$= 4.5\, a^2 M_A + 4.5\, a^3 V_A - 1.34\, Pa^3 - 0.167\, Pa^3$$

$$= 4.5\, a^2 M_A + 4.5\, a^3 V_A - 1.507\, Pa^3$$

$\therefore 4.5\, a^2 M_A + 4.5\, a^3 V_A - 1.507\, Pa^3 = 0$

$4.5a^2\, [M_A + a \times V_A - 0.334\, Pa] = 0$

$\therefore M_A + aV_A - 0.334\, Pa = 0$ ...(5)

From equation (4) and equation (5)

$M_A + 1.4\, aV_A - 0.834\, Pa = 0$

$M_A + a\, V_A - 0.334\, Pa = 0$

$- \quad - \quad +$

$0.5\ aV_A = 0.5\ Pa$

$$\therefore \quad V_A = \frac{0.5}{0.5}\ P = P$$

$$\therefore \quad V_A = P$$

$$V_B = P$$

∴ End moment

$M_A + a \times V_A = 0.334\ Pa$

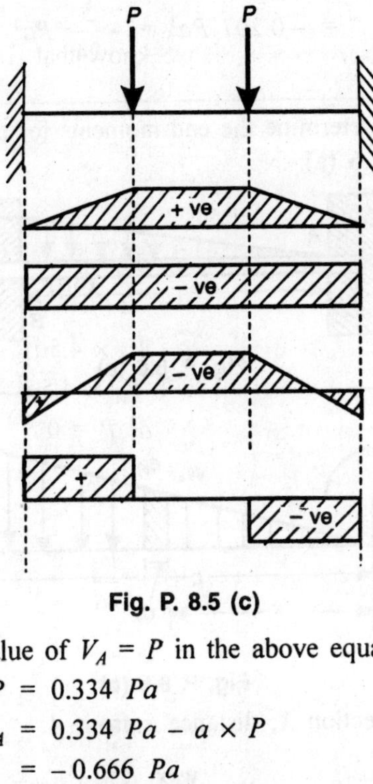

Fig. P. 8.5 (c)

Putting the value of $V_A = P$ in the above equation, then we get,

$$M_A + a \times P = 0.334\ Pa$$

$$\therefore \quad M_A = 0.334\ Pa - a \times P$$

$$= -0.666\ Pa$$

$$= -\left(\frac{2}{3}\right)\ Pa$$

So, assumed direction is not correct, actual direction of M_A is anticlockwise.

$$\therefore \quad M_B = M_A - P \times 2a - Pa$$

$$= -0.666\ Pa - 2Pa - Pa$$

$$= -3.666\ Pa$$

Mind span value of $EI\delta$

Value of L at mid span = 1.5 a

$$EIy = \frac{M_A}{2}x^2 + \frac{V_A}{6}x^3 - \frac{P}{6}(0.5a)^3$$

$$= \frac{-0.666\ Pa}{2} \times (1.5\ a)^2 + \frac{P}{6}(1.5\ a)^3 - \frac{P}{6}(0.5\ a)^3$$

$$= -0.749\ Pa^3 + 0.5625\ Pa^3 - 0.0208\ Pa^3$$

$$= -0.207\ Pa^3 = -\frac{5}{24}\ Pa^3$$

Problem 8.6: Determine the end moments for the restrained beam shown in Fig. 8.6 (a).

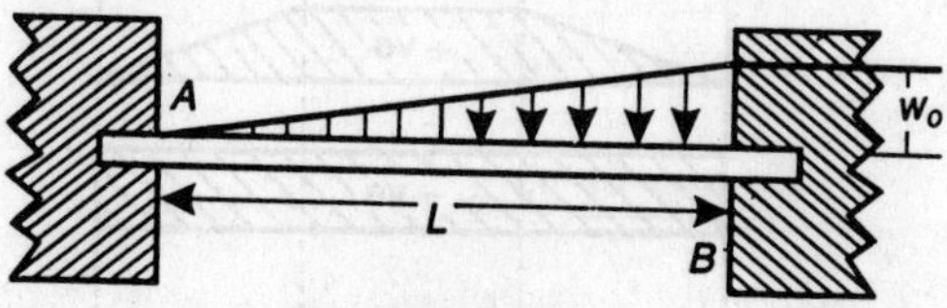

Fig. P. 8.6 (a)

Solution:

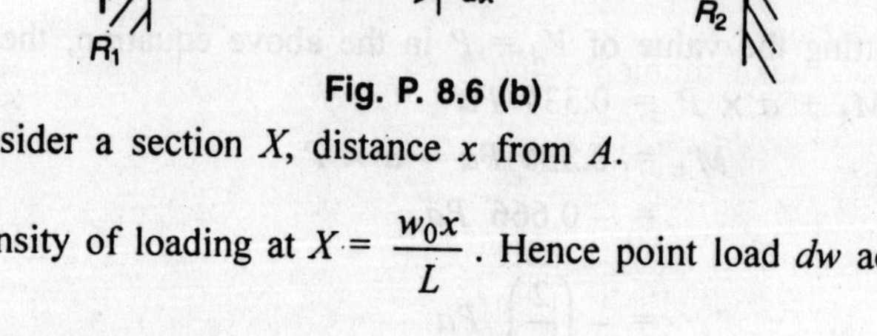

Fig. P. 8.6 (b)

Consider a section X, distance x from A.

Intensity of loading at $X = \dfrac{w_0 x}{L}$. Hence point load dw acting on an element at distance dx is $dw = \dfrac{w_0 x}{L} x\ dx$

Fixed end moments due to this elemental load are given by,

$$dM_1 = \left(\frac{w_0}{L} x\ dx\right)\frac{x(L-x)^2}{L^2} = \frac{w_0 x^2 (L-x)^2\ dx}{L^3}$$

and $$dM_2 = \left(\frac{w_0}{L}x\,dx\right)\frac{x^2\times(L-x)}{L^2}$$

$$= \frac{wx^3}{L^3}\,(L-x)\,dx,$$

$\therefore$ Total $$M_1 = \int_0^L \frac{w_0x^2}{L^3}\,(L-x)^2\,dx$$

$$= \frac{w_0}{L^3}\int_0^L (L^2x^2 + x^4 - 2L\,x^3)\,dx$$

$$= \frac{w_0}{L^3}\left[L^2\frac{L^3}{3} + \frac{L^5}{5} - 2L\left(\frac{L}{4}\right)^4\right]$$

$$= \frac{w_0L^2}{30}$$

Total $$M_2 = \int_0^L \frac{w_0x^3}{L^3}\,(L-x)\,dx$$

$$= \frac{w_0}{L^3}\int_0^L (Lx^3 - x^4)\,dx$$

$$= \frac{w_0}{L^3}\left[L\times\frac{L^4}{4} - \frac{L^5}{5}\right] = \frac{w_0L^2}{20}$$

For R_1, take moments about B

Thus $$\Sigma M_B = 0 = R_1L + M_2 - M_1 - \frac{w_0L^2}{6}$$

$$\therefore \quad R_1 = \frac{1}{L}\left[\frac{w_0L^2}{30} - \frac{w_0L^2}{20} + \frac{w_0L^2}{6}\right] = \frac{3}{20}\,w_0L$$

Similarly,

$$\Sigma M_A = 0$$

$$= R_2L + M_1 - M_2 - \frac{w_0L^2}{3}$$

From which,

$$R_2 = \frac{1}{L}\left[M_2 - M_1 + \frac{w_0 L^2}{3}\right]$$

$$= \frac{1}{L}\left[\frac{w_0 L^2}{20} - \frac{w_0 L^2}{30} + \frac{w_0 L^2}{3}\right] = \frac{7}{20}\, w_0 L$$

Problem 8.7: Determine the end moments and reactions for a fixed end beam loaded by a couple μ_0. Also, find the deflection at the point C where the couple acts.

Solution:

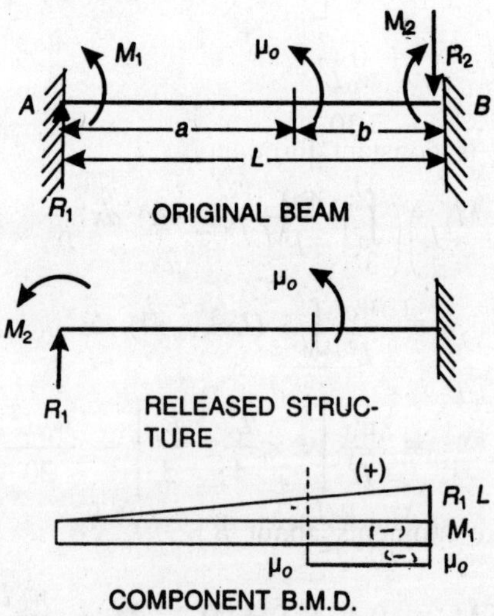

Fig. P. 8.7

Let M_1 and R_1 be the reaction component at end A. Release end A and apply M_1 and R_1 at A. Released structure shown in Fig. 8.7 (b).

The component B.M. diagram showing the effects of R_1, M_1 and μ_0 separately is shown in Fig. 8.7 (c). There are two unknowns: M_1 and R_1, and hence two condition equations are required. As a first condition, since both the ends of the beam have zero slopes, the change of slope between A and B is zero. Hence from area moment theorem.

$$\sum_A^B A_M = 0$$

$\because$ EI is constant throughout the beam length.

$$\therefore \frac{1}{2} R_1 L \times L - M_1 \times L - \mu_0 b = 0$$

$$\text{or } R_1 - \frac{2M_1}{L} = \frac{2\mu_0 b}{L^2} \qquad \ldots(1)$$

The second condition is obtained from the fact that tangent to the deflection curve at A passes through point B. This means that the moment of the area of B.M. diagram between A and B, taken about B, is zero.

$$\therefore \quad \sum_A^B A_M \bar{x} = 0$$

(since EI is constant throughout)

$$\therefore \left(\frac{1}{2} R_1 L \times L\right)\left(\frac{L}{3}\right) - M_1 L \times \frac{L}{2} - b\ \mu_0 \times \frac{b}{2} = 0$$

$$\text{or } R_1 - \frac{3M_1}{L} = \frac{3\mu_0 b^2}{L^3} \qquad \ldots(2)$$

Solving (1) and (2), we get

$$M_1 = \frac{\mu_0 b}{L^2}(2L - 3b)$$

$$= \frac{\mu_0 b}{L^2}(2a - b)$$

$$\text{and} \qquad R_1 = \frac{6\mu_0 ab}{L^3} \ (\uparrow)$$

Since $\Sigma V = 0$ for the whole beam, we get R_2

$$= -R_1 = -\frac{6\mu_0 ab}{L^3} \ (\text{i.e. } \downarrow)$$

Taking moments about A and assuming M_2 to be clockwise, we get

$$M_2 - M_1 - \mu_0 + \frac{6\mu_0 ab}{L^3} = 0$$

or
$$M_2 = \frac{\mu_0 b}{L^2}(2a - b) + \mu_0 - \frac{6\mu_0 ab}{L^3}$$

From which

$$M_2 = \frac{\mu_0 a}{L^2}(a - 2b)$$

Thus reaction components are completely known.

Now y_c is equal to the vertical offset from the tangent at A, since tangent at A is horizontal. Hence y_c is equal to moment of the area of $\frac{M}{EI}$ diagram between A and C, taken about C.

$$\therefore \quad y_c = \sum_C^A \frac{A_M}{EI}\bar{x}$$

$$= \left[\frac{1}{2}a \times R_1 a\left(\frac{a}{3}\right) - M_1 \times a \times \frac{a}{2}\right]\frac{1}{EI}$$

$$= \left(\frac{R_1 a^3}{6} - \frac{M_1 a^2}{2}\right)\frac{1}{EI}$$

$$= \frac{1}{EI}\left[\frac{6\mu_0 a^4 b}{6L^3} - \frac{\mu_0 b}{2L^2}(2a - b)a^2\right] = \frac{\mu_0 a^2 b^2 (b - a)}{2L^3 EI}$$

Problem 8.8: A fixed beam AB of span 4.5 m is subjected to a concentrated couple of 400 kN-m applied at a section 3 m from end A. Find the end moments from the first principles and draw the B.M. and S.F. diagrams.

Solution: Let M_1 and R_1 be the reaction components at A in the directions marked. Take section at a distance x from end A, then

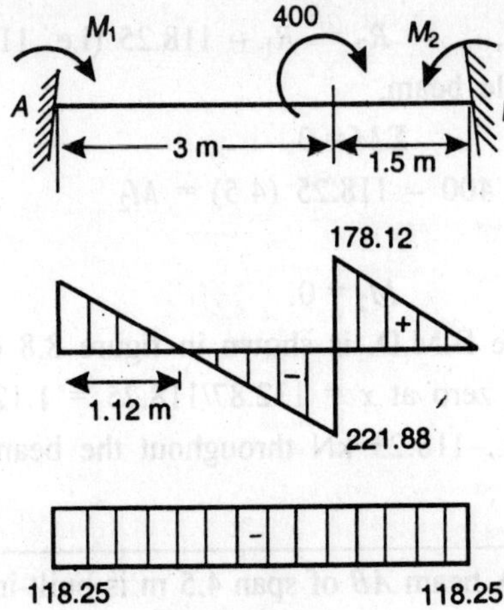

Fig. P. 8.8

$$EI\frac{d^2y}{dx^2} = -M_x = R_1x - M_1 \mid - 400\ (x-3)^0 \qquad \ldots(1)$$

$$\therefore \quad EI\frac{dy}{dx} = R_1\frac{x^2}{2} - M_1x + 0 \mid - 400\ (x-3) \qquad \ldots(2)$$

$$EI\ y = R_1\frac{x^3}{6} - \frac{M_1x^2}{2} + 0 \mid - \frac{400(x-3)^2}{2} \qquad \ldots(3)$$

At $x = 4.5$ m, $\dfrac{dy}{dx} = 0$

Hence from (2)

$$0 = \frac{R_1}{2}(4.5)^2 - M_1\ (4.5) - 400\ (4.5-3)$$

or $R_1 - 0.444\ M_1 = 59.259$ ...(4)

Also, at $x = 4.5$ m,

$y = 0$ Hence from (3),

$$0 = \frac{R_1}{6}(4.5)^3 - \frac{M_1}{2}(4.5)^2 - 200\ (4.5-3)^2$$

or $R_1 - 0.667\ M_1 = 29.63$ ...(5)

Solving (4) and (5), we get $M_1 = 132.87$ kN.m and $R_1 = 118.25$ kN. Plus signs with the values of M_1 and R_1 show that the assumed directions of M_1 and R_1 are correct.

From statics, $R_2 = -R_1 - 118.25$ (i.e. 118.25 ↑)

For the whole beam,

$$\Sigma M = 0$$

$$\therefore\ 132.87 + 400 - 118.25\,(4.5) = M_2$$

From which

$$M_2 = 0.$$

The complete B.M.D. is shown in figure 8.8 (b)

The B.M. is zero at $x = 132.87/118.25 = 1.125$ m from A.

S.F. constant −118.25 kN throughout the beam.

Problem 8.9: A beam AB of span 4.5 m is built-in at its both ends. It carries a gradually varying load from zero at A to 30 kN/m at B. Determine the fixed end moments and supports reactions at both ends of the beam, and hence plot the B.M. and S.F. diagrams for the beam.

Solution:

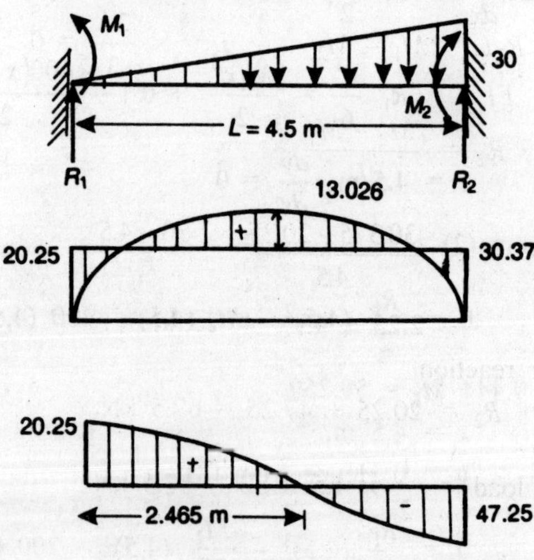

Fig. P. 8.9

$$M_1 = \frac{wL^2}{30} = \frac{30(45)^2}{30}$$

$$= 20.25 \text{ kN.m}$$

$$M_2 = \frac{wL^2}{20} = \frac{30(4.5)^2}{20}$$

$$= 30.375 \text{ kN.m}$$

For reaction R_1, take moments about B.

$$\text{or } R_1 \times L + M_2 - M_1 - \frac{1}{2} \times L \times w \times \frac{L}{3} = 0$$

$$R_1 = \frac{M_1 - M_2}{L} + \frac{wL}{6}$$

$$= \frac{20.25 - 30.375}{4.5} + \frac{30 \times 4.5}{6}$$

$$= -2.25 + 22.5$$

$$= 20.25 \text{ kN } (\uparrow)$$

Similarly, for finding reaction R_2, take moments about A.

$$\text{Thus} \quad R_2L = M_1 - M_2 - \frac{1}{2} L w \times \frac{2L}{3} = 0$$

$$\text{or} \quad R_2 = \frac{M_2 - M_1}{L} + \frac{wL}{3}$$

$$= \frac{30.375 - 20.25}{4.5} + \frac{30 \times 4.5}{3}$$

$$= 2.25 + 45 = 47.25 \text{ kN } (\uparrow)$$

Check for reaction

$$R_1 + R_2 = 20.25 + 47.25 = 67.5 \text{ kN}$$

$$\text{Total load} = \frac{1}{2} \times 4.5 \times 30 = 67.5 \text{ kN}$$

At any distance x from A, $w_x = \frac{30}{4.5} x = 6.667\, x$

$$F_x = R_1 - \frac{1}{2}(6.667\,x)\,x = 20.25 - 3.3333\,x^2$$

The variation is parabolic. S.F. is zero at $x = \sqrt{(20.25/3.3333)}$ $= 2.465$ m from A. The S.F.D. is shown in Fig. 8.9 (c).

Again, at any distance x from A.

$$M_x = R_1\,x - M_1 - \frac{1}{2}\,w_x \times x \times \frac{x}{3}$$

$$= 20.25\,x - 20.25 - 1.111\,x^3$$

This is a cubic variation. At $x = 2.465$ m, $M_x = +13.026$ kN.m

or $2\,x^2 - 3\,x\,L + L^2 = 0$

which gives $x = L/2$

Hence maximum deflection occurs at mid-span.

$\therefore$ at $x = L/2$,

$$y_{max.} = \frac{wL^4}{EI}\left[\frac{1}{384} - \frac{1}{96} + \frac{1}{96}\right] = \frac{wL^4}{384\,EI} \quad \textbf{Ans}$$

Problem 8.10: A fixed beam of length 4 m is loaded as shown in Fig. 8.10 (a). Find the end moments and reactions and plot the B.M. and S.F. diagrams.

Solution:

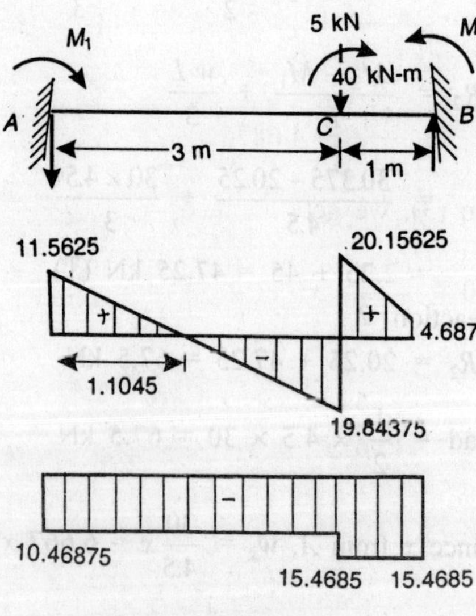

Fig. P. 8.10 (b)

Let us assume that M_1 acts on the clockwise direction and R_1 acts downwards.

Take section at a distance x from end A.

$$EI\frac{d^2y}{dx^2} = -M_1 + R_1x \mid$$

$$+ 5(x-3) - 40(x-3)^0 \qquad \ldots(1)$$

$$\therefore \quad EI\frac{dy}{dx} = -M_1x + \frac{R_1x^2}{2} + 0 \mid$$

$$+ \frac{5(x-3)^2}{2} - 40(x-3) \qquad \ldots(2)$$

and

$$EIy = -\frac{M_1x^2}{2} + \frac{R_1x^3}{6} + 0 \mid$$

$$+ \frac{5(x-3)^3}{2} - \frac{40(x-3)^2}{2} \qquad \ldots(3)$$

At $x = 4$ m,

$$\frac{dy}{dx} = 0 - M_1(4) + \frac{R_1}{2}(4)^2$$

$$+ 2.5(4-3)^2 - 40(4-3)$$

From which

$$R_1 = \frac{M_1}{2} + 4.6875.$$

Substituting in (3), we get

$$0 = -\frac{M_1}{2}x^2 + \frac{1}{6}\left(\frac{M_1}{2} + 4.6875\right)x^3 \mid$$

$$+ \frac{5}{6}(x-3)^3 - \frac{40(x-3)^2}{2} \qquad \ldots(4)$$

At $x = L = 4$ m, $y = 0$.

Hence from (4), we get

$$0 = -\frac{M_1}{2}(4)^2 + \frac{1}{6}\left(\frac{M_1}{2} + 4.6875\right)(4)^3 + \frac{5}{6}(4-3)^3 - 20(4-3)^2$$

From which

$$M_1 = +11.5625 \text{ kN.m}$$

The plus sign shows that the moment at A acts in the same direction as assumed.

$$\text{Also,} \quad R_1 = \frac{11.5625}{2} + 4.6875 = 10.4687 \text{ kN}$$

Here also, plus sign shows that R_1 acts in the same direction as assumed i.e., it acts in a downward direction.

Again, substituting

$x = L$ in (1), we get

$$(-M_x)_x = L = -M_2 = EI\frac{d^2y}{dx^2}$$

$$= -11.5625 + 10.46875(4) + 5(4-3) - 40(4-3)$$

or $\quad -M_2 = -4.6875$

or $\quad M_2 = 4.6875$ kN.m

(i.e. M_2 is of sagging nature, as marked.)

B.M. diagram

$$M_1 = +11.5625; \; M_2 = +4.6875$$

For $AC \quad M_x = 11.5625 - 10.46875\,x$

Hence M_x is zero at

$$x = -11.5625/10.46875$$

$$= 1.1045 \text{ m}$$

$$M_C \text{ (left)} = 11.5625 - 10.46875 \times 3$$

$$= -19.84375 \text{ kN–m}$$

$$M_C \text{ (right)} = 11.5625 - 10.46875 \times 3 + 40$$

$$= +20.15625$$

The B.M. diagram is shown in

S.F. diagram $\quad R_2 = (R_1 + 5)\ (\uparrow)$

$$= (10.46875 + 5) = 15.46875 \text{ kN } (\uparrow)$$

Problem 8.11: A horizontal shaft of length L is subjected at its centre to a vertical laod W. The shaft fits in bearings at its ends and when the slope of the shaft at the end is θ, the bearing exerts a bending moment on the shaft of magnitude $k\theta$. FInd (a) the B.M. at the ends and (b) deflection of the shaft at mid-span.

Solution:

The shaft is partially fixed at the ends. Let the support moment, exerted by the bearings be μ. Then $\mu = k\,\theta$.

or $$\theta = \frac{\mu}{k} = \frac{dy}{dx} \text{ at the end.}$$

Also, due to symmetry,

$$R_1 = R_2 = R = \frac{W}{2}, \text{ at each end.}$$

Take section of a distance x from left hand support.

$$EI\frac{d^2y}{dx^2} = -M_x = \mu - \frac{W}{2}x \mid + W\left(x - \frac{L}{2}\right) \quad \text{...(1)}$$

$$EI\frac{dy}{dx} = \mu\,x - \frac{W}{2}\frac{x^2}{2} + C_1 \mid + \frac{W}{2}\left(x - \frac{L}{2}\right)^2 \quad \text{...(2)}$$

At $$x = 0 \quad \frac{dy}{dx} = \theta = \frac{\mu}{k} \text{ (given)}$$

$$C_1 = EI\left(\frac{dy}{dx}\right)_{x=0} = EI \times \frac{\mu}{k}$$

Hence from (2)

$$EI\frac{dy}{dx} = \mu\,x - \frac{W}{4}x^2 + EI\frac{\mu}{k}\Bigg| + \frac{W}{2}\left(x - \frac{L}{2}\right)^2 \quad \text{...(3)}$$

Also, at $$x = \frac{L}{2},$$

$$\frac{dy}{dx} = 0, \text{ due to symmetry.}$$

$$\therefore \qquad 0 = \frac{\mu L}{2} - \frac{W}{4}\left(\frac{L}{2}\right)^2 + EI\frac{\mu}{k}$$

$$\text{or} \qquad \mu\left(\frac{L}{2} + \frac{EI}{k}\right) = \frac{WL^2}{16}$$

From which $\mu = \dfrac{WL^2 k}{8\,(Lk + 2\,EI)}$ **Ans**

Substituting this value of μ in (3), we get, for half the shaft,

$$EI\frac{dy}{dx} = \frac{WL^2 k}{8\,(Lk+2\,EI)} \times x - \frac{W}{4}\,x^2 + EI\frac{\mu}{k}$$

Integrating,

$$EIy = \frac{WL^2 k}{8\,(Lk+2\,EI)} \times \frac{x^2}{2} - \frac{W}{4} \times \frac{x^3}{3} + EI\frac{\mu}{k}\,x + 0$$

(Since $y = 0$ at $x = 0$)

At $\qquad x = \dfrac{L}{2}$,

we get the maximum deflection.

$$\therefore\; EIy_{\max} = \frac{WL^2 k}{16\,(Lk+2\,EI)}\left(\frac{L}{2}\right)^2 - \frac{W}{12}\left(\frac{L}{3}\right)^3 + EI\,\frac{\mu}{k}\left(\frac{L}{2}\right)$$

$$\text{or } EIy_{\max} = \frac{WL^4 k}{64\,(Lk+2\,EI)} - \frac{WL^3}{96} - \frac{EIL}{2\,k}\left[\frac{WL^2 k}{8\,(Lk+2\,EI)}\right]$$

$$\text{or} \qquad y_{\max} = \frac{WL^3}{192\,EI}\left[\frac{3\,kL}{kL+2\,EI} - 2 + \frac{12\,EI}{(kL+2\,EI)}\right]$$

$$= \frac{WL^3}{192\,EI}\left[\frac{3\,kL - 2\,kL - 4\,EI + 12\,EI}{kL + 2\,EI}\right]$$

$$= \frac{WL^3}{192\,EI}\left[\frac{kL + 8\,EI}{kL + 2\,EI}\right] \quad \textbf{Ans}$$

Problem 8.12: A fixed beam *AB* of length *L* carries a concentrated load *W* at the midspan. The moment of inertia of the beam from either end to a distance of *L*/4 is *I* and 2*I* for the remaining length, as shown in Fig. 8.12 (a). Determine the end moments and plot the B.M.D.

Solution:

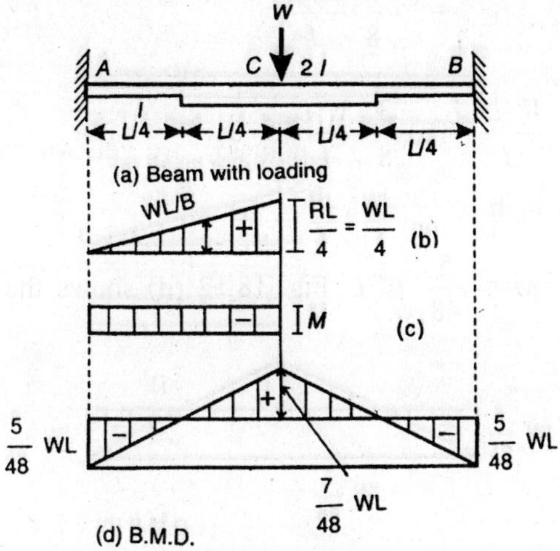

Fig. P. 8.12

Since beam is symmetrical

Hence $\quad M_1 = M_2 = M$ and $R_1 = R_2 = R = \dfrac{W}{2}$. The component B.M. diagrams due to *R* and *M* are shown in Fig. 8.12 (b) and (c) respectively. From Mohr's theorems,

$$E\theta_A^C = -\sum_A^C \frac{A_M}{I \text{ or } 2\,I} = 0$$

(since the tangent at *C* is horizontal.)

For B.M.D. of Fig. 8.12 (b):

$$\sum_A^C \frac{A_M}{I \text{ or } 2\,I} = \left(\frac{1}{2} \times \frac{L}{4} \times \frac{WL}{8}\right) \frac{1}{I} + \frac{1}{2}\left(\frac{WL}{8} + \frac{WL}{4}\right) \times \frac{L}{4} \times \frac{1}{2I}$$

$$= \frac{5}{128}\,\frac{WL^2}{I}$$

For B.M.D. of Fig. 8.12 (c)

$$\sum_A^C \frac{A_M}{I \text{ or } 2\,I} = -\left[\left(M \times \frac{L}{4} \times \frac{1}{I}\right) + \left(M \times \frac{L}{4} \times \frac{1}{2I}\right)\right]$$

$$= -\frac{3}{8}\frac{M\,L}{I}$$

$$\therefore \frac{5}{128}\frac{W\,L^2}{I} - \frac{3}{8}\frac{M\,L}{I} = 0$$

From which

$$M = \frac{5}{48}\,W\,L.$$ Fig. 18.12 (d) shows the B.M. diagram.

Chapter 9

Continuous Beams

Problem 9.1: Determine the B.M. in the beam over the supports

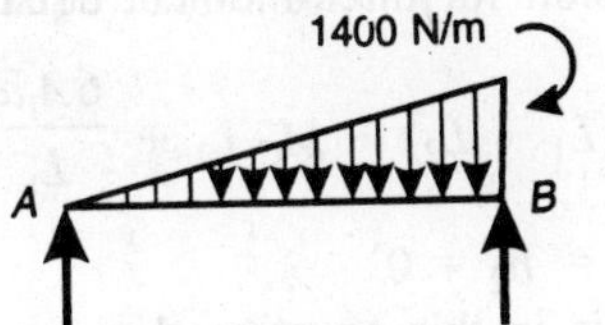

Fig. P. 9.1 (a)

Solution:

First we see from loading and support condition at beam, bending moment of the end of the beam is zero, i.e. $M_A = M_C = 0$

$$\therefore \quad M_A = 0$$

$$M_C = 0$$

For span AB

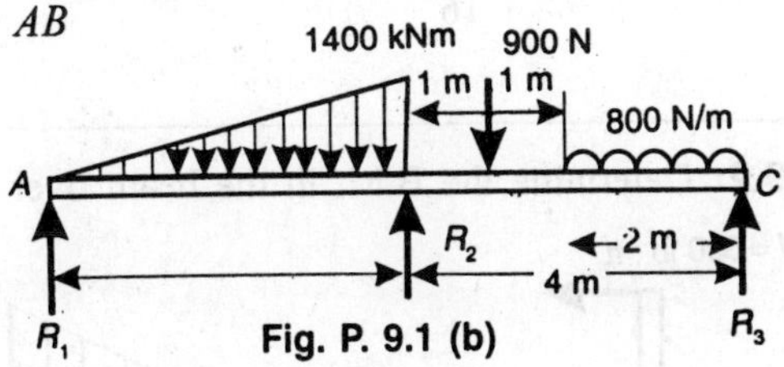

Fig. P. 9.1 (b)

$$\frac{6}{L} A \, \overline{a_1} = \frac{8}{60} w_0 L^3$$

= Moment of area of B.M.D. from end A

$$= \frac{8}{60} \times 1400 \times 4^3$$

$$= 11946.67 \text{ N.m}^2$$

For span BC

$$\frac{6}{L} A \bar{b}_2 = \frac{Pb}{L}(L^2 - b^2) + \frac{w_0}{4L} d^2 (2L^2 - d^2)$$

$$= \frac{900 \times 3}{4}[4^2 - 3^2] + \frac{800}{4 \times 4} \times 2^2 \times [2 \times 4^2 - 2^2]$$

$$= \frac{2700}{4}[16 - 9] + \frac{800}{4 \times 4} \times [32 - 4]$$

$$= \frac{2700}{4} \times 7 + \frac{800}{4} \times 28$$

$$= 10325 \text{ N.m}^2$$

General expression for three-moment equation,

$$M_A L_1 + 2M_B (L_1 + L_2) + M_C L_2 + \frac{6A_1\bar{a}}{L_1} + \frac{6A_2\bar{b}_2}{L_2} = 0$$

$$\because \quad h_1 = h_2 = 0$$

Putting the value in this equation then we get,

$$0 + 2 M_B (L_1 + L_2) + 0 + \frac{6A_1\bar{a}_1}{L_1} + \frac{6A_2\bar{b}_2}{L_2} = 0$$

$$\therefore 2M_B [4 + 4] + 11946.67 + 10325 = 0$$

$$16 M_B = -22271.67$$

$$\therefore \quad M_B = -\frac{22271.67}{16} = -1391.97 \text{ N.m}$$

Problem 9.2: Determine the B.M. in the beam over the supports.

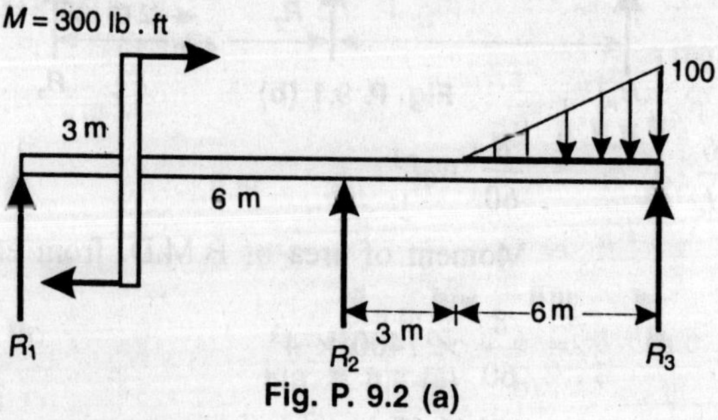

Fig. P. 9.2 (a)

Solution:

$$M_A = M_C = 0$$

For span AB

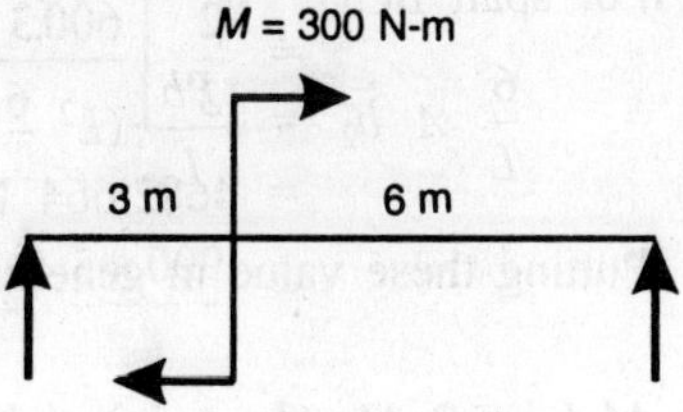

Fig. P. 9.2 (b)

$$\frac{6}{L} A\bar{a}_1 = -\frac{M}{L}(3a^2 - L^2)$$

$$= -\frac{300}{9}(3 \times 3^2 - 9^2)$$

$$= -33.33\ (27 - 81) = +1799.82 \text{ N.m}^2$$

For span BC

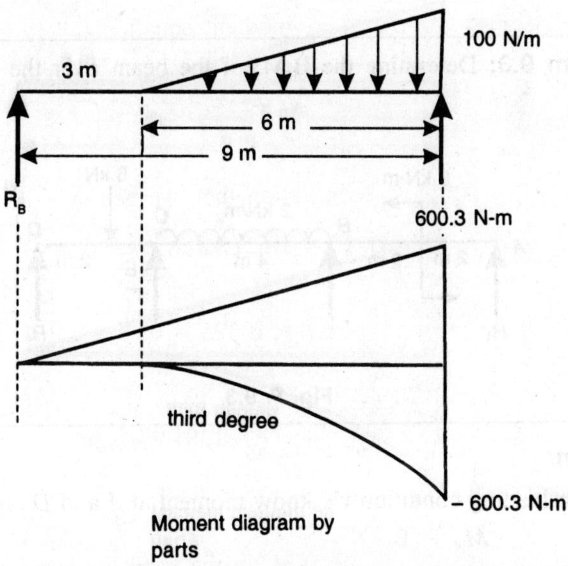

Fig. P. 9.2 (c)

$$R_B \times 9 = \frac{1}{2} \times 100 \times 6\text{m} \times \frac{6}{3} = 600$$

$$R_B = \frac{600}{9} = 66.67 \text{ N}$$

$$\frac{6A_2\bar{b}_2}{L_2} = \frac{6}{9} = \frac{2}{3}\left[\frac{1}{2} \times 600.3 \times 9 \times \frac{9}{3} - \frac{600.3 \times 6}{4} \times \frac{6}{5}\right]$$

$$= \frac{2}{3}\left[\frac{600.3 \times 81}{6} - \frac{600.3 \times 36}{20}\right]$$

$$= 4682.664 \text{ N.m}^2$$

Putting these value in general expression,

$$M_A L_1 + 2\ M_B\ (L_1 + L_2) + M_C\ L_2 + \frac{6A_1\overline{a_1}}{L_1} + \frac{6A_2\overline{b_2}}{L_2} = 0$$

$$0 + 2\ M_B\ (9 + 9) + 0 + 1799.82 + 4682.664 = 0$$

$$\therefore \quad 36\ M_B = -6482.484$$

$$\therefore \quad M_B = -180.069 \text{ N.m}$$

Problem 9.3: Determine the B.M. of the beam over the supports

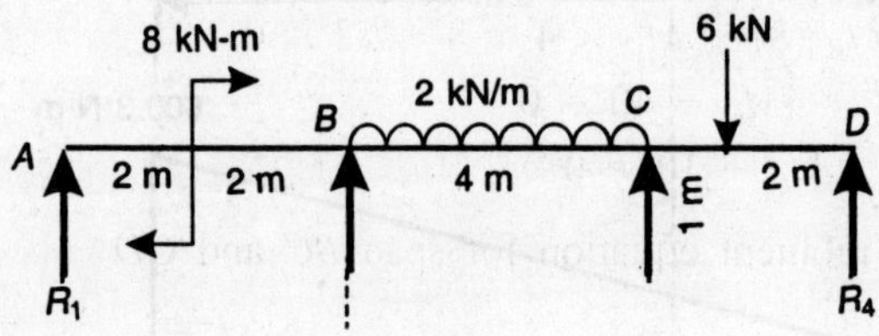

Fig. P. 9.3

Solution:

From loading condition we know moment at A and D are zero, so,

$$M_A = 0$$

$$M_D = 0$$

For span AB

$$\frac{6}{L}\ A\overline{a_1} = -\frac{M}{L}\ (3a^2 - L^2)$$

$$= -\frac{8}{4}\ [3 \times 2^2 - 4^2] = -2\ [12 - 16]$$

$$= -2 \times -4 = +8 \text{ kN.m}^2$$

For span BC

$$\frac{6}{L} A\overline{a_2} = \frac{w_0 L^3}{4} = \frac{2 \times 4^3}{4} = 32 \text{ kN.m}^2$$

$$\frac{6}{L} A\overline{b_2} = \frac{w_0 L^3}{4} = 32 \text{ kN.m}^2$$

For span *CD*

$$\frac{6}{L} A\overline{b_3} = \frac{Pb}{L}(L^2 - b^2)$$

$$= \frac{6 \times 2}{3}(9 - 4) = 4 \times 5 = 20 \text{ kN.m}^2$$

∴ Three moment equation for span *AB* and *BC*

$$M_A L_1 + 2M_B (L_1 + L_2) + M_C L_2 + \frac{6A_1\overline{a_1}}{L_1} + \frac{6A_2\overline{b_2}}{L_2} = 0$$

$0 + 2M_B (8) + M_C \times 4 + 8 + 32 = 0$

$16 M_B + 4M_C + 40 = 0$

$4 M_B + M_C + 10 = 0$ ...(1)

Three moment equation for span *BC* and *CD*

$$M_B L_1 + 2 M_C (L_1 + L_2) + M_D L_2 + \frac{6A\overline{a_2}}{L_1} + \frac{6A\overline{b_3}}{L_2} = 0$$

$M_B \times 4 + 2 M_C \times 7 + 0 + 32 + 20 = 0$

$4 M_B + 14 M_C + 52 = 0$ ...(2)

∴ From equation (1) and (2) we get

$4 M_B + M_C + 10 = 0$

$4 M_B + 14 M_C + 52 = 0$

$- \quad - \quad -$

$-13 M_C - 42 = 0$

$$\therefore \quad M_C = -\frac{42}{13} = -3.23 \text{ kN.m}$$

$$\therefore \quad 4 M_B = -52 + 14 \times 3.23$$

$$\therefore \quad M_B = -3.384 \text{ kN.m}$$

Problem 9.4: Determine the reaction and sketch the shear force diagram.

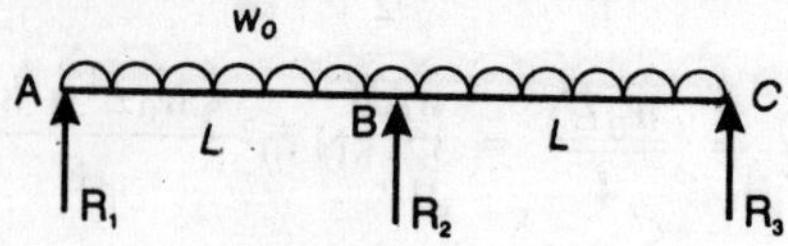

Fig. P. 9.4 (a)

Solution:

From loading diagram and support reaction we know bending moment at A and C are zero, i.e.

$$M_A = 0,$$
$$M_C = 0$$

$\therefore$ Three moment equation for this beam,

$$M_A L_1 + 2\, M_B\, (L_1 + L_2) + M_C L_2 + \frac{6A_1\overline{a_1}}{L_1} + \frac{6A_2\overline{b_2}}{L_2} = 0$$

$$0 + 2\, M_B\, (L + L) + 0 + \frac{6A_1\overline{a_1}}{L_1} + \frac{6A_2\overline{b_2}}{L_2} = 0$$

$$\therefore\ 4M_B \times L + \frac{6A_1\overline{a_1}}{L_1} + \frac{6A_2\overline{b_2}}{L_2} = 0 \qquad \ldots(1)$$

For span AB and BC

$$\frac{6A_1\overline{a_1}}{L_1} = \frac{w_0 L^3}{4}$$

$$\frac{6A_2\overline{b_2}}{L_2} = \frac{w_0 L^3}{4}$$

$$\therefore\quad 4\, M_B\, L + \frac{w_0 L^3}{4} + \frac{w_0 L^3}{4} = 0$$

$$4\, M_B + \frac{w_0 L^2}{4} + \frac{w_0 L^2}{4} = 0$$

$$\therefore\quad 4\, M_B = -\frac{2w_0 L}{4} = -\frac{w_0 L^2}{2}$$

$$\therefore\quad M_B = -\frac{w_0 L^2}{8}$$

For reaction R_1

$$-\frac{w_0 L^2}{8} = R_1 \times L - \frac{w_0 L^2}{2}$$

$$\therefore \quad R_1 L = \frac{w_0 L^2}{2} - \frac{w_0 L^2}{8} = \frac{4\, w_0 L^2 - w_0 L^2}{8} = \frac{3\, w_0 L^2}{8}$$

$$\therefore \quad R_1 = \frac{3\, w_0 L}{8}$$

For reaction R_3

$$-\frac{3\, w_0 L^2}{8} = R_3 \times L - \frac{w_0 L^2}{2}$$

$$\therefore \quad R_3 = \frac{3\, w_0 L}{8}$$

$$\therefore R_1 + R_2 + R_3 = 2 w_0 L$$

$$R_2 = 2\, w_0 L - \frac{3}{4}\, w_0 L = \frac{8\, w_0 L - 3\, w_0 L}{4} = \frac{5\, w_0 L}{4}$$

$$\therefore \quad R_1 = \frac{3\, w_0 L}{8}$$

$$R_2 = \frac{5\, w_0 L}{4}$$

$$R_3 = \frac{3\, w_0 L}{8}$$

S.F.D. of beam

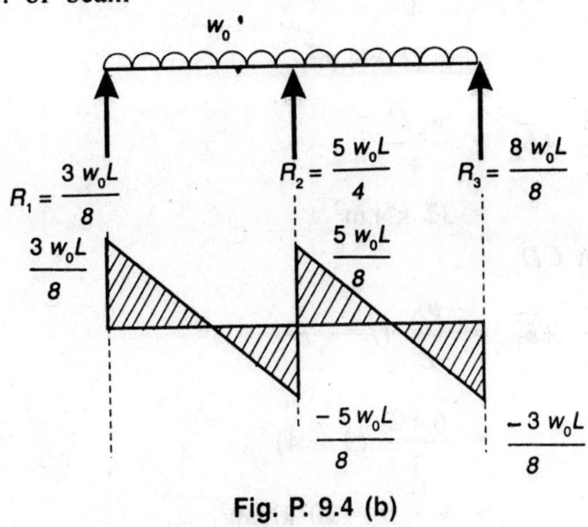

Fig. P. 9.4 (b)

Problem 9.5: Determine the reaction and sketch the shear force diagram.

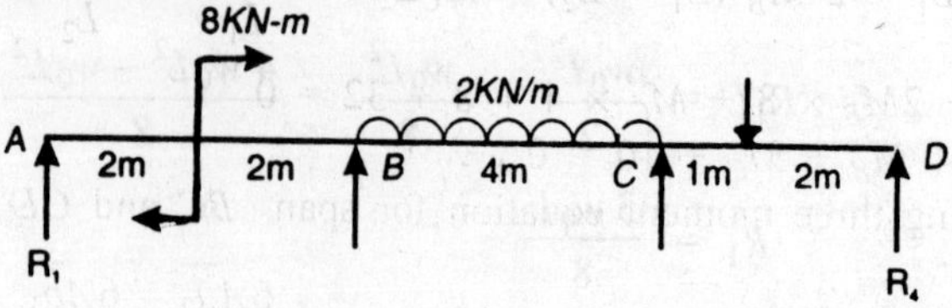

Fig. P. 9.5 (a)

Solution:

First draw end moment at the supports

From the loading condition we know moment at A and D are zero,

$$M_A = 0$$

$$M_D = 0$$

For span AB

$$\frac{6}{L} A\overline{a_1} = -\frac{M}{L}(3a^2 - L^2)$$

$$= -\frac{8}{4}[3 \times 2^2 - 4^2] = +8 \text{ kN.m}^2$$

For span BC

$$\frac{6}{L} A\overline{a_2} = \frac{w_0 L^3}{4}$$

$$= +32 \text{ kN.m}^2$$

$$\frac{6}{L} A\overline{b_2} = \frac{w_0 L^3}{4}$$

$$= 32 \text{ kN.m}^2$$

For span CD

$$\frac{6}{L} A\overline{b_3} = \frac{Pb}{L}(L^2 - b^2)$$

$$= \frac{6 \times 2}{3}(9 - 4)$$

$$= 4 \times 5 = 20 \text{ kN.m}^2$$

Using three moment equation for span *AB* and *BC*,

$$M_A L_1 + 2\ M_B\ (L_1 + L_2) + M_C L_2 + \frac{6A_1\overline{a_1}}{L_1} + \frac{6A_2\overline{b_2}}{L_2} = 0$$

$$0 + 2M_B \times 8 + M_C \times 4 + 8 + 32 = 0$$

$$\therefore\ 4\ M_B + M_C + 10 = 0 \qquad ...(1)$$

Using three moment equation for span *BC* and *CD*,

$$M_B\ L_1 + 2\ M_C\ (L_1 + L_2) + M_D L_2 + \frac{6A\overline{a_2}}{L_1} + \frac{6A\overline{b_3}}{L_2} = 0$$

$$\therefore\ M_B \times 4 + 2\ M_C \times 7 + 0 + 32 + 20 = 0$$

$$\therefore\ 4\ M_B + 14\ M_C + 52 = 0 \qquad ...(2)$$

From equation (1) and (2) we get

$$4\ M_B + M_C + 10 = 0$$

$$4\ M_B + 14\ M_C + 52 = 0$$

$$-\quad -\quad -$$

$$-13\ M_C - 42 = 0$$

$$\therefore\ M_C = -3.23\ \text{kN.m}$$

$$M_B = -3.384\ \text{kN.m}$$

For reaction

(a) For reaction R_1

$$M_B = \Sigma M_L$$

$$\therefore\ -3.384 = R_1 \times 4 + 8$$

$$\therefore\ R_1 \times 4 = -3.384 - 8$$

$$\therefore\ R_1 = -2.846\ \text{kN}\ (\downarrow)$$

(ii) For reaction R_B

$$M_C = \Sigma M_L$$

$$-3.23 = -2.848 \times 8 + 8 - 2 \times 4 \times 2 + R_B\ 4$$

$$= -22.768 + 8 - 16 + R_B \times 4$$

$$= -30.768 + R_B \times 4$$

$$\therefore\ R_B = +6.8845\ \text{kN}$$

(iii) For reaction R_D

$$-3.23 = R_4 \times 3 - 6 \times 1$$

$$R_4 = +0.9233\ \text{kN}$$

$\therefore$ For reaction R_C,

$$\Sigma F_y = 0$$

$R_1 + R_2 + R_3 + R_4 = 2 \times 4 + 6 = 8 + 6 = 14$

$-2.846 + 6.8846 + R_3 + 0.9233 = 14$

$\therefore \quad R_3 = 9.0381 \text{ kN}$

$\therefore \quad R_1 = -2.846 \text{ kN}$

$R_2 = 6.8846 \text{ kN}$

$R_3 = 9.0381 \text{ kN}$

$R_4 = 0.9233 \text{ kN}$

For S.F.D.

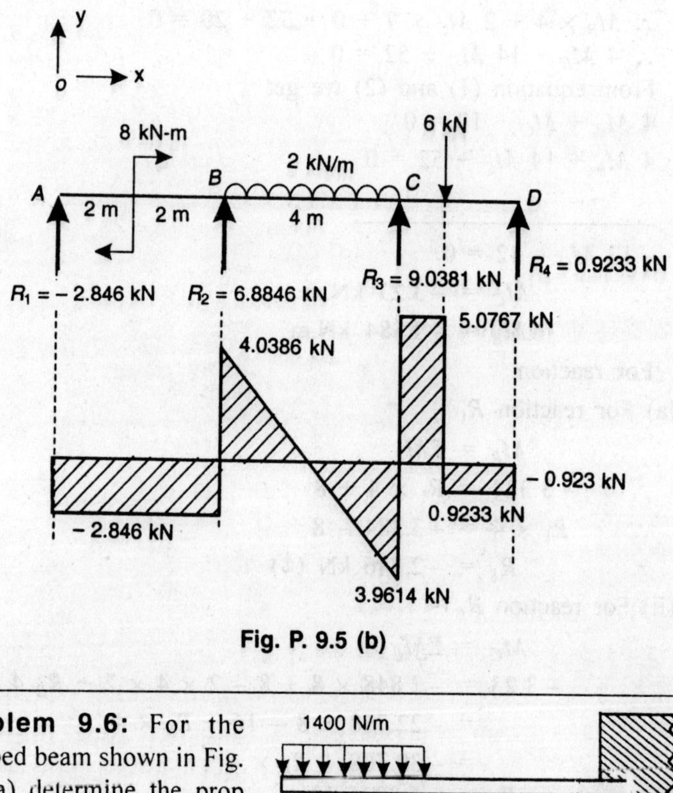

Fig. P. 9.5 (b)

Problem 9.6: For the propped beam shown in Fig. 9.6 (a) determine the prop reaction and the maximum positive bending moment.

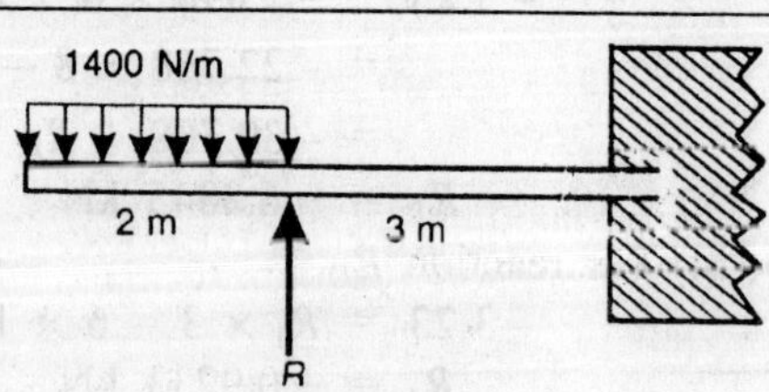

Fig. P. 9.6 (a)

Solution:

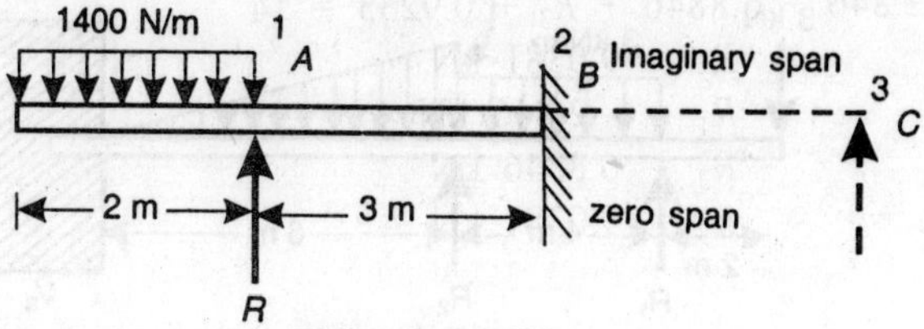

Fig. P. 9.6 (b)

$$M_A = -1400 \times 2 \times 1$$
$$= -2800 \text{ N.m}$$
$$M_C = 0,\ L_1 = 3 \text{ m},\ L_2 = 0$$

Apply three-moment equation,

$$M_A L_1 + 2\ M_B\ (L_1 + L_2) + M_C L_2 + \frac{6A_1\overline{a_1}}{L_1} + \frac{6A_2\overline{b_2}}{L_2} = 0$$

$$\text{or } M_A L_1 + 2\ M_B\ (L_1 + L_2) + \frac{6A_1\overline{a_1}}{L_1} = 0$$

$$\frac{6A_1\overline{a_1}}{L_1} = 0$$

Because no loading present between them

$\therefore\ M_A L_1 + 2\ M_B\ L_1 = 0$

$-2800 \times 3 + 2\ M_B \times 3 = 0$

$$6\ M_B = 8400$$
$$\therefore \quad M_B = 1400 \text{ N.m}$$

For reaction R

$$M_B = \Sigma M_L$$
$$1400 = R \times 3 - 1400 \times 2 \times 4 = 3\ R - 11200$$
$$\therefore \quad 3R = 11200 + 1400 = 12600$$
$$\therefore \quad R = 4200 \text{ N}$$

$\therefore$ Maximum positive bending moment at fixed support = 1400 N.m and support reaction at B = 4200 N.

> **Problem 9.7:** Compute the moments over the supports and sketch the shear diagram for the continuous beam shown in Fig. 9.7 (a).

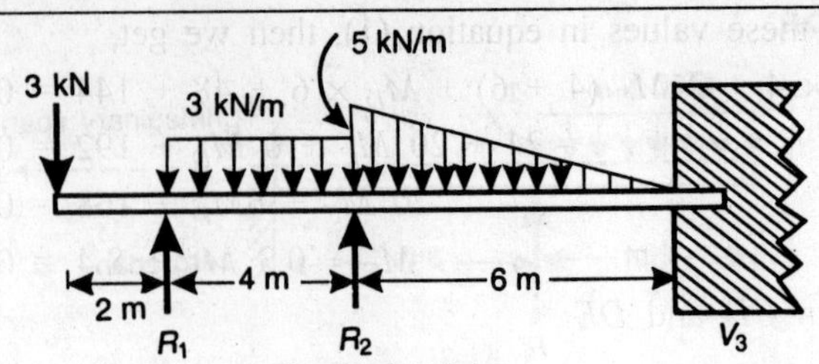

Fig. P. 9.7 (a)

Solution:

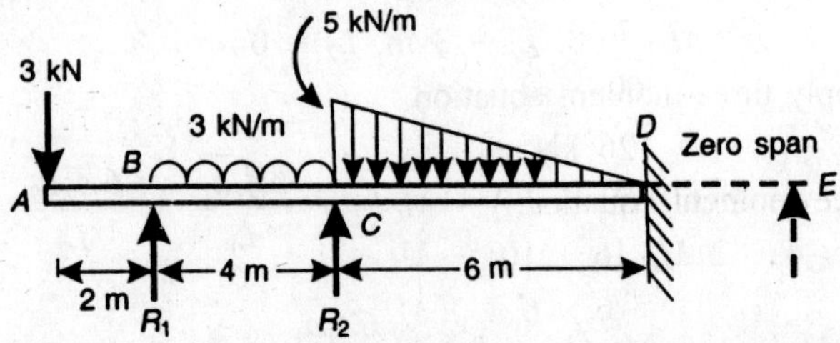

Fig. P. 9.7 (b)

$$M_B = -3 \times 2 = -6 \text{ kN.m}$$

For span *BC* and *CD*

$$M_B L_1 + 2\, M_C\, (L_1 + L_2) + M_D L_2 + \frac{6A_1\overline{a_1}}{L_1} + \frac{6A_2\overline{b_2}}{L_2} = 0 \ldots(1)$$

For span *BC*

$$\frac{6A_1\overline{a_1}}{L_1} = \frac{w_0 L^3}{4}$$

$$= \frac{3 \times 4^3}{4} = 48 \text{ kN.m}^2$$

$$\frac{6A_2\overline{b_2}}{L_2} = \frac{8}{60}\, w_0 L^3$$

$$= \frac{8}{60} \times 5 \times 6^3$$

$$= 144 \text{ kN.m}^2$$

Putting these values in equation (1), then we get,

$$-6 \times 4 + 2\ M_C\ (4 + 6) + M_D \times 6 + 48 + 144 = 0$$

$$-24 + 20\ M_C + 6\ M_D + 192 = 0$$

$$20\ M_C + 6M_D + 168 = 0$$

$$M_C + 0.3\ M_D + 8.4 = 0 \quad ...(2)$$

For span *CD* and *DE*

$$\frac{6A_2\overline{a_2}}{L_2} = \frac{7}{60}\ w_0 L_2^3$$

$$= \frac{7}{60} \times 5 \times 6^3$$

$$= 126\ \text{kN.m}^2$$

∴ Three moment equation,

$$M_B \times 6 + 2\ M_C\ [6 + 10] + M_D \times 0 + 126 + 0 = 0$$

$$-6 \times 6 + 2\ M_C \times 6 + 0 + 126 = 0$$

$$-36 + 12\ M_C + 126 = 0$$

$$12\ M_C = -\frac{90}{12} = -7.5\ \text{kN}$$

$$\therefore \quad M_C = -7.5\ \text{kN}$$

Putting this value in equation (2) then we get,

$$-7.5 + 0.3\ M_D + 8.4 = 0$$

$$M_D = -\frac{0.9}{0.3} = -3\ \text{kN.m}$$

For reaction R_1

$$M_C = \Sigma M_L$$

$$-7.5 = R_1 \times 4 - 3 \times 6 - 3 \times 4 \times 2$$

$$= 4\ R_1 - 18 - 24 = 4\ R_1 - 42$$

$$4\ R_1 = 34.5$$

$$\therefore \quad R_1 = 8.625\ \text{kN}\ (\uparrow)$$

For reaction R_2

$$M_D = \Sigma M_L$$

$$-3 = 8.625 \times 10 + R_2 \times 6 - 3 \times 4 \times 8$$

$$- \frac{1}{2} \times 5 \times 6 \times \frac{2}{3} \times 6$$

$$-3 = 86.25 + 6\ R_2 - 96 - 60 = 6\ R_2 - 69.75$$

$$6\ R_2 = 66.75$$

$$\therefore \quad R_2 = \frac{66.75}{6} = 11.125 \text{ kN}$$

$$\therefore \quad R_2 = 11.125 \text{ kN}$$

$$\therefore \quad R_1 = 8.625 \text{ kN}$$

$$R_2 = 11.125 \text{ kN}$$

Shear force diagram

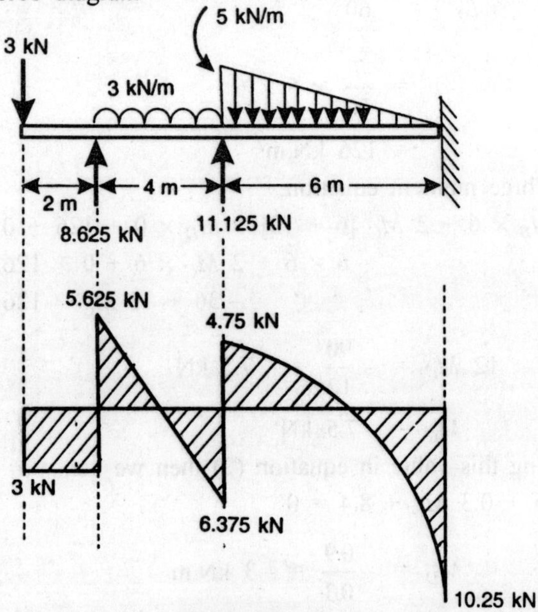

Fig. P. 9.7 (c)

Problem 9.8: Determine the midspan value of $EI\delta$ for the beam shown in Fig. 9.8 (a).

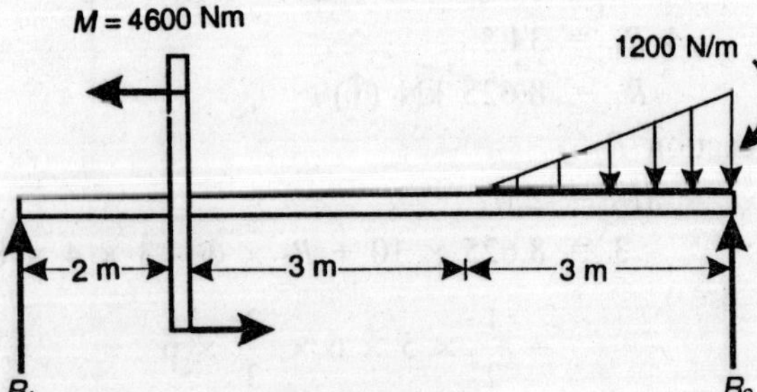

Fig. P. 9.8 (a)

Solution:

Selecting point 2 at the position of the designed deflection and points 1 and 3 at the reactions will make h_1 and h_3 each equal to the designed deflection. Also since 1 and 3 are above the horizontal line through 2, heights h_1 and h_3 will be positive.

The segments into which the beam is divided by the three points are designated as span (1) and (2)

For span (1)

$$L = 4 \text{ m}$$

For span (2)

$$L = 4 \text{ m}$$

For span *AB*

$$\frac{6}{L} A_1 \overline{a_1} = -\frac{M}{L}(3a^2 - L^2)$$

$$= -\frac{4600}{4}[3 \times 4 - 16]$$

$$= -\frac{4600}{4} \times -4 = +4600 \text{ N.m}^2$$

For span *BC*

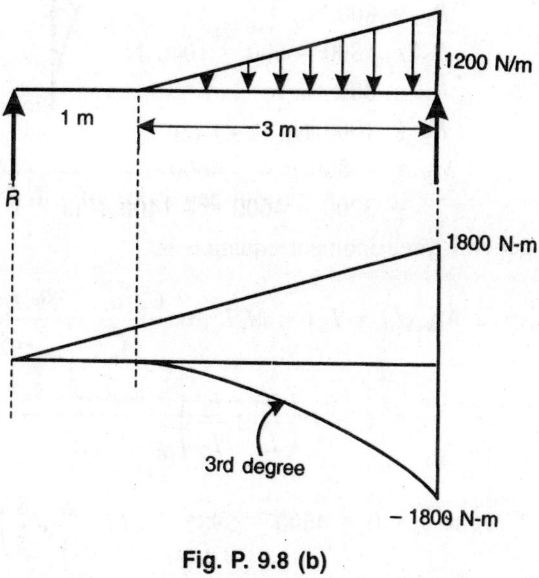

Fig. P. 9.8 (b)

$$R \times 4 = \frac{1}{2} \times 1200 \times 3 \times \frac{3}{3} = \frac{1}{2} \times 600 \times 3$$

$$\therefore \quad R = 450 \text{ N}$$

$$\frac{6A_2\overline{b_2}}{L_2} = \frac{6}{4}\left[\frac{1}{2} \times 1800 \times 4 \times \frac{4}{3} - \frac{1}{4} \times 1800 \times 3 \times \frac{3}{5}\right]$$

$$= \frac{6}{4}[4800 - 810] = \frac{6}{4} \times 3990 = 5985 \text{ N.m}^2$$

For Bending moment at B (M_B) and C

First calculate reaction at supports

$$R_1 + R_2 = \frac{1}{2} \times 1200 \times 3$$

$$R_1 + R_2 = 1800 \quad \ldots(1)$$

$$\Sigma M_C = 0$$

$$R_1 \times 8 - 4600 - \frac{1}{2} \times 1200 \times 3 \times \frac{3}{3} = 0$$

$$\therefore 8R_1 - 4600 - 1800 = 0$$

$$8R_1 = 6400$$

$$\therefore \quad R_1 = 800 \text{ N}$$

$$\therefore \quad R_2 = 1800 - 800 = 1000 \text{ N}$$

$$\therefore \quad R_1 = 800 \text{ N}$$

$$R_2 = 1000 \text{ N}$$

$$\therefore \quad M_B = +800 \times 4 - 4600$$

$$= 3200 - 4600 = -1400 \text{ N.m}$$

The general three moment equation is,

$$M_AL_1 + 2M_B(L_1 + L_2) + M_CL_2 + \frac{6A_1\overline{a_1}}{L_1} + \frac{6A_2\overline{b_2}}{L_2}$$

$$= 6EI\left(\frac{h_1}{L_1} + \frac{h_3}{L_2}\right)$$

$$0 + 2M_B \times 9 + 0 + 4600 + 5985 = 6EI\left(\frac{\delta}{4} + \frac{\delta}{5}\right)$$

$$18\ M_B + 4600 + 5985 = 6\ EI\delta\left(\frac{1}{4}+\frac{1}{5}\right)$$

$$-1400 \times 18 + 4600 + 5985 = \frac{6 \times 9}{20}\ EI\delta$$

$$-25200 + 4600 + 5985 = 2.7\ EI\delta$$

$$\therefore \qquad EI\delta = -\frac{14615}{2.7} = 5412.96\ \text{N.m}^3$$

> **Problem 9.9:** A beam *ABCD*, 16 m long is continuous over three spans: *AB* = 6 m, *BC* = 5 m and *CD* = 5 m, the supports being at the same level. There is a uniformly distributed load of 20 kN/m over *BC*. On *AB*, there is a point load of 80 kN at 2 m from *A*. On *CD*, there is a point load of 60 kN at 3 m from *D*. Calculate the moments and reactions at the supports.

Solution:

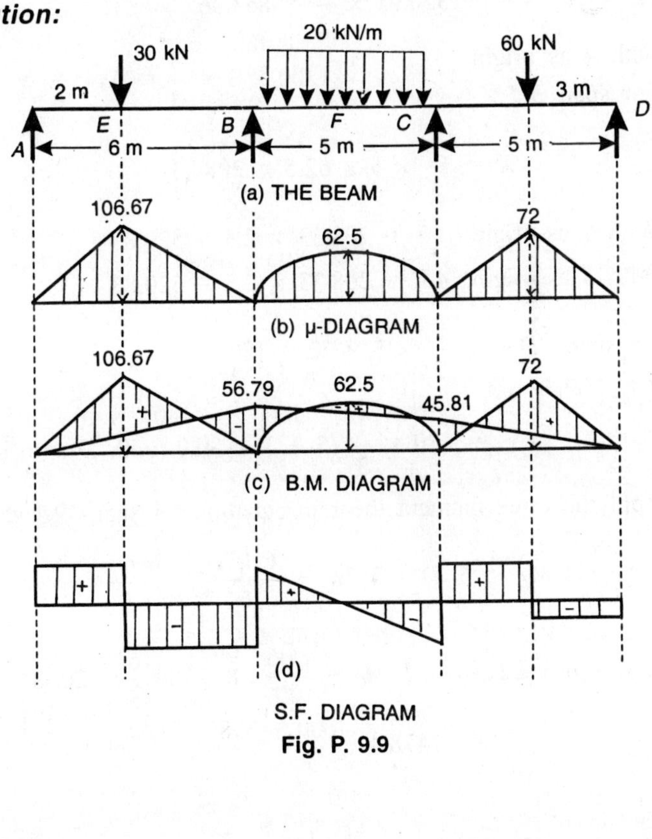

Fig. P. 9.9

The free B.M. diagrams for each span can be drawn as usual, as shown in Fig. 9.9 (b). Here, we have

$$M_E = \frac{80 \times 2 \times 4}{6} = 106.67 \text{ kN.m}$$

$$M_F = \frac{20(5)^2}{8} = 62.5 \text{ kN.m}$$

$$M_G = \frac{60 \times 2 \times 3}{5} = 72 \text{ kN.m}$$

For span *AB*

$$A = \frac{1}{2} \times 6 \times 106.67 = 320.01$$

$$\bar{x} = \frac{1}{3}(6 + 2) = \frac{8}{3}$$

$$\therefore \quad A\bar{x} = 320.01 \times \frac{8}{3} = 853.36$$

with *A* as origin.

For span *BC*

$$A = \frac{2}{3} \times 5 \times 62.5 = 208.33$$

With *C* as origin, $A\bar{x} = 208.33 \times 2.5 = 520.83$

With *B* as origin, $A\bar{x} = 208.33 \times 2.5 = 520.83$

For span *CDA* $= \frac{1}{2} \times 5 \times 72 = 180$

$$A\bar{x} = 180 \times \frac{1}{3}(5 + 3) = 480, \text{ with } D \text{ as origin.}$$

Applying three moment theorem equation for spans *AB–BC*

$$6M_A + 2\,M_B\,(6 + 5) + 5\,M_C + \frac{6A_1\bar{x}_1}{6} + \frac{6A_2\bar{x}_2}{5} - 0$$

$$\text{or } 6 \times 0 + 22\,M_B + 5\,M_C = -\left[\frac{6}{6} + 853.36 + \frac{6}{5} \times 520.83\right]$$

$$= -1478.4 \qquad \ldots(1)$$

Similarly, applying three moment theorem equation for spans *BC-CD*

$$5\ M_B + 2\ M_C\ (5+5) + 5\ M_D + \frac{6A_1\bar{x}_1}{5} + \frac{6A_2\bar{x}_2}{5} = 0$$

or $5\ M_B + 20\ M_C + 5 \times 0$

$$= -\left[\frac{6}{5} \times 520.83 + \frac{6}{5} \times 480\right] = -1200.1 \qquad \ldots(2)$$

Solving (1) and (2), we get

$$M_B = -56.79 \text{ kN.m}$$

and $\quad M_C = -45.81$ kN.m

For finding reaction R_A, write equation for B.M. at *B*. Thus

$$R_A \times 6 - 80 \times 4 = M_B = -56.79,$$

from which $\quad R_A = 43.87$ kN

For finding reaction R_B, write equation for B.M. at *C*. Thus

$$R_A \times 11 - 80 \times 9 + R_B \times 5 - 20 \times 5 \times 2.5 = M_C = -45.81$$

or $43.87 \times 11 - 720 + 5\ R_B - 250 = -45.81$

from which $\quad R_B = 88.32$ kN

For reaction R_D,

write equation for B.M. at *C*, considering R.H.S.

$\therefore\ R_D \times 5 - 60 \times 2 = M_C = -45.81$

from which $R_D = 14.84$ kN

Hence

$$R_C = (80 + 20 \times 5 + 60) - (43.87 + 88.32 + 14.84) = 92.97 \text{ kN}$$

The B.M. and S.F. diagrams are shown in Fig. 9.9 (c) and (d) respectively.

Problem 9.10: A continuous beam *ABCDE* has three supports with two over hangs as shown in Fig. 9.10. The overhangs are so selected that the reaction of the three supports are equal when the beam carries a uniformly distributed load over the whole length. Find the value of each overhang.

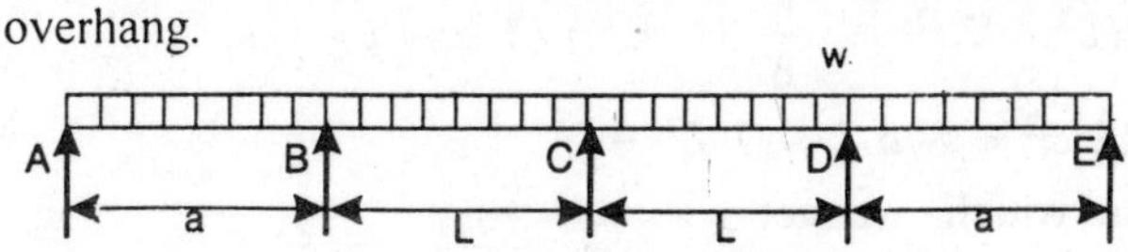

Fig. P. 9.10

Solution:

Total load on the beam = $w\ (2\ L + 2a) = 2\ w\ (L + a)$

Since all the reactions are equal, we have

$$R_B = R_C = R_D = \frac{2}{3}\ w\ (L + a)$$

Now $$M_B = -\frac{Wa^2}{2} = M_D$$

Applying three moments theorem equation for spans *BC-CD*, we get

$$M_B \times L + 2\ M_C\ (L + L) + M_D \times L = -\frac{WL^3}{4} - \frac{WL^3}{4}$$

or $$-\frac{Wa^2}{2} \times L + 4\ M_C \times L - \frac{Wa^2}{2} \times L = -\frac{WL^3}{2}$$

From which, we get

$$M_C = \frac{1}{4L}\left[-\frac{wL^3}{2} + \frac{wa^2 L}{2} + \frac{wa^2 L}{2}\right] = \frac{w}{4}\left(a^2 - \frac{L^2}{2}\right)$$

or $$M_C = -\frac{w}{8}\ (L^2 - 2a^2) \qquad \ldots(1)$$

But $$M_C = R_B \times L - \frac{w(a+L)^2}{6}$$

$$= \frac{2}{3}\ (L + a)\ L - \frac{w}{2}\ (L + a)^2$$

$$= -\frac{w(a+L)}{6}\ [3\ (L + a) - 4\ L]$$

$$= -\frac{w(a+L)}{6}\ (3\ a - L) \qquad \ldots(2)$$

Equating (1) and (2), we get

$$\frac{w}{8}\ (L^2 \quad 2a^2) - \frac{w}{6}\ (L + u)\ (3\ a - L)$$

or $18\ a^2 + 8\ a\ L - 7\ L^2 = 0$

From which, we get

$$a = 0.4398\ \text{L}$$

Problem 9.11: A continuous beam *ABC* of constant moment of inertia is simply supported at *A*, *B* and *C*. The beam carries a central point load of 40 kN in span *AB* and *a* central clockwise couple of 300 kN.m in the span *BC*, as shown in Fig. 9.11 (a). Find the support moments and plot the shear force and bending moment diagrams.

Solution:

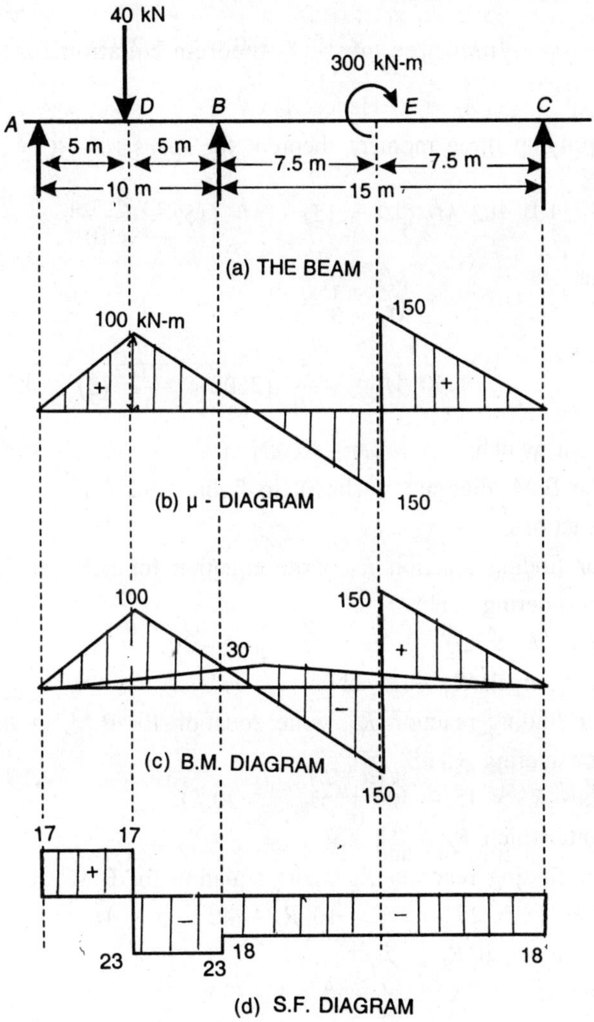

Fig. P. 9.11

For simply supported span *AB*,

$$M_D = \frac{40 \times 10}{4} = 100 \text{ kN.m}$$

For simply supported span *BC*,

$$\mu_E \text{ (left)} = -150 \text{ kN.m}$$

$$\mu_E \text{ (right)} = +150 \text{ kN.m}$$

$$A_1 \bar{x}_1 = \frac{1}{2} \times 10 \times 100 \times 5 = 2500$$

$$A_2 = 0 \text{ Hence } A_2 \bar{x}_2 = 0$$

Applying three moment theorem for spans *AB*–*BC* we get

$$M_A (10) + 2 M_B (10 + 15) + M_C (15) + \frac{6A_1\bar{x}_1}{10} + \frac{6A_2\bar{x}_2}{15} = 0$$

But $M_A = 0$;
$M_C = 0$

$$\therefore \quad 50 M_B = -\frac{6}{10} (2500) \quad -\frac{6}{15} (0)$$

From which $M_B = -30$ kN.m

The B.M. diagram is shown in figure 9.12 (c).

Reactions:

For finding reaction R_A, write equation for B.M. at *B*,

Considering L.H.S.

Thus, $R_A \times 10 - 40 \times 5 = M_B = -30$,

from which $R_A = 17$ kN

For finding reaction R_B, write equation for B.M. at *B*.

Considering R.H.S.

Thus $R_C \times 15 - 300 = M_B = -30$,

from which $R_C = 18$ kN

For finding reaction R_B, write equation for B.M. at *A*.

Thus $18 \times 25 - 300 + 10 R_B - 40 \times 5 = M_A = 0$

From which $R_B = 5$ kN

Check: Total vertical load = kN (↓);

Total reaction = 17 + 18 + 5 = 40 kN (↑)

The S.F. diagram is shown in Fig. 9.12 (d).

Problem 9.12: A three span beam *ABCD* has different moments of inertia and is loaded as shown in Fig. 9.12 (a). Find the reactions and support moments and draw the S.F. and B.M. diagrams.

Solution:

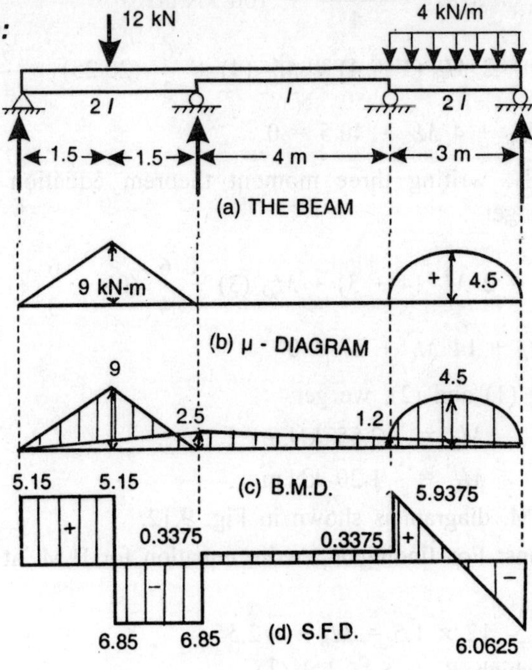

Fig. P. 9.12

The μ-diagram is shown in Fig. 9.12 (b).

For span *AB*,

$$M_{max} = \frac{12 \times 13}{4} = 9 \text{ kN.m}$$

$$A\bar{x} = \frac{1}{2} \times 9 \times 3 \times \frac{3}{2} = 20.25$$

For span *BC*

B.M. is zero throughout.

Hence $A\bar{x} = 0$

For span *CD*

$$M_{max} = \frac{4(3)^2}{8} = 4.5 \text{ kN.m}$$

here, $M_A = 0$

and $M_D = 0$

Writing three moment theorem equation for spans *AB–BC*, we get

$$M_A (3) + 2\ M_B (3 + 4) + M_C (4) + \frac{6}{3} (20.25) + \frac{6}{4} (0) = 0$$

or $14\ M_B + 4\ M_C + 40.5 = 0$ ...(1)

Similarly, writing three moment theorem equation for spans *BC–CD*, we get

$$M_B (4) + 2\ M_C (4 + 3) + M_D (3) + \frac{6}{4} (0) + \frac{6}{3} (13.5) = 0$$

or $4\ M_B + 14\ M_C + 27 = 0$ ...(2)

Solving (1) and (2), we get

$M_B = -2.55$ kN.m

and $M_C = -1.20$ kN.m

The B.M. diagram is shown in Fig. 9.12.

Reactions: For finding R_A, write equation for B.M. at *B*. Thus

$R_A \times 3 - 12 \times 1.5 = M_B = -2.55$

From which $R_A = 5.15$ kN (↑)

For finding R_B, write down equation for B.M. at *C*, Considering L.H.S.

$\therefore\ 5.15 (7) - 12 (5.5) + R_B \times 4 = M_C = -120,$

from which $R_B = 7.1875$ kN (↑)

For finding R_D, write equation for B.M. at *C*, Considering R.H.S.

$R_D (3) - 4 \times 3 \times 1.5 = M_C = -1.20$

from which $R_D = 5.6$ kN (↑)

Lastly, for finding R_C, write equation for B.M. at *B*, Considreing R.H.S.

$\therefore\ 5.6 \times 7 + R_C (4) - 4 \times 3 \times 5.5 = M_B = -2.55,$

From which $R_C = 6.0625$ kN (↑)

Check

Total load = $12 + 4 \times 3 = 24$ kN

Total reaction = $5.15 + 7.1875 + 5.6 + 6.0625 = 24$ kN

S.F. diagram: The S.F. diagram is shown in Fig. 9.12 (d).

Chapter 10

Strain Energy, Resistance and Impact of Loading

Problem 10.1: Using Costigliano's first theorem find the vertical and horizontal displacement of the point C. Neglect arial deformations.

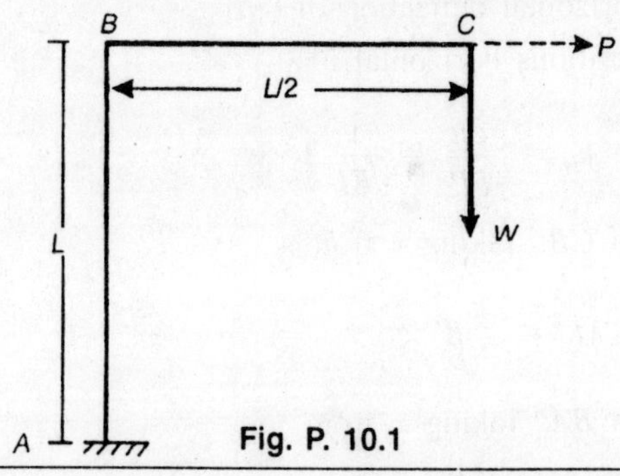

Fig. P. 10.1

Solution:

For

Vertical deflection of C

$$y_v = \frac{\partial u}{\partial w} = \frac{1}{E\,I} \int M_x \times \frac{\partial M_x}{\partial w}\, dx \qquad \ldots(A)$$

For portion CB: Taking x from C,

$$M_x = -wx; \quad \frac{\partial M_x}{\partial w} = x$$

For portion *BA*: Taking *x* from *B*,

$$M_x = -\frac{WL}{2};\ \frac{\partial M_x}{\partial w} = -\frac{L}{2}$$

Substituting these value in above equation (A)

$$\therefore \quad y_v = \frac{1}{EI}\left[\int_0^{\frac{L}{2}} W \times x \times x\, dx + \int_0^{L} \frac{WL}{2} \times \frac{L}{2} dx\right]$$

$$= \frac{1}{EI}\left[\frac{wx^3}{3}\right]_0^{\frac{L}{2}} + \frac{1}{EI}\left[\frac{WL^2 x}{4}\right]_0^{L}$$

$$= \frac{1}{EI}\left[\frac{WL^3}{24} + \frac{WL^3}{4}\right] = \frac{7\,WL^3}{24\,EI} \ (\downarrow)$$

<ii> for horizontal deflection of *C*,

Apply a fictitious horizontal load *P* at *C*. Then,

$$y_H = \frac{\partial u}{\partial P} = \frac{1}{EI}\int M_x \times \frac{\partial M_x}{\partial P}\, dx \qquad \ldots(B)$$

For portion *CB*: Taking *x* from *C*,

$$M_x = -W \times x;\ \frac{\partial M_x}{\partial P} = 0$$

For portion *BA*: Taking *x* from *B*

$$M_x = -\frac{WL}{2} - P \times x;$$

$$\frac{\partial M_x}{\partial P} = -x$$

Substituting these value in eq. (B) above, then we get

$$y_H = \frac{1}{EI}\left[\int_0^{L/2} (-w \times x)(0)dx + \int_0^{L}\left(\frac{WL}{2} + P_x\right)x dx\right]$$

Putting $P = 0$ before integrating

$$\therefore \quad y_H = \frac{1}{EI}\int_0^{L} \frac{WL}{2}\, x\, dx$$

$$= \frac{1}{EI}\frac{WL}{2}\left[\frac{x^2}{2}\right]_0^L$$

$$= \frac{WL^3}{4\,EI} \quad (\rightarrow)$$

Problem 10.2: Calculate the central deflection

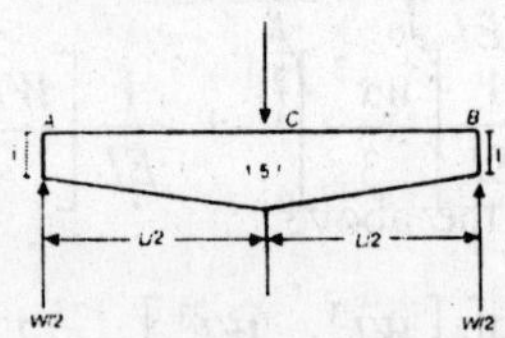

Fig. P. 10.2 (a)

Solution:

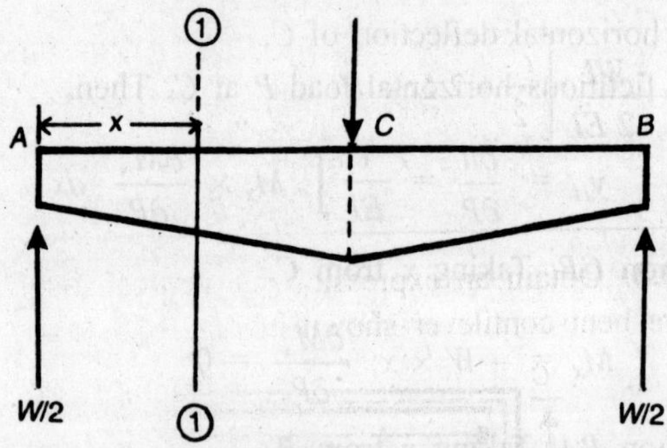

Fig. P. 10.2 (b)

The central deflection is given by

$$y_C = \frac{\partial u}{\partial w} = \frac{1}{E}\int_0^L \frac{M_x}{I_x}\frac{\partial M_x}{\partial w}\,dx$$

$$= \frac{2}{E}\int_0^{L/2} \frac{M_x}{I_x}\frac{\partial M_x}{\partial w}\,dx, \text{ due to symmetry}$$

At any distance x from A,

$$I_x = I + \frac{I}{2} \times \frac{2}{L} \times x = I\left(1 + \frac{x}{L}\right)$$

$$M_x = \frac{w}{2}x \text{ and } \frac{\partial M_x}{\partial w} = \frac{x}{2}$$

$$y_C = \frac{2}{E}\int_0^{L/2} \frac{1}{I\left(1+\frac{x}{L}\right)} \times \frac{w}{2} x \times \frac{x}{2}\, dx$$

$$= \frac{w}{2\,EI}\int_0^{L/2} \frac{x^2}{I\left(1+\frac{x}{L}\right)}\, dx$$

Putting $x + L = t$ in the above integral and simplying, we get

$$y_C = \frac{WL}{2\,EI}\int_L^{3L/2}\left[t - 2L + \frac{L^2}{t}\right] dt$$

$$= \frac{WL}{2\,EI}\left[\frac{t^2}{2} - 2\,Lt + L^2 \log_e t\right]_L^{3L/2} = 0.015\,\frac{WL^3}{EI}$$

Problem 10.3: Obtain an expression for vertical displacement of point A in the bent contilever shown in Fig. 10.3 (a).

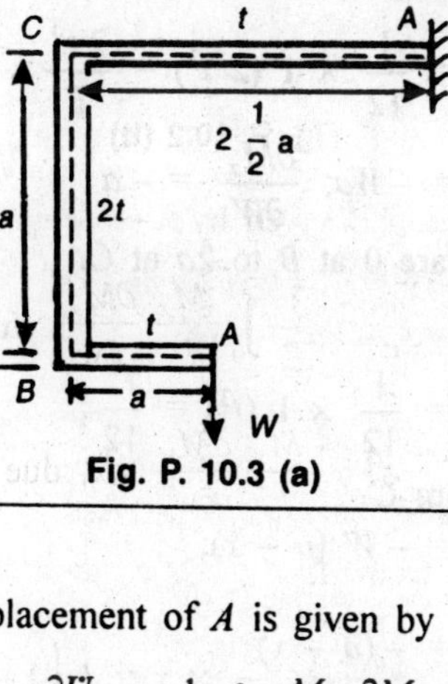

Fig. P. 10.3 (a)

Solution:

The vertical displacement of A is given by

$$y_A = \frac{\partial U}{\partial W} = \frac{1}{E}\int \frac{M_x}{I_x}\frac{\partial M_x}{\partial W}\, dx \qquad \ldots(1)$$

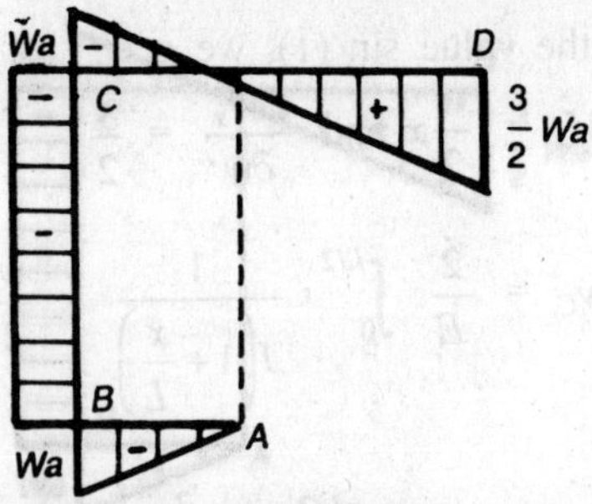

Fig. P. 10.3 (b)

Where the integration is carried over the whole frame. A bending moment is designated negative if it causes convexity to the inside of the frame. The B.M.D. for the whole frame is shown in Fig. 10.3 (b). Assume width of the frame as unity.

(i) For *AB*

$$I = \frac{1}{12} \times 1 \times t^3 = \frac{t^3}{12}$$

Measuring x from A,

$$M_x = -W \times x; \quad \frac{\partial M_x}{\partial W} = -x$$

The limits of x are 0 at A to a at B.

(ii) For *BC*:

$$I = \frac{1}{12} \times 1\,(2\,t^3) = \frac{2\,t^3}{3};$$

$$M_x = -Wa; \quad \frac{\partial M_x}{\partial W} = -a$$

The limits of x are 0 at B to $2a$ at C.

(iii) For *CD*:

$$I = \frac{1}{12} \times 1\,(t^3) = \frac{t^3}{12};$$

Measuring x from C,

$$M_x = -W\,(a - x);$$

$$\frac{\partial M_x}{\partial W} = -(a - x)$$

The limits of x are 0 at C to $5a/2$ at D.

Substituting the value sin (1), we get

$$y_A = \frac{1}{E} \times$$

$$\left[\int_0^a W\,x(x)\frac{12}{t^3}dx + \int_0^{2a} W\,a(a)\frac{3\,dx}{2\,t^3} + \int_0^{\frac{5a}{2}} W(a-x)\,(a-x)\frac{12}{t^3}dx\right]$$

$$= \frac{1}{E\,t^3} \times$$

$$\left[\left(4\,W\,x^3\right)_0^a + \left(\frac{3}{2}W\,a^2 x\right)_0^{2a} + 12\,W\left(a^2 x + \frac{x^3}{3} - \frac{2\,a\,x^2}{2}\right)_0^{\frac{5a}{2}}\right]$$

$$= \frac{1}{E\,t^3} \times$$

$$\left[4\,W\,a^3 + 3\,W\,a^3 + 12\,W\left(\frac{5}{2}a^3 + \frac{125}{24}a^3 - \frac{25}{4}a^3\right)\right] = \frac{24.5\,W\,a^3}{E\,t^3}$$

Problem 10.4: For a given cantilever of rectangular cross-section, length L, depth d and carrying a concentrated load at the free end, show that

$$\frac{\text{Deflection due to shearing strain}}{\text{Bending deflection}} = \text{constant} \times \left(\frac{d}{L}\right)^2$$

and find the value of the constant for a steel cantilever. Hence find the least value of (L/d) if the deflection due to shearing strain is not to exceed 1% of the total deflection.

Take $E = 2 \times 10^5\ \text{N/mm}^2$ and $N = 0.765 \times 10^5\ \text{N/mm}^2$

Solution:

At any layer distant y from the N.A., shear stress is given by

$$q = \frac{F}{2\,I}\left(\frac{d^2}{4} - y^2\right), \text{ where } F = W \text{ and } I = \frac{1}{12}\ bd^3$$

$$\therefore \qquad q = \frac{6W}{bd^3}\left(\frac{d^2}{4} - y^2\right)$$

The total strain energy stored is given below

$$U = \int_0^L \int_{-\frac{d}{2}}^{+\frac{d}{2}} \frac{b}{2N}\left\{\frac{6W}{b\,d^3}\left(\frac{d^2}{4} - y^2\right)\right\} dx\ dy$$

$$= \frac{b}{2N}\int_0^L \int_{-\frac{d}{2}}^{+\frac{d}{2}} \frac{36W^2}{b^2 d^6}\left(\frac{d^4}{16} - \frac{d^2y^2}{2} + y^4\right) dx\ dy$$

$$= \frac{18W^2}{Nb^2d^6}\int_0^L \left(\frac{d^5}{16} - \frac{d^2}{6}\times\frac{d^3}{4} + \frac{1}{5}\frac{d^5}{16}\right) dx$$

$$= \frac{18W^2}{Nb^2d^6} \times \frac{d^5}{30} \times L = \frac{3W^2L}{5Nbd} \qquad \ldots(1)$$

$$\text{Also, work done} = \frac{1}{2}Wy_s \qquad \ldots(2)$$

Equating the two, we get

$$\frac{1}{2}Wy_s = \frac{3}{5}\frac{W^2L}{Nbd}$$

$$\text{From which } y_s = \frac{6WL}{5Nbd}$$

Deflection due to bending

$$y_b = \frac{WL^3}{3EI} = \frac{4WL^3}{Ebd^3} \qquad \ldots(3)$$

$$\therefore \qquad \frac{y_s}{y_b} = \frac{6WL}{5Nbd} \times \frac{bd^3E}{4WL^3}$$

$$= \frac{3}{10}\frac{E}{N}\left(\frac{d}{L}\right)^2 = k\left(\frac{d}{L}\right)^2$$

Hence $k = \dfrac{3}{10}\dfrac{E}{N}$ (ans.)

Substituting numerical values,

$$k = \frac{3}{10} \times \frac{2 \times 10^5}{0.765 \times 10^5} = 0.7843$$

Again $y_s = 0.01\ (y_s + y_b)$ ...(Given)

or $\dfrac{y_s}{y_b} = \dfrac{1}{99} = \left(\dfrac{d}{L}\right)^2 k$

$$\therefore \quad \left(\frac{d}{L}\right)^2 = \frac{1}{99k} = \frac{1}{99 \times 0.7843}$$

From which $\dfrac{L}{d} = [99 \times 0.7843]^{1/2} = 8.81$

Problem 10.5: Two bars, each of length L and of different materials are each subjected to the same tensile force P. The first bar has a uniform diameter D and the second bar has a diametre $D/2$ for a length $L/4$ and a diameter D for the remaining length. Compare the strain energies of the two bars if (a) $\dfrac{E_1}{E_2} = \dfrac{4}{7}$ and (b) $E_l = E_2$.

Solution:

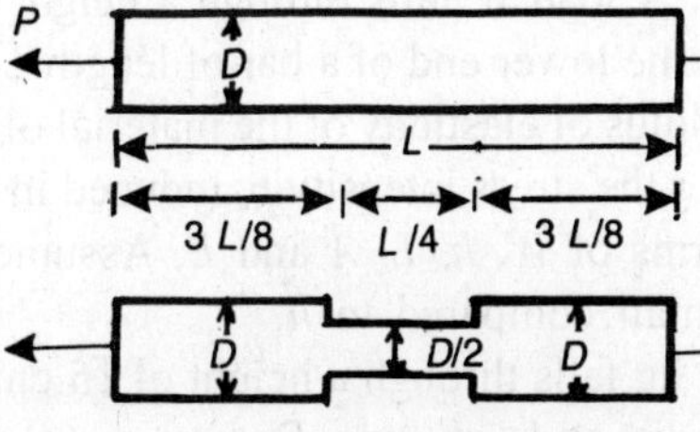

Fig. P. 10.5

Let us use suffix 1 for uniform bar and suffix 2 for non-uniform bar,

$$U_1 = \frac{P^2 L}{2\,A\,E_1} = \frac{P^2 L}{2\left(\dfrac{\pi}{4}D^2\right)E_1} = \frac{2\,P^2\,L}{\pi\,D^2\,E_1}$$

Also
$$U_2 = \frac{P^2}{2\,E_2}\left[\frac{\frac{3}{4}L}{\frac{\pi}{4}D^2}+\frac{\frac{1}{4}L}{\frac{\pi}{4}\left(\frac{D}{2}\right)^2}\right]$$

$$= \frac{P^2}{2\,E_2}\left[\frac{3\,L}{\pi\,D^2}+\frac{4\,L}{\pi\,D^2}\right] = \frac{7\,P^2 L}{2\,\pi D^2 E_2}$$

$$\therefore \quad \frac{U_1}{U_2} = \frac{2\,P^2 L}{\pi D^2 E_1} \div \frac{7\,P^2 L}{2\,\pi D^2 E_2} = \frac{4}{7}\frac{E_2}{E_1}$$

(a) If $\frac{E_1}{E_2} = \frac{4}{7}$ (or $\frac{E_2}{E_1} = \frac{7}{4}$),

$$\frac{U_1}{U_2} = \frac{4}{7} \times \frac{7}{4} = 1$$

(b) If $E_1 = E_2$, $\frac{U_1}{U_2} = \frac{4}{7}$

Problem 10.6: A load W falls through a height h before coming to rest on a stop at the lower end of a bar of length L and cross-sectional area A. The modulus of elasticity of the material of the rod is E. Derive an expression for the stress intensity p, induced in the rod as the result of impact in terms of W, h, L, A and E. Assume that the extension of the rod is small compared to h.

A mass of 10 kg falls through a height of 15 cm and then to stretch a bar 20 mm in dia. and 1 m long. Determine (a) the maximum stress induced in the bar, (b) max. elongation of the bar, (c) the energy stored in the bar when the elongation of the bar is at its maximum value. Take $E = 2 \times 10^5$ N/mm^2.

Solution:

Potential energy lost by weight = Strain energy stored in the bar

Hence, neglecting the extension of the rod, we have

$$Wh = \frac{p^2}{2E} \times V = \frac{p^2 AL}{2E}$$

$$\therefore \quad p = \sqrt{\frac{2WhE}{AL}}$$

(a) Mass = 10 kg

Hence Weight = $Mg = 10 \times 9.81 = 98.1$ N

$$A = \frac{\pi}{4}(20)^2 = 314.16 \text{ mm}^2;$$

$$L = 1000 \text{ mm}$$

$$h = 150 \text{ mm}$$

$$\therefore \quad p = \sqrt{\frac{2 \times 98.1 \times 150 \times 2 \times 10^5}{314.16 \times 1000}}$$

$$= 136.88 \text{ N/mm}^2$$

(b) $$\Delta = \frac{pL}{E} = \frac{136.88 \times 1000}{2 \times 10^5} = 0.6844 \text{ mm}$$

(c) Strain Energy = $U = P \times E$

lost $= W(h + \Delta)$

$= 98.1\ (150 + 0.6844) = 14782$ N.mm

$= 14.782$ N.m $= 14.782\ J$

Problem 10.7: Find the strain energy stored in a steel bar 50 cm long and 3 cm × 1 cm in cross-section, when it is subjected simultaneously to an axial pull of 50 kN and compressive stress of 100 N/mm² on its narrow edge.

Take $E = 2 \times 10^5$ N/mm² and $\frac{1}{m} = 0.28$.

Solution:

$$U = \frac{1}{2E}\left[\sigma_1^2 + \sigma_2^2 - \frac{2\sigma_1\sigma_2}{m}\right] V$$

Here $\sigma_1 = \dfrac{50 \times 10^3}{30 \times 10} = 166.67$ N/mm² (tensile)

Also $\sigma_2 = -100$ N/mm² (i.e. comp.)

$\therefore \quad U = \dfrac{1}{2 \times 2 \times 10^5} \times$

$[(166.67)^2 + (-100)^2 + 2 \times 166.67 \times 100 \times 0.28]\ (500 \times 30 \times 10)$

$= 17667$ N.mm $= 17.667$ N.m $= 17.667$ J

Problem 10.8: A member formed by connecting a steel bar to an aluminium bar is shown in figure. Assuming that the bars are prevented from buckling sidewise, calculate the magnitude of the force P that will cause the total length of the member to decrease by 0.20 m. Take $E_s = 2 \times 10^5$ N/mm² and $E_a = 0.7 \times 10^5$ N/mm². Also, compute the work done by force P.

Solution:

Area of steel bar,

$$A_s = \frac{\pi}{4}(20)^2$$

$$= 314.16 \text{ mm}^2$$

Area of aluminium bar,

$$A_a = \frac{\pi}{4}(50)^2$$

$$= 1963.5 \text{ mm}^2 = 6A_s$$

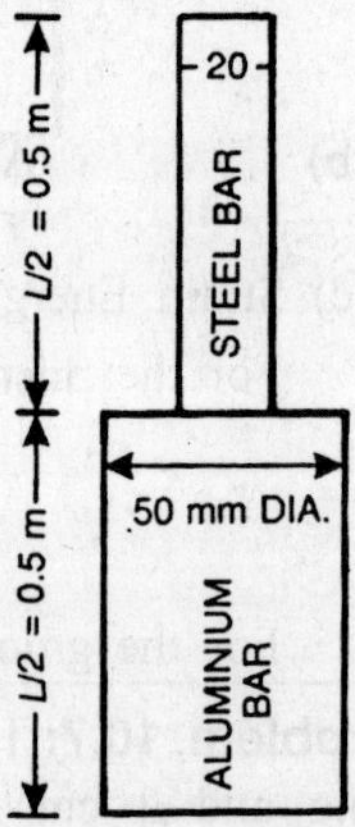

Fig. P. 10.8

Now $\dfrac{PL_s}{A_sE_s} + \dfrac{PL_a}{A_aE_a} = \Delta$

or $P\left[\dfrac{500}{314.16 \times 2 \times 10^5} + \dfrac{500}{1963.5 \times 0.7 \times 10^5}\right] = 0.2$

or $P\,[1.15955 \times 10^{-5}] = 0.2$ from which $P = 17248$ N

Total work done by $P = \dfrac{1}{2}$ load × deflection

$$= \frac{1}{2} \times 17248 \times 0.2 = 1724.8 \text{ N.mm}$$

$$= 1.725 \text{ N.m} = 1.725 \text{ J}$$

Problem 10.9: Compare the amount of strain energy in three circular bars shown in Fig. 10.9, assuming uniform distribution of stresses over the cross-section of the bars. Comment on the result, taking a large value of *n*.

Solution:

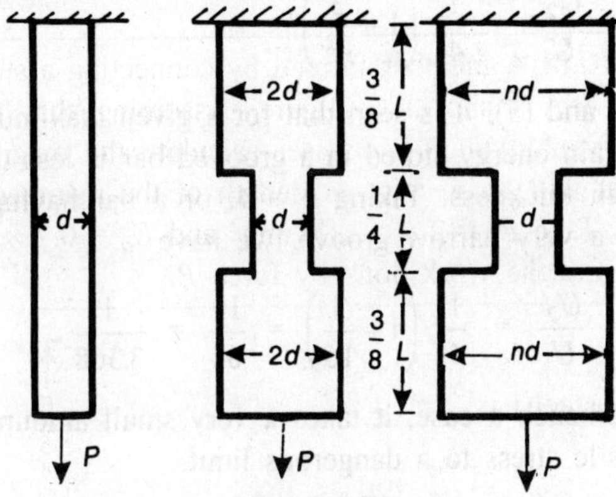

Fig. P. 10.9

For the prismatic bar of Fig. 10.9 (a)

$$U = \frac{P^2 L}{2 AE} \quad \ldots(1)$$

For the grooved bar of Fig. 10.9 (b),

$$U_1 = \frac{P^2(L/4)}{2 A E} + \frac{P^2(3L/4)}{2(4 A)E} = \frac{7}{16} \frac{P^2 L}{2 AE} \quad \ldots(2)$$

For the grooved bar of Fig. 10.9 (c),
Area of grooved portion = A and area of other portions

$$= \frac{\pi}{4} (nd)^2 = n^2 A$$

$$\therefore \quad U_2 = \frac{P^2(L/4)}{2 AE} + \frac{P^2(3L/4)}{2(n^2 A)E}$$

$$= \frac{P^2 L}{2\,AE}\left(\frac{1}{4}+\frac{3}{4n^2}\right)$$

or $$U_2 = \frac{1}{4}\frac{P^2 L}{2\,AE}\left(1+\frac{3}{n^2}\right) \quad \ldots(3)$$

Hence $$\frac{U_1}{U} = \frac{7}{16} \quad \ldots(4)$$

and $$\frac{U_2}{U} = \frac{1}{4}\left(1+\frac{3}{n^2}\right) \quad \ldots(5)$$

From (4) and (5), it is seen that for a given maximum stress, the quantity of strain energy stored in a grooved bar is less than that of a bar of uniform thickness. Taking $n = 4$, or a bar having large outer diameter and a very narrow groove, we have

$$\frac{U_2}{U} = \frac{1}{4}\left(1+\frac{3}{16}\right) = \frac{19}{64} = \frac{1}{3.368}$$

Hence for such a case, it takes a very small amount of work to bring the tensile stress to a dangerous limit.

Problem 10.10: A sliding block, having a mass of 40 kg, slides over a 40 mm dia 1200 mm long hoirzontal rod at a velocity of 2 m/s, is stopped by its impact, with a rigid collar provided at the end of the rod as shown in Fig. 10.10. Ignoring friction and bending of the bar, find the instantaneous stress induced in the bar. Take $E = 2 \times 10^5$ N/mm^2.

Solution:

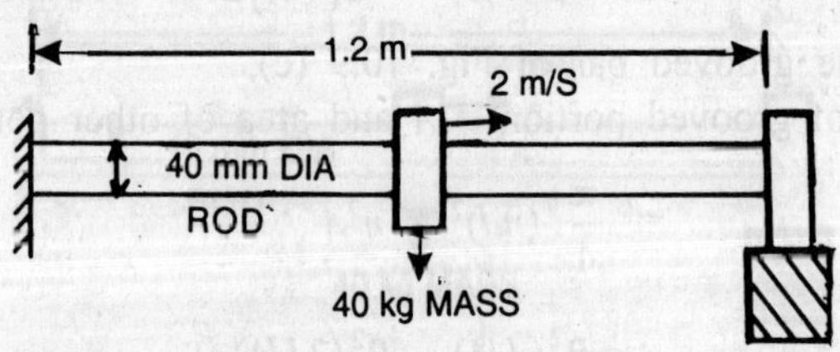

Fig. P. 10.10

Mass of sliding weight = M = 40 kg

$\therefore$ Weight $W = Mg = 40 \times 9.81 = 392.4$ N

$$\text{Area of bar} = \frac{\pi}{4}(40)^2 = 1256.6 \text{ mm}^2$$

Let the instantaneous stress induced in the bar = p

∴ Strain energy stored

$$= \frac{p^2}{2E} \times V = \frac{p^2}{2E} \times AL \qquad \text{...(a)}$$

$$\text{K.E. of sliding block} = \frac{Wv^2}{2g} \qquad \text{...(b)}$$

Equating (a) and (b), we have

$$\frac{p^2}{2E} AL = W\frac{v^2}{2g}$$

or

$$p^2 = \frac{2EWv^2}{2gAL}$$

$$= \frac{2 \times 2 \times 10^5 \times 392.4(2000)^2}{2 \times 9810 \times 1256.6 \times 1200} = 21221$$

From which p = 145.68 N/mm²

Problem 10.11: Find an expression for strain energy U stored in the bar shown in Fig. 10.11 cross-sectional area of the bar is A and modulus of elasticity is E.

Solution:

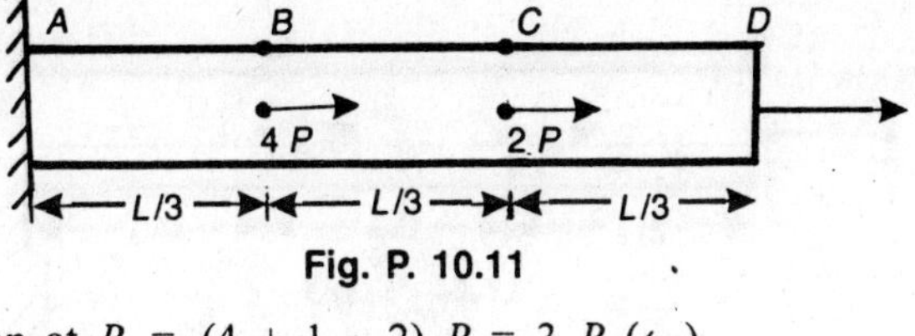

Fig. P. 10.11

Reaction at B = $(4 + 1 - 2)P = 3P$ (←)

∴ Force in AB = $3P$ (tension)

Force in BC = P (comp.)

Force in CD = P (tension)

Now strain energy in the bar = $\Sigma \dfrac{p^2 L}{2 AE}$

$$\therefore \qquad U = \frac{(3P)^2 (L/3)}{2AE} + \frac{(-P)^2 (L/3)}{2AE} + \frac{(P)^2 (L/3)}{2AE}$$

$$\text{or} \qquad U = \frac{11}{6} \frac{P^2 L}{AE}$$

Chapter 11

Torsion

Problem 11.1: What is the minimum diameter of a solid steel shaft that will not twist through more than 3° in a 6 m length when subjected to a torque of 12 kN.m? What maximum shearing stress is developed? Use $G = 83$ GPa

Solution: We know that

Given: $\theta = 3° = 3 \times \dfrac{\pi}{180}$ radian $\quad [\because 180° = \pi \text{ radian}]$

$= 0.0523$

$T = 12$ kN.m

$= 12 \times 1000$ N.m $= 12 \times 10^6$ N.mm

$G = 83$ GPa $= 83 \times 10^3$ N/mm²

$L = 6$ m $= 6 \times 1000$ mm

$$\therefore \quad \frac{\pi}{32} d^4 = \frac{12 \times 10^6 \times 6 \times 1000}{0.0523 \times 83 \times 10^3}$$

$$\therefore \quad d^4 = \frac{12 \times 10^6 \times 6 \times 32}{0.0523 \times 83 \times \pi}$$

$$\therefore \quad d = 114.008 \text{ mm}$$

$$\text{Max . } \tau = \frac{16T}{\pi d^3} = \frac{16 \times 12 \times 10^6}{\pi \times (114.008)^3}$$

$$= 41.242 \approx 41.3 \text{ N/mm}^2 \approx 41.3 \text{ MPa}$$

Problem 11.2: A solid steel shaft 5 m long is stressed to 80 MPa when twisted through 4°. Using $G = 83$ GPa, compute the shaft diameter. What power can be transmitted by the shaft at 20 Hz?

Solution: Given:

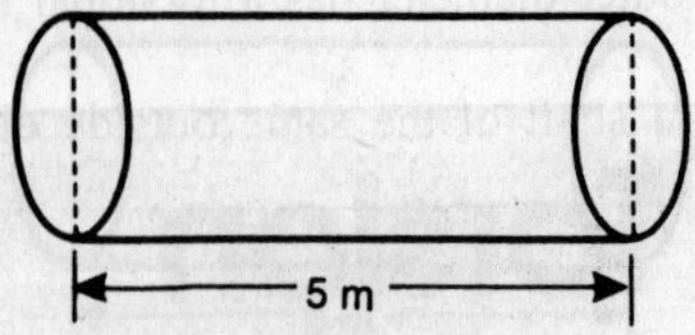

Fig. P. 11.2

$$\text{Length} = 5 \text{ m} = 5 \times 1000 \text{ mm}$$

$$\tau_{max.} = 80 \text{ MPa} = 80 \times 10^6 \text{ N/m}^2 = 80 \text{ N/mm}^2$$

$$\theta = 4° = \frac{4 \times \pi}{180} = 0.0698 \text{ radian}$$

$$G = 83 \text{ GPa} = 83 \times 10^9 \text{ N/m}^2$$

$$= 83 \times 10^3 \text{ N/mm}^2$$

$$n = f = 20 \text{ Hz} = \text{rotation per second}$$

We know that,

$$\tau_{max} = \frac{16\,T}{\pi d^3} \qquad \ldots(1)$$

$$\frac{T}{J} = \frac{N\theta}{l} = \frac{\tau}{R}$$

$$\therefore \quad R = \frac{\tau \times l}{N\,\theta}$$

$$= \frac{80 \times 5 \times 1000}{83 \times 10^3 \times 0.0698} = 69.031 \text{ mm}$$

$$\therefore \quad \text{Diameter} = 2\,R = 138.062 \approx 138 \text{ mm}$$

$\therefore$ Power transmitted to the shaft

$$= 2\,\pi\,\eta T \qquad \ldots(2)$$

From equation (1)

$$T = \frac{\tau_{max.} \times \pi d^3}{16} = \frac{80 \times \pi \times (138)^3}{16} = 41.28 \text{ kN.m}$$

$$\therefore \quad P = 2\,\pi\,\eta T$$

$$= 2\,\pi \times 20 \times 41.28 \times 10^3$$

$$= 5.1876 \text{ MW} \approx 5.19 \text{ MW}$$

Problem 11.3: Show that a hollow circular shaft whose inner diameter is half the outer diameter has a torsional strength equal to $\frac{15}{16}$ of that of a solid shaft of the same outside diameter.

Solution:

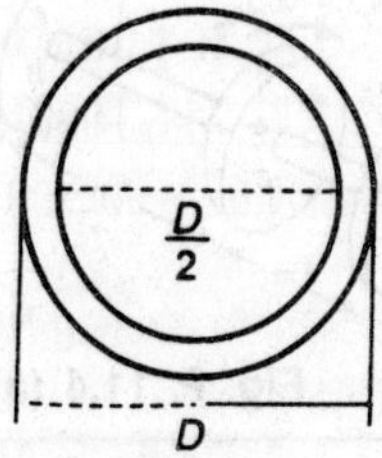

Fig. P. 11.3

Torsional strength of hollow shaft is

$$T = fs \times \frac{T}{R} = fs \times \frac{\pi}{16}\left[\frac{D^4 - d^4}{D}\right] \qquad ...(1)$$

But according to question, we know $d = \frac{D}{2}$

$\therefore$ Putting this value of $d = \frac{D}{2}$ in equation (1)

$$\therefore T = fs \times \frac{\pi}{16}\left[\frac{D^4 - \frac{D^4}{16}}{D}\right]$$

$$= fs \times \frac{\pi}{16}\left[\frac{16D^4 - D^4}{16D}\right]$$

$$= fs \times \frac{\pi}{16}\,\frac{15\,D^4}{16\,D} = \frac{15}{16}\left[fs \times \pi\frac{D^4}{D}\right] = \frac{15}{16}\,[fs \times \pi\ D^3]$$

$$= \frac{15}{16} \times \text{Torsional strength of solid shaft} \quad \textbf{Proved}$$

Problem 11.4: An aluminum shaft with a constant diameter of 50 mm is loaded by torques applied to gears attached to it as shown in Fig. P. 11.4 (a). Using G = 28 GPa, determine the relative angle of twist of gear D relative to gear A.

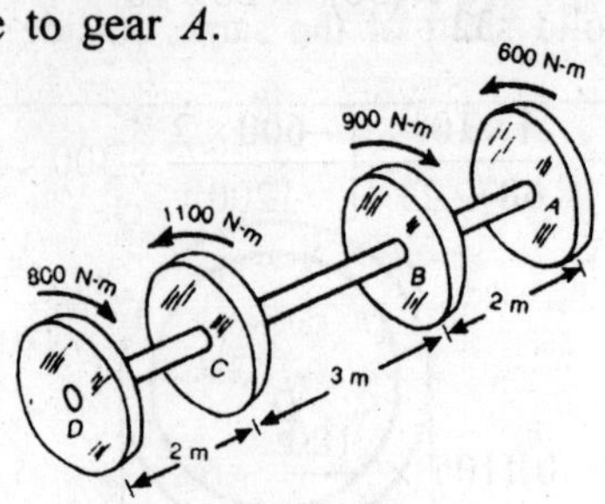

Fig. P. 11.4 (a)

Solution:

(→→) Torque denoted by double headed vector

→→ For clockwise torque of section,

←← For anticlockwise torque at section,

∴ Simplified figure

Fig. P. 11.4 (b)

∴ Torque in section AB

$$T_{AB} = -600 \text{ N.m}$$

Torque in section BC

$$T_{BC} = +300 \text{ N.m}$$

$$T_{CD} = -800 \text{ N.m}$$

Total angular deformation is the algebaric sum of the deformation in each segment.

$$\theta A/D = \frac{\Sigma TL}{JG}$$

$$= \frac{-600 \times 10^3 \times 2 \times 10^3}{\frac{\pi}{32} \times (50)^4 \times 28 \times 10^3} + \frac{300 \times 10^3 \times 3 \times 10^3}{\frac{\pi}{32}(50)^4 \times 28 \times 10^3}$$

$$= \frac{800 \times 10^3 \times 2 \times 10^3}{\frac{\pi}{32} \times (50)^4 \times 28 \times 10^3}$$

$$= \frac{1 \times 10^3}{6872.23}\left[\frac{-600 \times 2}{1200} + 300 \times 3 \frac{-800 \times 2}{1600}\right]$$

$$[\pi \text{ radian} = 180°] \therefore 1 \text{ radian} = \frac{180}{\pi}$$

$$= 0.1105 \times \frac{180}{\pi} = 6.336 \approx 6.34°$$

Problem 11.5: Determine the maximum torque that can be applied to a hollow circular steel shaft of 100-mm outside diameter and an 80-mm inside diameter without exceeding a shearing stress of 60 MPa or a twist of 0.5 deg/m. Use G = 83 GPa.

Solution: Given:

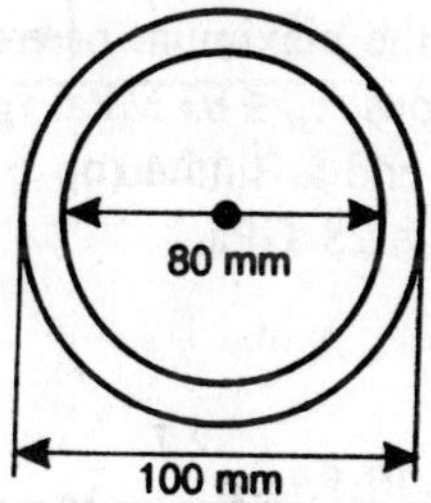

Fig. P. 11.5

$$\tau_{max.} = 60 \text{ MPa} = 60 \times 10^6 \text{ N/m}^2 = 60 \text{ N/mm}^2$$

Twist of 0.5 deg/m

i.e. $$\frac{\theta°}{L} = 0.5° = 0.5 \times \frac{\pi}{180} = 0.00872$$

$$G = 83 \text{ GPa} = 83 \times 10^9 \text{ N/m}^2 = 83 \times 10^3 \text{ N/mm}^2$$

We know that,

$$\frac{T}{J} = \frac{N\theta}{L} = \frac{q}{R}$$

$$\therefore \qquad T = \frac{N\,\theta}{L} \times J = \frac{N\,\theta}{L} \times \frac{\pi}{32}\,[D^4 - d^4]$$

$$= \frac{N\,\theta}{L} \times \frac{\pi}{32}\,[100^4 - 80^4]$$

$$= 83 \times 10^3 \times 0.00872 \times \frac{\pi}{32}\,[100^4 - 80^4]$$

$$= 4195.08 \text{ kN.m}$$

We know that,

$$\tau_{max.} = \frac{16\,TD}{\pi\left[D^4 - d^4\right]}$$

$$T = \frac{\tau_{max.} \times \pi\left[D^4 - d^4\right]}{16\,D}$$

$$= \frac{60 \times \pi\left[100^4 - 80^4\right]}{16 \times 100} = 6.955 \text{ kN.m}$$

$\therefore$ Maximum torque (T) = 4195.08 kN.m

Problem 11.6: A compound shaft consisting of a steel segment and an aluminium segment is acted upon by two torques as shown in Fig. P. 11.6 (a). Determine the maximum permissible value of T subject to the following conditions: $\tau_{st} \leq 83$ MPa, $\tau_{al} \leq 55$ MPa, and the angle of rotation of the free end is limited to 6°. For steel, G = 83 GPa and for aluminium, G = 28 GPa.

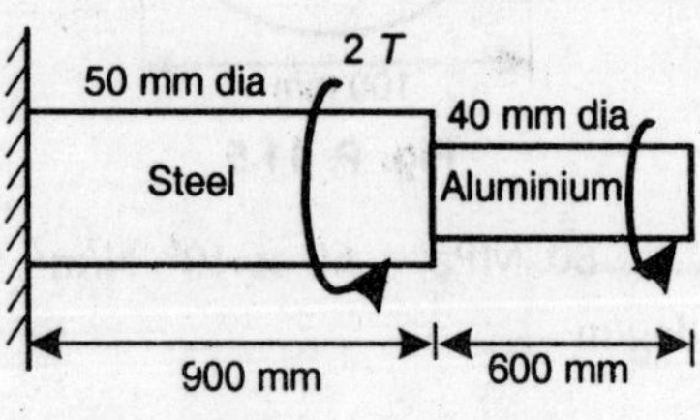

Fig. P. 11.6 (a)

Solution: Given:

Compound shaft

Steel $\qquad \tau_{st} \leq 83$ MPa

$\qquad \leq 83$ N/mm²

$$G_{st} = 83 \text{ GPa} = 83 \times 10^3 \text{ N/mm}^2$$

Aluminium $\tau_{al} \leq 55$ MPa

$$\leq 55 \text{ N/mm}^2$$

$$G_{al} = 28 \text{ GPa} = 28 \times 10^3 \text{ N/mm}^2$$

Angle at rotation of the free end is limited to 6°.

Determine maximum permissible value of T

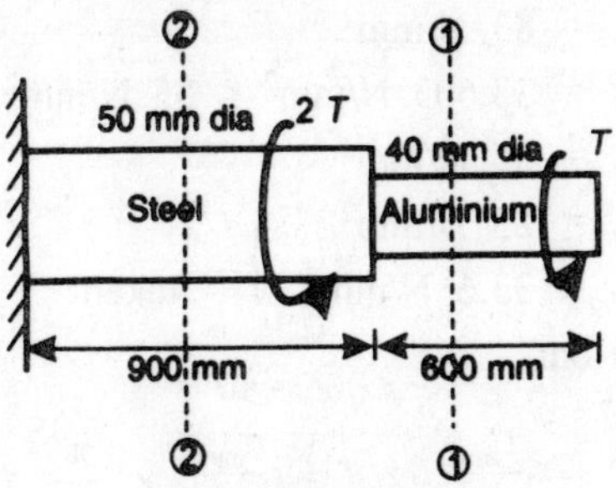

Fig. P. 11.6 (b)

F.B.D.

Section (1)-(2)

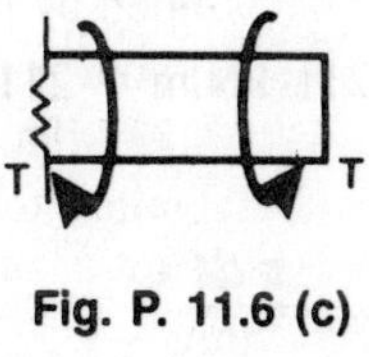

Fig. P. 11.6 (c)

∴ Torque in aluminium bar = T

Section (2)-(2)

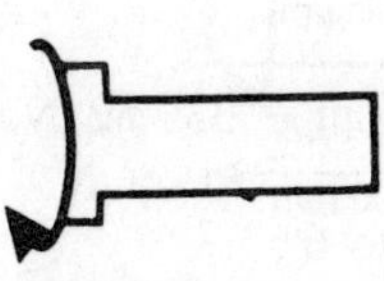

Fig. P. 11.6 (d)

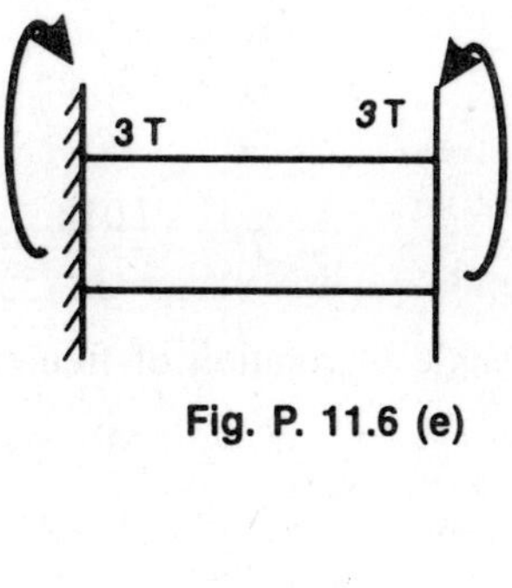

Fig. P. 11.6 (e)

∴ Torque in steel bar = $3T$

∵ Shaft is connected in parallel

$$\therefore \quad \theta_1 = \theta_2$$

$$\frac{\tau_{st} \times L_s}{R_{st} \times N_{st}} = \frac{\tau_{al} \times L_{al}}{R_{al} \times N_{al}}$$

$$\therefore \quad \tau_{st} = \tau_{al} \times \frac{R_{st} \times N_{st}}{R_{al} \times N_{al}} \times \frac{L_{al}}{L_{st}}$$

$$= \tau_{al} \times \frac{25 \times 83 \times 10^3}{20 \times 28 \times 10^3} \times \frac{600}{900}$$

$$\tau_{st} = 2.47\ \tau_{al} \qquad \ldots(1)$$

If $\tau_{st} = 83\ \text{N/mm}^2$

$$\tau_{al} = 33.603\ \text{N/mm}^2 < 55\ \text{N/mm}^2$$

If $\tau_{al} = 55\ \text{N/mm}^2$

$$\therefore \quad \left.\begin{aligned} \tau_{st} &= 83\ \text{N/mm}^2 \\ \tau_{al} &= 33.6\ \text{N/mm}^2 \end{aligned}\right\} \text{taken}$$

For aluminium bar

$$T = \frac{\tau_{max}}{r} \times J = \frac{\tau_{max}}{r} \times \frac{\pi}{32}\ d^4$$

$$= \tau_{max} \times \frac{\pi d^3}{32} = 33.6 \times \pi \times \frac{(40)^3}{32}$$

$$= 0.211\ \text{kN.m} = 211\ \text{N.m}$$

For steel bar

$$3T = \tau_{max}\ \frac{\pi\ d^3}{32}$$

$$= 83 \times \frac{\pi \times (50)^3}{32} = 1.018\ \text{kN.m}$$

$$\therefore \quad T = \frac{1.018}{3} = 0.3395\ \text{kN.m} = 339.52\ \text{N.m}$$

Angle of rotation of free end limited to 6°,

$$\frac{T}{J} = \frac{N\theta}{l} = \frac{q}{r}$$

$$\therefore \quad \theta = \frac{Tl}{NJ} = \frac{211 \times 10^3 \times 600 \times 32}{28 \times 10^3 \times \pi \times (40)^4}$$

$$= 0.01799\ \text{radian} = 1° < 6°$$

Problem 11.7: The compound shaft shown in Fig. P. 11.7 (a) is attached to rigid supports. For the bronze segment *AB*, the diameter is 75 mm. $\tau \leq 60$ MPa, and $G = 35$ GPa. For the steel segment *BC*, the diameter is 50 mm, $\tau \leq 80$ MPa, and $G = 83$ GPa. If $a = 2$ m and $b = 1.5$ m, compute the maximum torque *T* that can be applied.

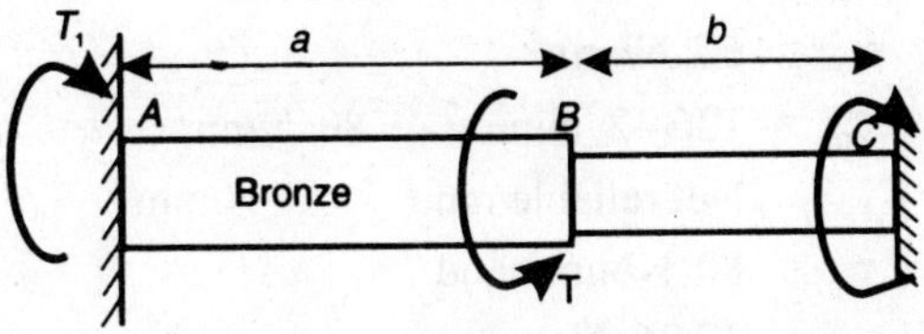

Fig. P. 11.7 (a)

Solution: Given:

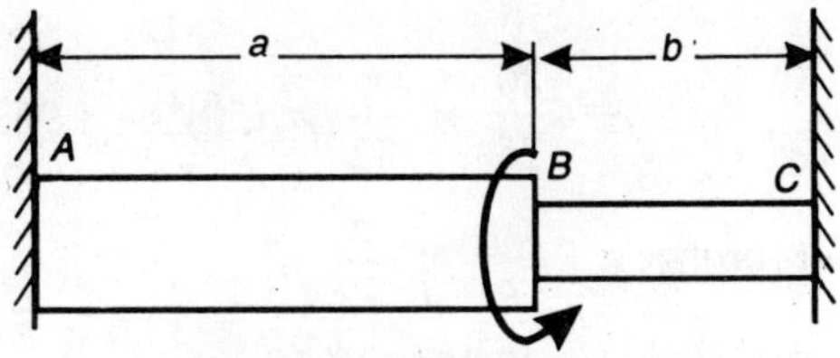

Fig. P. 11.7 (b)

(a) Bronze $\quad D = 75$ mm

$$\tau \leq 60 \text{ MPa} = 60 \text{ N/mm}^2$$
$$G = 35 \text{ GPa} = 35 \times 10^3 \text{ N/mm}^2$$
$$a = 2 \text{ m} = 2000 \text{ mm}$$

(b) Steel $\quad D = 50$ mm

$$\tau \leq 80 \text{ MPa} = 80 \text{ N/mm}^2$$
$$G = 83 \text{ Gpa} = 83 \times 10^3 \text{ N/mm}^2$$
$$b = 1.5 \text{ m} = 1500 \text{ mm}$$

We know $\dfrac{T}{J} = \dfrac{N\theta}{L} = \dfrac{\tau}{R}$

$\because$ Shaft is in parallel so,

$$T = T_1 + T_2 \qquad \ldots(a)$$

and

$$\theta_1 = \theta_2 \qquad \ldots(b)$$

From equation (b),

$$\frac{\tau_{st} \times L_{st}}{N_{st} R_{st}} = \frac{\tau_{br} \times L_{br}}{N_{br} \times R_{br}}$$

$$\therefore \quad \frac{\tau_{st} \times 1.5 \times 1000}{83 \times 10^3 \times 25} = \frac{\tau_{br} \times 2 \times 1000}{35 \times 10^3 \times 37.5}$$

$$\tau_{st} = \frac{\tau_{br} \times 2 \times 83 \times 25}{35 \times 37.5 \times 1.5}$$

$$\tau_{st} = 2.107\ \tau_{br} \qquad \ldots(c)$$

$\therefore$ $\tau_{st} = 80$ N/mm^2

and $\tau_{br} = 37.96$ N/mm^2 < 60 N/mm^2

If $\tau_{br} = 60$ N/mm^2

Then $\tau_{st} = 126.42$ N/mm^2 > 80 N/mm^2

Not reliable

So, $\tau_{st} = 80$ N/mm^2 and

$\tau_{br} = 37.96$ N/mm^2

$$\therefore T_2 \text{ (For steel)} = \frac{\tau_{st}}{R} \times \frac{\pi}{32} d^4$$

$$= \frac{80}{25} \times \frac{\pi}{32} \times (50)^4 = 1.963 \text{ kN.m}$$

$$T_1 \text{ (For bronze)} = \frac{\tau_{br}}{R} \times \frac{\pi}{32} d^4$$

$$= \frac{37.96}{37.5} \times \frac{\pi}{32} \times (75)^4 = 3.144 \text{ kN.m}$$

$$\therefore \quad T = T_1 + T_2$$

$$= 3.144 + 1.963 = 5.107 \text{ kN.m}$$

Problem 11.8: In problem 11.7 determine the ratio of lengths b/a so that each material will be stressed to its permissible limit. What torque T is required?

Solution:

If $\tau_{br} = 60$ N/mm^2

$\tau_{st} = 80$ N/mm^2

$$\text{then,} \quad \frac{b}{a} = \frac{\tau_{br}}{\tau_{st}} \times \frac{N_{st} \times R_{st}}{N_{br} \times R_{br}}$$

$$= \frac{60}{80} \times \frac{83 \times 10^3 \times 25}{35 \times 10^3 \times 375}$$

$$= 1.1857 \approx 1.19$$

$$\therefore \quad T = \frac{\tau_{br}}{R} \times \frac{\pi}{32} \times d^4 + \frac{\tau_{st}}{R} \times \frac{\pi}{32} d^4$$

$$= \frac{60}{37.5} \times \frac{\pi}{32} \times (75)^4 + \frac{80}{25} \times \frac{\pi}{32} \times (50)^4$$

$$= 6.933 \text{ kN.m}$$

Problem 11.9: A torque T is applied, as shown in Fig. P. 11.9, to a solid shaft with built-in ends. Prove that the resisting torques at the walls are $T_1 = Tb/L$ and $T_2 = Ta/L$. How would these values be changed if the shaft were hollow?

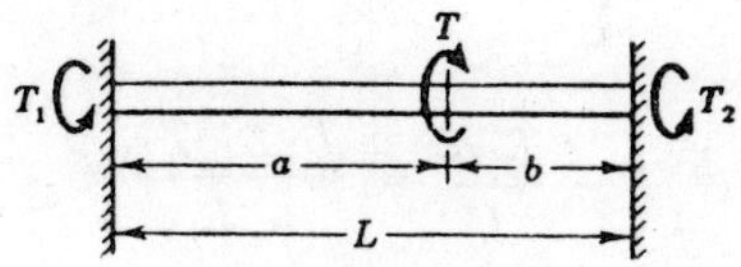

Fig. P. 11.9

Solution:

∴ Shaft connected in parallel so,

Applied torque = sum of resisting torque

i.e. $T = T_1 + T_2$ ...(1)

and $\theta = \theta_1 = \theta_2$

We know that,

$$\frac{T}{J} = \frac{N\theta}{L} = \frac{q}{\gamma}$$

$$T = \frac{N\theta}{L} \times J$$

$$\therefore \quad \theta = \frac{TL}{NJ}$$

$$\because \quad \theta_1 = \theta_2$$

$$\therefore \quad \frac{T_1 L_1}{N_1 J_1} = \frac{T_2 L_2}{N_2 J_2}$$

$$\because \quad N_1 = N_2 J_1 = J_2$$

$$L_1 = aL_2 = b$$

$$\therefore \quad \frac{T_1 a}{N_1 J_1} = \frac{T_2 a}{N_1 J_1}$$

$$\therefore \quad T_1 = \frac{b}{a} T_2 \qquad \text{...(2)}$$

Putting this value in eq. (1) we get,

$$T = \frac{b}{a} T_2 + T_2 = \left(\frac{b}{a} + 1\right) T_2 = \left(\frac{a+b}{a}\right) T_2$$

$$\therefore \quad T_2 = \frac{T \cdot a}{L}$$

$$\therefore \quad T_1 = T - \frac{T \cdot a}{L} = \frac{T \cdot L - T \cdot a}{L}$$

$$= \frac{T[L-a]}{L} = \frac{T \cdot b}{L} \quad \textbf{Proved}$$

Problem 11.10: A solid steel shaft is loaded as shown in Fig. P. 11.10 (a). Using G = 83 GPa, determine the required diameter of the shaft if the shearing stress is limited to 60 MPa and the angle of rotation at the free end is not to exceed 4 deg.

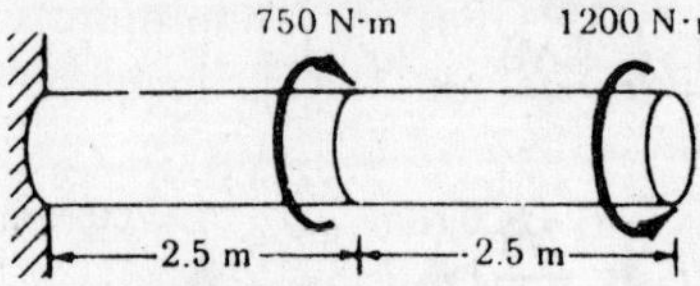

Fig. P. 11.10 (a)

Solution: Given

$$G = 83 \text{ GPa}$$

Shearing stress is limited to 60 MPa, i.e.

$$\tau_{max} = 60 \text{ MPa} = 60 \text{ N/mm}^2$$

The angle of rotation of the free end is not to exceed 4°.

$$4° = 4 \times \frac{\pi}{180} \text{ radian} = 0.06981317 \text{ radian}$$

We know that,

$$\frac{T}{J} = \frac{N\theta}{L} = \frac{q}{R}$$

F.B.D.

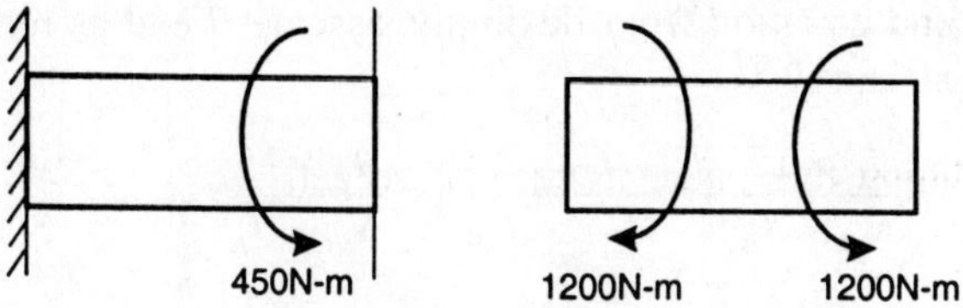

Fig. P. 11.10 (b)

$\therefore$ θ at the free end not exceed 4°

$$\frac{TL}{NJ} = \theta = 0.06981$$

$$\therefore \quad \frac{T\,L}{N \times 0.06981} = \frac{\pi}{32} \times d^4$$

$$\therefore \quad d^4 = \frac{32\,T\,L}{N\,\pi \times 0.06981} = \frac{32 \times 1200 \times 2.5 \times 1000 \times 1000}{83 \times 10^3 \times \pi \times 0.06981}$$

$$\therefore \quad d = 47.92 \text{ mm}$$

$\therefore$ Diameter of the shaft = 47.92 mm

Problem 11.11: A shaft composed of segments *AC, CD* and *DB* is fastened to rigid supports and loaded as shown in Fig. P. 11.11. For bronze, G = 35 GPa; for aluminium, G = 28 GPa, and for steel, G = 83 GPa. Determine the maximum shearing stress developed in each segment.

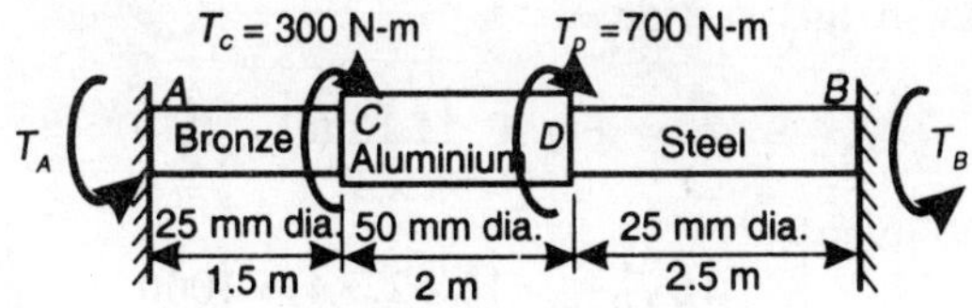

Fig. P. 11.11

Solution:

$$G_B = 35 \text{ GPa} = 35 \times 10^3 \text{ N/mm}^2$$

$$G_{al} = 28 \text{ GPa} = 28 \times 10^3 \text{ N/mm}^2$$

$$G_{steel} = 83 \text{ GPa} = 83 \times 10^3 \text{ N/mm}^2$$

This arrangement is statically indeterminate so two equation required to solve this problem,

First equation from equilibrium,

$$T_A + T_B = 1000 \quad ...(1)$$

Second equation from flexibility, assume B end as redundat. Then rotation at end B is,

$$\theta_B = \frac{T_C \times 1.5}{N_B \times J_B} + \frac{T_D \times 2}{N_{al} \times J_{al}} - \frac{T_B \times 1.5}{N_B \times J_B}$$

$$- \frac{T_B \times 2}{N_{al} \times J_{al}} - \frac{T_B \times 2.5}{N_{st} \times J_{steel}}$$

$$= \frac{300 \times 1000 \times 1.5 \times 1000 \times 32}{35 \times 10^3 \times \pi \times (25)^4} + \frac{700 \times 2 \times 1000 \times 1000 \times 32}{28 \times 10^3 \times \pi \times (50)^4}$$

$$- T_B \left\{ \left[\frac{1.5 \times 1000 \times 32}{35 \times 1000 \times \pi \times (25)^4} \right] + \frac{2 \times 1000 \times 32}{28 \times 10^3 \times \pi \times (50)^4} \right.$$

$$\left. + \frac{2.5 \times 1000 \times 32}{83 \times 1000 \times \pi \times (25)^4} \right\}$$

$$0 = 0.335 + 0.081 - T_B\,[0.000001862]$$

$$\therefore \quad T_B = 223.41 \text{ N-m}$$

$$\therefore \quad T_A = 1000 - 223.41 = 776.59 \text{ N.m}$$

$\therefore$ Torque in steel section,

$$T_B = 223.41 \text{ N.m}$$

$\therefore$ Stress in this section

$$= \frac{T_B}{J} \times R = \frac{T_B \times 32}{\pi \times d^4} \times \frac{d}{2}$$

$$= \frac{16\,T_B}{\pi\,d^3} - \frac{16 \times 223.41 \times 1000}{\pi \times (25)^3} - 72.82 \text{ N/mm}^2$$

Stress in Aluminium section,

$$T = 476.59 \text{ N.m}$$

$$\therefore \quad \tau = \frac{16 \times 476.59 \times 1000}{\pi \times (50)^3} = 19.41 \text{ N/mm}^2$$

Stress in Bronze section,

$$\tau = \frac{16\,T_A}{\pi\, d^3} = \frac{16 \times 776.59}{\pi \times (25)^3} \times 1000 = 253.128 \text{ N/mm}^2$$

Problem 11.12: A flanged bolt coupling consists of ten steel 20-mm diameter bolts spaced evenly around a bolt circle 400 mm in diameter. Determine the torque capacity of the coupling if the allowable shearing stress in the bolts is 40 MPa.

Solution: Given:

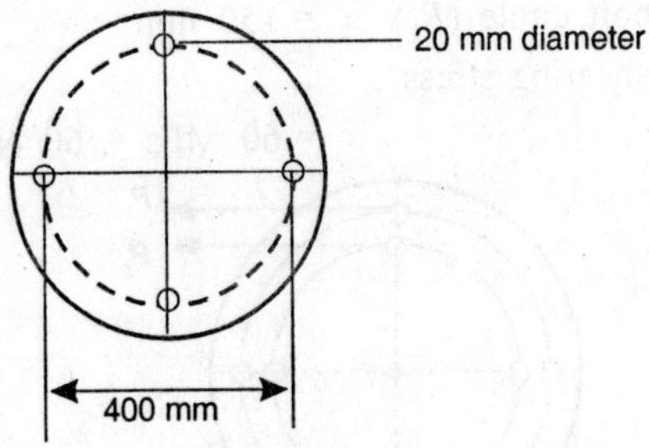

Fig. P. 11.12

Diameter of bolt (d_b) = 20 mm

Diameter of bolt circle = 400 mm

Allowable shearing stress in the bolts is = 40 MPa = 40 N/mm²

∴ Torque capacity of the coupling i.e.

$$T = q_b \times \frac{\pi}{4}\, d_b^2 \times n \times R_b$$

$$= 40 \times \frac{\pi}{4} \times (20)^2 \times 10 \times 200$$

$$= 25.13 \text{ kN.m}$$

Problem 11.13: A flanged bolt coupling consists of eight 10-mm diameter steel bolts on a bolt circle 400 mm in diameter, and six 10-mm diameter steel bolts on a concentric bolt circle 300 mm in diameter, as shown in Fig. P. 11.13. What torque can be applied without exceeding a shearing stress of 60 MPa in the bolts?

Solution:

First:

No. of bolt = 8

Diameter of bolt (d_b) = 10 mm

Diameter of bolt circle = 400 mm

Radius of bolt circle (R_b) = 200 mm

Second:

No. of bolt = 6

Diameter of bolt (d_b) = 10 mm

Diameter of bolt circle = 300 mm

Radius of bolt circle (R_b) = 150 mm

Allowable shearing stress

= 60 MPa = 60 N/mm^2

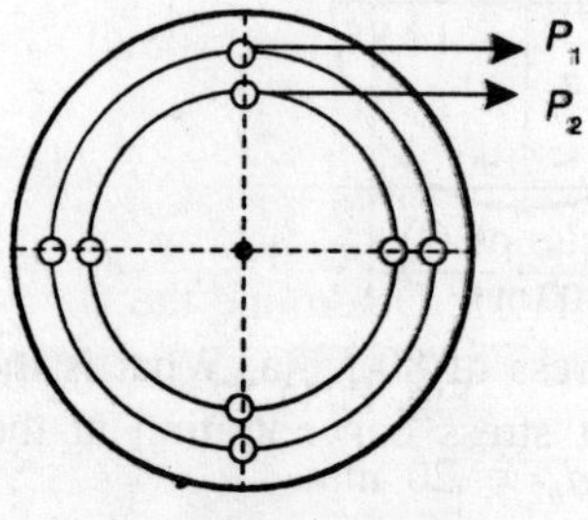

Fig. P. 11.13

Torque applied (T)

$$= P_1R_1\eta_1 + p_2R_2\eta_2$$

$$= 60 \times \frac{\pi}{4} (10)^2 \times 200 \times 8 + 60 \times \frac{\pi}{4} \times (10)^2 \times 150 \times 6$$

$$= 60 \times \frac{\pi}{4} (10)^2 [1600 + 900]$$

$$= 11.78 \text{ kN.m}$$

Problem 11.14: Determine the number of 10-mm diameter steel bolts that must be used on the 400-mm bolt circle of the coupling described in problem 11.13 to increase the torque capacity to 14 kN.m.

Solution:

Diameter of bolt = 10 mm

No. of bolts = n

Diameter of bolt circle = 400 mm

$\therefore \quad R = 200$ mm

Torque = 14 kN.m = 14×10^6 N.mm

Shear stress = 60 N/mm^2

$\therefore$ We know that,

$$T = P_1 R_1 n_1$$

$$14 \times 10^6 \times 10 = 60 \times \frac{\pi}{4} (10)^2 \times 200 \times n$$

$$140 = 6 \times \frac{\pi}{4} \times 2 \times n$$

$$\therefore \quad n = \frac{140}{3\pi} = 14.85 \approx 15 \text{ bolts}$$

Problem 11.15: A torque of 600 N.m is applied to the rectangular section shown in Fig. P. 11.15. Determine the wall thickness t so as not to exceed a shear stress of 80 MPa. What is the shear stress in the short sides? Neglect stress concentration at the corners.

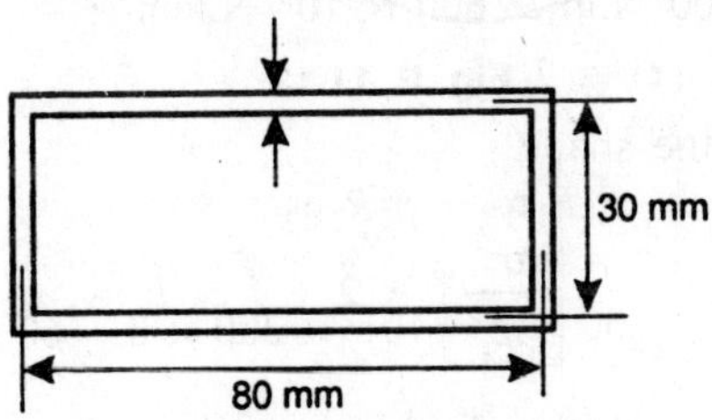

Fig. P. 11.15

Solution:

Torque (T) = 600 N.m = 600×10^3 N.mm

Wall thickness (t) = ?

Allowable shear stress (τ) = 80 MPa = 80 N/mm^2

We know that,

$$T = \tau \times 2\,A \times t$$

$$\therefore \quad t = \frac{T}{\tau\, 2\, A}$$

$$= \frac{600 \times 10^3}{80 \times 2 \times 80 \times 30} = \frac{100}{64} = 1.5625 \text{ mm}$$

∴ Shear stress in the shorter side

$$= \frac{T}{2A} = \frac{600 \times 10^3}{2 \times 1 \cdot 5625 \times 30} = 6400 \text{ N.mm}^2$$

Problem 11.16: A tube 2 mm thick has the shape shown in Fig. P.find shearing stress caused by a torque of 600 N.m.

Fig. P. 11.16

Solution:

Thickness of the shape = 2 mm

Torque = 600 N.m = 600 × 10^3 N.mm

Shear stress (τ) = ?

∴ Area of the shape

$$= \left[\frac{\pi r^2}{2}\right] \times 2 + L \times b$$

$$= \pi r^2 + Lb$$

$$= \pi (10)^2 + 80 \times 20 = 1914.159 \text{ mm}^2$$

$$\therefore \qquad \tau = \frac{T}{2At} = \frac{600 \times 10^3}{2 \times 1914.159 \times 2} = 78.36 \text{ N/ nm}^2$$

Problem 11.17: Compute the maximum shearing stress developed in a phosphor bronze spring having a mean diameter of 200 mm and consisting of 24 turns of 20-mm diameter wire when the spring is stretched 100 mm and G = 42 GPa.

Solution:

$$\text{Mean diameter} = 200 \text{ mm}$$

$$\text{Radius, } R = 100 \text{ mm}$$

$$\text{No. of turns} = 24$$

$$\text{Diameter of bar, } d = 20 \text{ mm}$$

$$\text{Deflection } (\delta) = 100 \text{ mm,}$$

$$m = \frac{2R}{d} = \frac{2 \times 100}{20} = 10$$

$$G = 42 \text{ GPa} = 42 \times 10^3 \text{ N/mm}^2$$

∴ We know deflection in spring

$$\delta = \frac{64\, P\, R^3 n}{Gd^4}$$

$$\therefore \quad p = \frac{Gd^4 \times \delta}{64\, R^3 n} = \frac{42 \times 10^3 \times (20)^4 \times 100}{64 \times (100)^3 \times 24} = 437.5 \text{ N}$$

∴ Maximum shearing stress (τ_{max})

$$= \frac{16\, P\, R}{\pi d^3}\left[\frac{4m-1}{4m-4} + \frac{0.615}{m}\right]$$

$$= \frac{16 \times 525 \times 100}{\pi \times 20^3}\left[\frac{39}{36} + \frac{0.615}{10}\right]$$

$$= 27.86 \times 1.1448 = 31.90 \text{ N/mm}^2$$

Problem 11.18: Determine the maximum shearing stress and elongation in a helical steel spring composed of 20 turns of 20-mm diameter wire on a mean radius of 90 mm when the spring is supporting a load of 1.5 kN. and G = 83 GPa.

Solution:

No. of turns = 20

Diameter of wire = 20 mm

Mean radius = 90 mm

Load on the spring = 1.5 kN = 1500 N

$$G = 83 \text{ GPa} = 83 \times 10^3 \text{ N/mm}^2$$

$$m = \frac{2R}{d} = \frac{180}{20} = 9$$

Maximum shearing stress (τ_{max})

$$= \frac{16\,PR}{\pi d^3}\left[\frac{4m-1}{4m-4}+\frac{0.615}{m}\right]$$

$$= \frac{16\times 1500\times 90}{\pi\times(20)^3}\left[\frac{35}{32}+\frac{0.615}{m}\right]$$

$$= 85.94 \times 1.16208$$

$$= 99.869 \text{ N/mm}^2 \approx 100 \text{ MPa}$$

$$\text{elongation } (\delta) = \frac{64\,PR^3 n}{Gd^4}$$

$$= \frac{64\times 1500\times(90)^3\times 20}{83\times 10^3\times(20)^4} = 105.39 \text{ mm} \approx 105 \text{ mm}$$

Problem 11.19: Two steel springs arranged in series as shown in Fig. P. 11.19 support a load P. The upper spring has 12 turns of 25-mm-diameter wire on a mean radius of 100 mm. The lower spring consists of 10 turns of 20-mm diameter wire on a mean radius of 75 mm. If the maximum shearing stress in either spring must not exceed 200 MPa. Compute the maximum value of P and the total elongation of the assembly. G = 83 GPa. Compute the equivalent spring constant by dividing the load by the total elongation.

Fig. P. 11.19

Solution:

$$G = 83 \text{ GPa} = 83 \times 10^3 \text{ N/mm}^2$$

$$\tau_{max} = 200 \text{ MPa} = 200 \text{ N/mm}^2$$

$$n = 12,\ d = 25 \text{ mm}$$

$$R = 100 \text{ mm},\ m = \frac{2R}{d} = 8$$

$$n = 10,\ d = 20 \text{ mm}$$

$$R = 75 \text{ mm}, \; m = \frac{2R}{d} = 7.5$$

[For both spring]

$$\delta_1 = \frac{64\, PR^3 n}{Gd^4} = \frac{64 \times P \times 100^3 \times 12}{83 \times 10^3 \times 25^4}$$

$$\delta_1 = 0.02368\ P \qquad \ldots(1)$$

$$\delta_2 = \frac{64 \times P \times (75)^3 \times 10}{83 \times 10^3 \times 20^4} = 0.02033\ P$$

$$\therefore \quad \delta_2 = 0.02033\ P \qquad \ldots(2)$$

$\therefore$ Maximum stress in first spring

$$= \frac{16\, PR}{\pi d^3}\left[\frac{4m-1}{4m-4} + \frac{0.615}{m}\right]$$

$$= \frac{16 \times 100 \times P}{\pi \times 25^3}\left[\frac{8 \times 4 - 1}{4 \times 8 - 4} + \frac{0.615}{8}\right]$$

$$= \frac{1600 \times P}{\pi \times 25^3}\,[1.184]$$

$$\therefore \quad P = 5182.36 = 42.229\ \delta_1$$

$$\therefore \quad \delta_1 = 122.71 \text{ mm}$$

Maximum stress in second spring

$$= \frac{16\, PR}{\pi d^3}\left[\frac{7.5 \times 4 - 1}{4 \times 7.5 - 4} + \frac{0.615}{7.5}\right]$$

$$200 = \frac{16 \times P \times 75}{\pi \times (20)^3}\left[\frac{29}{26} + 0.082\right]$$

$$\therefore \quad P = 3498.28 = 49.188\ \delta_2$$

$$\therefore \quad \delta_2 = 71.12 \text{ mm}$$

$\therefore$ Total deflection $(\delta) = \delta_1 + \delta_2$

$$= 122.71 + 71.12 = 193.83 \text{ mm}$$

$\therefore$ Total elongation $(\delta) = \Sigma \dfrac{64\, PR^3 n}{Gd^3}$

$$193.83 = 0.02368\ P + 0.02033\ P$$
$$= 0.044011\ P$$
$$\therefore \quad P = 4404.12\ \text{N} = 4.4\ \text{kN}$$

$$\therefore \text{Equivalent spring constant} = \frac{\text{Total load}}{\text{Total deflection}} = \frac{4404.12}{193.83} = 22.72$$

Problem 11.20: A rigid bar, hinged at one end, is supported by two identical springs as showin in Fig. P11.20 (a). Each spring consists of 20 turns of 10-mm wire having a mean diameter of 150 mm. Compute the maximum shearing stress in the springs.

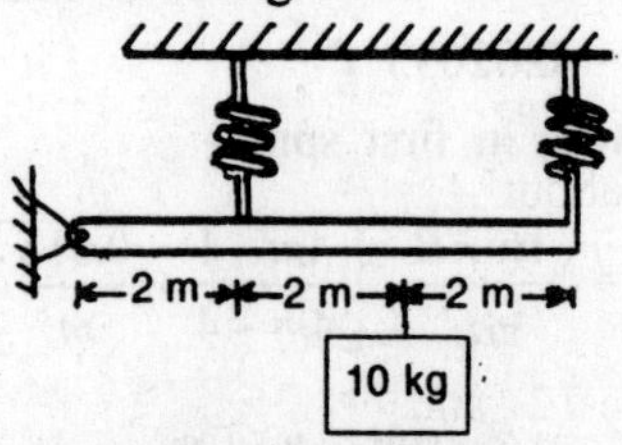

Fig. P. 11.20 (a)

Solution:

Mean radius (R) = 150 mm

F.B.D.

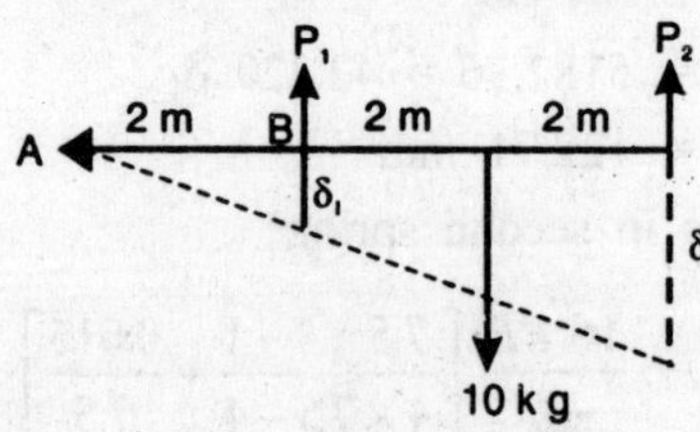

Fig. P. 11.20 (b)

$\therefore$ From similar triangle,

$$\frac{\delta_1}{2} = \frac{\delta_2}{6_3}$$

$$\therefore \quad \delta_2 = 3\delta_1 \qquad \ldots(1)$$

Where δ_1 = deflection in first spring and δ_2 is deflection in second spring,

But we know that deflection in spring is equal to,

$$\delta = \frac{64\,PR^3 \cdot n}{Gd^3}$$

$$\therefore \quad \frac{64\,P_2R_2^3 \cdot n}{G_2d_2^4} = \frac{3 \times 64\,P_1R_1^3 \times n}{G_1d_1^4}$$

$\because$ According to the problem

$$R_1 = R_2$$

$$G_1 = G_2$$

$$d_1 = d_2$$

$$n_1 = n_2$$

$$\therefore \quad P_2 = 3P_1 \qquad \ldots(2)$$

Taking moment about $A = 0$

$$P_1 \times 2 + P_2 \times 6 = 10 \times 4$$

$$P_1 + 3P_2 = 20 \qquad \ldots(3)$$

$\therefore$ From equation (2) and (3)

$$P_1 + 3 \times 3P_1 = 20$$

$$P_1 + 9\,P_1 = 20$$

$$\therefore \quad 10\,P_1 = 20$$

$$\therefore \quad P_1 = \frac{20}{10} = 2$$

$$\therefore \quad P_1 = 2 \text{ kg}$$

$$\therefore \quad 3P_2 = 20 - 2 = 18$$

$$\therefore \quad P_2 = \frac{18}{3} = 6 \text{ kg}$$

$$\therefore \quad P_1 = 2 \text{ kg}$$

$$P_2 = 6 \text{ kg}$$

Maximum shear stress (τ)

$$= \frac{16\,P\,R}{\pi d^3}\left(1 + \frac{0.5}{m}\right)$$

where $$m = \frac{2\,R}{d}$$

$$R = \frac{150}{2} = 75 \text{ mm}$$

$$d = 10 \text{ mm}$$

$$\therefore \quad m = \frac{2 \times 75}{10} = 15$$

∴ Maximum shearing stress in first spring is

$$= \frac{16 P_1 R}{\pi d^3}\left[1+\frac{0.5}{m}\right]$$

$$= \frac{16 \times 20 \times 75}{\pi \times (10)^3}\left[1+\frac{0.5}{15}\right] = 7.89 \text{ N/mm}^2$$

Maximum shearing stress in second spring

$$= \frac{16 P_2 R}{\pi d^3}\left[1+\frac{0.5}{m}\right]$$

$$= \frac{16 \times 60 \times 75}{\pi (10)^3}\left[1+\frac{0.5}{m}\right] = 23.68 \text{ N/mm}^2$$

Problem 11.21: As shown in Fig. P. 11.21 (a), a homogeneous 50-kg rigid block is suspended by the three springs whose lower ends were originally at the same level. Each steel spring has 24 turns of 10-mm diameter wire on a mean diameter of 100 mm, and G = 83 GPa. The bronze spring has 48 turns of 20-mm diameter wire on a mean diameter of 150 mm, and G = 42 GPa. Compute the maximum shearing stress in each spring.

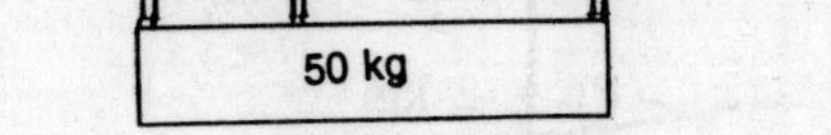

Fig. P. 11.21 (a)

Solution: Given:

For steel n = 24 turns

d = 10 mm

$$R = \frac{100}{2} = 50 \text{ mm}$$

G = 83 GPa = 83 × 10^3 N/mm²

For bronze n = 48 turns

d = 20 mm

$$R = \frac{150}{2} = 75 \text{ mm}$$

$$G = 42 \text{ GPa} = 42 \times 10^3 \text{ N/mm}^2$$

F.B.D.

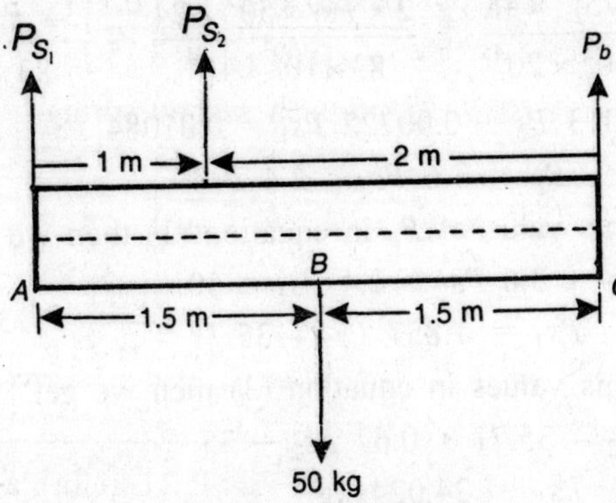

Fig. P. 11.21 (b)

From equilibrium equation,

$Ps_1 + Ps_2 + P_b = 50$ ...(1)

Taking moment about $C = 0$

$Ps_1 \times 3 + Ps_2 \times 2 = 50 \times 1.5$

$\therefore \quad Ps_1 = 25 - 0.67\ Ps_2$ ...(2)

Deflection diagram of beam

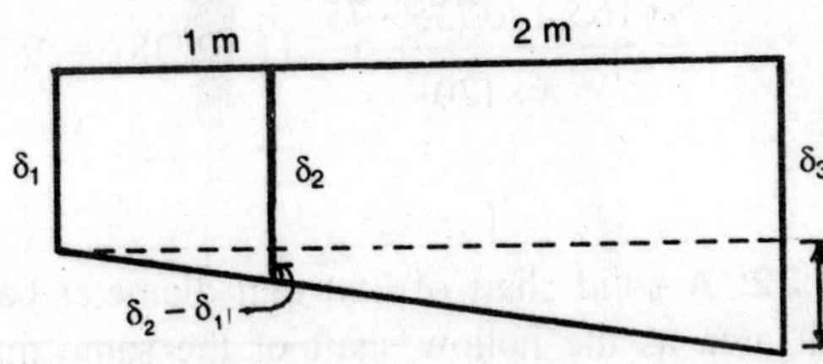

Fig. P. 11.21 (c)

$\therefore$ From properties of similar triangle

$$\frac{\delta_2 - \delta_1}{1} = \frac{\delta_3 - \delta_1}{3}$$

$\therefore \quad \delta_3 + 2\delta_1 = 3\delta_2$ ...(3)

Putting value of deflection in above equation,

$$\frac{P_b \times R_b^3 \times n_b}{G_b \times d_b^4} + 2 \times \frac{Ps_1 \times R_{s_1}^3 \times n_{s_1}}{Gs_1 \times ds_1^4} = 3\ \frac{Ps_2 \times R_{s_2}^3 \times n_{S_2}}{Gs_2 \times ds_2^4}$$

$$\Rightarrow \frac{P_b \times (75)^3 \times 48}{42 \times 10^3 \times 20^4} + \frac{2 \times Ps_1 \times 50^3 \times 24}{83 \times 10^3 \times 10^4} = \frac{3 \times Ps_2 \times 50^3 \times 24}{83 \times 10^3 \times 10^4}$$

or $0.003013\ P_b + 0.00722\ Ps_1 = 0.01084\ Ps_2$

$\therefore \quad P_b = 3.6\ Ps_2 - 2.4\ Ps_1$ ...(4)

Putting this value of P_b in equation (1) then we get,

$Ps_1 + Ps_2 + 3.6\ Ps_2 - 2.4\ Ps_1 = 50$

$\therefore \quad Ps_1 = 1.857\ Ps_2 - 35.71$ ...(5)

Putting this values in equation (3) then we get,

$1.857\ Ps_2 - 35.71 + 0.67\ Ps_2 = 25$

$\therefore \quad Ps_2 = 24.024$ kg

$Ps_1 = 8.9025$ kg

$P_b = 17.0735$ kg

$P_b = 17.0735 \times 10$ N $= 170.735$ N

$\therefore$ Maximum shearing stress in bronze spring

$$= \frac{16\ P_b\ R}{\pi d^3}\left[\frac{4m-1}{4m-4} + \frac{0.615}{m}\right]$$

$$= \frac{16 \times 170.735 \times 75}{\pi \times (20)^3}\ [1.19738] = 9.76\ \text{N/mm}^2$$

Problem 11.22: A solid shaft of 250 mm diameter has the same cross-sectional area as the hollow shaft of the same material with inside diameter of 200 mm. (a) Find the ratio of power transmitted by the two shafts for the same angular velocity, and (b) compare the angles of twist in equal lengths of these shafts, when stressed to the same intensity.

Solution:

Let the outside diameter of hollow shaft $= D$ mm

Area of cross-section of solid shaft $= A_s = \frac{\pi}{4}\ (250)^2$

Area of cross-section of hollow shaft $= A_H = \frac{\pi}{4}\ [D^2 - (200)^2]$

Since As $\quad A_s = A_H$, we have $\frac{\pi}{4}\ [D^2 - (200)^2] = \frac{\pi}{4}\ (250)^2$.

$$\therefore \quad D = [(250)^2 + (200)^2]^{1/2} = 320.16 \text{ mm}$$

(a) Ratio of power transmitted

We know that

$$P = \frac{2\,\pi n T}{60}$$

Shafts having same angular velocity will have same r.p.m. Hence ratio of H.P. transmitted will be equal to the ratio of torques transmitted. Hence if f_{sa} = allowable shear stress for the material, we have

$$\therefore \quad \frac{P_s}{P_H} = \frac{T_s}{T_H} = \frac{\frac{\pi}{16} f_{sa} \cdot D_5^3}{\frac{\pi}{16} f_{sa}\left[\frac{D^4 - d^4}{D}\right]} = \frac{(250)^3}{\frac{(320.16)^4 - (200)^4}{320.16}}$$

$$\text{or} \quad \frac{P_s}{P_H} = 0.56165$$

$$\text{or} \quad \frac{P_H}{P_s} = 1.78$$

(b) Angles of twists: Let θ_s angles of twist for the solid shaft and θ_h be the angle of twist for the hollow shaft, for the same length.

In general, we have

$$\frac{f_s}{R} = \frac{N\,\theta}{L}$$

$$\text{or} \quad \theta = \frac{f_s\,L}{R\,N}, \text{ where } R = \text{outside diameter.}$$

Now, for both the shafts, f_s, N and L are the same.

$$\therefore \quad \frac{\theta_s}{\theta_H} = \frac{R_H}{R_s} = \frac{D_H}{D_s}$$

$$= \frac{320.16}{250} = 1.2806$$

or $$\frac{\theta_H}{\theta_s} = 0.7809$$

Problem 11.23: A hollow shaft, subjected to pure torque, attains a maximum shear stress of f_s. Given that the strain energy per unit volume is $f_s^2/3$ N, calculate the ratio of shaft diameters. Determine the actual diameters of such a shaft to transmit 5 MW at 120 r.p.m. when energy stgored is 25000 N.m/m^3 and N = 80000 N/mm^2.

Solution:

We know from strain energy relation,

$$\frac{U}{\text{Volume}} = \frac{f_s^2}{4\,N}\left(\frac{D^2+d^2}{D^2}\right) \quad \text{for a hollow shaft}$$

But $$\frac{U}{\text{Volume}} = \frac{f_s^2}{3\,N} \text{ (Given)}$$

$$\therefore \frac{f_s^2}{4\,N}\left(\frac{D^2+d^2}{D^2}\right) = \frac{f_s^2}{3\,N}$$

or $$\frac{D^2+d^2}{D^2} = \frac{4}{3}$$

From which $$\frac{d}{D} = \sqrt{\frac{1}{3}} \quad \ldots(i)$$

or $$\frac{D}{d} = \sqrt{3} = 1.732$$

If $$\frac{f_s^2}{3\,N} = 25000 \text{ N.m/m}^3$$

$$= 25000 \times 10^{-6} \text{ N.mm/mm}^3 \text{ (Given)}$$

$$f_s = (25000 \times 10^{-6} \times 3 \times 80000)^{1/2} = 77.46 \text{ N/mm}^2$$

Also $$T = \frac{60\,P}{2\,\pi\,n} = \frac{60 \times 5 \times 10^3}{2\,\pi(120)} = 397.89 \text{ kN.m} = 397.89 \times 10^6 \text{ N.mm}$$

Now $$f_s = T\,\frac{R}{J} = T\,\frac{D/2}{\frac{\pi}{32}\left(D^4 - d^4\right)} = \frac{16\,D\,T}{\pi\left(D^4 - d^4\right)} \quad ...(2)$$

Substituting the values of f_s and T and noting that $\frac{d}{D} = \left(\frac{1}{3}\right)^{\frac{1}{2}}$, we get

$$77.46 = \frac{16\,D\left(397.89 \times 10^6\right)}{\pi\,D^4\left(1 - \frac{1}{9}\right)} = \frac{2280 \times 10^6}{D^3}.$$

From which

$$D^3 = 29.43 \times 10^6 \text{ or } D \approx 309 \text{ mm}$$

and $$d = \frac{309}{\sqrt{3}} \approx 178 \text{ mm}$$

Problem 11.24: A horizontal shaft fixed at each end, has a free length of 8 m. Viewed from one end of the shaft, axial couples of 40 kN.m clockwise and 50 kN.m counter clockwise act on the shaft at distance of 3.2 m and 5.5 m from the viewed end. Determine the end fixing couples in magnitude and direction and find the diameter of solid shaft for a maximum shearing stress of 65 N/mm². Also, find the position where the shaft suffers no angular twist.

Solution:

Let T = fixing torque at the viewed end.

∴ Torque in the middle portion of the shaft

$$= (T - 40) \text{ kN.m}$$

and torque at the far end of the shaft

$$= T - 40 + 50 = (T + 10) \text{ kN.m}$$

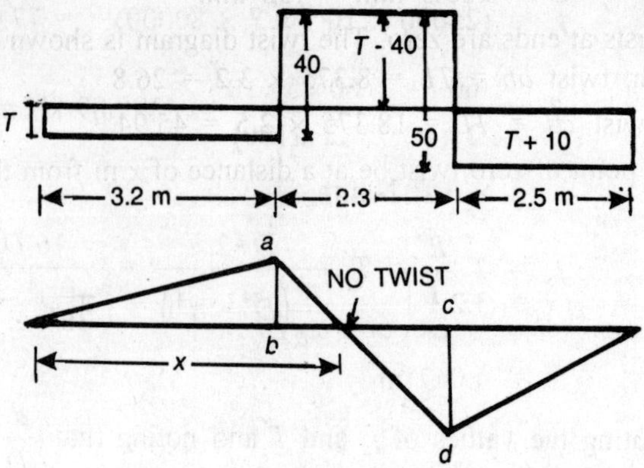

Fig. P. 11.24

In general, twist $\theta = \dfrac{TL}{NJ}$

Torsional rigidity (NJ) constant throughout

$$\text{Total twist} = \Sigma\theta = \frac{1}{NJ}\Sigma TL$$

For no resultant twist

$$\Sigma\theta = 0 = \Sigma TL = (T \times 3.2) + (T - 40) \times 2.3 + (T + 10) \times 2.5$$

$$\therefore \quad T = 8.375 \text{ kN.m}$$

Hence fixing couple at one end = T = 8.375 kN.m

Fixing couple at the other end = $T + 10 = 8.375 + 10$

$= 18.375$ kN.m

The torque diagram is shown in

Maximum torque = $T - 40 = 8.375 - 40 = 31.625$ kN.m

Maximum shear stress

$$F_s = \frac{16\,T_{max}}{\pi D^3} = \frac{16 \times 31.625 \times 10^6}{\pi D^3}$$

$$= \frac{161.06 \times 10^6}{D^3} \text{ N/mm}^2$$

But this is equal to 65 N/mm² (given).

$$\therefore \frac{161.06 \times 10^6}{D^3} = 65$$

$\therefore \quad D = 135.3 \text{ mm} \approx 136 \text{ mm}$

The twists at ends are zero. The twist diagram is shown in P.11.25 (b) in which, twist $ab = TL = 8.375 \times 3.2 = 26.8$

Also, twist $cd = TL = 18.375 \times 2.5 = 45.94$

Let the point of zero twist be at a distance of x m from the viewed end.

$$\therefore \quad x = 3.2 + \left(\frac{ab}{ab+cd}\right) 2.3 = 3.2 + \left(\frac{26.8}{26.8+45.94}\right) 2.3$$

$$= 4.047 \text{ m}$$

Problem 11.25: A circular bar AB, fixed at one end and free at the other is loaded by a uniformly distributed torque of intensity t per unit distance, along the axis of the bar, as shown Fig. P. 11.25. Determine the strain energy stored in the bar. Taking $L = 6$ m, diameter of bar = 200 mm, $t = 6$ kN.m/m and $N = 0.8 \times 10^5$ N/mm^2, find the value of the strain energy.

Solution:

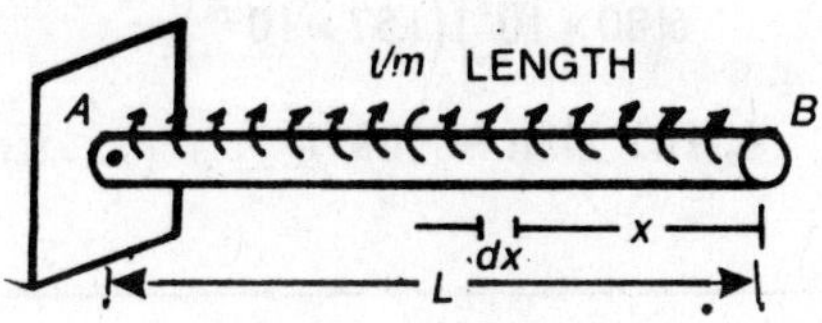

Fig. P. 11.25

Take a section x from the free end B

The torque T_x at that section = $t.x$. Consider a length dx, where the angle of twisting is θ_x.

$$\therefore \quad dU = \frac{1}{2} T_x \cdot \theta_x \quad \ldots(1)$$

But $\quad \theta = \dfrac{TL}{NJ}$, in general

$$\text{Hence} \quad \theta_x = \frac{T_x \cdot dx}{NJ_x} \quad \ldots(2)$$

Substituting in (1),

$$dU = \frac{T_x^2 dx}{2\ NJ_x}$$

Total strain energy

$$U = \int_0^L \frac{T_x^2 dx}{2\ NJ_x}$$

or $$U = \frac{1}{2\ NJ}\int_0^L (tx)^2\ dx = \frac{t^2 L^3}{6\ NJ}$$

Given $t = 6$ kN.m/m;

$L = 6$ m;

$N = 0.8 \times 10^5$ N/mm^2 = 80×10^6 kN/m^2

$$J = \frac{\pi}{32} d^4 = \frac{\pi}{32} (0.2)^4 = 1.57 \times 10^{-4} \text{ m}^4$$

$$\therefore \quad U = \frac{(6)^2 (6)^3}{6(80 \times 10^3)(1.57 \times 10^{-4})} = 10.3 \times 10^{-2} \text{ kN.m}$$

$$= 103 \text{ N.m} = 103 \text{ J}$$

Problem 11.26: A shaft is subjected to turques as shown in Fig. P. 11.26 (a). Plot the diagram showing the variation of torque along the length of the shaft.

Solution:

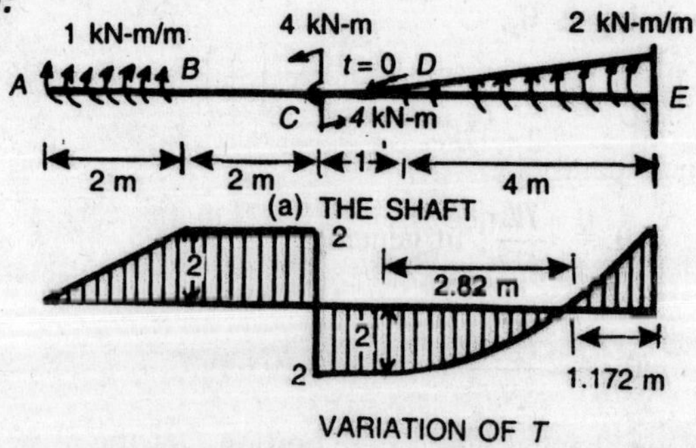

Fig. P. 11.26 (b)

The shaft carries varrying torque along its length

(i) For portion AB of the shaft

$$t = 1 \text{ kN.m/m.}$$

$\therefore$ $T_x = t \,.\, x = 1 \,.\, x$ (↶) (The variation being linear)

At A, $x = 0$ Hence, $T_A = 0$

At B, $x = 2$ m.

Hence $T_B = 1 \times 2 = 2$ kN.m (↶)

(ii) For portion BC of the shaft

This portion is subjected to a constant torsion

$$= 1 \times 2 = 2 \text{ kN.m}$$

(iii) For portion CD of the shaft

At C, the shaft is subjected to a concentrated torque of 4 kN.m in the anticlockwise direction.

Now T_c (left) = 2 kN.m (↶);

T_c (right) = 2 – 4 = –2, i.e. 2 kN.m (↷).

Thus, the torque changes sign at C.

The torque is constant from C to D.

(iv) For portion DE of the shaft

This portion is subjected to a uniformly variable torque, having intensity $t = 0$ at D to $t = 2$ kN-m/m at E. Consider a section X at a distance x metres from D.

The intensity of torque sat that section is

$$t_x = \frac{2}{4}\, x = \frac{x}{2} \text{ kN.m/m}$$

Hence total torque at X is

$$T_x = (1 \times 2) - 4 + \frac{1}{2}\, x \left(\frac{x}{2}\right) = -2 + \frac{x^2}{4} \ (↶)$$

The variation is parabolic.

At D, $x = 0$ Hence $T_D = -2$ kN.m (as earlier)

At E, $x = 4$ m.

Hence $T_E = -2 + \frac{1}{4}(4)^2 = 2$ kN.m (↶)

Thus, the torque changes sign in the portion DE; the torque is zero at $x = \sqrt{8} = 2.828$ m from D or at 1.172 m from E.

Problem 11.27: A propeller shaft 160 mm external diameter and 80 mm internal diameter transmits 1000 kW of power at 100 r.p.m. In addition, it subjected to a B.M. of 8 kN-m and an end thrust of 120 kN. Find (i) the principal stresses and their planes (ii) the maximum shear stress and its plane, (iii) the stress which acting alone would produce the same maximum strain. Tale 1/m = 0.3.

Solution:

From question, given data is,

$$D = 160 \text{ mm}$$

$$d = 80 \text{ mm}$$

$$P = 1000 \text{ kW}$$

$$n = 100 \text{ r.p.m.}$$

$$M = 8 \text{ kN.m} = 8 \times 10^6 \text{ N.mm}$$

$$C = 120 \text{ kN} = 120 \times 10^3 \text{ N}$$

$$A = \frac{\pi}{4}(160^2 - 80^2) = 15080 \text{ mm}^2$$

$$Z = \frac{\pi}{32}\left(\frac{D^4 - d^4}{D}\right)$$

$$= \frac{\pi}{32}\frac{(160)^4 - (80)^4}{160} = 376991 \text{ mm}^4$$

Torsional section modulus

$$= \frac{\pi}{16}\frac{D^4 - d^4}{d}$$

$$= \frac{\pi}{16}\frac{(160)^4 - (80)^4}{160} = 753982 \text{ mm}^4$$

$$f_o = \frac{C}{A} = \frac{120 \times 10^3}{15082} = 7.96 \text{ N/mm}^2$$

$$f_b = \frac{M}{Z} = \frac{8 \times 10^6}{376991} = 21.22 \text{ N/mm}^2$$

$\therefore$ Max. compressive direct stress

$$= f_c = 7.96 + 21.22 = 29.18 \text{ N/mm}^2$$

$$T = \frac{60\,P}{2\,\pi\,n} = \frac{60 \times 1000}{2\,\pi\,(100)}$$

$$= 95.493 \text{ kN.m} = 95.493 \times 10^6 \text{ N.mm}$$

$\therefore$ Max. shear stress

$$f_s = \frac{95.493 \times 10^6}{753982} = 126.65 \text{ N/mm}^2$$

(i) Principal stresses

$$\sigma_1 = \frac{f_c}{2} + \sqrt{\left(\frac{f_c}{2}\right)^2 + f_s^2}$$

$$= \frac{29.18}{2} + \sqrt{\left(\frac{29.18}{2}\right)^2 + (126.65)^2}$$

$$= 14.59 + 127.49 = 142.08 \text{ N/mm}^2 \text{ (comp.)}$$

$$\sigma_2 = 14.59 - 127.49 = -112.9 \text{ i.e. } 112.9 \text{ N/mm}^2 \text{ (tensile)}$$

(ii) Maximum shear stress intensity

$$= \sqrt{\left(\frac{29.18}{2}\right)^2 + (126.25)^2} = 127.49 \text{ N/mm}^2$$

(iii) Stress to give maximum strain

$$f = \sigma_1 - \frac{1}{m}\,\sigma_2 = 142.08 + 0.3 \times 112.9$$

$$= 175.95 \text{ N/mm}^2$$

Problem 11.28: A solid body shaft 60 mm diameter is to be coupled in series with a hollow steel shaft of the same external diameter. Find the internal diameter of the steel shaft if the angle of twist per unit length is to be 75% of that of the alloy shaft. Determine the speed at which the shafts are to be driven to transmit 400 kW, if the limits of shearing stress are to 60 N/mm² and 80 N/mm² in alloy and steel respectively. Take $N_{steel} = 2.2\ N_{alloy}$.

Solution:

Using s for steel shaft and a fer alloy shaft

$$\text{Angle of twist per unit length} = \frac{\theta}{L} = \frac{T}{NJ}$$

$$\therefore \quad \left(\frac{T}{NJ}\right)_s = 0.75\left(\frac{T}{NJ}\right)_a$$

$$\text{or} \quad \frac{32\,T}{\pi\left(60^4 - d^4\right)} = \frac{0.75 \times 2.2 \times 32\,T}{\pi(60)^4}$$

$$\text{or } 12960000 - d^4 = \frac{12960000}{0.75 \times 2.2} = 7854545$$

$$\therefore \quad d = 47.53 \text{ mm}$$

$$\text{Again,} \quad \frac{f_s}{D/2} = \frac{N\,\theta}{L}$$

$$\therefore \quad \frac{f_{ss}}{f_{sa}} = \left(\frac{N_s}{N_s}\right)\left(\frac{D_s}{D_a}\right)\frac{(\theta/L)_s}{(\theta/L)_a}$$

$$\text{But} \quad D_s = D_a, \left(\frac{\theta}{L}\right)_s = 0.7\left(\frac{\theta}{L}\right)_a$$

$$\text{and} \quad N_s = 2.2\,N_a$$

$$\therefore \quad \frac{f_{ss}}{f_{sa}} = 2.2 \times 1 \times 0.7 = 1.54$$

The limits of shearing stresses are 80 N/mm^2 in steel and 60 N/mm^2 in alloy, but the actual maximum stresses must be 80 N/mm^2 in steel and 80/1.54 = 51.95 N/mm^2 in alloy.

For the solid shaft,

$$T = f_s \times \frac{\pi D^3}{16} = \frac{51.95 \times \pi(60)^3}{16}$$

$$= 2203277 \text{ N.mm} = 2.203 \text{ kN.m}$$

$$\text{Now} \quad P = \frac{2\,\pi NT}{60}$$

or $$n = \frac{60\,P}{2\,\pi T} = \frac{60 \times 400}{2\,\pi\,(2.203)} = 1734 \text{ r.p.m.}$$

Problem 11.29: A solid circular shaft of diameter d = 100 mm isloaded with torques as shown in figure P. 11.29 (a). Plot the diagrams showing variation of T and θ.

Solution:

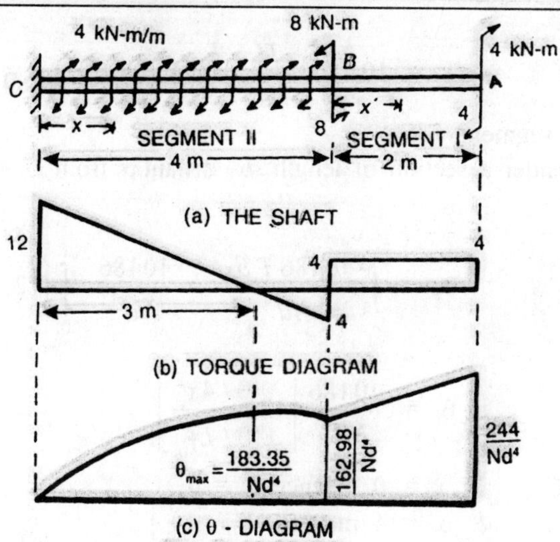

Fig. P. 11.29 (b)

(a) Torque diagram

For CB, measuring x from C,

$$T_x = 4 - 8 + 4\,(4 - x) = 12 - 4\,x \qquad \ldots(1)$$

The variation is linear.

At C, $\quad x = 0.$

Hence $\quad T_c = 12$ kN.m (↺)

At B, $\quad x = 4$ m.

Hence $\quad T_B$ (left) $= 12 - 4 \times 4 = -4$ kN.m (↻)

The torsion changes sign from C to B and is zero at x

$$= \frac{12}{4} = 3 \text{ m from } C.$$

For BA: $\quad T = 4$ kN.m (↻)

This is constant from B to A.

(b) Rotation diagram:

The left hand section is fixed and cannot turn; hence the angle of twist should be calculated from the left hand end of the rod.

In general $\theta = \dfrac{TL}{NJ} = \dfrac{TL}{N\dfrac{\pi}{32}d^4}$

$$= 10.186\,\frac{TL}{Nd^4}$$

For segment CB

Consider a section of length dx, distant x from C, where torsion is T_x.

$$\theta_x = \int_0^x \frac{10.186\,T_x dx}{Nd^4} = \frac{10.186}{Nd^4}\int_0^x (12 - 4\,x)\,dx$$

or $$\theta_x = \frac{10.186}{Nd^4}\left[12x - \frac{4x^2}{2}\right] \qquad \ldots(2)$$

At C, $x = 0$ Hence $\theta_c = 0$

At B, $x = 4$ m.

$$\theta_b = \frac{10.186}{Nd^4}\left[12\times 4 - \frac{4}{2}(4)^2\right] = \frac{162.98}{Nd^4}$$

For max. value of θ_x, we have

$$\frac{d\theta_x}{dx} = 0$$

$$= 12 - 4\,x, \text{ giving } x = 3 \text{ m.}$$

Hence θ is maximum where T is zero.

Hence $$\theta_{max.} = \frac{10.186}{Nd^4}\left[12\times 3 - \frac{4}{2}(3)^2\right] = \frac{183.35}{Nd^4}$$

For segment BA

Measuring x' from B,

$$\theta_x = \theta_B + \int_0^x 10.186\,\frac{T_x dx'}{Nd^4}$$

$$= \frac{162.98}{Nd^4} + \frac{10.186}{Nd^4}\int_0^x 4\,dx'$$

or $$\theta_x = \frac{162.98}{Nd^4} + \frac{10.186}{Nd^4}(4\,x')$$

At B, $x' = 0.$

Hence $$\theta_x = \frac{162.98}{Nd^4}, \text{ as before}$$

At A, $x' = 2$ m.

Hence $$\theta_x = \frac{162.98}{Nd^4} + \frac{10.186}{Nd^4}(4 \times 2) = \frac{244.47}{Nd^4}$$

Problem 11.30: A close-coiled helical spring has to absorb 60 N.m of energy when compressed 60 mm. The coil diameter is eight times the wire diameter. If there are ten coils, determine (a) diameter of the coil, (b) diameter of the wire, and (c) the maximum shear stress. Take $N = 0.86 \times 10^5$ N/mm^2.

Solution:

$$U = 60 \text{ N.m} = 60 \times 10^3 \text{ N.mm}$$

$$\delta = 60 \text{ mm}$$

$$D = 8\,d$$

Where D = coil dia and d = wire dia

R = coil radius

$n = 10$

Now $$U = \frac{1}{2} W\delta$$

or $$60 \times 10^3 = \frac{1}{2} W(60),$$

from which $W = 2000$ N

Also, $$\delta = \frac{64\,WR^3 n}{Nd^4} = \frac{64\,W(4d)^3 n}{Nd^4} = \frac{64 \times 64\,Wn}{Nd}$$

$$\text{or} \qquad d = \frac{64 \times 64\, Wn}{N\delta} = \frac{64 \times 64 \times 2000 \times 10}{0.86 \times 10^5 \times 60}$$

$$= 16 \text{ mm}$$

$$\therefore \qquad D = 8\text{ d} = 6 \times 16 = 128 \text{ mm}$$

$$f_s = \frac{16\, WR}{\pi d^3}\left(1 + \frac{d}{4\, R}\right)$$

$$= \frac{16 \times 2000\,(64)}{\pi (16)^3}\left(1 + \frac{16}{4 \times 64}\right)$$

$$= \frac{16 \times 2000\,(64)}{\pi (16)^3}\,(1 + 0.0625)$$

$$= 169.1 \text{ N/mm}^2$$

Problem 11.31: A open coiled helical spring, made out of 20 mm diameter steel rod, has 10 complete turns at a mean diameter of 150 mm, the angle of helix being 15°. An axial load of 400 N is applied. Compute (a) deflection under the load and (b) maximum intensities of direct and shear stresses, induced in the section of the wire. Take $N = 0.84 \times 10^5$ N/mm² and $E = 2 \times 10^5$ N/mm².

Solution: Given

$n = 10$ $\qquad D = 150$ mm

$R = 75$ mm $\qquad \alpha = 15°$

$W = 400$ N

$$T = WR\cos\alpha = 400 \times 75 \cos 15° = 28978 \text{ N.mm}$$

$$M = WR\sin\alpha = 400 \times 75 \sin 15° = 7764.6 \text{ N.mm}$$

(a) We know $\delta = \dfrac{64\, WR^3 n \sec\alpha}{d^4}\left[\dfrac{\cos^2\alpha}{N} + \dfrac{2\sin^2\alpha}{E}\right]$

$$= \frac{64 \times 400\,(75)^3 \times 10 \sec 15°}{(20)^4}\left[\frac{\cos^2 15°}{0.84 \times 10^5} + \frac{2\sin^2 15°}{2 \times 10^5}\right] = 8.23 \text{ mm}$$

(b) Direct and shearing stresses

$$f_b = \frac{M}{Z} = \frac{7764.6}{\frac{\pi}{32}(20)^3} = 9.89 \text{ N/mm}^2$$

$$f_s = \frac{T}{J} \times \frac{d}{2} = \frac{28978}{\frac{\pi}{16}(20)^3} = 18.45 \text{ N/mm}^2$$

$$\left(\text{Neglecting } \frac{d \tan\alpha}{4R}\right)$$

But $\dfrac{d \tan\alpha}{4R} = \dfrac{20 \tan 15^\circ}{4 \times 75} = 0.0179$ which is too small

$\therefore \quad f_s = 18.45\ (1 + 0.0179) = 18.78 \text{ N/mm}^2$

Principal stresses,

$$\sigma = \frac{f_b}{2} \pm \sqrt{\frac{f_b^2}{4} + f_s^2}$$

$$= \frac{9.89}{2} \pm \sqrt{\left(\frac{9.89}{2}\right)^2 + (18.45)^2} = 4.945 \pm 19.101$$

$\therefore \quad \sigma_1 = 24.05 \text{ N/mm}^2$ and

$\sigma_2 = -14.16 \text{ N/mm}^2$

Maximum shear stress,

$$f_s' = \sqrt{\left(\frac{f_b}{2}\right)^2 + (f_s)^2} = 19.101 \text{ N/mm}^2$$

Problem 11.32: A stiff bar of negligible weight transmits a load W to a combination of 3 springs, as shown in Fig. P. 11.32. The three springs are made of the same material and out of rods of equal diameters. They are of the same length before loading. The number of coils in the three springs are 10, 12 and 15 respectively while the mean radii of the coils are in the proportion 1 : 1.2 : 1.4 respectively. Find the distance x such that the stiff bar remains horizontal after applying the load. [AMIE, CSE 95]

Solution:

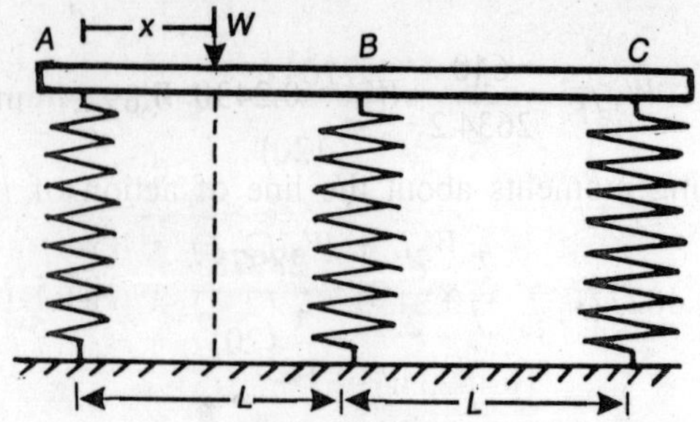

Fig. P. 11.32

Since bar is rigid, so deflection is same in each spring.

$$\therefore \quad \delta_1 = \delta_2 = \delta_3 = \delta \text{ (say)}$$

Also, the springs are made of the same material and out of the rods of equal diameter, we have

$$N_1 = N_2 = N_3 = N$$

and $$d_1 = d_2 = d_3 = d$$

Also Given

$$n_1 = 10$$

$$n_2 = 12 \text{ and}$$

$$n_3 = 15 \text{ and}$$

$$R_1 : R_2 : R_3 = 1 : 1.2 : 1.4$$

Hence $$R_2 = 1.2\ R_1$$

and $$R_3 = 1.4\ R_1$$

Now $$W_1 = \frac{Nd^4\delta}{64\ R_1^3 n_1} = \frac{Nd^4\delta}{64\ R_1^3(10)} = \frac{Nd^4\delta}{640\ R_1^3} \quad \text{...(1)}$$

$$W_2 = \frac{Nd^4\delta}{64\ R_2^3 n_2} = \frac{Nd^4\delta}{64\ (1.2\ R_1)^3(12)} = \frac{Nd^4\delta}{1327.1\ R_1^3} \quad \text{...(2)}$$

and $$W_3 = \frac{Nd^4\delta}{64\ R_3^3 n_3} = \frac{Nd^4\delta}{64\ (1.4\ R_1)^3(15)} = \frac{Nd^4\delta}{2634.2\ R_1^3} \quad \text{...(3)}$$

From (1) and (2), we get

$$W_2 = \frac{640}{1327.1} W_1 = 0.4823\ W_1$$

From (1) and (3), we get

$$W_3 = \frac{640}{2634.2} W_1 = 0.2430 \; W_1$$

Now, taking moments about the line of action of W_1, we get

$$W_2 L + W_3 \,(2\,L) = Wx$$

or $$0.4823 \; W_1 L + 0.2430 \; W_1 \,(2\,L) = Wx$$

$\therefore$ $$x = \frac{(0.4823\,W_1 + 0.4860\,W_1)\,L}{W} \qquad \text{...(1)}$$

But $$W = W_1 + W_2 + W_3$$
$$= W_1 + 0.4823 \; W_1 + 0.2430 \; W_1$$
$$= 1.7253 \; W_1 \qquad \text{...(2)}$$

From (1) $$x = \frac{0.4823\,W_1 + 0.4860\,W_1}{1.7253\,W_1} L = 0.5612 \; L$$

> **Problem 11.33:** An open coiled helical spring and a close-coiled helical spring are similar in every respect, except that the elongation of the open coiled spring is 2.5% greater than that of the close-coiled spring, under the same load. Determine the angle of inclination of the coils of the open coil springs. Take E/N = 2.5.

Solution:

Suffix 1 for open-coiled spring and 2 for closed-coiled spring.

We have $\delta_1 = 1.025 \; \delta_2$

$\therefore$ $$\frac{64\,WR^3 \times n}{d^4 \cos\alpha} \left[\frac{\cos^2\alpha}{N} + \frac{2\sin^2\alpha}{E}\right] = 1.025 \times \frac{64\,WR^3 n}{Nd^4}$$

or $$\frac{1}{N\cos\alpha}\left[\cos^2\alpha + \frac{2\,N}{E}\sin^2\alpha\right] = \frac{1.025}{N}$$

or $$\cos^2\alpha + \frac{2}{2.5}\,(1 - \cos^2\alpha) = 1.025 \;\cos\alpha$$

or $$\cos^2\alpha + 0.8\,(1 - \cos^2\alpha) = 1.025 \;\cos\alpha$$

or $$0.2\,\cos^2\alpha - 1.025\,\cos\alpha + 0.8 = 0$$

or $$\cos^2\alpha - 5.125\,\cos\alpha + 4 = 0$$

$$\therefore \quad \cos\alpha = \frac{5.125 \pm \sqrt{(5.125)^2 - 4 \times 4 \times 1}}{2 \times 1}$$

$$= 4.102 \text{ and } 0.9604$$

$$\therefore \quad \cos\alpha = 0.9604$$

$$\therefore \quad \alpha = 16.16^\circ$$

> **Problem 11.34:** An open-coiled helical spring of a given length of wire, 5 mm is diameter is coiled to a mean diameter of 40 mm, and is loaded with an axial load. If the spring is assumed to be a closely coiled, it would cause an error of 2.5% in axial deflection. Determine the pitch of the open-coiled spring. Take μ = 0.3.

Solution: Given

$$d = 5 \text{ mm}$$

$$D = 40 \text{ mm}$$

$$R = 20 \text{ mm}$$

$$\mu = 0.3$$

Using suffix 1 for open-coiled spring and 2 for closed-coiled spring,

We have $\quad E = 2\,N\,(1 + \mu)$

$$\therefore \quad \frac{2\,N}{E} = \frac{1}{1+\mu} = \frac{1}{1+0.3} = 0.7692 \qquad \ldots(1)$$

We have $\quad L_1 = 2\,\pi Rn \sec\alpha$

and $\quad L_2 = 2\,\pi Rn$

$\therefore \quad L_1 > L_2$ and hence $\delta_1 > \delta_2$

$$\therefore \quad \frac{\delta_1 - \delta_2}{\delta_1} = 0.025$$

$$\text{or} \quad 1 - \frac{\delta_2}{\delta_1} = 0.025$$

$$\therefore \quad \delta_2 = 0.975\,\delta_1 \qquad \ldots(2)$$

$$\text{or} \quad \frac{64\,WR^3 n}{Nd^4} = 0.975 \times \frac{64\,WR^3 n}{d^4 \cos\alpha}\left[\frac{\cos^2\alpha}{N} + \frac{2\sin^2\alpha}{E}\right]$$

or $$1 = \frac{0.975}{\cos\alpha}\left[\cos^2\alpha + \frac{2N}{E}\sin^2\alpha\right]$$

or $$1 = \frac{0.975}{\cos\alpha}\left[\cos^2\alpha + 0.7692\,(1 - \cos^2\alpha)\right], \text{ from (1)}$$

$$1.0256\,\cos\alpha = \cos^2\alpha + 0.7692 - 0.7692\,\cos^2\alpha$$

or $$0.2308\,\cos^2\alpha - 1.02556\,\cos\alpha + 0.7692 = 0$$

$$\cos^2\alpha - 4.4437\,\cos\alpha + 3.3328 = 0$$

$$\therefore \quad \cos\alpha = \frac{4.4437 \pm \sqrt{(4.4437)^2 - 4 \times 3.3328}}{2}$$

$$= 3.4883,\ 0.9554$$

$$\therefore \quad \cos\alpha = 0.9554$$

From which $\alpha = 17.17°$

Now, if p is the pitch, we have

$$\frac{p}{\pi D} = \tan\alpha$$

$$\therefore \quad p = \pi D \tan\alpha = \pi\,(40)\tan 17.17° = 38.84 \text{ mm}$$

Chapter 12

Special Topics

Problem 12.1: The B.M. M applied to a solid shaft carries a maximum direct stress f_y at elastic failure. Determine the numerical relationships between M and twisting moment T which, acting alone on the shaft, will produce elastic failure according to each of the following theories of failure. (a) Maximum principal stress, (b) Maximum principal strain, (c) Maximum shear stress, (d) Maximum strain energy, and (e) shear strain energy. Take $\mu = 0.3$.

Solution:

The elastic limit stress f_y caused by bending moment M is given by

$$f_y = \frac{32\,M}{\pi\,d^3}$$

Similarly, the shear stress τ produced by the torque T acting alone is

$$\tau = \frac{16\,T}{\pi\,d^3}$$

This shear stress is accompanied by complimentary shear stress, giving a state of shear stress, due to which the induced principal stress are

$$\sigma_1 = -\sigma_2 = \frac{16\,T}{\pi\,d^3}$$

Hence the principal stresses acting on the shaft are

$$\frac{16\,T}{\pi\,d^3},\ 0,\ -\frac{16\,T}{\pi\,d^3}$$

(a) From maximum principal stress theory

$$\frac{16\,T}{\pi\, d^3} = f_y = \frac{32\,M}{\pi\, d^3} \text{ at elastic failure}$$

$$\therefore \qquad \frac{M}{T} = 0.5$$

(b) From maximum principal strain theory

$$\sigma_1 - \mu\sigma_2 = f_y$$

or $$\frac{16\,T}{\pi\, d^3} + \frac{0.3 \times 16\,T}{\pi\, d^3} = \frac{32\,M}{\pi\, d^3}$$

From which

$$\frac{M}{T} = \frac{16 + 4.8}{32} = 0.65$$

(c) From maximum shear stress theory

$$\sigma_1 - \sigma_2 = f_y$$

or $$\frac{16\,T}{\pi\, d^3} + \frac{16\,T}{\pi\, d^3} = \frac{32\,M}{\pi\, d^3}$$

From which

$$\frac{M}{T} = 1$$

(d) From Maximum strain energy theory

$$\sigma_1^2 + \sigma_2^2 - 2\,\mu\sigma_1\sigma_2 = f_y^2$$

or $$\left(\frac{16\,T}{\pi\, d^3}\right)^2 + \left(\frac{16\,T}{\pi\, d^3}\right)^2 + 2 \times 0.3 \left(\frac{16\,T}{\pi\, d^3}\right)^2 = \left(\frac{32\,M}{\pi\, d^3}\right)^2$$

or $$2.6 \left(\frac{16\,T}{\pi\, d^3}\right)^2 = \left(\frac{32\,M}{\pi\, d^3}\right)^2$$

or $$1.612 \left(\frac{16\,T}{\pi\, d^3}\right)^2 = \frac{32\,M}{\pi\, d^3}$$

From which

$$\frac{M}{T} = \frac{1.612}{2} = 0.806$$

(e) Maximum shear strain energy theory

$$\sigma_1^2 + \sigma_2^2 - \sigma_1\sigma_2 = f_y^2$$

or $$\left(\frac{16\,T}{\pi\,d^3}\right)^2 + \left(\frac{16\,T}{\pi\,d^3}\right)^2 + \left(\frac{16\,T}{\pi\,d^3}\right)^2 = \left(\frac{32\,M}{\pi\,d^3}\right)^2$$

or $$3\left(\frac{16\,T}{\pi\,d^3}\right)^2 = \left(\frac{32\,M}{\pi\,d^3}\right)^2$$

$$\therefore \qquad \frac{M}{T} = \frac{\sqrt{3}}{2} = 0.866$$

Problem 12.2: A circle of 400 mm diameter is scribed on a mild steel plate before it is subjected to stresses as shown in Fig. P. 12.2. In stressing, the circle deforms to an ellipse. Calculate the lengths of the major and minor axes of the ellipse and also find their directions.

Take $\frac{1}{m}$ = 00.286 and E = 205 kN/mm^2.

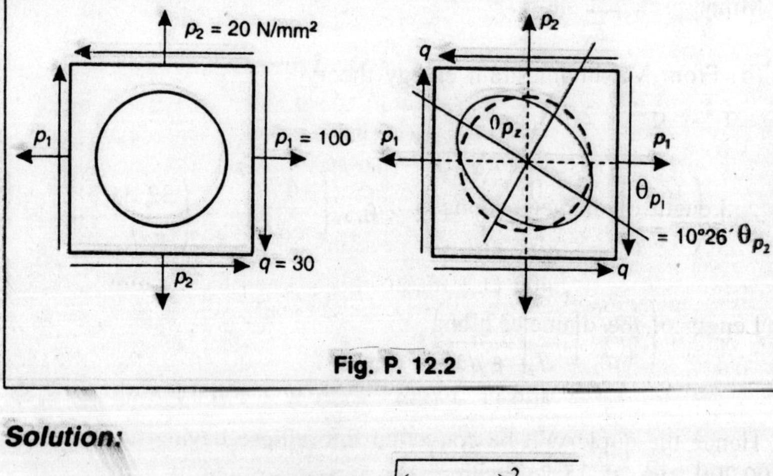

Fig. P. 12.2

Solution:

$$\sigma = \frac{p_1 + p_2}{2} \pm \sqrt{\left(\frac{p_1 - p_2}{2}\right)^2 + q^2}$$

$$= \frac{100+20}{2} \pm \sqrt{\left(\frac{100-20}{2}\right)^2 + (30)^2} = 60 \pm 50$$

$\therefore \quad \sigma_1 = +110 \text{ N/mm}^2$

and $\quad \sigma_2 = +10 \text{ N/mm}^2$

$$\tan 2\theta = \frac{2q}{p_1 - p_2}$$

$$= -\frac{2 \times 30}{100-20} = -0.75;$$

$$2\theta_p = -36.87^\circ = -36^\circ 52'$$

$\theta_{p1} = -18^\circ 26'$ (i.e. in the clockwise direction)

and $\quad \theta_{p2} = -18^\circ 26' + 90^\circ = 71^\circ 34'$

Major principal strain,

$$\varepsilon_1 = \frac{\sigma_1}{E} - \frac{\sigma_2}{mE}$$

$$= (110 - 0.286 \times 10)\frac{1}{205 \times 1000}$$

$$= +5.226 \times 10^{-4} \text{ (increase)}$$

Minor principal strain,

$$\varepsilon_2 = \frac{\sigma_2}{E} - \frac{\sigma_1}{mE} = (10 - 110 \times 0.286)\frac{1}{205 \times 1000}$$

$$= -1.047 \times 10^{-4} \text{ (increase)}$$

$$= 1.047 \times 10^{-4} \text{ (decrease)}$$

$\therefore$ Length of diameter along σ_1

$$= d + \varepsilon_1 d = d(1 + \varepsilon_1)$$

$$= 400(1 + 5.226 \times 10^{-4}) = 400.209 \text{ mm}$$

Length of the diameter along σ_2

$$= d - \varepsilon_2 d = d(1 - \varepsilon_2)$$

$$= 400(1 - 1.047 \times 10^{-4}) = 399.958 \text{ mm}$$

Hence the circle will be converted into ellipse having major axis = 400.209 mm at 18°26′ with the x-axis (measured in clock wise direction) and minor axis = 399.958 mm at 71°34′ with x-axis (measured in the anticlockwise direction).

Problem 12.3: A flat brass plate was stretched by tensile force acting in direction x and y at right angle. Strain gauges show that strains in x-direction was 0.00108 and in the y-direction 0.00024. Find (a) stresses acting in x and y directions, and (b) direct and shearing strains on a plane inclined at 40° to the x-direction (c) normal and shearing stresses on that plane. Take E = 80 kN/mm² and μ = 0.3.

Solution:

(a) Stresses in x and y directions

For two dimensional stress system, we have,

$$p_1 = \frac{E(e_1 + \mu e_2)}{1 - \mu^2}$$

$$= \frac{80 \times 10^3 (0.00108 + 0.3 \times 0.00024)}{1 - (0.3)^2}$$

$$= 101.27 \text{ N/mm}^2$$

$$p_2 = \frac{E(e_2 + \mu e_1)}{1 - \mu^2}$$

$$= \frac{80 \times 10^3 (0.00024 + 0.3 \times 0.00108)}{1 - (0.3)^2}$$

$$= 49.58 \text{ N/mm}^2$$

(b) Direct and shearing strains at 40° plane

We know value of eq. when $\phi = 0$,

$$e_\theta = \frac{1}{2}(e_1 + e_2) + \frac{1}{2}(e_1 - e_2)\cos 2\theta$$

$$= \frac{1}{2}(0.00108 + 0.00024)$$

$$+ \frac{1}{2}(0.00108 - 0.00024)\cos 80^\circ$$

$$= 7.33 \times 10^{-4}$$

Also,

$$\frac{1}{2}\varphi_\theta = \frac{1}{2}(e_1 - e_2)\sin 2\theta \text{ (numerically)}$$

or $$\frac{1}{2}\varphi_\theta = \frac{1}{2}(0.00108 - 0.00024)\sin 80^\circ$$

$$= 4.14 \times 10^{-4} \text{ radian}$$

or $$\varphi_\theta = 8.28 \times 10^{-4} \text{ radian}$$

The values of e_θ and φ_θ can also be obtained from Mohr's circle for strain shown in Fig. P. 12.3.

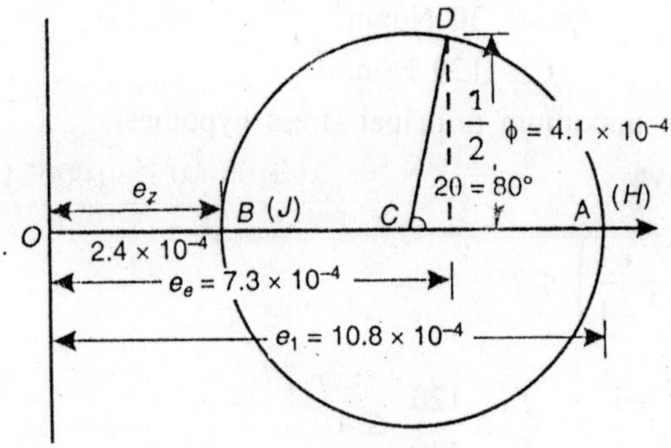

Fig. P. 12.3

From graphical construction, we get

$$e_\theta = 7.3 \times 10^{-4}$$

and $$\varphi_\theta = 8.2 \times 10^{-4}$$

(c) The normal and shearing stresses at 40° plane

The normal stress is given below noting that in that equation, θ is the angle made with y direction. Hence $\theta = 90^\circ - 40^\circ = 50^\circ$.

$$\therefore \quad p_n = p_1 \cos^2\theta + p_2 \sin^2\theta$$

$$= 101.27 \cos^2 50^\circ + 49.58 \sin^2 50^\circ$$

$$= 70.94 \text{ N/mm}^2$$

Also, $$p_t = \frac{p_1 - p_2}{2}\sin 2\theta \text{ (numerically)}$$

$$= \frac{101.27 - 49.58}{2}\sin 100^\circ$$

$$= 25.45 \text{ N/mm}^2$$

Problem 12.4: A thick cylindrical shell 150 mm internal diameter has to withstand an internal pressure of 30 N/mm^2. Calculate the thickness of the metal necessary for the cylinder on the basis of (i) maximum principal stress (ii) maximum principal strain, (iii) maximum shear, and (iv) maximum stain energy. Neglect the longitudinal direct stress. The permissible tensile stress for the shell material is 120 N/mm^2 and 1/m = 0.3.

Solution: Given

$$r = 75 \text{ mm}$$
$$p_r = 30 \text{ N/mm}^2$$
$$f = 120 \text{ N/mm}^2$$

(1) From maximum principal stress hypothesis

We have

$$p_r\left(\frac{K^2+1}{K^2-1}\right) \le f$$

$$\text{or} \quad \frac{K^2+1}{K^2-1} = \frac{f}{p_r} \le \frac{120}{130} \le 4$$

From which

$$K^2 \ge 5/3$$

$$\text{or} \quad K = 1.291$$

$$\text{Hence} \quad R = K \times r = 1.291 \times 75 = 96.82 \text{ mm}$$

$\therefore$ Metal thickness

$$= R - r = 96.82 - 75 = 21.82 \text{ mm}$$

(2) From maximum principal strain hypothesis

We have

$$p_r\left[\frac{K^2+1}{K^2-1}+\frac{1}{m}\right] \le f$$

$$\text{or} \quad \left[\frac{K^2+1}{K^2-1}+\frac{1}{m}\right] \le \frac{f}{p_r} \le \frac{120}{30} \le 4$$

$$\text{or} \quad \frac{K^2+1}{K^2-1} \le (4-0.3) \le 3.7$$

From which $K^2 = 4.7/2.7 = 1.7407$

or $K = 1.3194$

Hence $R = K \times r$

$= 1.3194 \times 75 = 98.95$ mm

$\therefore$ Hence thickness

$= R - r$

$= 98.95 - 75 = 23.95$ mm

(3). We have from maximum shear stress hypothesis

$$p_r \frac{2K^2}{K^2 - 1} \leq f \text{ or } \frac{K^2}{K^2 - 1} < \frac{f}{2p_r} \leq \frac{120}{2 \times 30} \leq 2$$

From which $K^2 = 2$ or $K = 1.4142$

$\therefore$ $R = K \times r$

$= 1.4142 \times 75 = 106.07$ mm

$\therefore$ Metal thickness $= R - r = 106.07 - 75 = 31.07$

(4) From maximum strain energy hypothesis

We have $$\frac{p_r^2\left(2\left(1 + \frac{1}{m}\right)K^4 + 2\left(1 - \frac{1}{m}\right)\right)}{\left(K^2 - 1\right)^2} \leq f^2$$

or $$\frac{2(1 + 0.3)K^4 + 2(1 - 0.3)}{\left(K^2 - 1\right)^2} \leq \left(\frac{f}{p_r}\right)^2 \leq \left(\frac{120}{30}\right)^2 \leq 16$$

or $2.6\ K^4 + 1.4 = 16\ (K^4 + 1 - 2K^2)$

or $K^4 - 2.388\ K^2 + 1.0896 = 0$

or $(K^2 - 1.194)^2 = 0.336$

or $K^2 - 1.194 = \pm 0.5797$

$K^2 = 1.194 + 0.5797 = 1.7737$

$K = 1.3318$

$R = K.r$

$= 1.3318 \times 75 = 99.89$

Metal thickness

$= R - r = 99.89 - 75$

$= 24.89$ mm

Problem 12.5: A steam boiler, 3 m in diameter, is made of 25 mm thick mild steel plates. The efficiency of longitudinal riveted joint is 88% and that of circumferential riveted joint is 70%. Determine the permissible steam pressure if the maximum tensile stress in the plate section through the rivets is not to exceed 120 N/mm^2. Also calculate (a) circumferential stress in the solid plate section, and (b) longitudinal stress in the plate section through the rivets.

Solution: Given

$$d = 3 \text{ m}$$

$$t = 25 \text{ mm}$$

$$\eta_1 = 88\%$$

$$\eta_2 = 70\%$$

$$f_j = 120 \text{ N/mm}^2$$

$$f_j = \frac{pd}{2t\eta}$$

or $$p = \frac{2 f_j t\eta}{d}$$

$$= \frac{2 \times 120 \times 25 \times 0.88}{3000} = 1.76 \text{ N/mm}^2$$

$$f = \frac{pd}{2t} = \frac{1.76 \times 3000}{2 \times 25} = 105.6 \text{ N/mm}^2;$$

Also, $$f_{oj} = \frac{pd}{4t\eta} = \frac{1.75 \times 3000}{4 \times 25 \times 0.7} = 75 \text{ N/mm}^2$$

Problem 12.6: A cylindrical shell 2 m long and 90 cm internal diameter and 12 mm metal thickness is subjected to an internal pressure of 1.6 N/mm^2. Determine (a) maximum intensity of shear stress, and (b) changes in the dimensions of the shell. Take E = 2 × 10^5 N/mm^2 and 1/m = 0.3.

Solution: Given

$$L = 2 \text{ m}$$

$$d = 900 \text{ mm}$$
$$t = 12 \text{ mm}$$
$$p = 1.6 \text{ N/mm}^2$$

(a) From maximum shear stress

$$f = \frac{pd}{2t} = \frac{1.6 \times 900}{2 \times 12} = 60 \text{ N/mm}^2 \text{ (tensile)}$$

$$f_0 = \frac{pd}{4t} = \frac{f}{2} = 30 \text{ N/mm}^2 \text{ (tensile)}$$

$$\tau_{max} = \frac{f+p}{2} = \frac{60+1.6}{2} = 30.8 \text{ N/mm}^2$$

or $$\tau_{max} = \frac{f}{2} = \frac{60}{2} = 30 \text{ N/mm}^2, \text{ if } p \text{ is neglected}$$

(b) From changes in the dimensions of the shell

$$e = \frac{f}{E} - \frac{f_0}{mE}$$

$$= \frac{1}{E}(60 - 30 \times 0.3) = \frac{51}{E} \text{ (increase)}$$

$$e_0 = \frac{f_0}{E} - \frac{f}{mE}$$

$$= \frac{1}{E}(30 - 60 \times 0.3) = \frac{12}{E} \text{ (increase)}$$

∴ Change in the diameter

$$= \delta d = e.d = \frac{51 \times 900}{2 \times 10^5} = 0.2295 \text{ mm (increase)}$$

Change in the length

$$= \delta L = e_0.L = \frac{12 \times 2000}{2 \times 10^5}$$

$$= 0.12 \text{ mm (increase)}$$

$$\frac{\delta V}{V} = e_v = e_0 + 2e = \frac{12}{E} + 2 \times \frac{51}{E} = \frac{114}{E}$$

$$\therefore \quad \delta V = \frac{114}{2 \times 10^5}\left[\frac{\pi}{4}(900)^2 \times 2000\right]$$

$$= 725.2 \times 10^3 \text{ mm}^3 \text{ (increase)}$$

Problem 12.7: A steel tube having outside and inside diameters of 40 mm and 35 mm is firmly plugged at both the ends, leaving an internal length of 200 mm between the flat ends of the plugs. The plugs are designed so that water can be admitted to the inner space and also so that an axial pull can be applied to the tube. If the tube is subjected to an axial pull of 20 kN and in addition is filled with water at a guage pressure of 2 N/mm², find the volume of water which will escape from the tube if the axial load is removed and the inner space is open to the atmosphere.
For steel, take $E = 2.1 \times 10^5$ N/mm² and $\mu = 0.286$. For water, take $K = 2.1 \times 10^3$ N/mm².

Solution: Given:

$$d = 35 \text{ mm}$$

$$D = 40 \text{ mm}$$

$$L = 200 \text{ mm}$$

$$P = 20 \text{ kN}$$

$$p = 2 \text{ N/mm}^2$$

$$V = \frac{\pi}{4} d^2 L = \frac{\pi}{4} (35)^2 \times 200$$

$$= 19.24 \times 10^4 \text{ mm}^3$$

Initial conditions:

Longitudinal stress

$$f_0 = \frac{1}{A}\left(\frac{\pi}{4} d^2\ p + p\right)$$

$$= \frac{\frac{\pi}{4}(35)^2 \times 2 + 20000}{\frac{\pi}{4}\left(40^2 - 35^2\right)}$$

$$= 74.44 \text{ N/mm}^2$$

$$\text{Hoop stress } f = \frac{pd}{2t} = \frac{2 \times 35}{2(20 - 17.5)} = 14 \text{ N/mm}^2 \text{ (tensile)}$$

Transverse compressive stress = water pressure, $p = 2$ N/mm²

$$\therefore \quad e = \frac{1}{E}\,[14 - 0.286 \times 74.44 + 0.286 \times 2] = -\frac{6.72}{E}$$

$$e_0 = \frac{1}{E}\,[74.44 - 0.286 \times 14 + 0.286 \times 2] = +\frac{71.01}{E}$$

∴ Volumetric strain

$$e_v = e_0 + 2\,e = \frac{71.01}{E} - \frac{2 \times 6.72}{E} = \frac{57.57}{E} \text{ (increase)}$$

Now δV = expansion of tube caused by water pressure and external force

$$= \frac{57.57}{2.1 \times 10^5} \times 19.24 \times 10^4 = 52.745 \text{ mm}^3$$

δV = compressoin of water under compressive stress p

$$= \frac{pV}{K} = \frac{2 \times 19.24 \times 10^4}{2.1 \times 10^3}$$

$$= 183.238 \text{ mm}^3$$

∴ Volume of water escaping when both external load and internal pressure is removed

$$= \delta V + \delta V' = 52.745 + 183.238 \approx 236 \text{ mm}^3.$$

Problem 12.8: Calculate Euler's critical stresses for a series of columns having slenderness ratio of 50, 100, 150, 200 and 250 under the following conditions: (a) both ends hinged, and (b) both ends fixed. Take $E = 2 \times 10^5$ N/mm².

Solution:

(a) When both ends hinged: the Euler's cripling load

$$p_E = \frac{\pi^2 E r^2}{L^2} = \frac{\pi^2 E}{\left(\frac{L}{r}\right)^2} = \frac{19.739 \times 10^5}{\left(\frac{L}{r}\right)^2}$$

Substituting the values of $\frac{L}{r}$ in the above formula, we obtain following values.

$\frac{L}{r} \rightarrow$	100	150	200	250
P_E (N/mm²)	197.4	87.7	49.3	31.6

(b) When both ends fixed: the Euler's cripling load,

$$L_E = \frac{L}{2}$$

$$\therefore \quad P_E = \frac{4\pi^2 E}{\left(\frac{L}{r}\right)^2} = \frac{78.9568 \times 10^5}{\left(\frac{L}{r}\right)^2}$$

Substituting various values of $\frac{L}{r}$, we obtain as follows,

$\frac{L}{r} \rightarrow$	100	150	200	250
P_E (N/mm²)	789.6	350.9	197.4	126.3

Problem 12.9: Compare the crippling loads given by Euler's and Rankine's formulae, for a tubular steel strut 2.5 m long, having outer and inner diameters as 40 mm and 30 mm respectively, loaded through pin joints at the ends. Take yield stress as 320 N/mm², the Rankine's constant $= \frac{1}{7500}$ and $E = 2 \times 10^5$ N/mm². For what length of the strut of this cross-section does the Euler's formula cease to apply?

Solution:

Moment of Inertia of the section

$$I = \frac{\pi}{64}(40^4 - 30^4) = 85903 \text{ mm}^4$$

Area of cross-section

$$A = \frac{\pi}{4}(40^2 - 30^2) = 549.8 \text{ mm}^2$$

Radius of zyration

$$r = \sqrt{\frac{I}{A}} = \sqrt{\frac{85903}{549.8}} = 12.5 \text{ mm};$$

$$L = 2.5 \text{ m} = 2500 \text{ mm}$$

Euler's cripling load

$$P_E = \frac{\pi^2 EI}{L^2} = \frac{\pi^2 \times 2 \times 10^5 \times 85903}{(2500)^2}$$

$$= 27131 \text{ N} = 27.131 \text{ kN}$$

$$P_R = \frac{f_c \times A}{1 + a\left(\frac{L}{r}\right)^2} = \frac{320 \times 549.8}{1 + \frac{1}{7500}\left(\frac{2500}{12.5}\right)^2}$$

$$= 27779 \text{ N} = 27.779 \text{ kN}$$

$$\therefore \quad \frac{P_E}{P_R} = \frac{27.131}{27.779} = 0.977$$

Now $\quad p_E = \frac{P_E}{A} = 320 = \frac{\pi^2 E}{\left(\frac{L}{r}\right)^2}$

$\therefore$ Validity limit of

$$\frac{L}{r} = \sqrt{\frac{\pi^2 E}{320}} = \sqrt{\frac{\pi^2 (2 \times 10^5)}{320}} = 78.54$$

or $f_0^2 - 694.5 f_0 + 28104 = 0$

From which we get

$$f_0 = 43.147 \text{ N/mm}^2$$

$\therefore$ Safe load $P = f_0 A$

$$= 43.147 \times 2827.4$$

$$= 121994 \text{ N} \approx 122 \text{ kN}$$

Problem 12.10: A column section, shown in Fig. P 12.10, consists of three plates, each of thickness t welded together. It carries an axial load of 1000 kN over an effective length of 3.6 m. Taking $f_y = 250$ N/mm² and $E = 2 \times 1^5$ N/mm², determine the thickness t.

Solution:

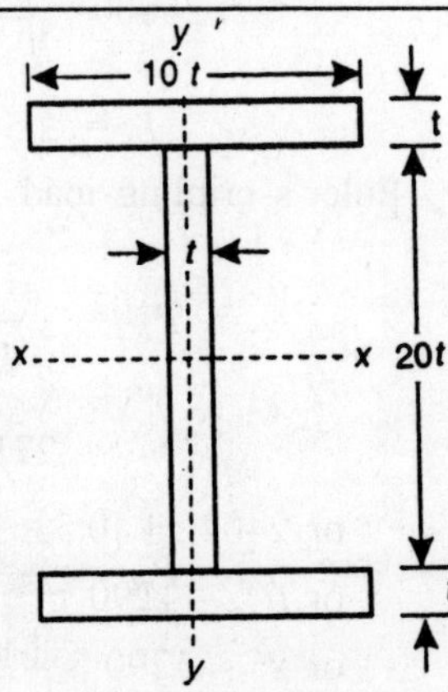

Fig. P. 12.10

M.O.I. about x-x axis

$$(I_X) = \frac{1}{12} t\,(20\,t)^3 + 2$$

$$\left[\frac{1}{12} \times (10\,t)\,t^3 + 10\,t^2\left(10\,t + \frac{t}{2}\right)^2\right]$$

$$= 2273.33\ t^4$$

M.O.I. about y-y axis

$$(I_Y) = 2 \times \frac{1}{12} \times t\,(10\,t)^3 + \frac{1}{12}\,(20\,t)\,t^3$$

$$= 168.33\ t^4$$

Area of section

$$(A) = 2\,[t\,(10\,t)] + 20\,t \times t = 40\,t^2$$

Minimum radius of zyration

$$(r_{\min}) = \sqrt{\frac{I_{\min}}{A}} = \sqrt{\frac{168.33\,t^4}{40\,t^2}} = 2.0514\ t$$

Slenderness ratio (λ)

$$= \frac{l}{r_{\min}} = \frac{3600}{2.0514\,t} = \frac{1754.9}{t}$$

$$\sigma_{ac} = \frac{P}{A} = \frac{1000 \times 10^3}{40\,t^2} = \frac{25000}{t^2}$$

$$f_{cc} = \frac{\pi^2 E}{\lambda^2} = \frac{\pi^2 \times 2 \times 10^5 \times t^2}{(1754.9)^2} = 0.64095\ t^2$$

Now $\sigma_{ac} = \dfrac{0.6\, f_{cc} \cdot f_y}{\left[f_{cc}^n + f_y^n\right]^{\frac{1}{n}}}$ where $n = 1.4$

$$\therefore \quad \frac{25000}{t^2} = \frac{0.6\left(0.64095\, t^2\right)250}{\left[\left(0.64096\, t^2\right)^{1.4} + (250)^{1.4}\right]^{\frac{1}{1.4}}}$$

$$\frac{260.03}{t^4} = \frac{1}{\left[0.5365\, t^{28} + 2275.7\right]^{1/1.4}}$$

or $2404.54\ [0.5365\ t^{2.8} + 2275.7] = t^{5.6}$

or $t^{5.6} - 1290\ t^{2.8} - 5.47 \times 10^6 = 0$

or $x^2 - 1290\ x - 5.47 \times 10^6 = 0$ (where $x = t^{2.8}$)

From which $x = 3071.5 = t^{2.8}$

$$t = (3071.5)^{1/2.8} = 17.6 \text{ mm} = 18 \text{ mm}$$

Problem 12.11: A uniform bar of cross-section area A and flexural stiffness EI is heated so that its temperature varies linearly from $1/2t$ at one end to t at the other end. One end is pin jointed to a rigid foundation. The other end is pin jointed so that it can slide in the direction of the length of the bar, the thermal expansion of which is resisted by a compression spring of stiffness k. If there is no load in the spring when $t = 0$, obtain an expression for the stress in the bar when it is heated and show that it buckles in flexure when

$$t = \frac{4\pi^2 I}{3\alpha L^2 A}\left(1 + \frac{EA}{rL}\right)$$

where α = coefficient of linear thermal expansoin.

Solution:

Average temperature along the length of bar

$$= \frac{1}{2}\left(\frac{1}{2}t + t\right) = \frac{3}{4}\ t$$

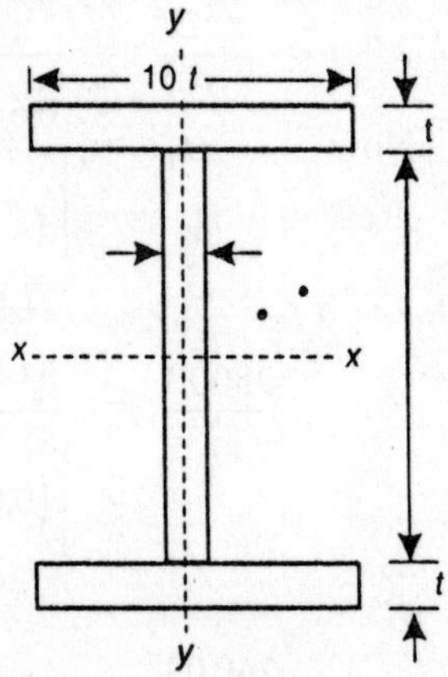

Fig. P. 12.11

$\therefore$ Thermal expansion of bar due to change in temp.

$$= \Delta_t = \frac{3}{4}\alpha Lt \qquad \ldots(1)$$

Let $\quad P$ = force exerted by the spring on the bar.

Then, compression produced in the bar

$$= \frac{PL}{AE} \qquad \ldots(2)$$

Also, compression of the spring

$$= \frac{P}{k} \qquad \ldots(3)$$

Now, net expansion of the bar = compression of spring

$$\therefore \quad \frac{3}{4}\alpha Lt - \frac{PL}{AE} = \frac{P}{k}$$

$$\text{From which } P = \frac{\frac{3}{4}\alpha Lt}{\frac{L}{AE}+\frac{1}{k}} \qquad \ldots(4)$$

and stress in the bar

$$= p = \frac{P}{A} = \frac{\frac{3}{4}\alpha Lt}{\frac{L}{E}+\frac{A}{k}} \quad \textbf{Ans} \qquad \ldots(5)$$

Now, the bar will buckle when buckling load

$$P = \frac{\pi^2 EI}{L^2} \qquad \ldots(6)$$

Equating (4) and (6), we get

$$\frac{\pi^2 EI}{L^2} = \frac{\frac{3}{4}\alpha L t}{\frac{L}{AE} + \frac{1}{k}}$$

Rearranging, we get

$$t = \frac{4\pi^2 I}{3\alpha L^2 A}\left(1 + \frac{AE}{KL}\right)$$

Chapter 13

IES Examination Papers with Solutions

Problem 13.1: *A* rigid beam *ABC* is hinged at *A* and suspended at two points *B* and *C* as shown by two bars *BD* and *CE* made of aluminum and steel respectively. The beam ABC carries a load of 1000 kg at midpoint of BC. The cross-sectional area of the aluminum bar *BD* is 6 mm^2 and that of steel bar that of steel bar *CE* is 4 mm^2. Determine the loads taken up by the two bars BD and CE. Also find the stresses developed in each bar. Modulus of elasticity is 7000 kg/mm^2 for aluminium and 20000 kg/mm^2 for steel. [IES 1987]

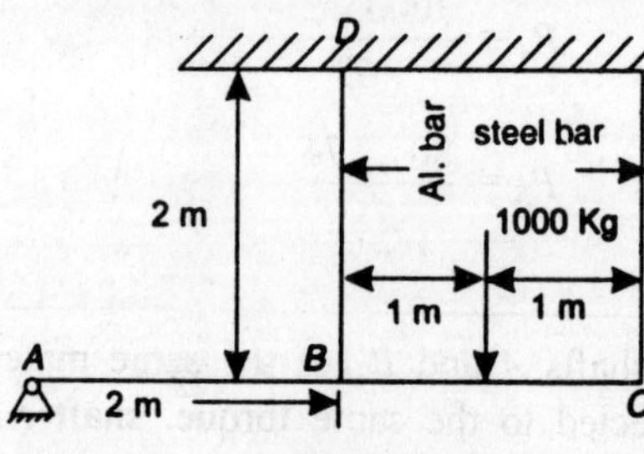

Fig. P. 13.1

Solution:

Since beam ABC is rigid, so it will undergo rigif body rotation, i.e. slope is constant

$$\frac{(\text{deflection})_{\text{Al}}}{2} = \frac{(\text{deflection})_{\text{steel}}}{4} \quad \ldots(1)$$

Deflection in steel bar is 2 times the deflection in A1. bar.

Let forces in seel and aluminium bar be P_S and P_A.

Deflection in a bar loaded with axial load is $\dfrac{PL}{AE}$

Hence, deflection in steel bar $= \dfrac{P_S\, L_S}{A_S\, E_S} = \dfrac{P_S \times 2000}{4 \times 20000} = \dfrac{P_S}{40}$

Deflection in Aluminium bar $= \dfrac{P_A\, L_A}{A_A\, E_A} = \dfrac{P_A \times 2000}{6 \times 7000} = \dfrac{P_A}{21}$

Putting above values in equation (1) we get $= \dfrac{P_S}{40} = \dfrac{2\, P_A}{21}$

$$P_S = \frac{80\, P_A}{21} \qquad \ldots(2)$$

Taking moments about hinge A we get

$$4\, P_S + 2\, P_A = 1000 \times 3 = 3000$$

Putting equation (2) in above equation we get

$$\frac{4 \times 80 P_A}{21} + 2\, P_A = 3000$$

$$\Rightarrow \qquad P_A = \frac{3000 \times 21}{362} = 174 \text{ kg}$$

$$P_S = \frac{80 \times 174}{21} = 662.9 \text{ kg}$$

Problem 13.2: Two shafts A and B of the same material and of same lengths are subjected to the same torque. shaft A is a solid circular section, while shaft B is a hollow circular section whose internal diameter is 0.7 time its outside diameter. If the maximum shear stress in each shaft is to be the same, compare the weight of the two shafts. [IES 1987]

Solution:

Basic Torque equation:

$$\frac{T}{I_P} = \frac{f_S}{r}$$

Since, torque and $(f_s)_{max}$ is same in both hollow and solid shaft. Hence,

$$\left(\frac{I_P}{r}\right)_{hollow} = \left(\frac{I_P}{r}\right)_{solid} \quad \ldots(1)$$

$$(I_P)_{hollow} = \frac{\pi}{64}\cdot\left(D_o^4 - D_i^4\right)$$

$$D_i = 0.7\ D_o$$

$$(I_P)_{hollow} = \frac{\pi}{64}\cdot D_o^4\,(1 - 0.7^4) = \frac{0.7599}{64}\ \pi\, D_o^4$$

$$(I_P)_{solid} = \frac{\pi}{64}D^4$$

Putting above values in equation (1) we get

$$\frac{0.7599}{64}\cdot\frac{\pi\, D_o^4}{(0.5\, D_o)} = \frac{\pi}{64}\cdot\frac{D^4}{(0.5\, D)}$$

$$D_o^3 = \frac{D^4}{0.7599}$$

$$D_o = \frac{D}{0.9125} = 1.096\ D \quad \ldots(2)$$

Since, length and material is same of two shafts. So weight of two shafts is directly proportional to area

$$\frac{W_{solid}}{W_{hollow}} = \frac{\pi\, D^2}{4\cdot\frac{\pi}{4}\cdot\left(D_o^2 - D_i^2\right)}$$

$$= \frac{D^2}{D_o^2\left(1 - 0.7^2\right)} = \frac{D^2}{0.51\, D_o^2}$$

Putting equation (2) in above equation we get

$$\frac{W_{solid}}{W_{hollow}} = \frac{D^2}{0.51(1.096\ D)^2} = 1.632$$

Problem 13.3: Following figure shows a pin connected framework. Determine the axial force in each bar caused by the vertical load P. The two outer bars are identical and have the same cross-sectional area A_i. While the middle bar has an area A_V. All bars have the same modulus of elasticity E.

[IES 1987]

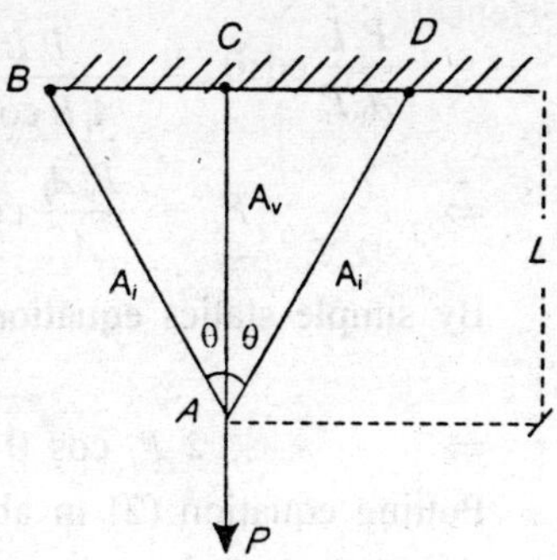

Fig. P. 13.3

Solution:

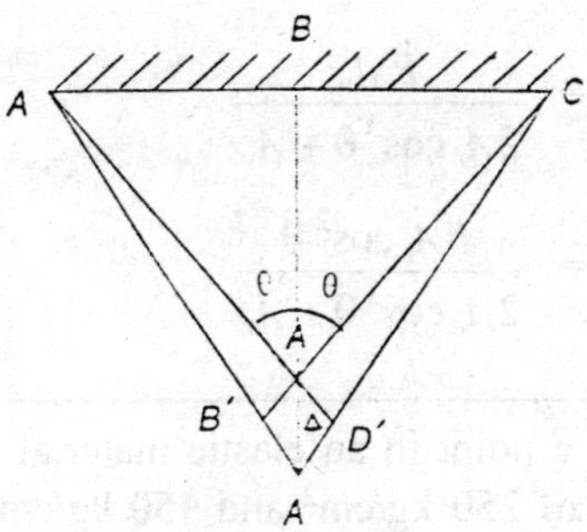

Fig. P. 13.3 (a)

Since the above problem is symmetrical, hence axial force in two outer bars will be same say P_i

Let axial force in vertical bar be P_v and vertical displacement of A be Δ

$AA' = \Delta$

Since, change in length is very small, angle θ will remain same.

$A'B' = A'D' = \Delta \cos \theta$ = elongation, in outer bar.

Elongation in a bar due to axial force P is given by $\dfrac{PL}{AE}$

$$\text{Elongation in vertical bar, } AC = \Delta = \frac{P_v L}{A_v E} \qquad ...(1)$$

$$\text{Length of outer bars} = \frac{L}{\cos\theta}$$

$$\text{Elongation in outer bars} = \Delta\cos\theta = \frac{P_i L / \cos\theta}{A_i E}$$

Putting equation (1) in above equation, we get

$$\frac{P_v L}{A_v E}\cos\theta = \frac{P_i L}{A_i E \cos\theta}$$

$$\Rightarrow \quad P_i = \frac{P_v A_i}{A_v}\cos^2\theta \quad \text{...(2)}$$

By simple statics equation

$$\Sigma V = 0$$

$$\Rightarrow \quad 2\,P_i \cos\theta + P_v = P$$

Putting equation (2) in above equation, we get

$$\frac{2P_v A_i}{A_v}\cos^3\theta + P_v = P$$

$$P_v = \frac{P A_v}{2A_i\cos^3\theta + A_v}$$

$$P_i = \frac{P A_i \cos^2\theta}{2A_i\cos^3\theta + A_v}$$

Problem 13.4: At a point in an elastic material under stress, there are normal stresses of 250 kg/cm^2 and 150 kg/cm^2 on planes A and B respectively at right angles to each other. In addition, there are shear stresses of 25 kg/cm^2 clockwise on the plane A and anti-clockwise on the plane B. Find the principal stresses, maximum shear stress and orientation of principal planes with respect to the plane A and indicate on a sketch. When (i) 250 kg/cm^2 is tensile and 150 kg/cm^2 is compressive, and (ii) both the normal stresses are tensile. [IES 1987]

Solution:

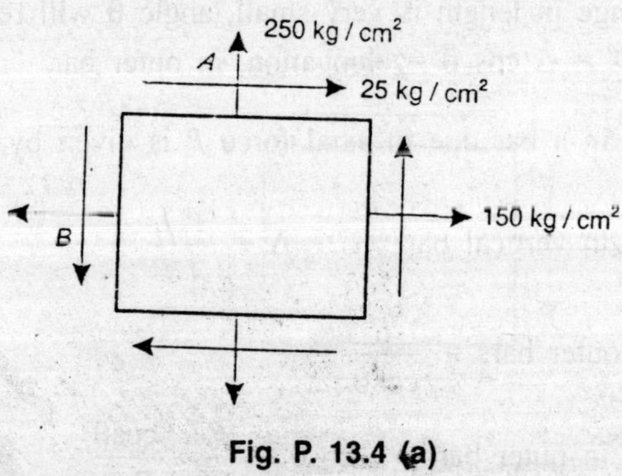

Fig. P. 13.4 (a)

Case (i) $\tau_{xx} = 250 \text{ kg/cm}^2$

$\tau_{yy} = -150 \text{ kg/cm}^2$

$\tau_{xy} = 25 \text{ kg/cm}^2$

Principal stresses in given by

$$\sigma_{1,2} = \left(\frac{\tau_{xx} + \tau_{yy}}{2}\right) \pm \sqrt{\left(\frac{\tau_{xx} - \tau_{yy}}{2}\right)^2 + \tau_{xy}^2}$$

$$\sigma_{1,2} = \frac{250 - 150}{2} \pm \sqrt{\left(\frac{250 + 150}{2}\right)^2 + 25^2}$$

$$= 50 \pm \sqrt{40625} = 50 \pm 201.56$$

$$\sigma_1 = 50 \pm 201.56 = 251.56 \text{ kg/cm}^2$$

$$\sigma_2 = 50 - 201.56 = -151.56 \text{ kg/cm}^2$$

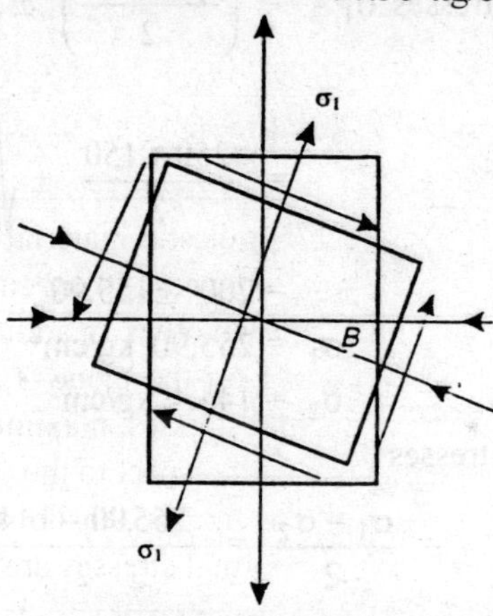

Fig. P. 13.4 (b)

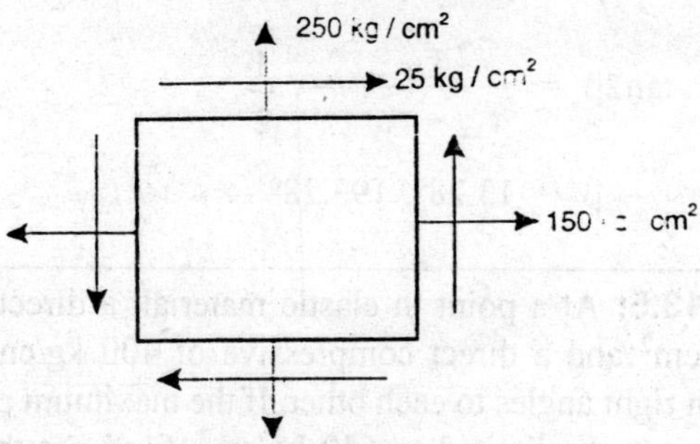

Fig. P. 13.4 (c)

Maximum shear stress

$$= \frac{\sigma_1 - \sigma_2}{2} = \frac{251.56 + 151.56}{2}$$

$$= 201.56 \text{ kg/cm}^2$$

$$\tan 2\beta = \frac{2\tau_{xy}}{\tau_{xx} - \tau_{yy}} = \frac{2 \times 25}{250 + 150}$$

$$\beta = 3.56^\circ \text{ and } 183.56^\circ$$

Case (ii) $\tau_{xx} = 250 \text{ kg/cm}^2$

$\tau_{yy} = 150 \text{ kg/cm}^2$

$\tau_{xy} = 25 \text{ kg/cm}^2$

$$\text{Principal stresses } \sigma_{1,2} = \left(\frac{\tau_{xx} + \tau_{yy}}{2}\right) \pm \sqrt{\left(\frac{\tau_{xx} - \tau_{yy}}{2}\right)^2 + \tau_{xy}^2}$$

$$= \frac{250 + 150}{2} \pm \sqrt{\left(\frac{250 - 150}{2}\right)^2 + 25^2}$$

$$= 200 \pm 55.90$$

$$\sigma_1 = 255.90 \text{ kg/cm}^2$$

$$\sigma_2 = 144.1 \text{ kg/cm}^2$$

Max, shear stresses

$$= \frac{\sigma_1 - \sigma_2}{2} = \frac{255.90 - 144.1}{2}$$

$$= 55.90 \text{ kg/cm}^2$$

Direction of principal stress is given by:

$$\tan 2\beta = \frac{2\tau_{xy}}{\tau_{xx} - \tau_{yy}} = \frac{1}{2}$$

$$\beta = 13.28^\circ, 193.28^\circ$$

Problem 13.5: At a point in elastic material, a direct tensile stress of 600 kg/cm^2 and a direct compressive of 400 kg/cm^2 are applied on planes at right angles to each other. If the maximum principal stress in material is to be limited to 650 kg/cm^2. Find out the shear stress that may be alowed on the planes. Also determine the magnitude and the direction of the minimum principal stress and the maximum shear stress. [IES 1988]

Solution:

The principal stresses are given by formula $\sigma_{1,\,2}$

$$\sigma_{1,\,2} = \frac{\tau_{xx}+\tau_{yy}}{2} \pm \sqrt{\left(\frac{\tau_{xx}-\tau_{yy}}{2}\right)^2 + \tau_{xy}^2} \qquad \text{...(1)}$$

$$\tau_{xy} = 600 \text{ kg/cm}^2$$
$$\tau_{yy} = -400 \text{ kg/cm}^2$$
$$\sigma_1 = 650 \text{ kg/cm}^2$$

Equating above values in equations (1) we get

$$650 = \frac{600-400}{2} \pm \sqrt{\left(\frac{600+400}{2}\right)^2 + \tau_{xy}^2}$$

$$500^2 + \tau_{xy}^2 = 550^2$$
$$\tau_{xy} = 229.13 \text{ kg/cm}^2$$

Min. principal stress

$$= \left(\frac{600-400}{2}\right) - \sqrt{\left(\frac{600+400}{2}\right)^2 + 229.13^2}$$

$$\sigma_2 = -450 \text{ kg/cm}^2$$

Direction of principal stress is given by

$$\tan 2\beta = \frac{2\tau_{xy}}{\tau_{xx}-\tau_{yy}} = \frac{2\times 229.13}{600+400} = 0.45826$$

$$2\beta = \tan^{-1}(0.45826)$$
$$2\beta = 24.62$$
$$\beta = 12.62$$

or $= 180 + 12.62 = 12.62,\ 192.62$

$$\text{Max shear stress} = \left(\frac{\sigma_1 - \sigma_2}{2}\right) = \frac{650-(-450)}{2}$$

$$= \frac{650+450}{2} = 550 \text{ kg/cm}^2$$

Problem 13.6: Two identical steel canilever beams (EI constant) of span L are built into a wall at left with a vertical distance 'a' between them and are connected together by a steel rod at the right as shown in following figure. A load W is applied at the midpoint of the bottom beam. If P is the pull in the steel rod, show that

$$P = \frac{5WL^3}{32\left[L^3 + \frac{6aI}{\pi d^2}\right]}$$

where, d is the diameter of the steel rod and I is moment of inertia of each beam [IES 1988]

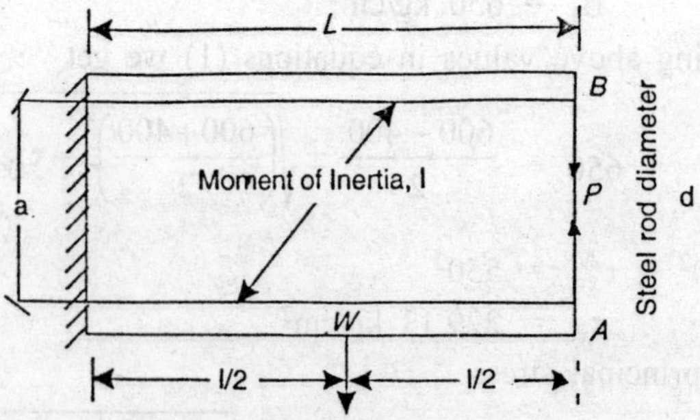

Fig. P. 13.6

Solution:

Deflection at 'A' due to load W can be found by using unit load method:

$$\Delta = \int_0^{L/2} \frac{Wx\cdot(L/2+x)dx}{EI}$$

$$= \frac{W}{EI}\int_0^{L/2}\left(\frac{xL}{2}+x^2\right)dx$$

$$= \frac{W}{EI}\left[\frac{L}{2}\cdot\frac{x^2}{2}+\frac{x^3}{3}\right]_0^{L/2}$$

$$= \frac{W}{EI}\left[\frac{L}{4}\cdot\frac{L^2}{4}+\frac{1}{3}\frac{L^3}{8}\right] = \frac{5WL^3}{48EI}$$

Net deflection at 'A'

$$= \frac{5\,WL^3}{48\,EI} - \frac{PL^3}{3\,EI}$$

where, $\frac{PL^3}{3\,EI}$ is upward deflection due to reaction P.

Deflection at 'B' due to load P at end is

$$\Delta_B = \frac{PL^3}{3\,EI}$$

Net deflection will be axial elongation in steel rod.

Elongation in steel rod due to force $P = \frac{P.a}{AE} = \frac{4\,Pa}{\pi d^2 E}$

$$\Delta_A - \Delta_B = \frac{4\,Pa}{\pi d^2 E}$$

$$\frac{5}{48}\frac{WL^3}{EI} - \frac{2\,PL^3}{3\,EI} = \frac{4\,Pa}{\pi d^2 E}$$

$$\frac{5}{48}\frac{WL^3}{I} = P\left[\frac{2L^3}{3I} + \frac{4a}{\pi d^2}\right]$$

Multiply throughout by $3I$ we get

$$\frac{5WL^3}{16} = P\left[2L^3 + \frac{12\,a\,I}{\pi d^2}\right]$$

$$P = \frac{5WL^3}{32\left[L^3 + \frac{6aI}{\pi d^2}\right]}$$

Problem 13.7: A solid circular shaft 200 mm in diameter has the same cross sectional area as a hollow circular shaft of the same material with inside diameter of 150 mm. For the same maximum shear stress, determine the ratio of torque transmitted by the hollow shaft to that by the solid shaft. Also, compare the angle of twist in the above shafts for equal length and same maximum shear stress. From the above results, what will be your conclusions regarding strength and stiffnes of the two shafts? [IES 1988]

Solution:

Let outer diameter of hollow shaft be D_o

Since areas of two shafts are same

$$\frac{\pi \cdot 200^2}{4} = \frac{\pi}{4}(D_o^2 - 150^2)$$

$$D_o^2 = 200^2 + 150^2$$

$$D_o = 250 \text{ mm}$$

Basic torsion equation is

$$\frac{T}{I_P} = \frac{f_s}{r} = \frac{G\theta}{L}$$

For same max, shear stress, ratio of torque transmitted will be

$$\frac{T_S}{T_h} = \frac{\dfrac{(I_P)_S}{D/2}}{\dfrac{(I_P)_h}{D_0/2}}, \quad \frac{T_S}{T_h} = \frac{\left(\dfrac{\pi \cdot 200^4}{64 \cdot \dfrac{200}{2}}\right)}{\dfrac{\pi}{64}\dfrac{\left(250^4 - 150^4\right)}{250/2}}$$

$$= \frac{200^3 \times 250}{250^4 - 150^4} = 0.588$$

$$\frac{T_{\text{solid}}}{T_{\text{hollow}}} = 0.588$$

Ratio of angle of twist

$$\frac{\theta_s}{\theta_h} = \frac{1}{D/2}\frac{D_o}{2} = \frac{250}{200} = 1.25$$

$$\theta_s = 1.25\ \theta_h$$

From the above result it can be concluded that holow shaft is more stronger and stiffer than solid shaft.

Problem 13.8: A rigid *ABCD* is hinged at *D* and supported by two springs at A and B as shown. The beam carries a vertical load *W* at the point C at a distance of c from the hinged end. The flexibilities (deflection/unit load) of springs at *A* and *B* are d_1 and d_2 respectively. Determine the forces in the springs and also the reaction at the hinged support. [IES 1988]

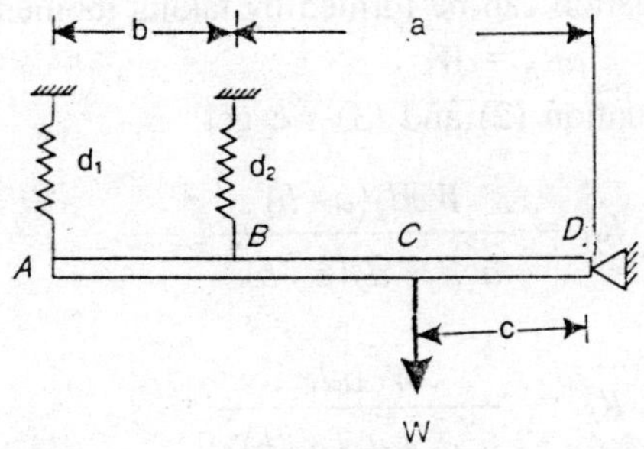

Fig. P. 13.8

Solution:

The above problem is indeterminate to one degree. So a compatibility equation apart from simple statics equation is also to be created. Since rod *AD* is rigid so it will under go rigid body rotation as shown

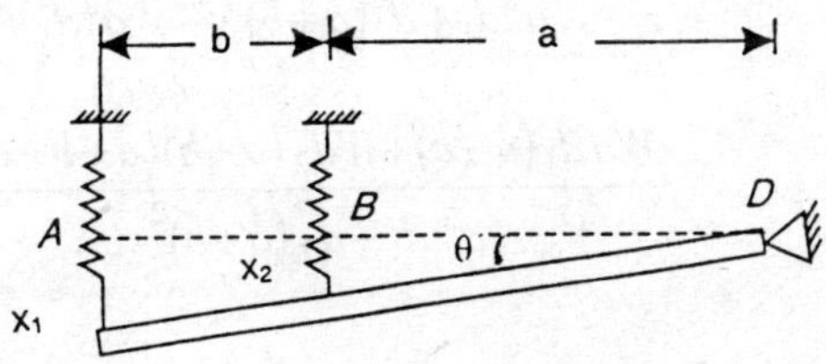

Fig. P. 13.8 (a)

Since, slope θ is constant

$$\frac{x_2}{a} = \frac{x_1}{(a+b)} \qquad \ldots(1)$$

where,

x_1 and x_2 are deflection at A and B.

Let the forces in springs be R_A and R_B

So deflection in springs will be

$$x_1 = R_A d_1$$

$$x_2 = R_B d_2$$

Putting above values in equation (1) we get

$$\frac{R_B d_2}{a} = \frac{R_A d_1}{(a+b)}$$

$$R_B d_2 \ (a + b) = R_A d_1 \ (a) \qquad \text{...(2)}$$

Second equation can be formed by taking moments about hinge D

$$(a + b) \ R_A + aR_B = Wc \qquad \text{...(3)}$$

Solving equation (2) and (3) we get

$$R_A = \frac{Wcd_2(a+b)}{a^2 d_1 + d_2(a+b)^2}$$

$$R_B = \frac{Wcad_1}{a^2 d_1 + d_2(a+b)^2}$$

Now by simply statics

$$\Sigma V = 0$$

R_D = Reaction at hinged support

$$= W - \frac{Wcd_2(a+b)}{a^2 d_1 + d_2(a+b)^2} - \frac{Wcad_1}{a^2 d_1 + d_2(a+b)^2}$$

$$= \frac{Wad_1(a-c) + Wd_2(a+b)(a+b-c)}{a^2 d_1 + d_2(a+b)^2}$$

Problem 13.9: A rigid bar with negligible weight is hinged at 0 and attached to two vertical load P that can be applied at A so that the maximum stresses in the steel and bronze rods do not exceed 150 MPa and 70 MPa respectively? Area of steel bar is 900 mm²; E for steel is 210 GPa and length is 1.5 m, Area of bronze bar is 300 mm², *E* for bronze is 84 GPa and length is 2 m. [IES 1989]

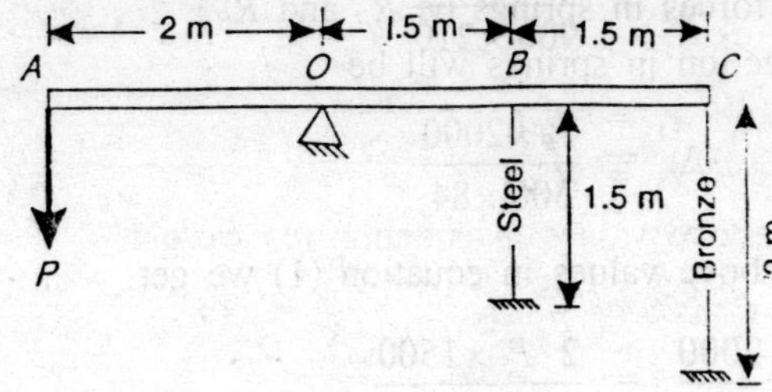

Fig. P. 13.9

Solution:

Let elongation in steel rod be Δ_S and that in bronze bar be Δ_b

Let stress in steel bar be P_S and that in bronze bar be P_b.

Since bar ABC is rigid so it will undergo rigid body rotation, i.e.

$$\frac{\Delta_S}{1.5} = \frac{\Delta_b}{3.0}$$

$$\Rightarrow \quad \Delta_b = 2\,\Delta_S \quad \ldots(1)$$

Deflection in bar due to axial force of $P = \dfrac{PL}{AE}$

So, elongation in steel and bronze bars due to force of P_S and P_b respectively is

$$\Delta_S = \frac{P_S L_S}{A_S E_S}$$

and $$\Delta_b = \frac{P_b L_b}{A_b E_b}$$

$$A_S = 900 \text{ mm}^2$$
$$E_S = 210 \text{ GPa}$$
$$L_s = 1.5 \text{ m} = 1500 \text{ mm}$$
$$A_b = 300 \text{ mm}^2$$
$$E_b = 84 \text{ GPa}$$
$$L_b = 2 \text{ m} = 2000 \text{ m}$$

Hence, $\Delta_S = \dfrac{P_s \cdot 1500}{900 \times 210}$

$$\Delta_b = \frac{P_b \times 2000}{300 \times 84}$$

Putting above values in equation (1) we get

$$\frac{P_b \times 2000}{300 \times 84} = \frac{2 \cdot P_S \times 1500}{900 \times 210}$$

$$P_S = \frac{P_b \cdot (2000) \times 3 \times 210}{2 \times 1500 \times 84} = 5\ P_b$$

Stress in steel bar $= \dfrac{P_S}{A_S} = \dfrac{P_S}{900}$ should be limiteed to 150 MPa

This implies max. value of P_S = 900 × 150 N = 135 kN

Similarly max. value of Pb = 70 × 300 N = 21 kN

$$\frac{\text{Max. value of } P_S}{\text{Max. value of } P_b} = \frac{135}{21}$$

= 6.43 > 5 (as per compatibility requirement since $P_S = 5\ P_b$)

Hence, bronze will reach its maximum stress first than steel bar

So, P_b = 21 kN

$P_s = 5\ P_b = 5 \times 21 = 105$ kN

Taking moment about hinge '0' to find maximum value of load, P, we get

$$2\ P = 1.5\ P_S + 3.0\ P_b$$
$$= 1.5 \times 105 + 3 \times 21$$
$$= 220.5$$
$$P - 110.25 \text{ kN}$$

Problem 13.10: The shear force diagram in a statically determinate beam is shown in Fig. Determine the loading diagram and draw the bending moment diagram. [IES 1989]

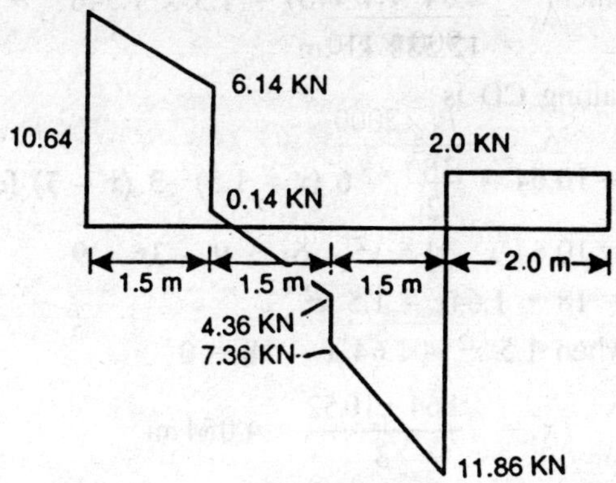

Fig. P. 13.10

Solution:

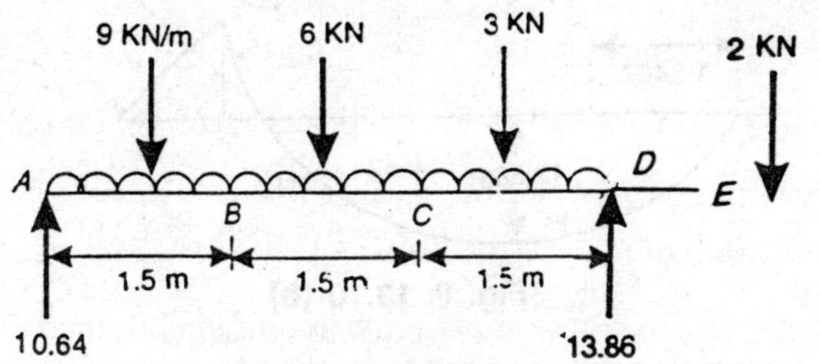

Fig. P. 13.10 (a)

$$\text{Moment along } A = M_x = 10.64\ x - \frac{3x^2}{2} = 10.64\ x - 1.5\ x^2$$

$$\text{Moment at} \quad B = 10.64 \times 1.5 - 1.5^3 = 12.585 \text{ kN.m}$$

$$\text{Moment along } BC = M_x = 10.64x - \frac{3x^2}{2} - 6\ (x - 1.5)$$

$$= 10.64x - 1.5x^2 - 6x + 9 \qquad \ldots\text{(i)}$$

$$\text{Moment at } C = 4.64 \times 3 - 1.5 \times 9 + 9 = 9.42 \text{ kN.m}$$

Since moment 'C' is less than moment at 'B' so max. moment will occur along BC only

Taking differention of equation (i) and equating to zero

$$\frac{dM}{dx} = 4.64 - 3\ x = 0$$

$$x = 1.5467 \text{ m}$$

Max. Moment $= 4.64 \times 1.5467 - 1.5 \times 1.5467^2 + 9$

$= 12.588$ kN.m

Moment along CD is

$$M_x = 10.64 - \frac{3x^2}{2} - 6(x - 1.5) - 3(x - 3) \text{ for } x > 3$$

$$= 10.64x - 1.5x^2 - 6x + 9 - 3x + 9$$

$$= 18 + 1.64x - 1.5x^2$$

$M_x = 0$ when $1.5x^2 - 1.64x - 18 = 0$

$$x = \frac{1.64 \pm 10.52}{3} = 4.054 \text{ m}$$

Plotting the moment diagram

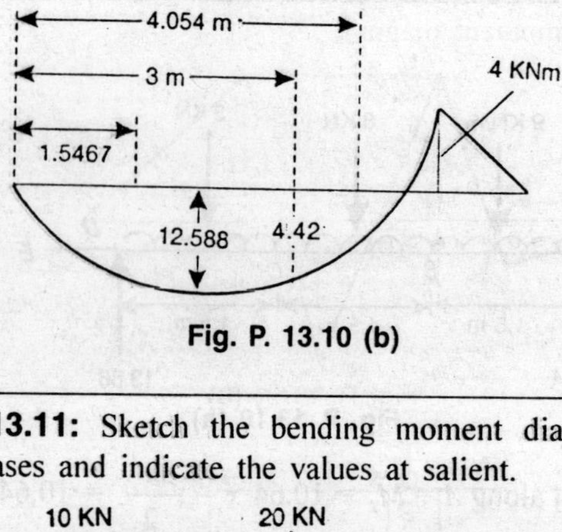

Fig. P. 13.10 (b)

Problem 13.11: Sketch the bending moment diagrams in the following cases and indicate the values at salient. [IES 1989]

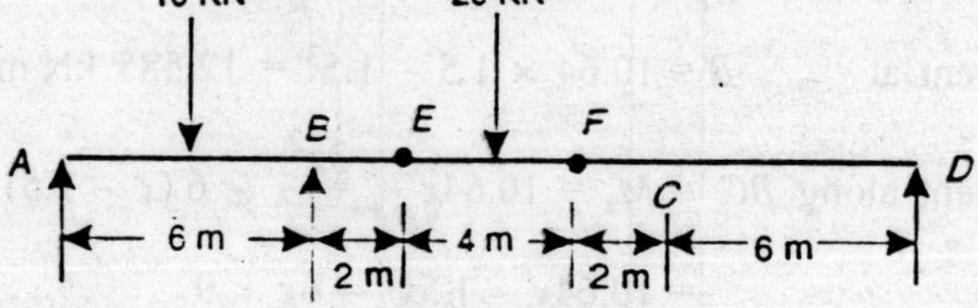

Fig. P. 13.11

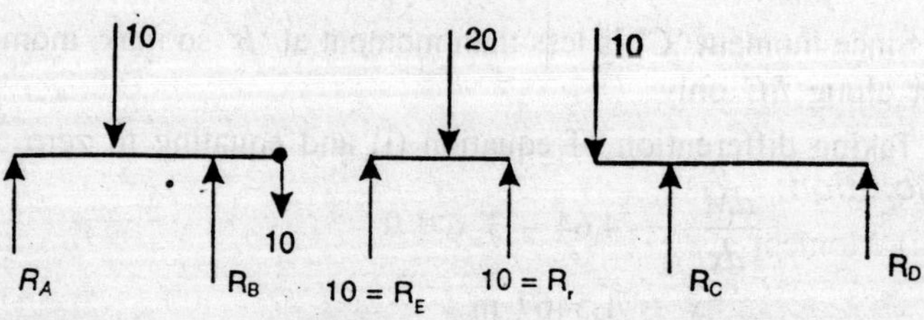

Fig. P. 13.11 (a)

Solution:

(i) Taking moment about A

$$R_B = \frac{10\times 8}{6} + \frac{10\times 3}{6}$$

$$= 18.33$$

$$\Sigma V = 0$$

$$R_A = 20 - 18.33$$

$$= 1.67$$

Taking moment about 'D'

$$R_C = \frac{10\times 8}{6} = 13.33$$

Ploting the moment diagram

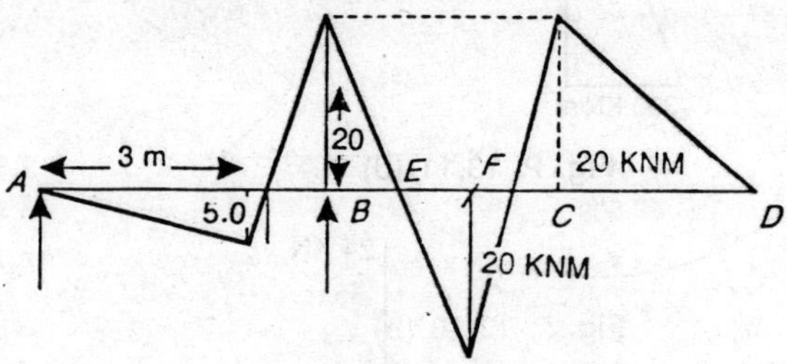

Fig. P. 13.11 (b)

(ii)

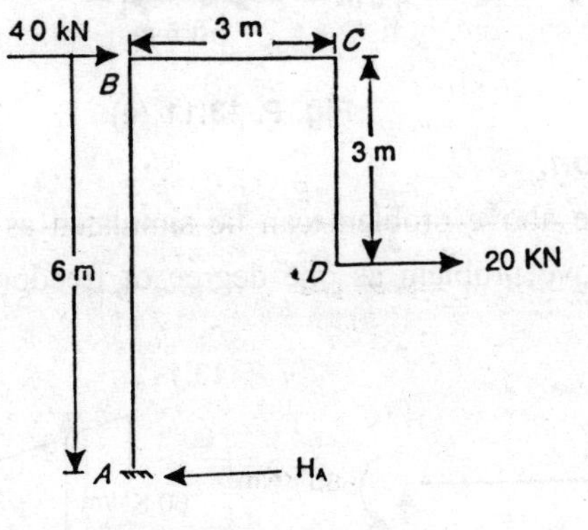

Fig. P. 13.11 (c)

Solution:

(ii) Moment at

$$A = M_A = 40 \times 6 + 20 \times 3 = 300 \text{ kN.m}$$

$$\Sigma H = 0$$

$$H_A = 40 + 20 = 60 \text{ kN}$$

$$\Sigma V = 0$$

$$V_A = 0$$

Plotting moment diagram, we get

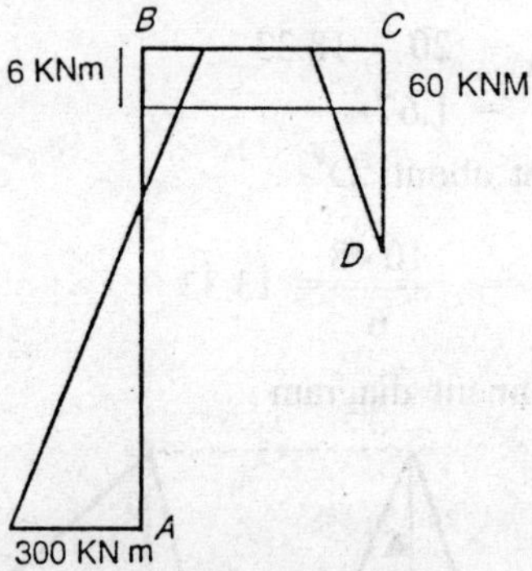

Fig. P. 13.11 (d)

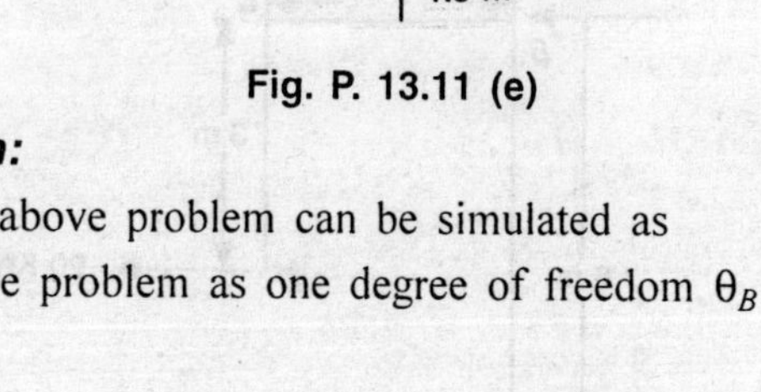

Fig. P. 13.11 (e)

Solution:

(iii) The above problem can be simulated as

The above problem as one degree of freedom θ_B

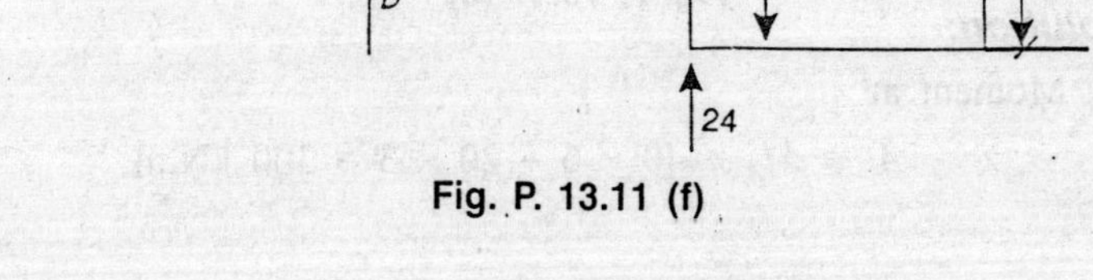

Fig. P. 13.11 (f)

$$M_{AB} = \frac{2\,EI}{4}\,\theta_B$$

$$M_{BA} = \frac{4\,EI\theta_B}{4.0} = 60 \qquad \ldots(1)$$

$$\theta_B = \frac{60}{EI}$$

Putting above value of θ_B in (1) we get

$$M_{AB} = \frac{2}{4} \times 60 = 30 \text{ kN.m}$$

Ploting moment diagram, we get

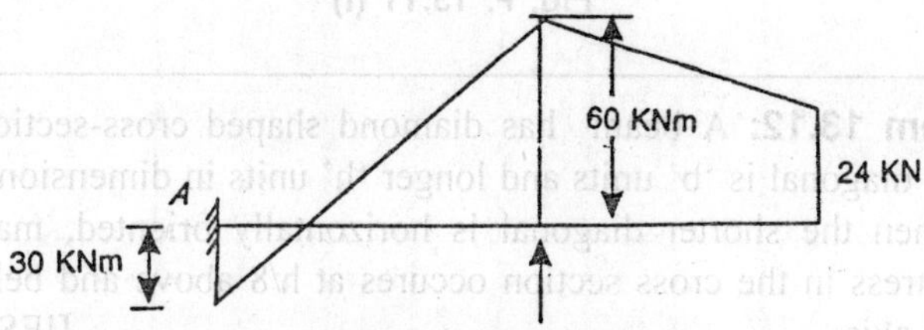

Fig. P. 13.11 (g)

(iv)

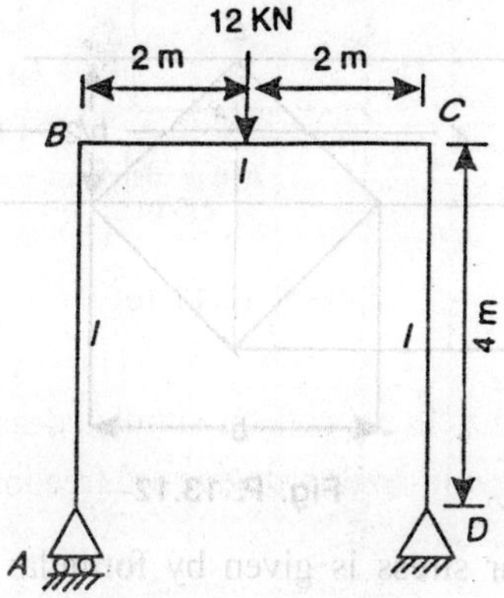

Fig. P. 13.11 (h)

Solution:

(iv) Taking moments about D, we get

$$V_A = \frac{12}{2} = 6 \text{ kN}$$

$V_D = 6$ kN

Plotting moment diagram

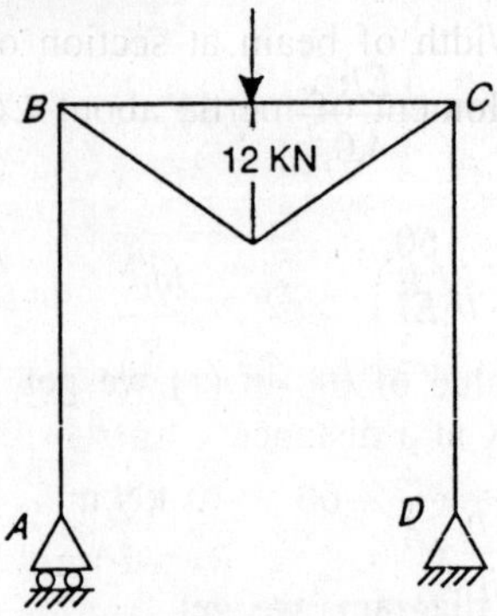

Fig. P. 13.11 (i)

Problem 13.12: A beam has diamond shaped cross-section. The smaller diagonal is 'b' units and longer 'h' units in dimension. Prove that when the shorter diagonal is horizontally oriented, maximum shear stress in the cross section occures at h/8 above and below the neutral axis. [IES 1989]

Solution:

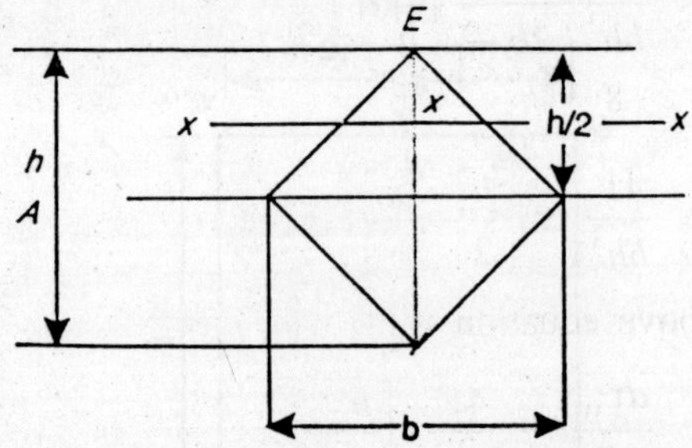

Fig. P. 13.12

The basic shear stress is given by formula:

$$\tau_{xy} = \frac{V}{Ib_i} \Sigma A_i \bar{y}_i$$

where, V = Shear force at vertical section

A_i = Area of an element between section of interest and exterme fibre.

$\bar{y}_i$ = Distance of centroid of this area, A_i, from the neutral axis.

b = Width of beam at section of interest.

I = Moment of inertia about CG.

Moment of Inertia,

$$I = \frac{1}{2} b \left(\frac{h}{2}\right)^3 \times 2 = \frac{bh^3}{8}$$

Taking a section X-X at a distance x from extreme fibre E, we get

$$b_i = \frac{2b}{h} x$$

$$A_i = \frac{1}{2} \cdot \left(\frac{2b}{h} x\right) \cdot x = \frac{b}{h} x^2$$

$$y_i = \frac{h}{2} - \frac{2x}{3}$$

Putting above values in basic shear stress formula we get

$$\tau_{xy} = \frac{V}{\frac{bh^3}{8}\left(\frac{2b}{h} x\right)} \cdot \frac{b}{h} x^2 \left[\frac{h}{2} - \frac{2x}{3}\right]$$

$$= \frac{4V}{bh^3} x \left[\frac{h}{2} - \frac{2x}{3}\right]$$

Differentiating above equation w.r.t. x we get

$$\frac{d\tau_{xy}}{dx} = \frac{4V}{bh^3} \left[\frac{h}{2} - \frac{4}{3} x\right]$$

τ_{xy} is maximum when

$$\frac{d\tau_{xy}}{dx} = 0$$

$$\Rightarrow \quad \frac{h}{2} - \frac{4x}{3} = 0$$

$$x = \frac{3h}{8}$$

Shear stress is maximum at a distance at distance of $\left(\frac{h}{2}-\frac{3h}{8}\right)$ from neutral axis.

Hence, shear stress is max, at a distance of $h/8$ from neutral axis.

> **Problem 13.13:** A rectangular beam 100 × 300 mm deep carries a UDL of 24 kN/m over a simply supported span of 4 m. Determine the principal stresses at a point 75 mm below the neutral axis at quarter span section. Neglect the self weight of thet beam
> Determine moments and shear at quarter section X-X. [IES 1989]

Solution:

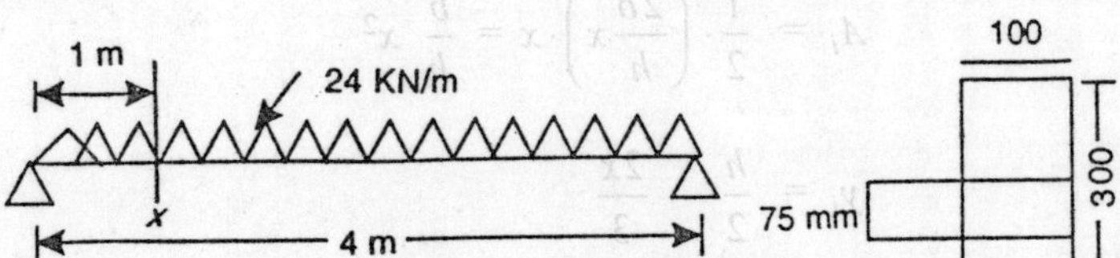

Fig. P. 13.13

$$\text{Shear} = \frac{24\times4}{2} - 24 = 24 \text{ kN.}$$

$$\text{Moment} = 48 \times 1 - 24 \times 1 \times 0.5$$

$$= 48 - 12 = 36 \text{ kN}$$

$$I = \frac{bd^3}{12} = \frac{100\times300^3}{12}$$

$$= 225 \times 10^6 \text{ mm}^4$$

Bending stress at 75 mm below neutral axis

$$= \sigma_b = \frac{M}{I}\cdot y$$

$$\sigma_b = \frac{36\times10^6}{225\times10^6}\cdot 75$$

$$= 12 \text{ N/mm}^2$$

Shear stress at 75 mm below neutral axis

$$\tau_{xy} = \frac{V}{Ib} \cdot \Sigma A_i y_i$$

$$= \frac{24 \times 10^3}{225 \times 10^6 \times 100} \cdot 75 \times 100 \left(150 - \frac{75}{2}\right)$$

$$= \frac{24 \times 75 \times 112.5}{225 \times 10^3}$$

$$= 0.9 \text{ N/mm}^2$$

Principal stress, $\sigma_{1,2}$

$$= \frac{\sigma_b}{2} \pm \sqrt{\left(\frac{\sigma_b}{2}\right)^2 + \tau_{xy}^2}$$

$$= \frac{12}{2} \pm \sqrt{5.915^2 + 0.9^2}$$

$$= 6 \pm 5.983$$

$$\sigma_1 = 6 + 5.983 = 11.983 \text{ N/mm}^2$$

$$\sigma_2 = 6 - 5.983 = -0.0.017 \text{ N/mm}^2$$

Problem 13.14: A beam *ABC* of length *L* is simply supported at one end *A* and at end B, distance *x* from the free end C. the beam carries a UDL of intensity *w* per unit length over its entire length. Determine the value of *x* so that the distribution of bending moment in span will be most economical [IES 1989]

Solution:

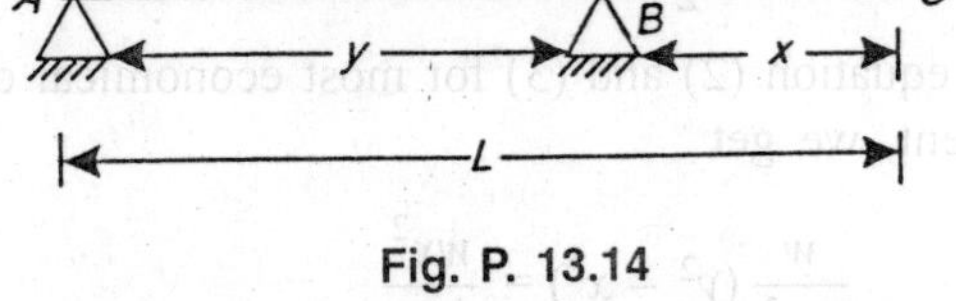

Fig. P. 13.14

The distribution of bending moment in the span will be most economical when the max. sagging moment along AB is equal to hogging moment at support B.

Taking moment about 'B' to find reaction at 'A' we get

$$R_A\, y = \frac{wy^2}{2} - \frac{wx^2}{2} = \frac{w}{2}(y^2 - x^2)$$

$$R_A = \frac{w}{2y}(y^2 - x^2)$$

Moment along AB at a distaance 'a' from A can be written as

$$M_a = R_A\, a - \frac{wa^2}{2}$$

$$= \frac{w}{2y}(y^2 - x^2)\, a - \frac{wa^2}{2} \quad \text{...(1)}$$

M_a is maximum where

$$\frac{dMa}{da} = 0$$

$$\frac{w}{2y}(y^2 - x^2) - wa = 0$$

$$a = \left(\frac{y^2 - x^2}{2y}\right)$$

Putting above value of 'a' in equation (1) to find maximum sagging moment we get

$$M_{\max} = \frac{w}{4y^2}(y^2 - x^2)^2 - \frac{w\left(y^2 - x^2\right)^2}{8y^2} = (y^2 - x^2)^2 \quad \text{...(2)}$$

Maximum hogging moment at support 'B' is

$$= \frac{wx^2}{2} \quad \text{...(3)}$$

Equating equation (2) and (3) for most economical distribution of bending moment, we get

$$\frac{w}{8y^2}(y^2 - x^2) = \frac{wx^2}{2}$$

$$\Rightarrow \quad (y^2 - x^2)^2 = 4x^2 y^2$$

$$\Rightarrow \quad y^4 + x^4 - 6x^2 y^2 = 0$$

Dividing by x^4 we get

$$\left(\frac{y}{x}\right)^4 + 1 - \frac{6y^2}{x^2} = 0$$

Let $\frac{y^2}{x^2}$ be α

$$\Rightarrow \qquad \alpha^2 - 6\alpha + 1 = 0$$

Solving above quadratic equation,

$$\alpha = \frac{6 \pm \sqrt{32}}{2} = 5.828,\ 0.1716$$

$$\frac{y}{x} = \sqrt{\alpha} = 2.414,\ 0.414$$

$$\Rightarrow \qquad y = 2.414\ x \text{ or } 0.414\ x$$

Since $\quad x + y = L$

Putting above values of y, we get

$$x = 0.293\ L \text{ or } x = 0.707\ L$$

When $x = 0.707\ L$ there will be hogging moment in AB so 'x' can't be equal to $0.707\ L$.

Hence $\qquad x = 0.293\ L$

Problem 13.15: Determine the stiffness of the beam at the end is of the beam shown in Fig. P. 13.15 and the carry over moment at A.C is an internal hinge. [IES 1990]

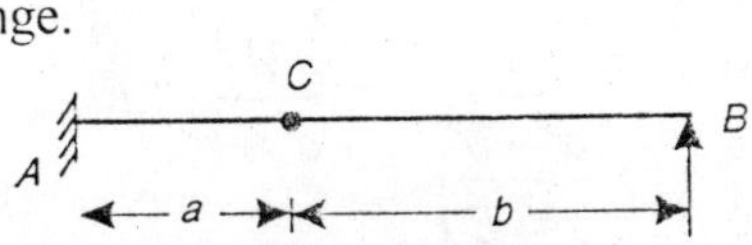

Fig. P. 13.15

Solution:

Applying a unit moment at B

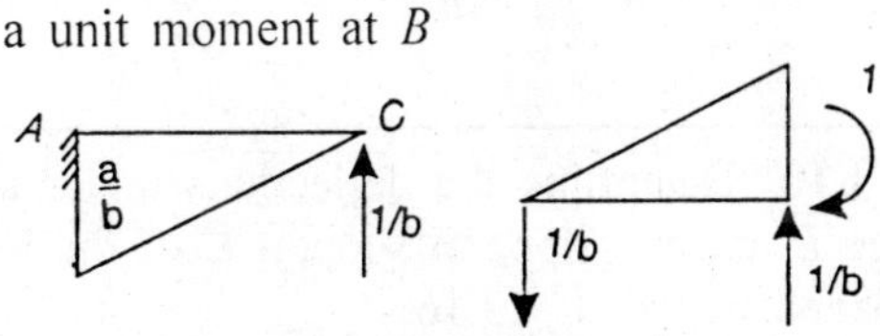

Fig. P. 13.15 (a)

Rotation at 'B' can be found using unit load method

$$\theta_B = \int \frac{Mm\,dx}{EI} = \int_0^{(a+b)} \left(1-\frac{x}{b}\right)^2 \frac{dx}{EI}$$

$$= \frac{1}{EI}\int_0^{1+b} \left(1+\frac{x^2}{b^2}-\frac{2x}{b}\right)\frac{dx}{EI}$$

$$= \frac{1}{EI}\left(x+\frac{x^3}{3b^2}-\frac{x^2}{b}\right)\Bigg|_0^{a+b}$$

$$= \left[(a+b)+\frac{(a+b)^3}{3b^2}-\frac{(a+b)^2}{b}\right]\frac{1}{EI}$$

$$= \left[a+b+\frac{a^3+b^3+3a^2b+3ab^2}{3b^2}-\frac{a^2+b^2+2ab}{b}\right]\frac{1}{EI}$$

$$= \left[a+b+\frac{a^3}{3b^2}+\frac{b}{3}+\frac{a^2}{b}+a-\frac{a^2}{b}-b-2a\right]\frac{1}{EI}$$

$$= \frac{1}{EI}\,\frac{a^3+b^3}{3b^2} = \left(\frac{a^3+b^3}{3b^2}\right)\cdot\frac{1}{EI}$$

So moment a B due to unit rotation at B

$$= \frac{1}{\dfrac{a^3+b^3}{3b^2\,EI}} = \frac{3b^2}{a^3+b^3}\cdot EI$$

Carry over moment at

$$A = \frac{a}{b}\cdot\frac{3b^2}{\left(a^3+b^3\right)}EI = \frac{3ab}{a^3+b^3}\cdot EI$$

Problem 13.16: Determine the deflections under the load, at the internal hinge B and the slope at support C of the beam from ABC shown as loaded in Fig. P. 13.16. [IES 1990]

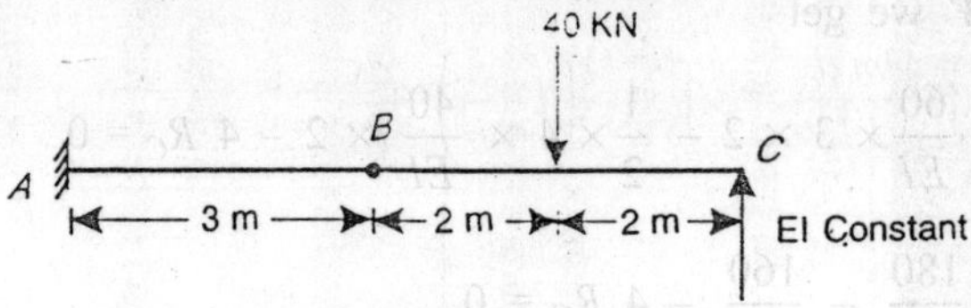

Fig. P. 13.16

Solution:

Conjugate beam method will be more appropriate as shear at a point loaded with $\frac{M}{EI}$ diagram in conjugate beam gives the slope and moment at that point gives deflection at that point.

Conjugate beam will be

Fig. P. 13.16 (a)

Bending Moment diagram of original beam is

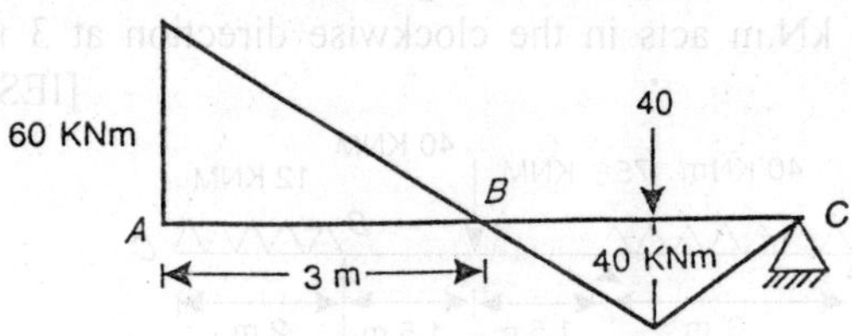

Fig. P. 13.16 (b)

BC behaves as simply supported at B

Loading conjugate beam with $\frac{M}{EI}$ we get

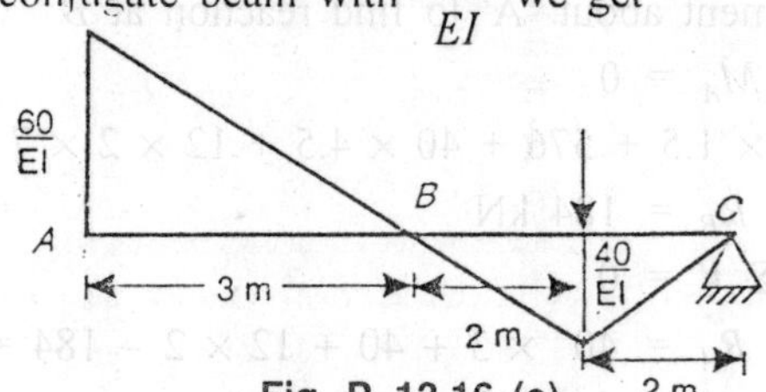

Fig. P. 13.16 (c)

Slope at '*C*' reaction at '*C*' in conjugate beam. Taking moments about 'B' we get

$$\frac{1}{2}\frac{60}{EI}\times 3\times 2-\frac{1}{2}\times 4\times\frac{40}{EI}\times 2-4\,R_C=0$$

$$\Rightarrow \frac{180}{EI}-\frac{160}{EI}-4\,R_C=0$$

$$R_C=\frac{5}{EI}=\text{Slope at } C$$

Deflection under load is moment in conjugate beam

$$\Rightarrow \frac{5}{EI}\times 2+\frac{1}{2}\times\frac{40}{EI}\times 2\times\frac{2}{3}=\frac{110}{3\,EI}$$

Deflection at internal hinge

$$B=\frac{5\times 4}{EI}+\frac{1}{2}\times\frac{4\times 40\times 2}{EI}=\frac{180}{EI}$$

Problem 13.17: Draw the bending moment and shear force diagram for the beam shown as loaded in Fig. P. 13.17. The concentrated moment of 576 kN.m acts in the clockwise direction at 3 m from end A. [IES 1990]

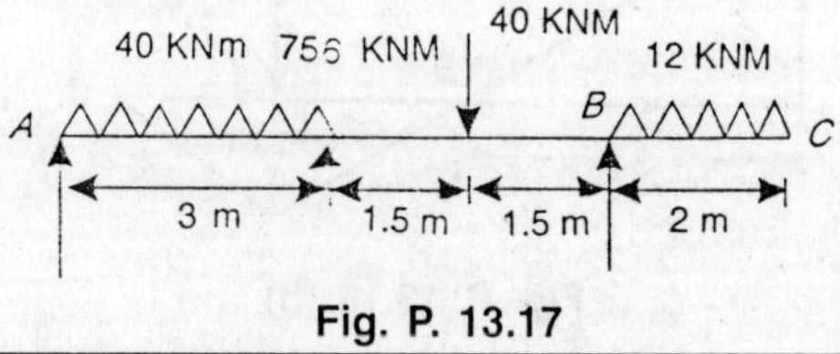

Fig. P. 13.17

Solution:

Taking moment about 'A' to find reaction at *B*

$$\Sigma M_A=0$$

$$\Rightarrow 40\times 3\times 1.5+576+40\times 4.5+12\times 2\times 7=6\,R_B$$

$$R_B=184\text{ kN}$$

$$\Sigma V=0$$

$$\Rightarrow \quad R_A=40\times 3+40+12\times 2-184=0$$

Plotting shear force diagram

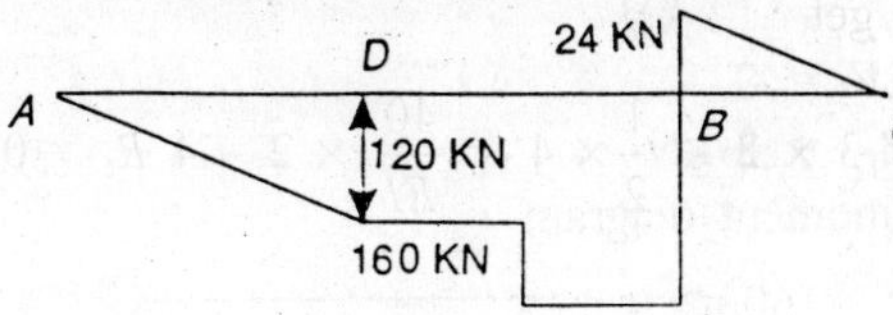

Fig. P. 13.17 (a)

Plotting the moment diagram

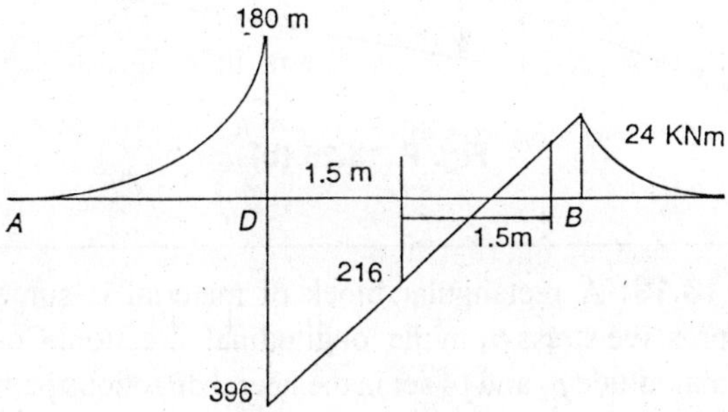

Fig. P. 13.17 (b)

Problem 13.18: (a). Sketch the bending moment and or other moments acting on the members in the following cases shown in Fig. P. 13.18 (a). Identify the other type of moments wherever applicable. Indicate the value at salient points.

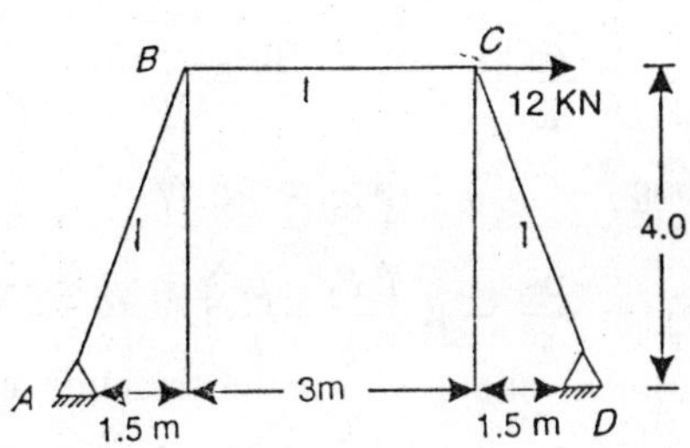

Fig. P. 13.18 (a)

Solution:

Taking moments about D

$$R_A = \frac{12 \times 4}{6} = 8 \text{ kN}$$

$$\Sigma H = 0$$

$$\Rightarrow \quad H_A = 12 \text{ kN}$$

$$\Sigma V = 0$$

$$\Rightarrow \quad R_D = 8 \text{ kN}$$

Plotting the moment diagram

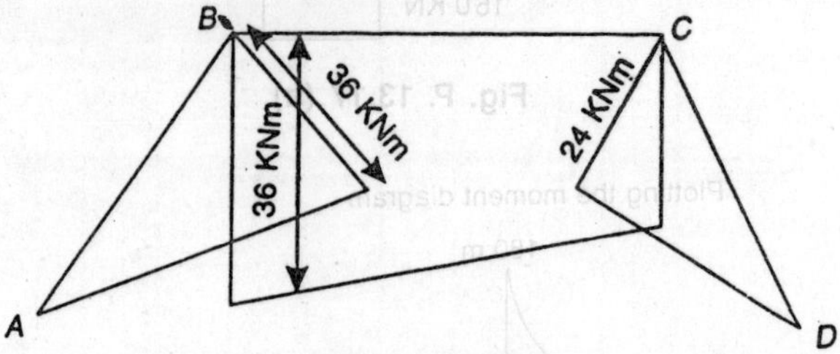

Fig. P. 13.18 (b)

> **Problem 13.19:** A rectangular block of material is subjected to a direct compressive stress p_1 in the longitudinal direction. Compressive stresses of magnitude p_2 and p_3 act in the lateral directions perpendicular to the longitudinal direction so that the net transverse strain along the p_2 and p_3 directions will be 1/2 and 1/4 the strains that would have occurred if p_1 alone were acting. Determine p_2 and p_3 in terms of p_1 and the net strain in the longitudinal direction. [IES 1990]

Solution:

Let modulus of elasticity of material be E and poissons ratio be μ.

$$p_2 = \frac{p_2}{E} - \frac{\mu p_1}{E} - \frac{\mu p_3}{E} \qquad \ldots(1)$$

Net strain along

$$p_3 = \frac{p_3}{E} - \mu\left(\frac{P_2}{E} + \frac{p_1}{E}\right) \qquad \ldots(2)$$

Net strain along p_2 when only p_1 is acting $= -\dfrac{\mu p_1}{E}$ ...(3)

Net strain along p_3 when only p_1 is acting $= -\dfrac{\mu p_1}{E}$ ...(4)

From equations (1) to (4) we get

$$\frac{p_2}{E} - \frac{\mu p_1}{E} - \frac{\mu p_3}{E} = \frac{1}{2}\left(-\frac{\mu p_1}{E}\right) \quad ...(5)$$

$$\frac{p_3}{E} - \mu\left(\frac{p_2}{E} + \frac{p_1}{E}\right) = \frac{1}{4}\left(-\frac{\mu p_1}{E}\right) \quad ...(6)$$

From equation (5) we get

$$2\,p_2 - 2\,\mu p_3 = \mu p_1 \quad ...(7)$$

From equation (6) we get

$$4\,p_3 - 4\,\mu p_2 = 3\,\mu p_1 \quad ...(8)$$

Solving equations (7) and (8) we get

$$p_3 = \frac{\mu p_1 (2\mu + 3)}{4(1-\mu^2)}$$

$$p_2 = \frac{\mu p_1 (2 + 3\mu)}{4(1-\mu^2)}$$

Problem 13.20: A beam *ABCD* of span *L* is fixed at *A* and *B*. *A* a concentrated load *W* acts at *B* and *C*.

$$AB = BC = CD = \frac{L}{3}$$

Moment of inertia of segments *AB* and *CD* is I and that of *BC* is 2*I*. Determince the fixec end moments. [IES 1991]

Solution:

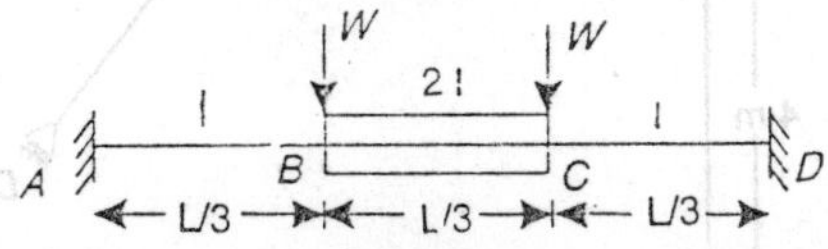

Fig. P. 13.20 (a)

Due symmetry, fixed end moments and Reactions at A and D are equal. The above problem figure can be divided into two parts.

Plotting $\frac{M}{EI}$ diagram.

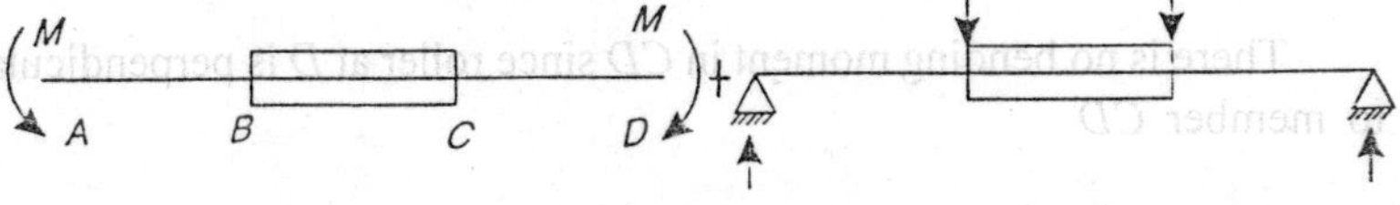

Fig. P. 13.20 (b)

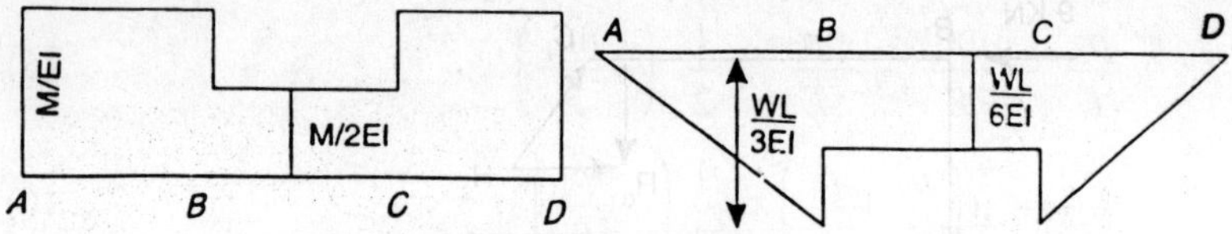

Fig. P. 13.20 (c)

Since slopes at A and D is zero

$$\frac{\Sigma M}{EI} = 0$$

$$\Rightarrow \frac{M}{EI}\cdot\frac{L}{3} + \frac{M}{2EI}\cdot\frac{L}{3} + \frac{M}{EI}\cdot\frac{L}{3} = \frac{1}{2}\frac{WL}{3EI}\cdot\frac{L}{3}\times 2 + \frac{WL}{6EI}\cdot\frac{L}{3}$$

$$\Rightarrow \frac{5ML}{6} = \frac{WL^2}{6}$$

$$\Rightarrow M = \frac{WL}{5}$$

Problem 13.21: Draw bending diagrams for the structures shown in following Fig. P. 13.21 and indicate the magnitude of bending moment at salient points. [IES 1991]

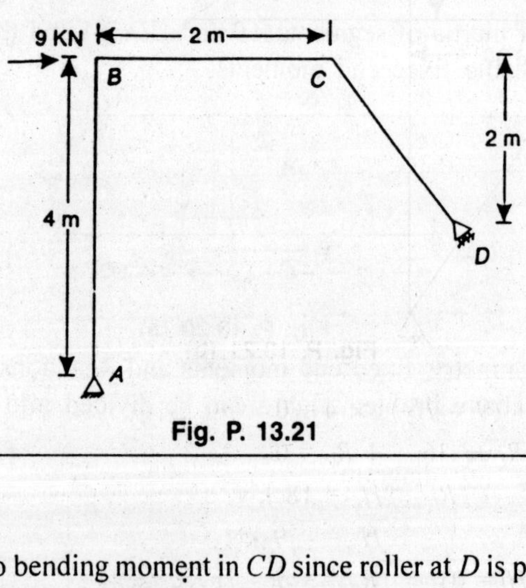

Fig. P. 13.21

Solution:

There is no bending moment in CD since roller at D is perpendicular to member CD

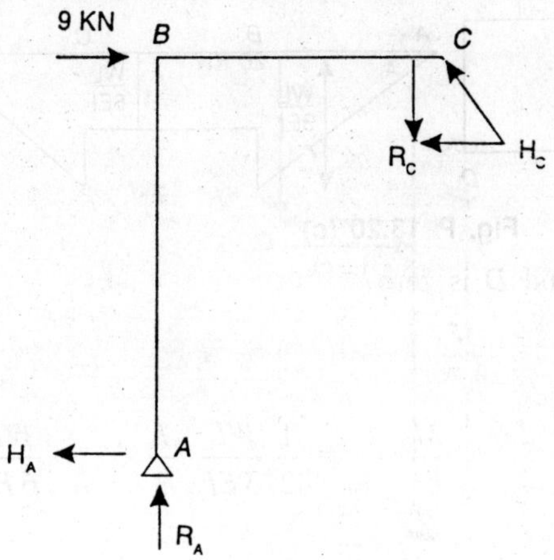

Fig. P. 13.21 (a)

Since CD is along 45° angle

So $\qquad R_C = H_C$

Taking moments about A, we get

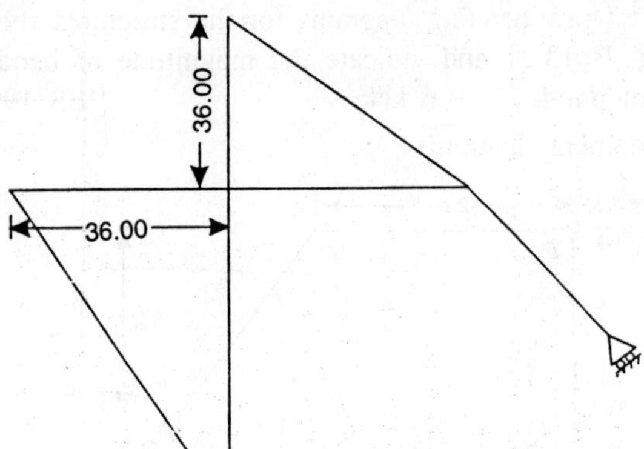

Fig. P. 13.21 (b)

$$R_C \times + (9 - H_C) \times 4 = 0$$

$$\Rightarrow \quad 2\,R_C + 36 - 4\,R_C = 0 \qquad (\therefore R_C = H_C)$$

$$\Rightarrow \quad R_C = H_C = 18 \text{ kN}$$

$$H_A = -\ 9 \text{ kN}$$

(Since, $\Sigma H = 0$) $\quad R_A = +\ R_C$

$$= 18 \text{ kN}$$

(ii)

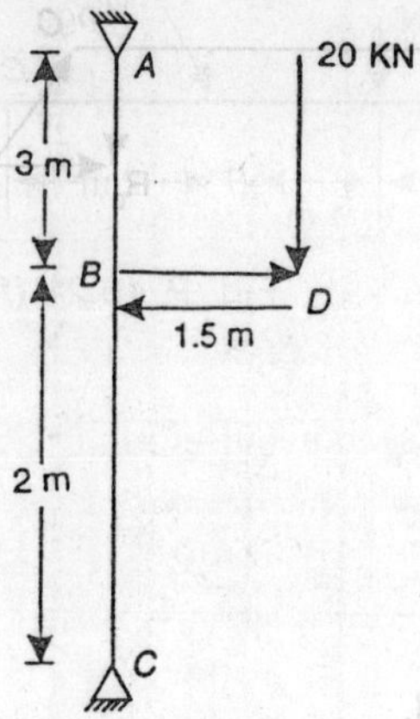

Fig. P. 13.21 (c)

Solution:

Taking moments about A

$$5\ H_C = 20 \times 1.5$$

$$H_C = 6\ \text{kN}$$

$$H_A = H_C = 6\ \text{kN}$$

Plotting moment diagram

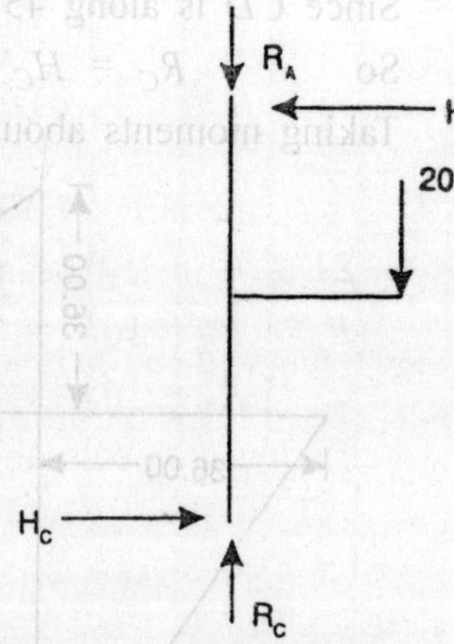

Fig. P. 13.21 (d)

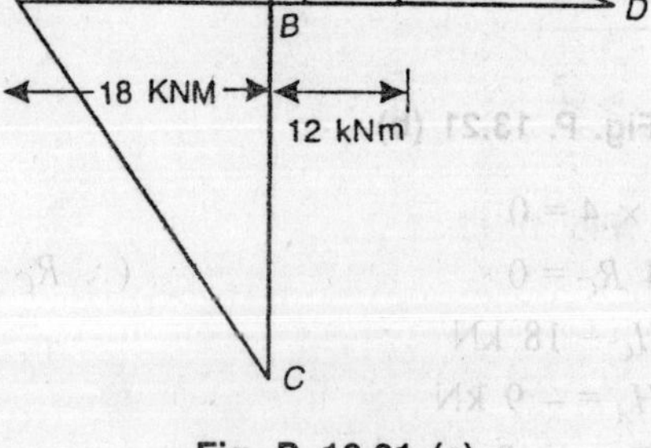

Fig. P. 13.21 (e)

(iii)

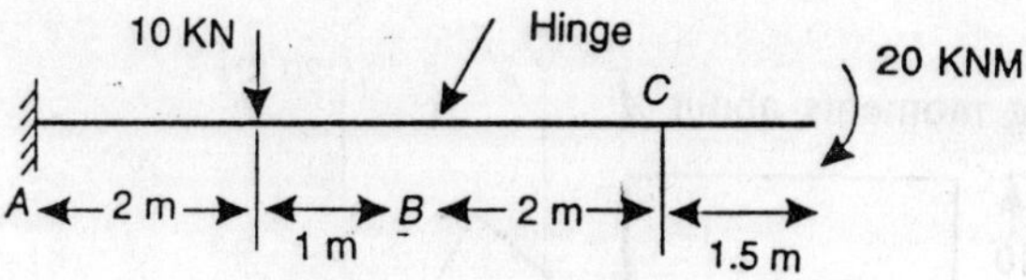

Fig. P. 13.21 (f)

Solution:

Taking moments about C in right beam

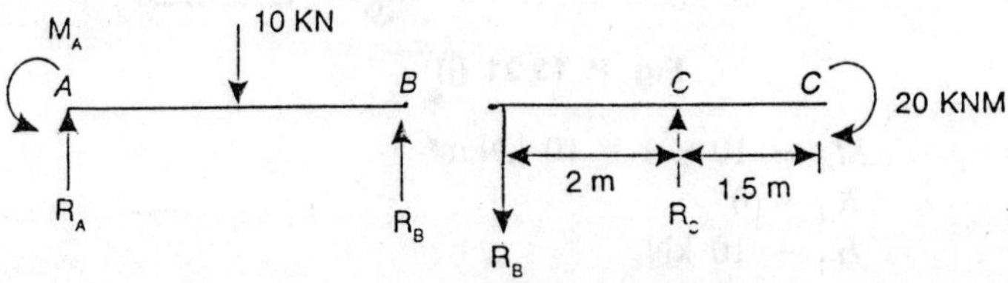

Fig. P. 13.21 (g)

$$2\ R_B = 20$$

$$\Rightarrow \quad R_B = 10 \text{ kN}$$

$$R_C = R_B = 10 \text{ kN}$$

$\Sigma V = 0$ in left beam AB

$$R_A + R_B = 10$$

$$\Rightarrow \quad R_A = 10 - R_B = 0$$

Taking moments about 'A' we get

$$M_A + 3 \times 10 = 10 \times 2$$

$$M_A = -10 \text{ kN.m}$$

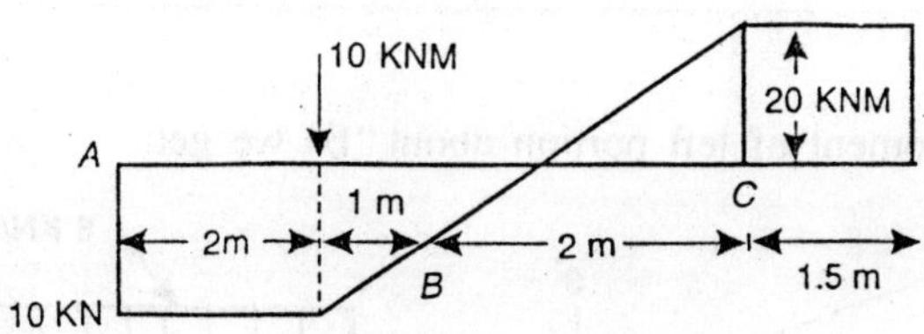

Fig. P. 13.21 (h)

(iv)

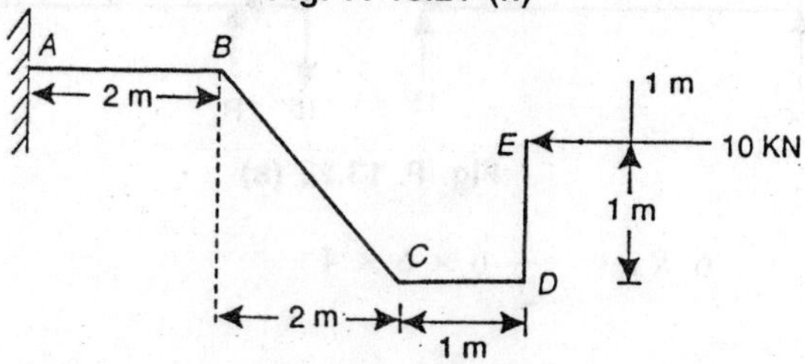

Fig. P. 13.21 (i)

Solution:

Taking moments about A

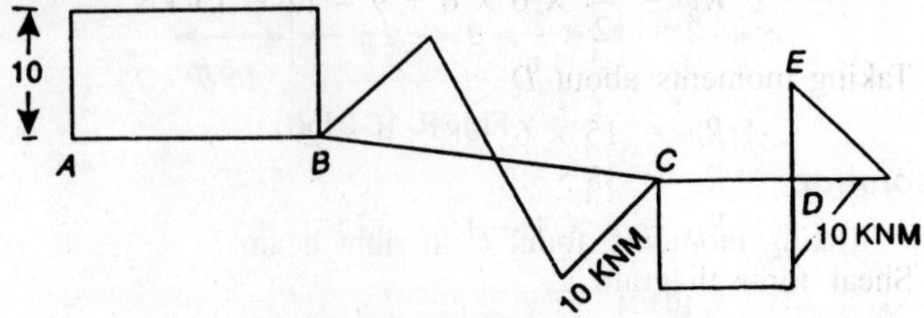

Fig. P. 13.21 (j)

$$M_A = 10 \times 1 = 10 \text{ kN.m}$$
$$R_A = 0$$
$$H_A = 10 \text{ kN}$$

Problem 13.22: Analyse the beam *ABCD* shown in Fig. P. 13.22 and bending moment and shear force diagrams. *B* is an internal hinge. Determine the positions and values of maximum positive moments in the beam. [IES 1991]

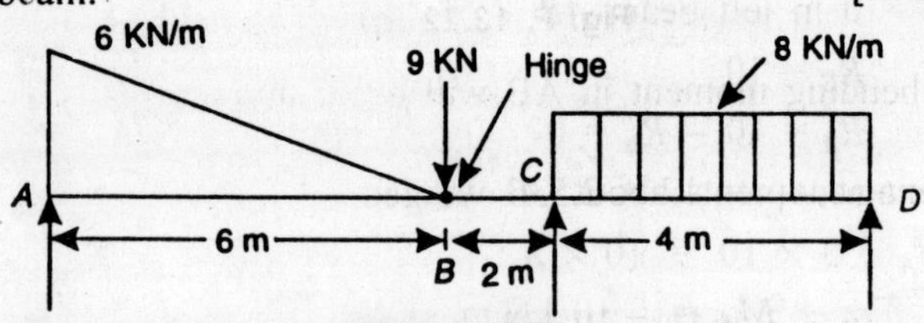

Fig. P. 13.22

Solution:

Taking moment of left portion about 'B' we get

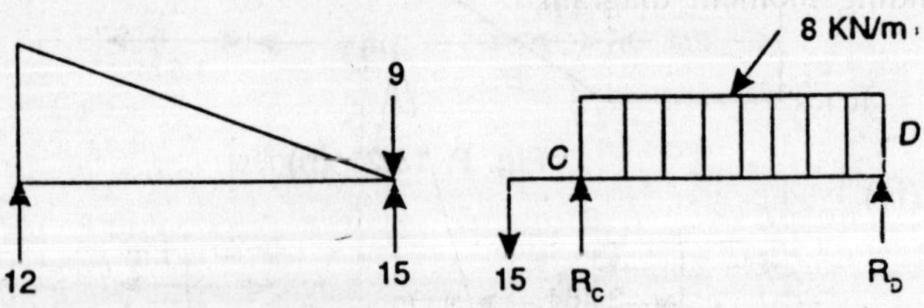

Fig. P. 13.22 (a)

$$6\ R_A = \frac{1}{2}\ 6 \times 6 \times 4$$

$$R_A = 12 \text{ kN}$$

$$R_B = \frac{1}{2} \times 6 \times 6 + 9 - 12 = 15 \text{ kN}$$

Taking moments about D

$$4\,R_C = 15 \times 6 + 8 \times 4 \times 2$$

$$R_C = 38.5$$

$$R_D = 8 \times 4 + 15 - 32.5 = 8.5$$

Shear force diagram

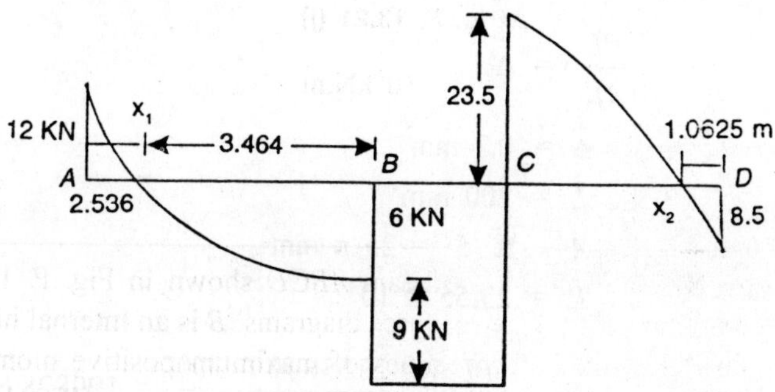

Fig. P. 13.22 (b)

Max. bending moment in AB will occur at x_1

$$\text{Moment at } x_1 = 12 \times 2.536 - 3.464 \times \frac{2.536^2}{2} - \frac{1}{2} \times 2 \times \frac{2.536^2}{2}(6 - 3.464) = 30.432 - 11.139 - 5.4366 = 13.8564$$

$$\text{Moment at } x_2 = 8.5 \times 1.0625 - \frac{8 \times 1.0625^2}{2} = 4.516 \text{ kN.m}$$

Bending moment diagram

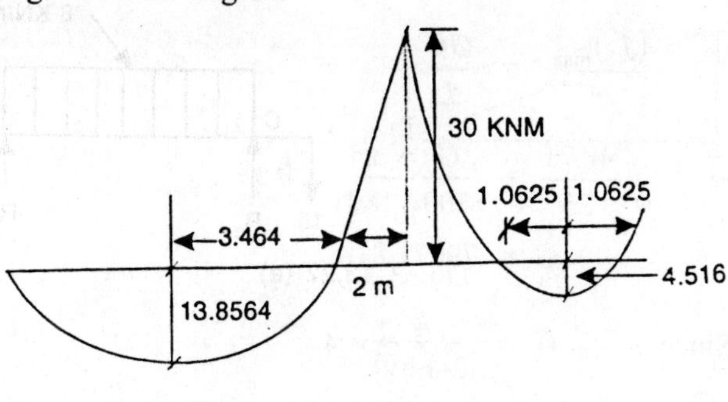

Fig. P. 13.22 (c)

Problem 13.23: A metal bar of 10 mm diameter when subjected to a pull of 23.55 kN gave an elongation of 0.3 mm on a gauge length of 200 mm. In a torsion test on the same material, a maximum shear stress of 40.71 N/mm^2 was measured on bar of 50 mm diameter, The angle of twist measured over a length of 300 mm was 0 deg 21 minutes. Determine poisson's ratio of the material. [IES 1991]

Solution:

Elongation in a bar due to pull '*P*' is given by

$$\frac{PL}{AE} = \Delta$$

$$\Delta = 0.3 \text{ mm}$$

$$L = 200 \text{ mm}$$

$$A = \pi \,.\, 5^2 = 25\,\pi \text{ mm}^2$$

$$P = 23.55 \times 10^3 \text{ N}$$

Hence, $$E = \frac{PL}{A\Delta} = \frac{23.55\times10^3\times200}{0.3\times25\times\pi} = 199898.6 \text{ N/mm}^2$$

Torsion test

Shaar stress (max.) = 40.71 N/mm^2

$$r = 25 \text{ mm}$$

$$\theta = 0 \text{ deg } 21 \text{ minutes} = 0.35^\circ$$

$$= \frac{\pi}{180}\times 0.35 \text{ radians}$$

$$L = 300 \text{ mm}$$

From torsion equation

$$\frac{(f_s)_{max}}{r} = \frac{G\theta}{L}$$

$$\Rightarrow \quad \frac{40.71}{25} = \frac{G}{300}\cdot\frac{\pi}{180}\times 0.35$$

$$G = 79971.81$$

Since $$G = \frac{E}{2(1+\nu)}$$

or $$1 + \nu = \frac{E}{2G}$$

$$\Rightarrow \quad \nu = \frac{E}{2G} - 1 = \frac{199898.6}{2 \times 79971.81} - 1$$

$$= 0.2498 = 0.25 \text{ (Say)}$$

Problem 13.24: A square pin is required to resist a pull of 40 kN and a shear force of 15 kN. Drive a suitable section according to strain energy theory. Maximum elastic stress in tension is 350 N/mm^2. Poisson's ratio is 0.3 [IES 1991]

Solution:

Pull = 40 kN

Shear force = 15 kN

$$\nu = 0.3 = \mu = \frac{1}{m}$$

$$\sigma_0 = 350 \text{ N/mm}^2$$

Let 'A' be the cross-sectional area of pin.

Actual stesses are :

$$\sigma_x = \frac{40,000}{A} \text{ N/mm}^2$$

$$\tau = \frac{15000}{A} \text{ N/mm}^2$$

Principal stresses are:

$$\sigma = \frac{\sigma_x}{2} \pm \frac{1}{2}\sqrt{\sigma_x^2 + 4\tau^2}$$

$$= \frac{20000}{A} \pm \frac{1}{2}\sqrt{\left(\frac{40000}{A}\right)^2 + \left(\frac{2 \times 15000}{A}\right)^2}$$

$$= \frac{20,000}{A} \pm \frac{50,000}{2A}$$

$$\sigma_1 = \frac{45000}{A}$$

$$\sigma_2 = -\frac{5000}{A}$$

$$\sigma_3 = 0$$

$$\text{Strain energy} = \frac{1}{2E}\left[\sigma_1^2+\sigma_2^2+\sigma_3^2-2\nu\left(\sigma_1\sigma_2+\sigma_2\sigma_3+\sigma_3\sigma_1\right)\right]\times\text{volume}$$

$$= \frac{1}{2EA^2}[45000^2 + 5000^2 - 2\times 0.3\,(45,000\times -5000)]\times\text{Volume}$$

$$= \frac{2185\times 10^6}{2EA^2}$$

Strain energy in simple tension

$$= \frac{\sigma_o^2}{2E}\times\text{volume}$$

$$\therefore \quad \frac{2185\times 10^6}{2EA^2}\times\text{volume} = \frac{350^2}{2E}\times\text{volume}$$

$$\Rightarrow \quad A = \frac{\sqrt{2185\times 10^6}}{350} = 133.55$$

$$\text{Size of square pin} = \sqrt{133.55}$$

$$= 11.56\text{ mm}$$

Problem 13.25: A solid conical bar of circular cross-section is suspended vertically. If the length of bar is L and the weight per unit volume of material of the bar is w, determine the total elongation of the bar due to its own weight. [IES 1991]

Solution:

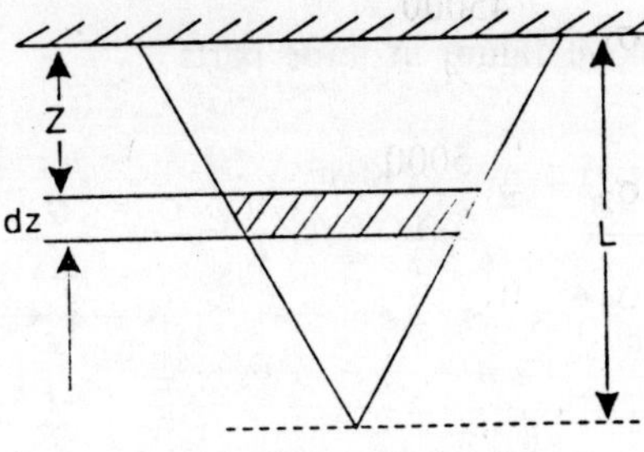

Fig. P. 13.25

Consider an elements of thickness dz at a distance z from top.

$$\text{Stress} = \frac{\text{force}}{\text{area}} = \frac{w \times \text{volume}}{Area}$$

$$\text{Volume of cone} = \frac{1}{3} \cdot A \cdot (L - z)$$

where, A = Cross-sectional area.

$$\text{Hence, stress} = w \cdot \frac{1}{3} (L - z)$$

$$\text{Strain} = \frac{stress}{E}$$

$$\text{Elongation of the element} = \text{strain} \times dz = \frac{w}{3E} (L - z)\, dz$$

$$\text{Total elongation} = \int_0^L \frac{w}{3E} (L - z)\, dz$$

$$= \frac{w}{3E}\left[L^2 - \frac{L^2}{2}\right] = \frac{wL^2}{6E}$$

Problem 13.26: Neatly sketch the bending moment diagrams for the following cases. [IES 1992]

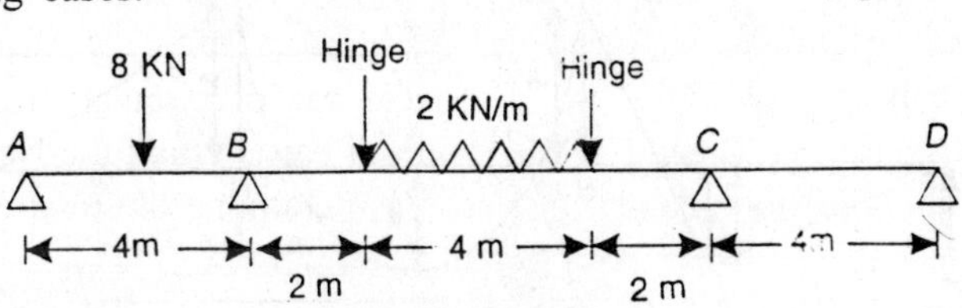

Fig. P. 13.26

Solution:

Breaking the above figure in three parts

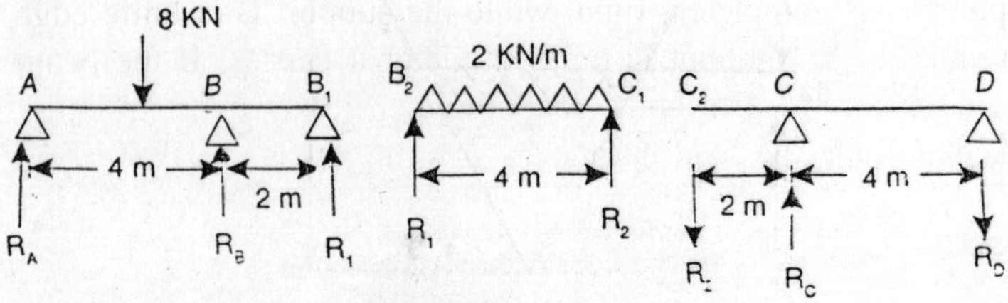

Fig. P. 13.26 (a)

Taking moment about *D*,

$$R_2 \times 6 = 4R_C => R_2 = \frac{2}{3} R_C \qquad ...(1)$$

Taking moment about *C*

$$4 R_1 = 2 \times 4 \times 2 = 16$$

$$R_1 = 4 \text{ kN}$$

$$R_1 + R_2 = 2 \times 4 = 8$$

$$\Rightarrow \quad R_2 = 4 \text{ kN}$$

Putting values of R_2 in eqn (*i*), we get $R_C = 6$ kN

$$\Sigma v = D$$

i.e. $$R_C = R_2 + R_D$$

$$\Rightarrow \quad R_D = 2 \text{ kN (–ve direction)}$$

Taking moment about *A*

$4R_B - 8 \times 2 - 6R_1 = 0$

$$R_B = 4 + \frac{3}{2} R_1 = 4 + \frac{3}{2} \times 4 = 10 \text{ kN}$$

$$\Sigma V = 0$$

i.e. $$R_A = 8 + R_1 - R_B = 8 + 4 - 10 = 2 \text{ kN}$$

Plotting the moment diagram

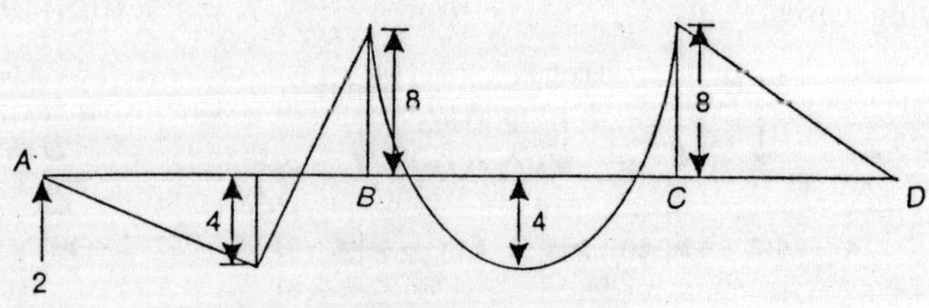

Fig. P. 13.26 (b)

Problem 13.27: A beam *ABC* of uniform flexural rigidity and of length 2L is fixed at A and overhangs the support B. AB = BC = L. Support A is completely rigid, while the support B is knife edge which sinks by an amount K times the load it carries. If the beam carries a uniformly distributed load of intensity w over its entire length, determine the load carried by the support B. [IES 1992]

Solution:

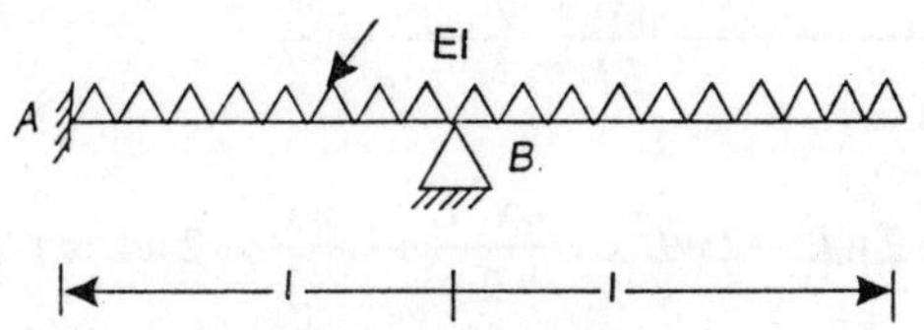

Fig. P. 13.27 (a)

The support B will sink by K times the load it carries Hence, reaction at support B will be K Δ if 'Δ' is total sinking at B.

The above figure can be shown as

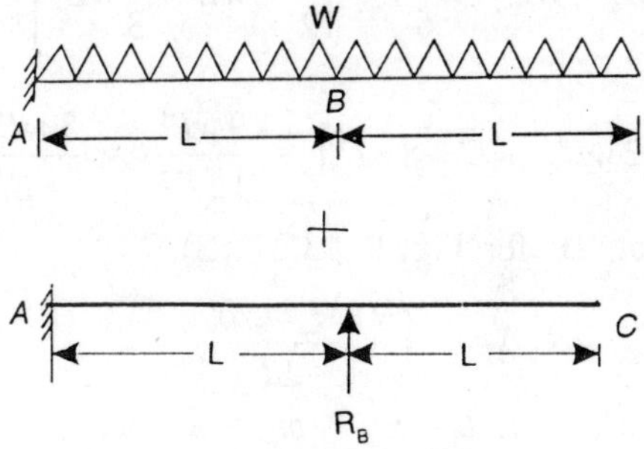

Fig. P. 13.27 (b)

Applying unit load method for calculation of deflection at B in both cases

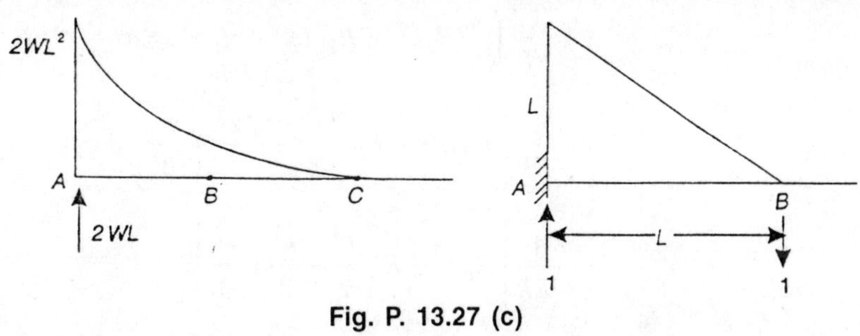

Fig. P. 13.27 (c)

$$M_x = 2\,wL^2 + \frac{wx^2}{2} - 2\,wLx \qquad m_x = L - x$$

Deflection at 'B' for Fig. P. 13.27 (a)

$$= \Delta_1 = \int_0^L \frac{M_x m_x d_x}{EI}$$

$$= \int_0^L \frac{\left(2wL^2 + \frac{wx^2}{2} - 2wlx\right)(L-x)dx}{EI}$$

$$= \frac{1}{EI}\int_0^L \left(2\,wL^3 - 2\,wL^2 x + \frac{wx^2 L}{2} + - \frac{wx^3}{3} - 2\,wL^2 x + 2\,wL\,x^2\right)dx$$

$$= \frac{1}{EI}\left[2\,wL^3 x - wL^2 x^2 + \frac{wx^3 L}{6} - \frac{wx^4}{12} - wL^2 x^2 + \frac{2}{3} wLx^3\right]$$

$$= \frac{1}{EI}\left[2\,wL^4 - wL^4 + \frac{wL^4}{6} - \frac{wL^4}{12} - wL^4 + \frac{2}{3} wL^4\right]$$

$$= \frac{1}{12\,EI}\left[2\,wL^4 - wL^4 + 8\,wL^4\right] = \frac{9\,wL^4}{12\,EI} = \frac{3\,wL^4}{4\,EI}$$

Deflection at 'B' for Fig. P. 13.27 (B)

$$= \Delta_2 = \int_0^L \frac{M_x m_x dx}{EI}$$

$$M_x = R_B L - x \qquad m_x = L - x$$

$$\Delta_2 = \frac{1}{EI}\int_0^L (R_B L - x)\,(L - x)\,dx$$

$$= \frac{1}{EI}\int_0^L (R_B L^2 - R_B\,Lx - xL + x^2)dx$$

$$= \frac{1}{EI}\left[R_B L^2 x - R_B \frac{Lx^2}{2} - \frac{Lx^2}{2} + \frac{x^3}{3}\right]$$

$$= \frac{1}{EI}\left[R_B L^3 - R_B \frac{L^3}{2} - \frac{L^3}{2} + \frac{L^3}{3}\right]$$

$$= \frac{1}{6EI}[3\,R_B\,L^3 - L^3] = \frac{L^3}{6\,EI}[3\,R_B - 1]$$

Net deflection at 'B'

$$= \Delta = \Delta_1 - \Delta_2 = \frac{3}{4\,EI}wL^4 - \frac{L^3}{6\,EI}[3\;R_B - 1]$$

Since, $\Delta = K\,R_B$

$$KR_B = \frac{3}{4\,EI}wL^4 - \frac{R_B L^3}{2\,EI} + \frac{L^3}{6\,EI}$$

$$\Rightarrow \quad R_B\left[K + \frac{L^3}{2\,EI}\right] = \frac{3}{4\,EI}wL^4 + \frac{L^3}{6\,EI}$$

$$R_B = \left[\frac{\dfrac{3\,wL^4}{4\;\;EI} + \dfrac{L^3}{6\,EI}}{K + \dfrac{L^3}{2\,EI}}\right] = \left[\frac{9\,wL + 2}{\dfrac{EIK}{12\,L^3} + 6}\right]$$

Problem 13.28: A bar of square is used as a beam with one of its diagonal horizontal. Calculate the value of shear stress at the neutral axis and position and magnitude of the maximum shear stress. Sketch the shear distribution curve along the vertical diagonal. [IES 1992]

Solution:

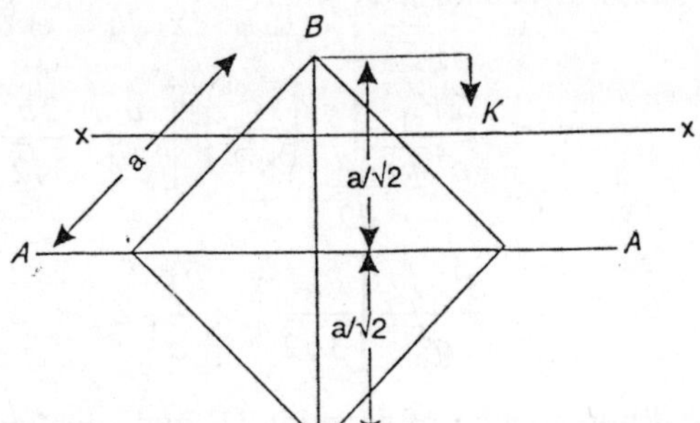

Fig. P. 13.28

Moment of Inertia about $AA = \sqrt{2a}\,\dfrac{(a/\sqrt{2})^3}{12}\cdot 2$

$$I = \sqrt{2a}\cdot\frac{a^3}{2\sqrt{2}}\,\frac{2}{12} = \frac{a^4}{12}$$

The basic formula for shear—stress is given by

$$\tau_{xy} = \frac{V}{Ib}\Sigma\, A_i y_i$$

where, V = Shear force at vertical section.

A_i = Area of an element between section of interest and extreme fibre.

y_i = Distance of centroid of this area, A_i, from the neutral axis.

b = Width of beam at section of interest.

I = Moment of Inertia about CG.

Taking a section at a distance of x from extreme fibre 'B'

$$b = 2\,x$$

$$A_i = \frac{1}{2}\cdot 2\,x\,\cdot x = x^2$$

$$y_i = \frac{a}{\sqrt{2}} - \frac{2x}{3}$$

$$\tau_{xy} = \frac{12V}{a^4}\cdot\frac{1}{2x}\,x^2\left[\frac{a}{\sqrt{2}}-\frac{2x}{3}\right] \qquad \ldots(1)$$

Shear stress at neutral axis where $x = \dfrac{a}{\sqrt{2}}$ we get

$$\tau = \frac{12V}{a^4}\cdot\frac{1}{\frac{2a}{\sqrt{2}}}\cdot\left(\frac{a}{\sqrt{2}}\right)^2\left[\frac{a}{\sqrt{2}}-\frac{2a}{3\sqrt{2}}\right]$$

$$= \frac{3V\sqrt{2}}{a^3}\left[\frac{a}{3\sqrt{2}}\right] = \frac{V}{a^2}$$

Taking differentiation of equation (1) and equating to zero to find section of maximum stress we get

$$\frac{d\tau_{xy}}{dx} = \frac{6V}{a^4}\left[\frac{a}{\sqrt{2}} - \frac{4x}{3}\right] = 0$$

$$\Rightarrow \frac{a}{\sqrt{2}} - \frac{4x}{3} = 0$$

$$\Rightarrow \frac{4x}{3} = \frac{a}{\sqrt{2}}$$

$$x = \frac{3a}{4\sqrt{2}}$$

$$(\tau_{xy})_{\max} = \frac{6V}{a^4} \cdot \frac{3a}{4\sqrt{2}}\left[\frac{a}{\sqrt{2}} - \frac{a}{2\sqrt{2}}\right]$$

$$= \frac{9V}{2\sqrt{2}a^2}\left[\frac{1}{2\sqrt{2}}\right] = \frac{9}{8}\frac{V}{a^2}$$

Plotting shear stress distribution

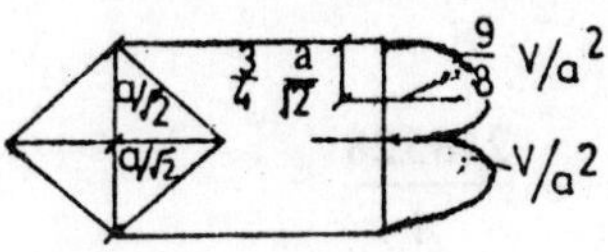

Fig. P. 13.28 (a)

> **Problem 13.29:** A bolt is subjected to an axial pull of 12 kN together with transverse shear of 6 kN. Estimate the diameter of bolt required if it is to be safe according to (i) maximum principal stress theory, (ii) maximum shear stress theory and (iii) maximum strain energy theory.
> Elastic limit in simple tension = 300 N/mm^2
> Factor of safety = 3
> Poisson's ratio = 0.3 [IES 93]

Solution:

$$\text{Permissible stress} = \frac{300}{3} = 100 \text{ B/mm}^2$$

Let 'A' be the cross-sectional area of bolt required.

Actual stresses are :

$$\sigma_x = \frac{12{,}000}{A}$$

Shear stress $\tau = \dfrac{6{,}000}{A}$

Principal stresses

$$\sigma_{1,2} = \frac{\sigma_x}{2} \pm \frac{1}{2}\sqrt{\sigma_x^2 + 4\tau^2}$$

$$= \frac{12{,}000}{2A} \pm \frac{1}{2}\sqrt{\left(\frac{12000}{A}\right)^2 + 4\left(\frac{6000}{A}\right)^2}$$

$$= \frac{12000}{2A} \pm \frac{16970.56}{2A}$$

$$\sigma_1 = \frac{14485.28}{A}$$

$$\sigma_2 = -\frac{2485.28}{A}$$

$$\sigma_3 = 0$$

(i) Maximum principal stress theory

$$\sigma_2 = 100$$

$$\Rightarrow \quad \frac{14485.28}{A} = 100$$

$$A = 144.85$$

$$\Rightarrow \quad \frac{\pi}{4}d^2 = 144.85$$

$$d = 13.58 \text{ mm}$$

(ii) Maximum shear stress theory

$$\text{Max. shear stress} = \frac{\sigma_1 - \sigma_2}{2}$$

$$= \frac{14485.28 - (-2485.28)}{2A} = \frac{8485.14}{A}$$

$$\therefore \qquad \frac{8485.14}{A} = \frac{100}{2}$$

$$\Rightarrow \qquad A = 169.70$$

$$d = 14.70 \text{ mm}$$

(iii) Maximum strain energy theory

$$\left[\sigma_1^2 + \sigma_2^2 + \sigma_3^2 - 2\nu(\sigma_1\sigma_2 + \sigma_2\sigma_3 + \sigma_3\sigma_1)\right] = \sigma_o^2 = 100^2$$

$$\Rightarrow \qquad \frac{1}{A^2}[14485.28^2 + 2485.28^2 - 2 \times 0.3\ (14485.28 \times -2485.28)]$$

$$= 100^2$$

$$A = 154.14$$

$$d = 14 \text{ mm}$$

Problem 13.30: A mild steel plate is stressed as shown in the figure. Before stressing, a circle of 300 mm diameter is drawn on the plate. Determine the length and directions of the major and minor area of the ellipse into which the circle deforms ofter stressing.
Poisson's ratio = 0.3, stressing poisson's ratio = 0.3
Modulus of elasticity = 210 kN/mm² [IES 93]

Solution:

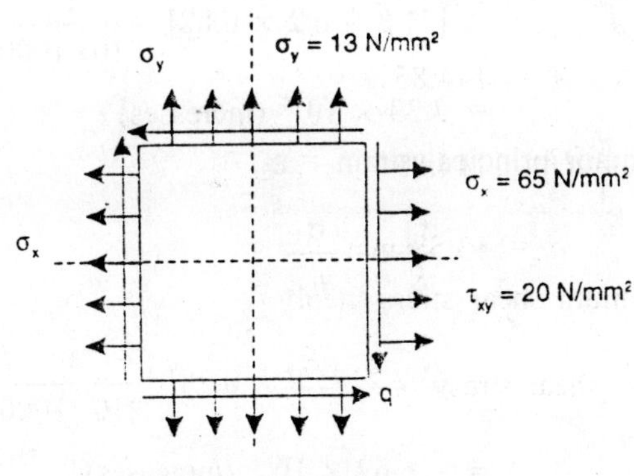

Fig. P. 13.30

$$\sigma_y = 13 \text{ N/mm}^2$$
$$\sigma_x = 65 \text{ N/mm}^2$$
$$\tau_{xy} = 20 \text{ N/mm}^2$$

The principal stresses are given by

$$\sigma_1 = \frac{\sigma_x + \sigma_y}{2} \pm \sqrt{\left(\frac{\sigma_x - \sigma_y}{2}\right)^2 + \tau^2{}_{xy}}$$

$$= \frac{65+13}{2} \pm \sqrt{\left(\frac{65-13}{2}\right)^2 + (20)^2}$$

$$= 39 \pm 32.8$$

$$\sigma_1 = +71.8 \text{ N/mm}^2, \quad \sigma_2 = +6.2 \text{ N/mm}^2$$

$$\tan 2\theta = -\frac{2\tau_{xy}}{\sigma_x - \sigma_x}$$

$$= -\frac{2 \times 20}{65-13} = -0.76923$$

$$\therefore \quad Q_1 = 18°48' \text{ and } Q_2 = 108°48'$$

Then major principal strain,

$$\varepsilon_1 = \frac{\sigma_1}{EI} - \frac{\sigma_2}{mE}$$

$$= [71.8 - 6.2 \times 0.82] \frac{1}{210 \times 1000}$$

$$= 3.33 \times 10^{-4} \text{ (incresses)}$$

The minor principal strain = ε_2

$$= \frac{\sigma_2}{E} - \frac{\sigma_1}{mE}$$

$$= [6.2 - 71.8 \times 0.28] \frac{1}{210 \times 1000}$$

$$= -6.62 \times 10^{-5} \text{ (increases)}$$
$$= 6.62 \times 10^{-5} \text{ (decreases)}$$

Length of the diameter along the direction of

$$\sigma_1 = d + \varepsilon_1 d$$
$$= 300\ (1 + \varepsilon_1)$$
$$= 300\ (1 + 3.33 \times 10^{-4})$$
$$= 300.099 \text{ mm}$$

Length of the diameter along the direction of

$$\sigma_2 = d - \varepsilon_2 d = d\ (1 - \varepsilon_2)$$
$$= 300\ (1 - 6.62 \times 10^{-5})$$
$$= 299.98 \text{ mm}$$

Hence, the circule will be converted into an ellips having major axis = 300.099 mm at 18°48′ and the minor axis = 299.98 mm at 108°48′ with the vertieal centre line (measural in the clockwise direction)

Problem 13.31: A rigid block ABC, weighing 200 kN us supported by three rods sysmetrieally placd, as shown in figure below. Assuming the block to remain horizontal, determine the stess in each rod after a temperature rise of 20°C. The lower ends of the rods are assumed to have been at the same level before the block was attached and the temperature changed, Given Area of steel rod 900 mm^2. area of bronze rod 1500 mm^2, E for steel 205 kN/ mm^2, E for bronze 82 kN/mm^2, $\alpha_s = 12 \cdot 10^{-6}$/°C, $\alpha_{bronze} = 20 \times 10^{-6}$/°C

Solution:

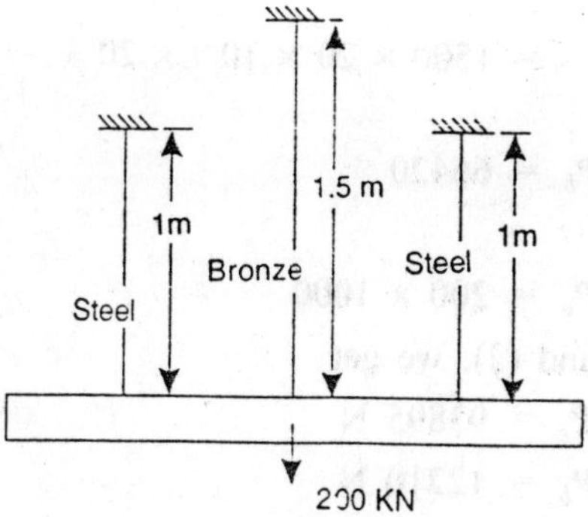

Fig. P. 13.31 (a)

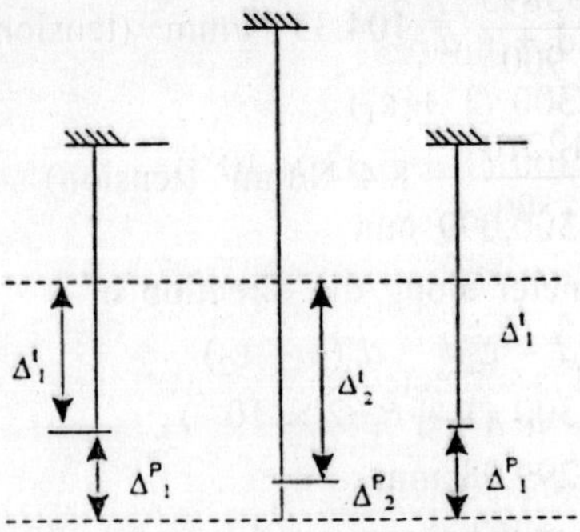

Fig. P. 13.31 (b)

It is clear that the sum of the two defromations of steel rod is equal to the sum of the deformations of the bronze rod.

Hence, we get

$$\Delta_1' + \Delta_1^p = \Delta_2' + \Delta_2^p$$

or $$L_S\,\alpha_s' + \frac{P_S L_S}{A_S E_S} = L_b\,\alpha_b' + \frac{P_b L_b}{A_b E_b}$$

P_s us the load carried by each of the steel rods and P_b is the load carried by bronze rod.

Substituting the values, we get

$$[1000 \times 12 \times 10^{-6} \times 20] + \frac{P_s \times 1000}{900 \times 205 \times 10^3}$$

$$= 1500 \times 20 \times 10^{-6} \times 20 + \frac{P_b \times 1500}{1500 \times 82 \times 10^3}$$

or $P_s - 2.25\,P_b = 66420$ ...(1)

from statics

$$2\,P_s + P_b = 200 \times 1000 \quad ...(2)$$

Solving (1) and (2), we get

$$P_s = 93895 \text{ N}$$

and $$P_b = 12210 \text{ N}$$

Hence the stresses are

$$p_s = \frac{93895}{900} = 104.33 \text{ N/mm}^2 \text{ (tension)}$$

and $$p_b = \frac{12210}{1500} = 8.4 \text{ N/mm}^2 \text{ (tension)}$$

Problem 13.32: A bolt is subjected to an axial pull to 12 kN together with a transverse shear of 6 kN. Estimate the diameter of the bolt required if it is to be safe according to
(a) maximum principal stress theory
(b) Maximum shear stress theory, and
(c) Maximum strain energes theory
Elastic limit in simple tension = 300 N/mm^2
Factor of safters = 3 , posson's ratio = 0.3 [IES 1993]

Solution:

Permissible tensile stress $= \frac{300}{3} = 100 \text{ N/mm}^2$

Let d be the required diameter of the bolt, in mm,
Hence the applied stress are

$$p = \frac{12000}{\frac{\pi}{4}d^2} = \frac{15278.9}{d^2} \text{ N/mm}^2$$

and $$q = \frac{6000}{\frac{\pi}{4}d^2} = \frac{7639.4}{d^2} \text{ N/mm}^2$$

(a) *Maximum principal stess theory*
The maximum principal stress in the bolt

$$\sigma = \frac{p}{2} + \sqrt{\left(\frac{p}{2}\right)^2 + q^2}$$

where p_1 = p and $p_2 = 0$

$$= \frac{1}{2} \times \frac{15278.9}{d^2} + \sqrt{\left(\frac{15278.9}{2d^2}\right)^2 + \left(\frac{7639.4}{d^2}\right)^2}$$

$$= \frac{18443.3}{d^2}$$

Permissible simple stress in tension

$$= 100 \text{ N/mm}^2$$

$$\therefore \quad \frac{18443.2}{d^2} = 100$$

$$\therefore \quad d = 13.58 \text{ mm}$$

(b) *Maximum shear stress theory*

Max. shear stress

$$= \sqrt{\left(\frac{p}{2}\right)^2 + q^2} = \sqrt{\left(\frac{15278.9}{2d^2}\right)^2 + \left(\frac{7639.4}{d^2}\right)^2}$$

$$= \frac{10803.8}{d^2}$$

$$\text{Max. shear stess} = \frac{100}{2} = 50 \text{ N/mm}^2$$

$$\therefore \quad \frac{10803.8}{d^2} = 50$$

$$\therefore \quad d = 14.7 \text{ mm}$$

(c) *Maximum strain energy theory*

The principal stressses are

$$\sigma_1 = \frac{18443.2}{d^2},$$

$$\sigma_2 = -\frac{3164.4}{d^2} \text{ and } \sigma_3 = 0$$

$\therefore$ Strain energy

$$= \frac{1}{2E}\left[\left(\frac{18443.2}{d^2}\right)^2 + \left(-\frac{3164.4}{d^2}\right)^2 - 2\times 0.3\left(\frac{1843.2}{d^2}\right)\times\left(-\frac{3164.4}{d^2}\right)\right]$$

$$= \frac{1}{2E}\cdot\frac{3.8518\times 10^8}{d^4}$$

Strain energy in simple tension

$$\frac{f^2}{2E} = \frac{(100)^2}{2E}$$

$$\therefore \quad \frac{1}{2E} \times \frac{3.8518 \times 10^8}{d^4} = \frac{(100)^2}{2E}$$

$$\therefore \quad d = 14.01 \text{ mm}$$

Problem 13.33: A beam of uniform section and length 2L is freely supported by rigid props at its ends and by an elastic prop at its centre. If the prop deffects by an amount which is *K* times the load it carries, determine the load carried by the prop due to a total uniformly distributed load 'w' on the beam. [IES 1993]

Solution:

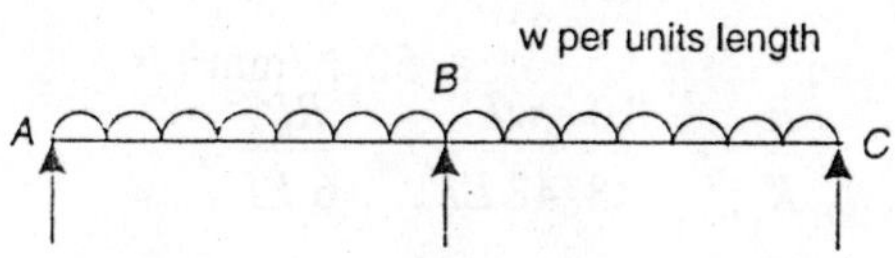

Fig. P. 13.33

Let the reaction at central prop be 'R'and net deffection be Δ

$$\therefore \quad R = K\Delta$$

$$\Delta = \delta_1 - \delta_2$$

where, δ_1 = deflection at centre due to u.d.l '*w*' and δ_2 = is deflection of centre due to load R at centre.

Applying unit load method for calculation of diflection

Deflection at centre due to u.d.l

$$\delta_1 = 2\int_0^l \left(wLx - \frac{wx^2}{2}\right) \cdot \frac{x}{2} \cdot \frac{dx}{EI}$$

$$= 2\int_0^l \left(\frac{wLx^2}{2} - \frac{wx^3}{4}\right) \frac{dx}{EI}$$

$$= \frac{2}{EI}\left[\frac{wLx^3}{6} - \frac{wx^4}{16}\right]_0^L$$

$$= \frac{2}{EI}\left[\frac{wL^4}{6} - \frac{wL^4}{16}\right] = \frac{2\,wL^4}{EI}\left[\frac{8-3}{48}\right] = \frac{10}{48}\frac{wL^4}{EI}$$

Deflection at centre due to central reaction

$$R = \delta_2 = 2\int_0^L \frac{R}{2}\,x \cdot \frac{x}{2} \cdot \frac{dx}{EI}$$

$$= \frac{R}{2} \cdot \frac{L^3}{3\,EI} = \frac{RL^3}{6\,EI}$$

$$\Delta = \delta_1 - \delta_2$$

$$= \frac{10\,wL^4}{48\,EI} - \frac{RL^3}{6\,EI} \qquad \ldots(1)$$

$\therefore \quad \Delta = \frac{R}{K}$ putting this value in equation (1) we get

$$\frac{R}{K} = \frac{10}{48}\frac{wL^4}{48\,EI} - \frac{RL^3}{6\,EI}$$

$$\Rightarrow \quad R\left[\frac{1}{K} + \frac{L^3}{6\,EI}\right] = \frac{10}{48}\frac{wL^4}{EI}$$

$$\Rightarrow \quad \frac{R}{K\,6\,EI} \cdot (6\,EI + EL^3) = \frac{10}{48}\frac{wL^4}{EI}$$

$$\therefore \quad R = \frac{10}{8}\left[\frac{wL^4 K}{6\,EI + KL^3}\right]$$

Problem 13.34: Find the maximum torque, that can be applied safely to shaft of 300 mm diameter. The permissible angle of twist is 1.5 degree in a length of 7.5 m and the shearing stress in not to exceed 42 N/mm^2. Take G = 84.4 kN/mm^2. [IES 1994]

Solution:

Using torsion equation

$$\frac{T}{I_P} = \frac{(f_s)_{max}}{R} = \frac{G\theta}{L}$$

$$\theta = 1.5° = \frac{\pi}{180} \times 1.5 \text{ radian} = 0.02168 \text{ radians}$$

$$R = 150 \text{ mm}$$

$$L = 7.5 \text{ m} = 7500 \text{ mm}$$

$$G = 84.4 \text{ kN/mm}^2$$

$$(f_s)_{max} = 42 \text{ N/mm}^2$$

$$I_p = \frac{\pi D^4}{64} = \pi \times \frac{300^4}{64} = 3.976 \times 10^8 \text{ mm}^4$$

From basic torsion equation

$$\frac{T}{I_P} = \frac{(f_s)_{max}}{R}$$

$$T = \frac{(f_s)_{max} \times I_P}{R}$$

$$T = \frac{42 \times 3.976 \times 10^8}{150} \text{ N.m} = 111.328 \text{ kN.m}$$

$$T = \left(\frac{G\theta}{L}\right) \cdot I_p$$

$$= \frac{84.4(0.02168)}{7500} \cdot \frac{3.976 \times 10^8}{1000} \text{ kN.m}$$

$$= 97 \text{ kN.m}$$

So maximum torque that can be applied safely to a shaft = 97 kN.m.

Problem 13.35: The diagram given in below figure is the shearing force diagram for a beam which rests on two supports, one being on the left hand end.

Deduce directly from the shearing force diagram:

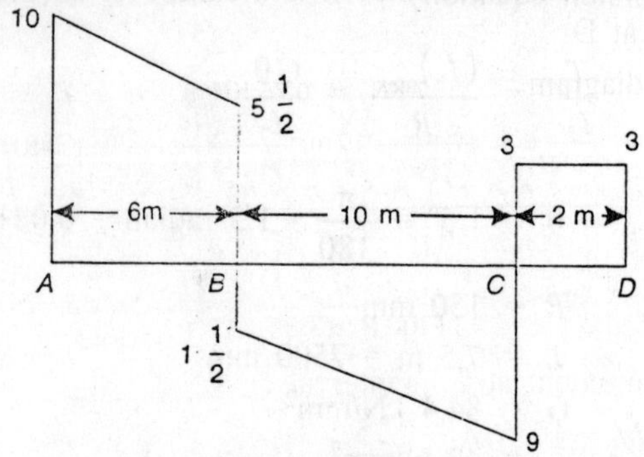

Fig. P. 13.35

(i) The bending moment at 2 m interval along the beam.
(ii) The loading on the beam.
Show clearly how the bending moment and the loading has been obtained. Draw loading diagram and the bending moment diagram.

[IES 1994]

Solution:

'A' is one of the support as given. other support will be at 'C' because reaction at support will be upwards and at 'C' shear force is increasing in + ve direction.

Along AB shear force is linearly varying hence 'AB' will have u.d.l. on it of value

$$\left(\frac{10-5\frac{1}{2}}{6}\right) = 0.75 \text{ kN/m.}$$

At 'B' shear force is decreasing, hence there will be a point load of value

$$\left(5\frac{1}{2}+1\frac{1}{2}\right) = 7 \text{ kN}$$

Along AB shear force is again linearly varying so, it will be loaded by u.d.l. of value

$$\left(\frac{9-1\frac{1}{2}}{10}\right) = 0.75 \text{ kN/m}$$

At 'C' there will be support with reaction of (9 +3) = 12 kN. Along CD the S.F. is constant but since it is a cantilever so there will point load of 3 kN at D.

Loading diagram

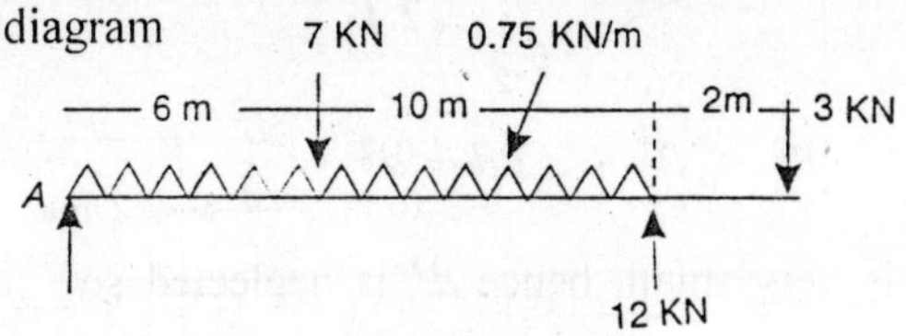

Fig. P. 13.35 (a)

Bending moment at 2 m interval from A:

At 2 m, $M = 10 \times 2 - 0.75 \times 2 \times 1 = 18.5$ kN.m

At 4 m, $M = 10 \times 4 - 0.75 \times 4 \times 2 = 34$ kN.m

At 6 m, $M = 10 \times 6 - 0.75 \times 6 \times 3 = 46.5$ kN.m

At 8 m, $M = 10 \times 8 - 7 \times 2 - 0.75 \times 8 \times 4 = 42$ kN.m

At 10 m, $M = 10 \times 10 - 7 \times 4 - 0.75 \times 10 \times 5 = 34.5$ kN.m

At 12 m, $M = 10 \times 12 - 7 \times 2 - 7 \times 6 - 0.75 \times 12 \times 6 = 24$ kN.m

At 14 m, $M = 10 \times 14 - 7 \times 8 - 0.75 \times 14 \times 7 = 10.5$ kN.m

At 16 m, $M = 3 \times 2 = -6$ kN.m

Moment diagram

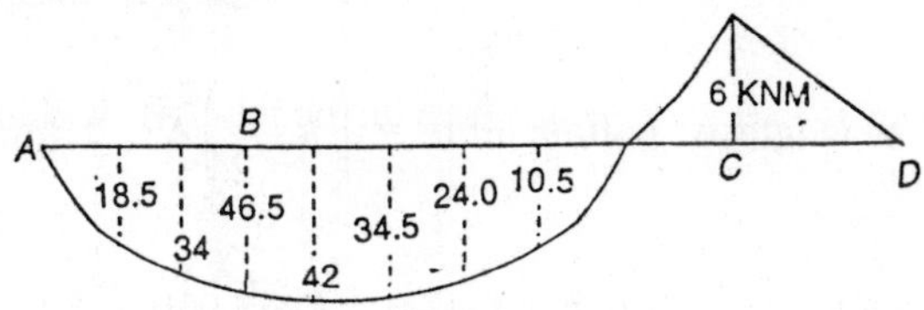

Fig. P. 13.35 (b)

Problem 13.36: A long of uniform rectangular section and thickness t, originally straight, is bent into the form of a circular are and the displacement d of the mid-point of length L is measured by means of a dial gauge. The displacement 'd' may be regarded as small compared with length L.

Show that the longitudinal surface strain ε in the rod is given by

$$\varepsilon = \frac{4td}{L^2}$$ [IES 1994]

Solution:

$$R^2 = (R - d)^2 + \left(\frac{L}{2}\right)^2$$

$$R^2 = R^2 - 2\,Rd + d^2 + \frac{L^2}{4}$$

Since d is very small, hence d^2 is neglected so

$$R = \frac{L^2}{4d}$$

Since, plane sections remain plane, the variation of strain is linear.

Considering an small element of length dx subtending an angle $\Delta\theta$ at center of curvature. Before bending, the vertical planes are parallel. After bending they rotate and subtend an angle $\Delta\theta$ with each other. The rotation will be about neutral axis.

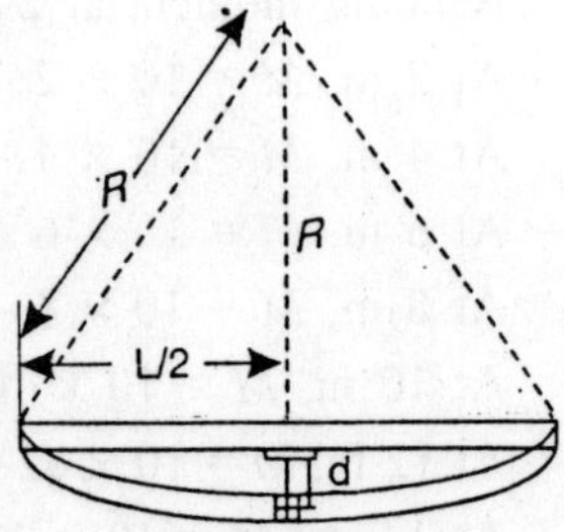

Fig. P. 13.36

So, change in length of top fibre $= \left(R - \frac{t}{2}\right)\Delta\theta - R\Delta\theta$

Change in length of bottom fibre $= \left(R + \frac{t}{2}\right)\Delta\theta - R\Delta\theta$

So net change over the depth 't' $= \left(R + \frac{t}{2}\right)\Delta\theta - \left(R - \frac{t}{2}\right)\Delta\theta$

$$= t\Delta\theta$$

Surface strain $\varepsilon = \dfrac{t\,\Delta\theta}{R\,\Delta\theta} = \dfrac{t}{R} = \dfrac{t}{\frac{L^2}{4d}} = \dfrac{4\,dt}{L^2}$

Problem 13.37 : The strain measurements from a rectangular strain rosette were $e_0 = 600 \times 10^{-6}$, $e_{45} = 500 \times 10^{-6}$ and $e_{90} = 200 \times 10^{-6}$. Find the magnitude and direction of principal strains. If $E = 2 \times 10^5$ N/mm^2 and $\mu = 0.3$ find the principal stresses.

[IES 1995]

Solution:

Using the analytical method,

$$\varepsilon_x = \varepsilon_0 = 600 \times 10^{-6}$$

$$\varepsilon_y = \varepsilon_{90} = 200 \times 10^{-6}$$

$$\varepsilon'_x = \varepsilon_x \cos^2\theta + \varepsilon_y \sin^2\theta + r_{xy} \sin\theta \cos\theta$$

$$\varepsilon'_x = 500 \times 10^{-6} \text{ for } \theta = 45°$$

$$\Rightarrow 500 \times 10^{-6} = 600 \times 10^{-6} \cdot \frac{1}{2} + 200 \times 10^{-6} \cdot \frac{1}{2} + \frac{r_{xy}}{2}$$

$$r_{xy} = 200 \times 10^{-6}$$

Principal strains

$$= \varepsilon_{1,2} = \frac{\varepsilon_x + \varepsilon_y}{2} \pm \frac{1}{2}\sqrt{\left(\varepsilon_x - \varepsilon_y\right)^2 + r_{xy}^2}$$

$$= \frac{600 \times 200}{2} \times 10^{-6} \pm \frac{1}{2}$$

$$\sqrt{\left(\frac{600-200}{2}\right)^2 \times 10^{-12} + (200 \times 10^{-6})^2}$$

$$= 400 \times 10^{-6} \pm \frac{1}{2} \times 200\sqrt{2} \times 10^{-6}$$

$$= (400 \pm 100\sqrt{2})\, 10^{-6}$$

$$\varepsilon_{max} = \varepsilon_1 = 541.42 \times 10^{-6}$$

$$\varepsilon_{min} = \varepsilon_2 = 258.58 \times 10^{-6}$$

If ϕ is angle of principal strain with ε_x, then

$$\tan 2\phi = \frac{r_{xy}}{\varepsilon_x - \varepsilon_y} = \frac{200}{600-200} = \frac{1}{2}$$

$$2\phi = 26.565° \text{ or } 206.565°$$

$$\phi = 13.283° \text{ or } 103.283°$$

$$E = 2 \times 10^5 \text{ N/mm}^2$$

$$\mu = 0.3$$

$$\varepsilon_1 = \frac{\sigma_1}{E} - \frac{\mu\sigma_2}{E}$$

$$\varepsilon_1 = \frac{\sigma_2}{E} - \frac{\mu\sigma_1}{E}$$

where, σ_1 and σ_2 are principal stresses

Hence, $$\sigma_2 = \frac{E(\varepsilon_2 + \mu\varepsilon_1)}{1-\mu^2}$$

$$\sigma_1 = \frac{E(\varepsilon_1 + \mu\varepsilon_2)}{(1-\mu^2)}$$

Hence, principal stresses are :

$$\sigma_1 = \frac{2\times10^5}{1-0.3^2}[541.42 + 258.58 \times 0.3] \times10^{-6}$$

$$= 136.04 \text{ N/mm}^2$$

$$\sigma_2 = \frac{2\times10^5}{1-0.3^2}[258.58 + 0.3 \times 541.42]\times10^{-6}$$

$$= 92.53 \text{ N/mm}^2$$

Problem 13.38: Draw bending moment and shear force diagramss for the beam loaded as shown in Fig. P. 13.38. [IES 1995]

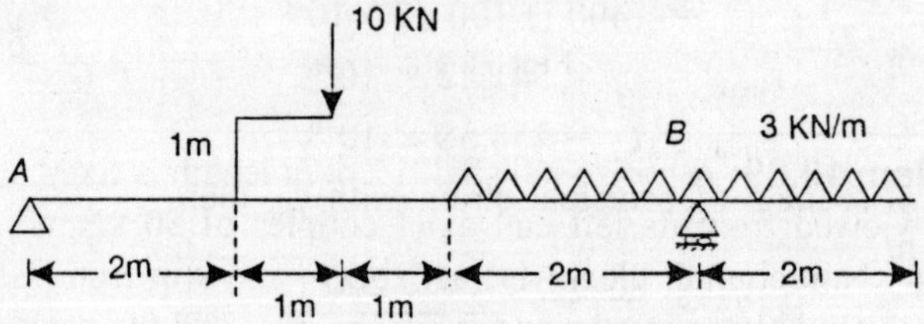

Fig. P. 13.38

Solution:

Taking moments about 'B' to find reaction at A we get

$$6\, R_A = 10 \times 3 + 3 \times 2 \times 1 - 3 \times 2 \times 1$$

$$\Rightarrow \quad R_A = 5 \text{ kN}$$

By simple statics

$$\Sigma V = 0$$

$$\Rightarrow \quad R_B = 10 + 3 \times 4 - 5 = 17 \text{ kN}$$

Plotting shear force diagram

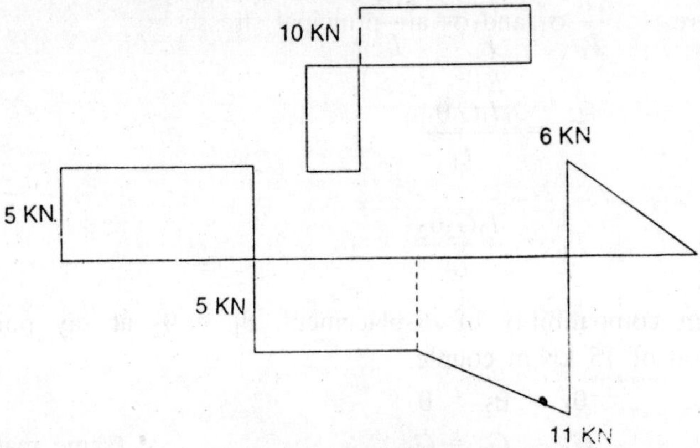

Fig. P. 13.38 (a)

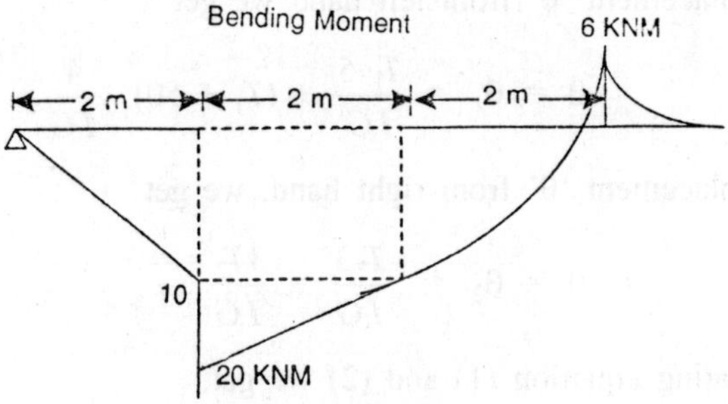

Fig. P. 13.38 (b)

> **Problem 13.39:** A horizontal shaft 12 m in length is fixed at its ends. when viewed from its left end axial couples of 50 kN.m clockwise and 75 kN.m counter clockwise act at 5 m and 9 m from the left end respectively. Determine the end fixing couples and the position where the shaft suffers no angular twist. [IES 1995]

Solution:

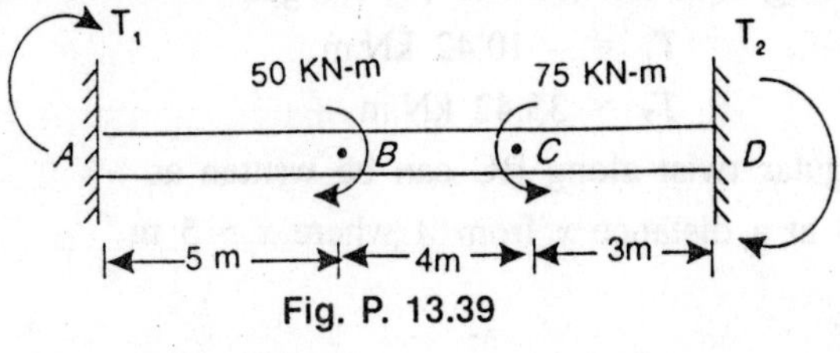

Fig. P. 13.39

$$T_1 + T_2 = 75 - 50 = 25$$

Using torsion equation

$$\frac{T}{I_P} = \frac{f_s}{r} = \frac{G\theta}{L}$$

$$T_1 = \frac{I_1 G_1 \theta_1}{L_1}$$

and $$T_2 = \frac{I_2 G_2 \theta_2}{L_2}$$

From compatibility of displacement, $\theta_1 = \theta_2$ at say point of application of 75 kN.m couple.

So $$\theta_1 = \theta_2 = \theta$$

$$G_1 = G_2 = G \qquad \text{(same material)}$$

$$I_1 = I_2 = I$$

Displacement 'θ' from left hand we get

$$\theta = \theta_1 = \frac{T_1 \cdot 5}{IG} + (T_1 + 50) \cdot \frac{4}{IG} \qquad \text{...(1)}$$

Displacement 'θ' from right hand, we get

$$\theta = \theta_2 = \frac{T_2 3}{IG} = \frac{3T_2}{IG} \qquad \text{...(2)}$$

Equating equation (1) and (2) we get,

$$\frac{9T_1}{IG} + \frac{200}{IG} = \frac{3T_2}{IG}$$

$$\Rightarrow \quad 3T_1 - T_2 + \frac{200}{3} = 0 \qquad \text{...(3)}$$

$$T_1 + T_2 - 25 = 0 \text{ (by simple statics equation)} \qquad \text{...(4)}$$

Solving equation (3) and (4) we get.

$$T_1 = -10.42 \text{ kN.m}$$

$$T_2 = 35.42 \text{ kN.m}$$

Angular twist along BC can be written as

Say at a distance x from A where $x > 5$ m

$$\theta = \frac{-10.42 \times 5}{IG} + \frac{(50-10.42)}{IG}(x-5)$$

[For $x \geq 5$ m and $x \leq 9$ m]

Equating this to zero we get

$$-52.1 + 50x - 250 - 10.42x + 52.1 = 0$$

$$39.58x = 250$$

$$x = 6.316 \text{ m}$$

Problem 13.40: The principal stresses at a point in an elastic material are 1.5 σ (tensile), σ (tensile) and 0.5 σ (compressive).The elastic limit in tension is 210 MPa and μ = 0.3. what would be the value of σ at failure when computed by the different theroies of failure? [IES 1995]

Solution:

$$\sigma_1 = 1.5\,\sigma$$

$$\sigma_2 = \sigma$$

$$\sigma_3 = -0.5\,\sigma$$

$$\sigma_0 = 210 \text{ MPa}$$

$$\mu = 0.3$$

(a) *Maximum principal stress theory*

Maximum stress = 1.5 σ

$$\sigma \text{ at failure} = \frac{210}{1.5} = 140 \text{ MPa}$$

(b) *Maximum shear stress theory*

$$\text{Maximum shear stress} = \left(\frac{\sigma_1 - \sigma_3}{2}\right)$$

$$= \frac{1.5\sigma - (-0.5\sigma)}{2} = \sigma$$

$$\text{In simple tension maximum shear stress } \frac{\sigma_0}{2} = \frac{210}{2} = 105 \text{ MPa}$$

$$\sigma = 105 \text{ MPa}$$

(c) Strain energy theory

$$\frac{1}{2E}\left[\sigma_1^2+\sigma_2^2+\sigma_3^2-2\mu\left(\sigma_1\sigma_2+\sigma_2\sigma_3+\sigma_3\sigma_1\right)\right] = \frac{\sigma_0^2}{2E}$$

$\Rightarrow [(1.5\ \sigma)^2 + \sigma^2 + (0.5\ \sigma)^2 - 0.6\ (1.5\ \sigma^2 - 0.5\sigma^2 - 1.5 \times 0.5\ \sigma^2)] = 210^2$

$\Rightarrow (3.5\ \sigma^2 - 0.15\ \sigma^2\ ) = 210^2$

$$3.35\ \sigma^2 = 210^2$$

$$\sigma = 114.74 \text{ MPa}$$

(d) shear strain energy theory

Shear stain energy

$$= \frac{1}{12G}[(\sigma_1 - \sigma_2)^2 + (\sigma_2 - \sigma_3)^2 + (\sigma_3 - \sigma_1)^2] \times \text{Volume}$$

$$= \frac{\sigma_0^2}{6G} \times \text{Volume}$$

$$\therefore \quad \frac{1}{12G}[(1.5 - 1)^2\ \sigma^2 + [1 + 0.5]^2\ \sigma^2 + [-\ 0.5 - 1]^2\ \sigma^2\ ] = \frac{\sigma_0^2}{6G}$$

$$\Rightarrow \quad \frac{4.75}{2}\sigma^2 = 210^2$$

$$\Rightarrow \quad \sigma = 136.26 \text{ MPa}$$

(e) Maximum Strain theory

$$\text{Maximum strain} = \frac{1}{E}[\sigma_1 - \mu\ \sigma_2 - \mu\ \sigma_3] = \frac{\sigma_0}{E}$$

$\Rightarrow [1.5\ \sigma - 0.3 \times \sigma + 0.3 \times 0.5\ \sigma\ ] = 210$

$$\sigma = 155.55 \text{ MPa}$$

Problem 13.41: Analyse the frame loaded as shown in figure and draw the bending moment diagram. The frame is fixed at A and hinged at D. The relative second moment of the areas are also indicated in the figure. [IES 1995]

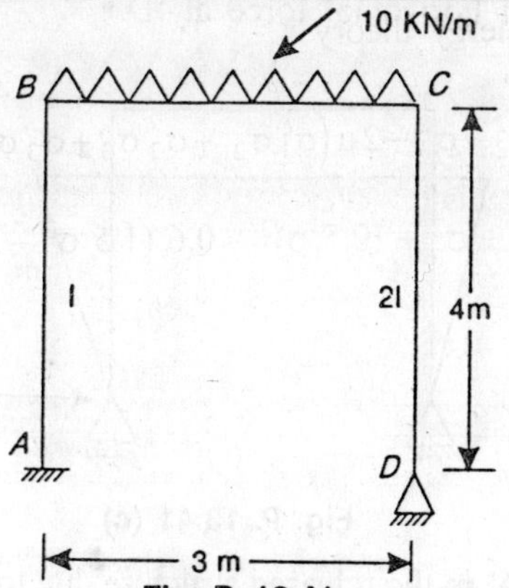

Fig. P. 13.41

Solution:

Since, above problem is indeterminate to two degrees it is better to use flexibility method. Let two redundants be moment at A and horizontal reaction at 'D'

Basic system with loading

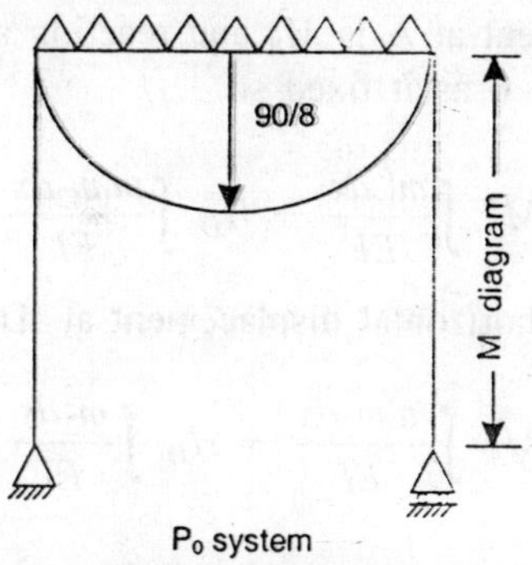

P_0 system

Fig. P. 13.41 (a)

Applying unit anticlockwise moment at 'A'

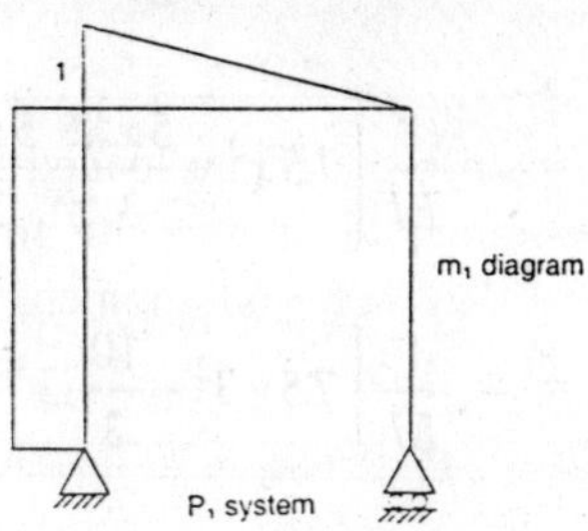

P_1 system

Fig. P. 13.41 (b)

Applying unit horizontal force at 'D'

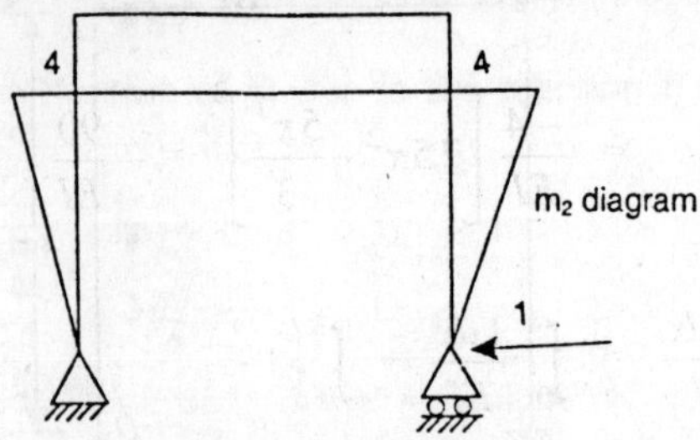

Fig. P. 13.41 (c)

Rotation at 'A' in the released structure due to loading = $\int \frac{Mm_1\, dx}{EI}$

Rotation at 'A' due to unit moment at 'A' = $\int \frac{m_1 m_1 dx}{EI}$

Rotation at A due to unit horizontal force at D = $\int \frac{m_1 m_2 dx}{EI}$

If actual moment at A is M_A and reaction at D is H_D then, since net rotation at A = 0 as it fixed so

$$\int \frac{Mm_1\, dx}{EI} + M_A \int \frac{m_1^2\, dx}{EI} + H_D \int \frac{m_1 m_2 dx}{EI} = 0 \qquad \ldots(1)$$

Similarly net horizontal displacement at 'D' is zero so

$$\int \frac{Mm_2 dx}{EI} + M_A \int \frac{m_1 m_2 dx}{EI} + H_D \int \frac{m_2^2\, dx}{EI} = 0 \qquad \ldots(2)$$

$$\int \frac{Mm_1\, dx}{EI} = -\int_0^3 \frac{\left(15x-5x^2\right)\left(1-\frac{x}{3}\right) dx}{EI}$$

$$= -\frac{1}{EI}\left[7.5x^2 - \frac{5x^3}{3} - \frac{5x^3}{3} + \frac{5x^4}{12}\right]_0^3$$

$$= -\frac{1}{EI}\left[7.5\times 3^2 - \frac{10}{3}\cdot 3^3 ++ \frac{5}{12}\cdot 3^4\right] = -\frac{11.25}{EI}$$

$$\int \frac{Mm_2}{EI}dx = -\int_0^3 (15x - 5x^2)\frac{4dx}{EI}$$

$$= \frac{-4}{EI}\left[7.5x^2 - \frac{5x^3}{3}\right]_0^3 = -\frac{90}{EI}$$

$$\int \frac{m_1^2 dx}{EI} = \int_0^4 \frac{1\,dx}{EI} + \int_0^3 \left(1-\frac{x}{3}\right)^2 \frac{dx}{EI}$$

$$= \frac{4}{EI} + \frac{1}{EI}\left[\left(1-\frac{x}{3}\right)^3 \times -1\right]_0^3 = \frac{5}{EI}$$

$$\int \frac{m_2^2 dx}{EI} = 2\int_0^4 \frac{x^2 dx}{EI} + \int_0^3 \frac{4^2 dx}{EI}$$

$$= \frac{2.4^3}{3\,EI} + \frac{3 \times 16}{EI} = \frac{90.667}{EI}$$

$$\int \frac{m_1 m_2 dx}{EI} = \int_0^4 \frac{x\,dx}{EI} + \int_0^3 \frac{4\,x\,dx}{EI}$$

$$= \frac{16}{2\,EI} + \frac{2 \times 9}{EI} = \frac{26}{EI}$$

Putting these value of integrals in equation (1) and (2) we get

$$-\frac{11.25}{EI} + \frac{5\,M_A}{EI} + \frac{26}{EI}H_D = 0$$

$$-\frac{90}{EI} + \frac{26}{EI}M_A + \frac{90.667}{EI}H_D = 0$$

Solving above equating we get

$$H_D = -0.707 \text{ kN}$$

$$M_A = 5.93 \text{ kN.m}$$

Plotting moment diagram

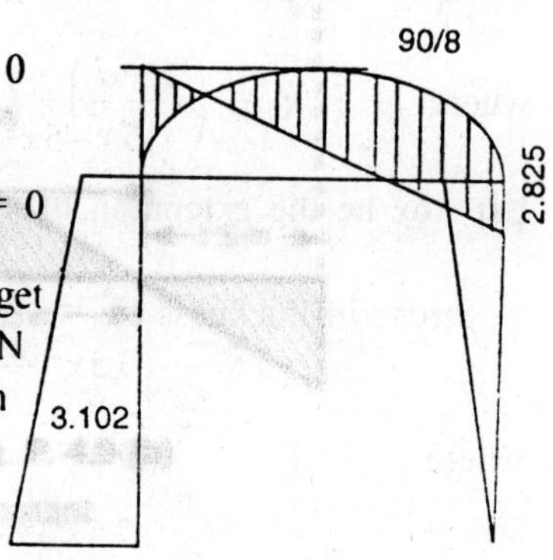

Fig. P. 13.41 (d)

Problem 13.42: A vertical tapered rod of 'L' has its diameter varying linearly from 'd' at lower. end to 'D' at the upper end which has fixed support. Young's modulus of the material is 'E'. show that the elongation to the rod at its lower end when subjected to à longitudinal force 'F' is given by $\delta = \dfrac{4FL}{\pi dDE}$

(b) What do you understand by SHEAR CENTRE ? Determine the location of shear centre of an are section shown in the Fig. P. 13.42.

(c) A steel shaft is subjected to a torque of 2000 kg a and a bending moment of 1000 kg.m. Determine of the shaft = 10 cm. Calculate maximum, minimum principal stresses and maximum shear stress in the shaft at its surface. [IES 1996]

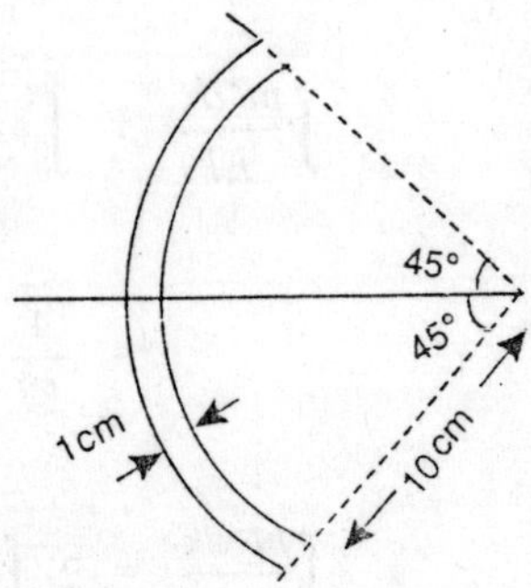

Fig. P. 13.42

Solution:

(a) A vertical tapered rod of which is fixed at top and free at bottom having D as diameter as top and d as diameter at bottom of length 'L' subjected to the force F.

Let 'D' be the diameter of bar at a distance X from the bottom

$$\therefore \qquad D' = d + \left(\frac{D-d}{L}\right)x = d + k.x$$

where $k = \left(\dfrac{D-d}{L}\right) = \text{Constant.}$

Let $\Delta\delta x$ be the extemtion of small thickness dx

$$\therefore \text{ Stress in thickness } dx = \frac{F}{\pi}$$

where $A' = \dfrac{\pi}{4}(D')^2$

$$= \frac{\pi}{4}(d + kx)^2$$

$$\therefore \qquad \Delta\delta x = \frac{F.dx}{\frac{\pi}{4}(d+kx)^2 - E} = \frac{4F.dx}{\pi E(d+kx)^2}$$

Total elangation

$$\Delta = \int_0^L \Delta\delta x = \int_0^L \frac{4F.dx}{\pi E(d+kx)^2}$$

$$= \frac{4F}{\pi E}\int_0^L \frac{1}{(d+kx)^2}dx = \frac{4F}{\pi E}\frac{1}{k}\left(\frac{1}{d+kx}\right)_0^L$$

$$= \frac{4F}{\pi E} - \frac{1}{k}\left[\frac{1}{d+kL} - \frac{1}{d}\right]$$

Putting , $\quad k = \left(\frac{D-d}{L}\right)$, We get

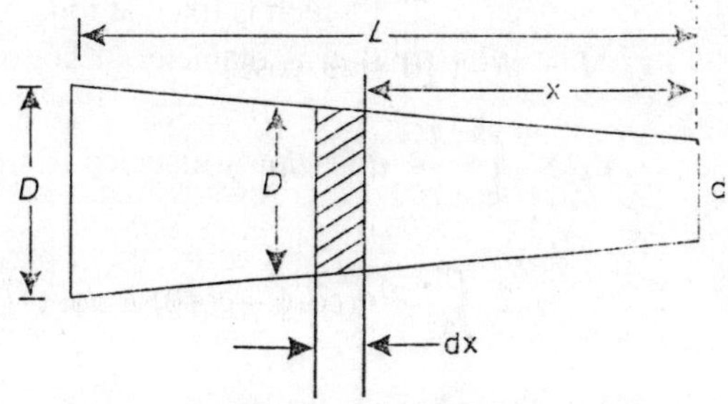

Fig. P. 13.42 (a)

$$\Delta = \frac{4F}{\pi E}\cdot\frac{L}{(D-d)}\left[\frac{1}{d+D-d} - \frac{1}{d}\right]$$

$$= \frac{4F}{\pi E}\cdot\frac{L}{(D-d)}\left(\frac{1}{D} - \frac{1}{d}\right)$$

$$= \frac{4F}{\pi E}\cdot\frac{L}{(D-d)}\frac{D-d}{D.d} = \frac{4FL}{\pi E.D.d}$$

$\therefore$ Hence extension of rod $\Delta = \dfrac{4FL}{\pi dDE}$.

(b) Given $\theta = 45°$, $R = 10$ cm, $t = 1$ cm.

Let 'e' be the shear centre.

Let V is acting at a distance 'e'

$$V_1 = \int_0^\theta \frac{V}{tI} \cdot d\bar{y} \cdot da$$

where
$$d\bar{y} = \int_\phi^\theta R d\theta.\ t.\ R \text{ Sin } \phi$$
$$= R^2 t + [\cos\phi - \cos\theta]$$

Moment of Inertia,

$$I = 2 \int_0^\theta R^3\ .\ d\phi.\ t\ (\sin\phi)^2$$
$$= 2\ R^3\ t \int_0^\theta \sin^2\phi . d\phi$$
$$= 2\ R^3 t \int_0^\theta \left(\frac{1-\cos 2\phi}{2}\right) d\phi$$
$$I = R^3 \text{ t } [\theta \sin\theta \cos\theta]$$

$$\therefore \quad V_1 = \int_0^\theta \frac{V}{tI}\ d\bar{y}.\ da$$
$$= 2 \int_0^\theta R^2 t\,(\cos\phi - \cos\theta)\, R\, d\phi.\text{ t} \quad (\because da = R.d\phi.t)$$

$$\Rightarrow \quad V_1 = \frac{2R^3.V.t}{I}\left[\sin\phi - \phi\cos\phi\right]_0^\theta$$

$$V_1 = \frac{2R^3.V.t}{I}\ [\sin\theta - \theta \cos\theta]$$
$$= \frac{2R^3 V.t}{2R^3 t}\ \frac{[\sin\theta - \theta\cos\theta]}{[\theta - \sin\theta\cos\theta]}$$

Now, $\quad V \times e = V_1 \times R$

$$\therefore \quad e = \frac{V_1 R}{V} = 2R\left(\frac{\sin\theta - \theta\cos\theta}{\theta - \sin\theta\cos\theta}\right)$$

Putting $\theta = 45°, R = 10$ cm

$$\therefore\ e = 2 \times 10 \left[\frac{\sin 45° - \frac{\pi}{4}\cos 45°}{\frac{\pi}{4} - \sin 45° \cos 45°}\right] = \frac{20 \times 0.1517}{0.28} = 10.645 \text{ cm}$$

(c) Torque,

$T = 2000$ kg m

Bending moment, M = 1000 kg m

Diameter of shaft = 10 cm

Due to torsion, shear stress will be developed and shear stress will be maximum at the surface of the shaft.

we know, $\frac{\tau}{r} = \frac{T}{J}$

$$\Rightarrow \quad \tau = \frac{T.r}{J} = \frac{T \cdot \frac{d}{2}}{\frac{\pi}{32} d^4} = \frac{16T}{\pi d^3}$$

$$= \frac{10 \times 2000 \times 100}{\pi (10)^3} = 1018.6 \text{ kg/cm}^2$$

Due to Bending moment, bending stresses are developed

$$\sigma_b = \frac{M}{Z} = \frac{M}{I} . R = \frac{32M}{\pi d^3}$$

$$= \frac{16 \times 2000 \times 100}{\pi \times (10)^3} = 1018.6 \text{ kg/cm}^2$$

Principal stressses are

$$\sigma_{1,2} = \frac{\sigma_b}{2} \pm \sqrt{\frac{\sigma_b^2}{4} + \tau^2}$$

$$= \frac{1018.6}{2} \pm \sqrt{\frac{(1018.6)^2}{4} + (1018.6)^2}$$

$$= 509.3 \pm \sqrt{259382.22 + 1037545-9} = 509.3 \pm 1138.8$$

$$\Rightarrow \quad \sigma_1 = +\ 1648.1 \text{ kg/cm}^2$$

$$\sigma_2 = -\ 629.5 \text{ kg.cm}^2$$

Maximum shear stress at the surface $\tau = 1018.6$ kg/cm^2.

Problem 13.43: Shear force diagram of beam is shown in the Fig. P. 13.43. Identify the location and nature of supports. Draw loading and bending moment diagram showing values of bending moment every 2 m and also showing position and magnitude of maximum bending moment.

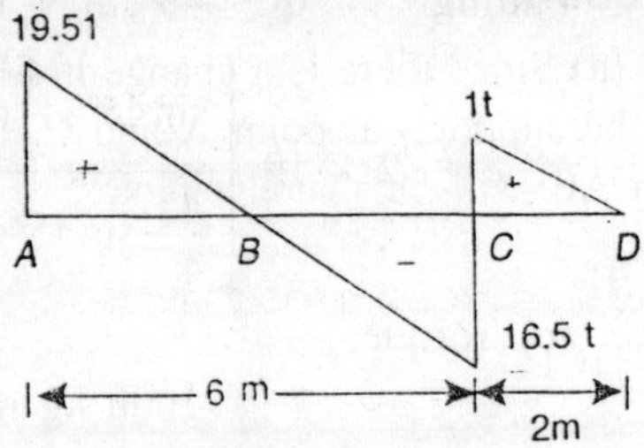

Fig. P. 13.43

(b) Analyse the following strectures (frames and beam) shown in Fig. P. 13.43 (a) and draw bending moment diagrams. [IES 1996]

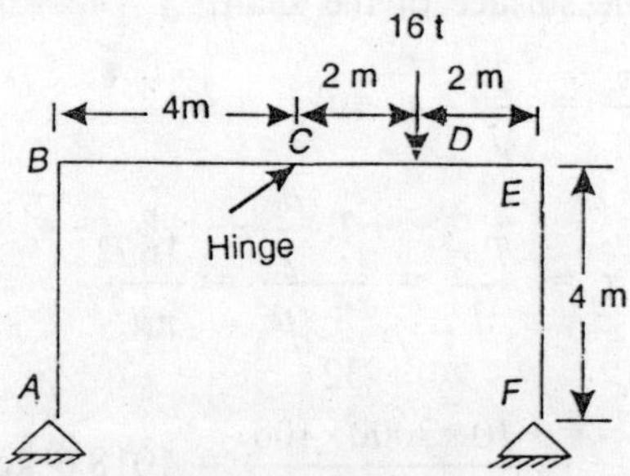

Fig. P. 13.43 (a) (i)

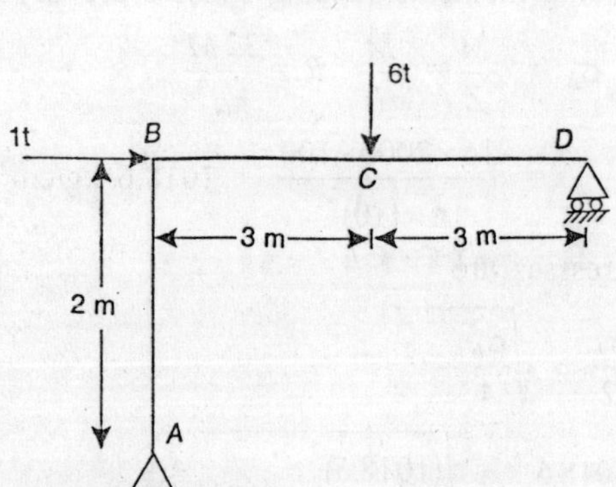

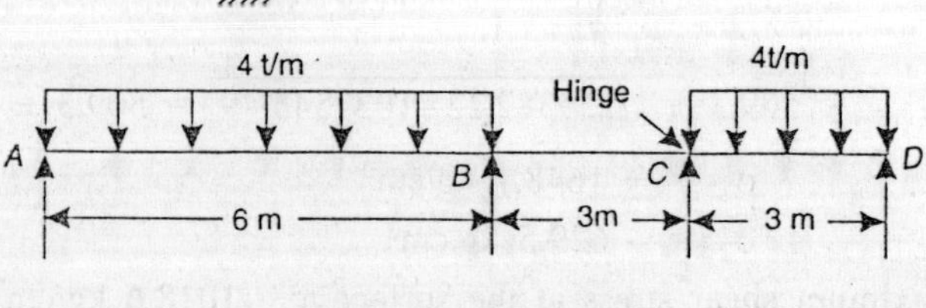

Fig. P. 13.43 (a) (ii)

Solution:

(a) Since there is a change in SF diagram at point A and C. There will be supports at point A and C. Let the intensity of loading on the span AC will be *w*/unit length

$$R_A \times 6 = w \times 6 \times \frac{6}{2} - w' \times 2. x \times 1$$

$$\Rightarrow \qquad R_A = 3w - \frac{w'}{3} = 19.5 \qquad ...(1)$$

$$R_C = 3w + \frac{7w'}{3} = 28.5\ t \qquad ...(2)$$

Solving eqns. (1) and (2), we get

$$w = 6.875 \text{ t/m and } w' = 3.375 \text{ t/m}$$

Let at a distance 'x' from support A, SF will be zero.

Then

$$19.5 - 6.875\ x = 0$$

$$x = 2.836 \text{ m}$$

$$\text{B.M.} = 19.5x - 6.875. \frac{x}{2}$$

At point A, $x = 0$, B.M. = 0

At point C, $x = 6$ m

$$\text{B.M.} = 19.5 \times 6 - 6.875 \times \frac{6^2}{2}$$

$$= -6.75 \text{ t.m}$$

At $\qquad x = 2.836$ m, B.M = 27.654 t.m

Let B.M. will be zero at a distance equal to x.

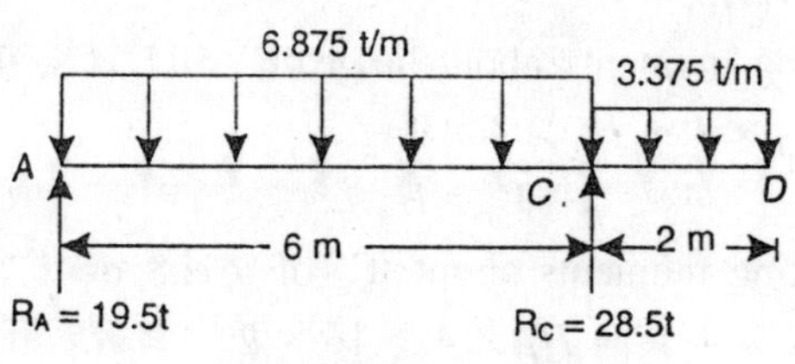

Fig. P. 13.43 (b)

$$\therefore \quad 19.5x - 6.875 \frac{x^2}{2} = 0$$

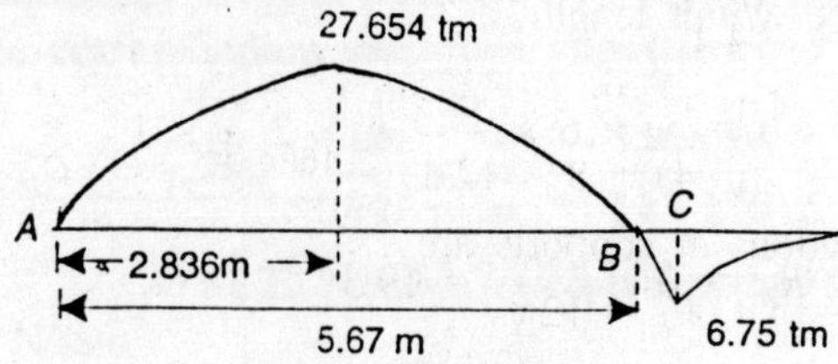

Fig. P. 13.43 (c)

$$\text{or} \quad 6.875x^2 = 39\,x$$

$$\therefore \quad x = \frac{39}{6.875}$$

$$= 5.67 \text{ m}$$

The bending moment diagram is shown.

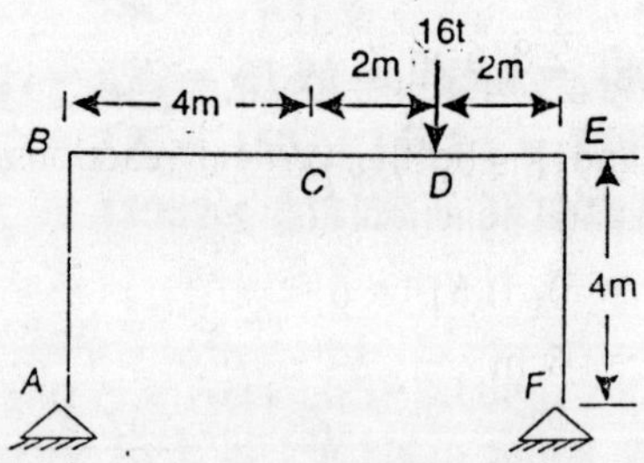

Fig. P. 13.43 (d)

(b) (i) Let V_A and V_F be the horizontal reactions:

Now, $\Sigma V = 0$

$\therefore \quad V_A + V_F = 16$ t

and $H_A + H_F = 0$

Now taking moments about hinge 'C' of L.H.S, then

$$-V_A \times 4 + H_A \times 4 = 0$$

$$\therefore \quad V_A = H_A$$

Again taking moments about 'C' of R.H.S of 'C'

$$V_F \times 4 = -H_F \times 4 + 16 \times 2$$

$$\text{or} \quad V_F = -H_F + 8$$

or $\quad V_A + V_F = H_A - H_F + 8$

or $\quad 16 = H_A - H_F + 8$

or $\quad H_A - H_F = 8$

$H_F = -4$ t

$V_A = 4$ t

$V_F = -(-4) + 8 = 12$ t

Hence reactions at supports are

$V_A = 4$ t, $V_F = 12$ t

$H_A = 4$ t, $H_F = -4$ t

B.M. diagram

BM at $A = 0$

BM at $B = 4 \times 4 = 16$ t.m

BM at $C = 4 \times 4 - 4 \times 4 = 0$

BM at $F = 0$

BM at $E = 4 \times 4 = 16$ t.m

BM at $D = 12 \times 2 + 4 \times 2 = 32$ t.m

BM at $C = 4 \times 4 - 4 \times 4 = 0$

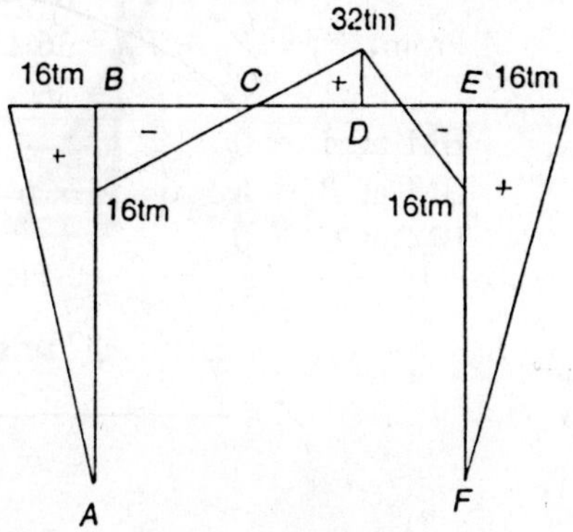

Fig. P. 13.43 (e)

(ii) Let V_A and V_D be the vertical reactions at A and D while H_A be the horizontal reaction at A.

$\Sigma H = 0$

or $H_A + 1 = 0$

$\therefore \quad H_A = 1\ t$

$\Sigma H = 0,\ V_A + V_D = 6\ t$

Taking moment about A,

$V_D \times 6 = 6 \times 3 + 1 \times 2$

or $\quad V_D = 3.33 = 2.67$ t.

B.M. diagram

BM at $A = 0$

BM at $D = 0$

BM at $B = 1 \times 2 = 2$ t.m

BM at $C = 2.67 \times 3 + 1 \times 2 = 10$ t.m

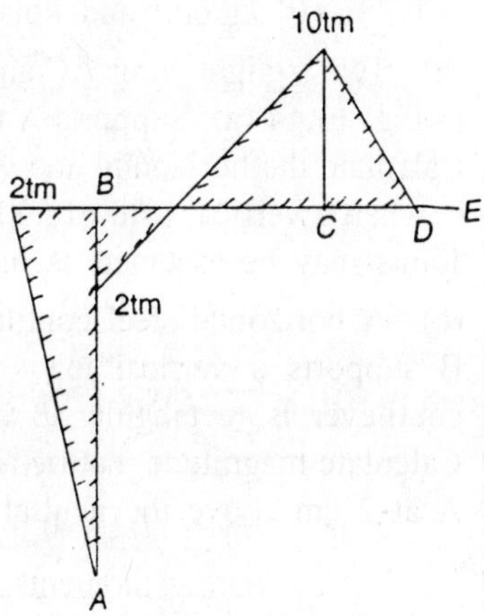

Fig. P. 13.43 (f)

(b) (iii) Let V_A, V_B and V_D be the vertical reactions at support A, B and D respectively,

$\therefore \quad V_A + V_B + V_D = 4 \times 6 + 4 \times 3 = 36$ t

Taking moment about hinge 'C' on R.H.S of C

$$V_D \times 3 = 4 \times 3$$

$\therefore$ $$V_D = 4 \text{ t}$$

Now, $$M_B = 0$$

or $$V_A \times 6 + 8 \times 3 = 4 \times 6 \times 3$$

$$V_A = 8 \text{ t}$$

From $$V_A + V_B + V_D = 36 \text{ t}$$

$$V_B = 36 - 12 = 24 \text{ t}$$

BM at $A = 0$

BM at $B = 8 \times 6 - 4 \times 6 \times 3 = -24$ t.m

BM at $C = 0$

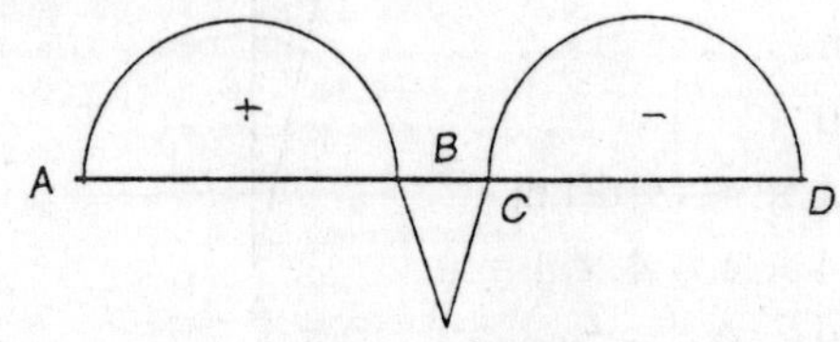

Fig. P. 13.43 (g)

Problem 13.44 : (a) A cylindrical piece of steel 80 mm dia and 120 mm long is subjected to an axial compressive force of 50000 kg. Calculate the change in the volume of the piece if bulk modulus = 1.7×10^5 kg/cm^2 and Poisson's ration = 0.3.

(b) Two straight wires AC and BC, 8 mm dia meet at joint C as shown in Fig. 13.44 (a). Supports A and B (at the same level) are unyielding. Calculate the horizontal and vertical components of deflection of joint C when a vertical load of 500 kg is applied at C.E. = 2×10^6 kg/cm^2. Joints may be assumed as hinged.

(c) A horizontal steel contilever is fixed at end A and its free end B supports a vertical load of 6000 kg. The cross-section of the contilever is rectangular, 3 cm wide, 12 cm deep. AB = 30 cm. Calculate magnitude, nature and direction of principal stresses at end A at 2 cm above the neutral axis. [IES–97]

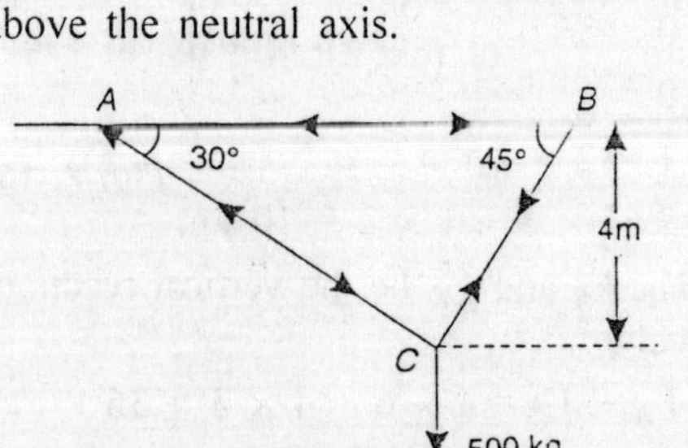

Fig. P. 13.44 (a)

Solution :

(a) Given :

$$d = 80 \text{ mm}$$
$$L = 120 \text{ mm}$$
$$P = 50{,}000 \text{ kg}$$
$$v = 0.3$$
$$K = 1.7 \times 106 \text{ kg/cm}^2$$

Volume of pipe,

$$V = \frac{\pi d^2 \cdot L}{4}$$

$$\therefore \quad \frac{\delta V}{V} = \frac{\pi \cdot L}{4} \cdot \frac{2d\delta d}{\frac{\pi}{4} d^2 L} + \frac{\pi d^2 \delta L}{4 \frac{\pi d^2 L}{4}} = \frac{2\delta d}{d} + \frac{\delta L}{L}$$

$$= -2\, v \frac{P}{E} + \frac{P}{E}$$

$$\frac{\delta V}{V} = \frac{P}{E}[1 - 2\, v] \qquad \text{...(i)}$$

$$p = \frac{P}{A} = \frac{50{,}000}{\pi \times \frac{(8)^2}{4}} \text{ kg/cm}^2 = 994.72 \text{ kg/cm}^2$$

$$E = 3\, K\,(1 - 2v)$$
$$= 3 \times 1.7 \times 10^6\,(1 - 2 \times 0.3) = 2.04 \times 10^6 \text{ kg/cm}^2$$

From equation (i), we get

$$\frac{\delta V}{V} = \frac{994.72}{2.04 \times 10^6}\,(1 - 2 \times 0.3)$$

$$\frac{\delta V}{V} = 1.95 \times 10^{-4}$$

Thus change in volume;

$$= \delta V = 1.95 \times 10^{-4} \times V$$

$$= 1.95 \times 10^{-4} \times \frac{\pi}{4} \times (8)^2 \times 12$$

$$= 0.1176 \text{ cm}^3 = 117.6 \text{ mm}^3$$

(b) At joint C

$$\Sigma H = 0$$

$$P_{CA} \cos 30^\circ = P_{CB} \cos 45^\circ \text{ and } \Sigma V = 0$$

$$P_{CA} \sin 30^\circ + P_{CB} \sin 45^\circ = 500$$

but $P_{CB} \sin 45^\circ = P_{CB} \cos 45^\circ = P_{CA} \cos 30^\circ$

$\therefore$ $P_{CA} \sin 30^\circ + P_{CA} \cos 30^\circ = 500$

$\Rightarrow$ $P_{CA} (\sin 30^\circ + \cos 30^\circ) = 500$

$$1.3660\ P_{CA} = 500$$

$\Rightarrow$ $P_{CA} = 366.025$ kg

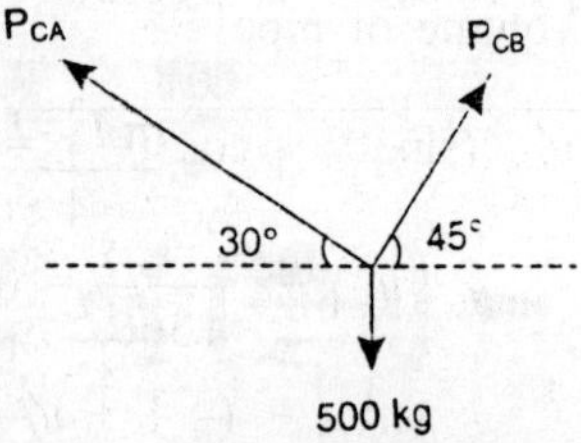

Fig. P. 13.44 (b)

and $P_{CA} = \dfrac{P_{CA} \cos 30^\circ}{\cos 45^\circ}$

$= 448.287$ kg

$$x = \frac{4}{\tan 45^\circ} = 4 \text{ m}$$

and $y = \dfrac{4}{\tan 30^\circ} = 6.9282$ m

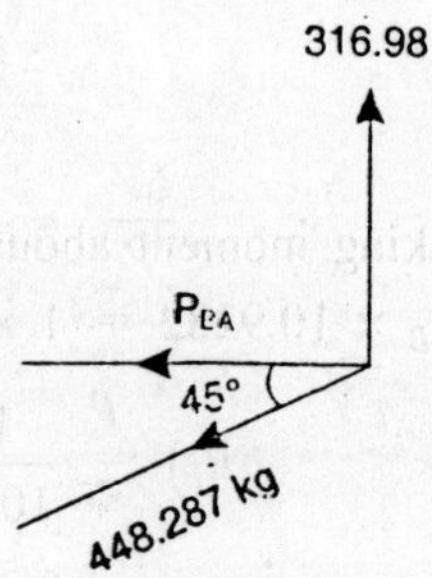

Fig. P. 13.44 (c)

Taking moment about B

$$R_B \times 10.9282 + 500 \times 6.9282 = 0$$

or $R_B = -\dfrac{500 \times 6.9282}{10.9282}$

$= -316.98$ kg $= 316.98$ kg ($\uparrow$)

$\therefore$ Reaction at joint

$A = 500 - 316.98$

$= 183.012$ kg ($\uparrow$)

Consider the equilibrium of point B

$$\Sigma H = 0$$

$P_{BA} = -448.287 \cos 48^\circ$

$= -316.98$ kg

$= 316.98$ kg (comp.)

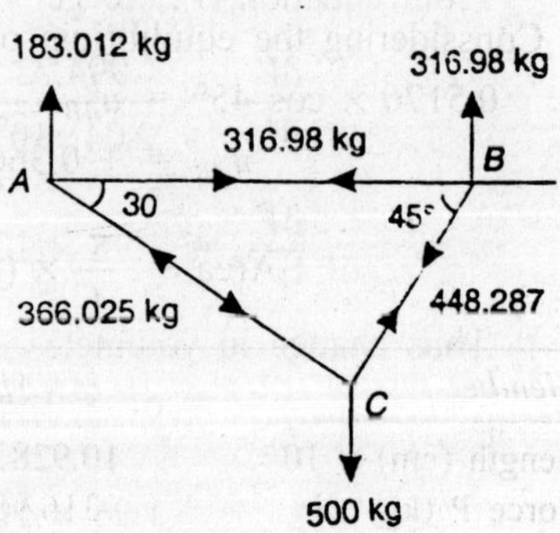

Fig. P. 13.44 (d)

When horizontal load of 1 kg will act

Considering the equilibrium of

point *C*.

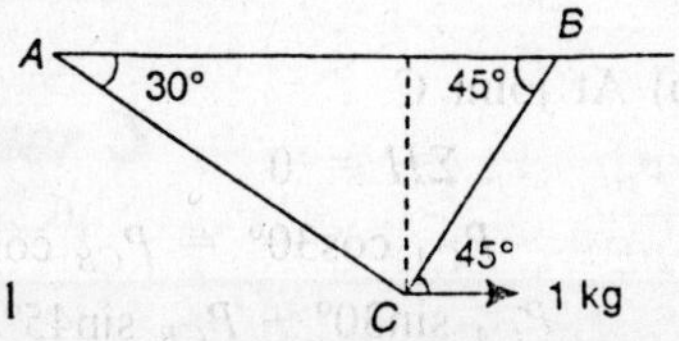

Fig. P. 13.44 (e)

$$\Sigma H = 0$$

$$u'_{CB}\cos45° + 1 = u'_{CA}\cos30° = 0$$

$$u'_{CB}\cos45° - u'_{CA}\cos30° = -1$$

$$\Sigma V = 0$$

$$u'_{CB}\sin45° + u'_{CA}\sin30° = 0$$

$$u'_{CB}\sin45° + u'_{CA}\sin30° = 0$$

$$-u'_{CB}\cos45° - u'_{CA}\cos30° = -1$$

$$+ \qquad +$$

$$u'_{CA}(\sin30° + \cos30°) = 1$$

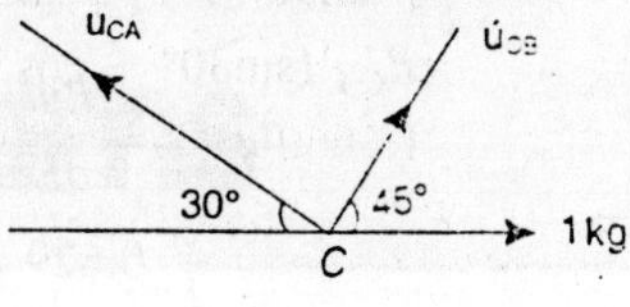

Fig. P. 13.44 (f)

$$\Rightarrow \qquad u'_{CA} = \frac{1}{1.366} = 0.7320$$

$$\therefore \qquad u'_{CB} = (-1 + u'_{CA}\cos30°)/\cos45°$$

$$= (-1 + 1 \times 0.86)/\cos45°$$

$$= \frac{-1 + 0.732 \times 0.86}{\cos45°} = -\frac{0.366}{\cos45°} = -0.5176 \text{ kg}$$

Taking moment about *A*

$$R_B \times 10.9282 = 1 \times 4$$

$$\Rightarrow \qquad R_{BV} = \frac{1 \times 4}{10.9282}$$

$$= 0.366\ (\downarrow)$$

as no vertical force is acting

$$R_{AV} = 0.366\ (\uparrow)$$

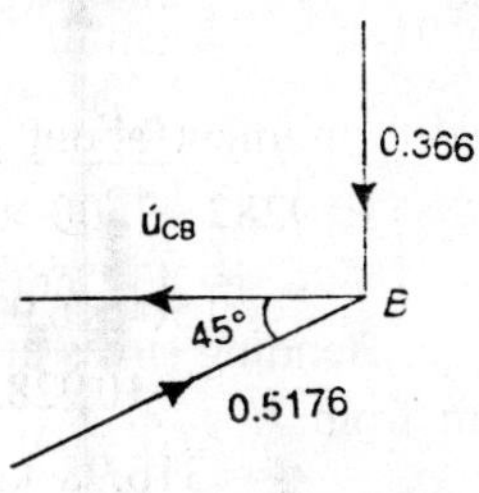

Fig. P. 13.44 (g)

Conssidering the equilibrium of joint *B*

$$0.5176 \times \cos 45° = u'_{AB}$$

$$u'_{AB} = +0.366 \text{ kg (tensile)}$$

$$\text{Area} = \frac{\pi}{4} \times (8 \times 10^{-1})^2 = 0.502 \text{ cm}^2$$

Member	*AB*	*BC*	*CA*
Length (cm) × 10^2	10.9282	5.656	6.9282
Force P (kg)	–316.98	448.287	366.025
Area *a*	0.502	0.502	0.502
u'	0.366	–0.5176	0.732

u	–0.634	0.8965	0.732
$\dfrac{Pul}{aE}$	0.2187	0.2264	0.1848
$\dfrac{Pu'l}{aE}$	–0.1262	–0.1307	+0.1848

$$\delta_V = \Sigma \frac{Pul}{aE} = 0.629$$

$$= 0.629 \text{ cm} = 6.29 \text{ mm}$$

$$\delta_H = \Sigma \frac{Pu'l}{aE}$$

$$= -0.072 \text{ cm} = 0.72 \text{ mm } (\leftarrow)$$

(c) Principal stress required at a distance 2 cm for the neutral axis.

Maximum B.M.= 6000 × 30 = 1, 80, 000 kg.cm

Shear force is maximum at *A*.

$$F = 6000 \text{ kg}$$

$$I = \frac{bd^3}{12} = \frac{3\times(12)^3}{12}$$

$$= 432 \text{ cm}^3$$

∴ Bending stress at distance 2 cm from A,

$$\sigma_b = \frac{M}{I} y_1$$

$$= \frac{1,80,000}{432} \times 2$$

$$= 833.33 \text{ kg/cm}^2$$

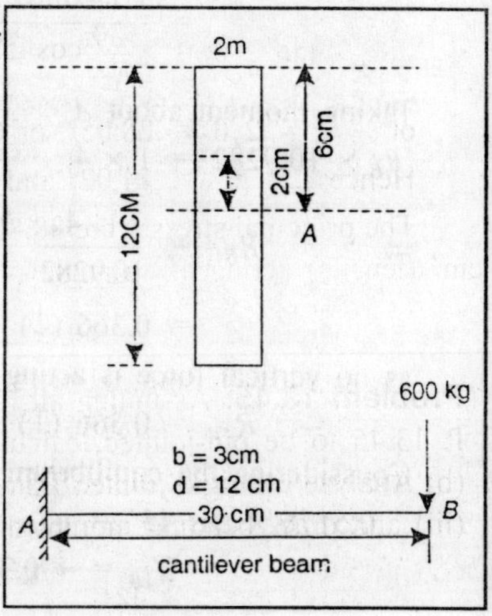

Fig. P. 13.44 (h)

Shear stress distribution along rectangular cross-section

$$\tau = \frac{F}{2I}\left[\frac{d^2}{4} - y^2\right] \quad \text{where } y = 2 \text{ cm}$$

∴ Shear stress,

$$\sigma_s = \frac{6000}{2 \times 432}\left[\frac{(12)^2}{4} - (2)^2\right]$$

$$= 222.22 \text{ kg/cm}^2$$

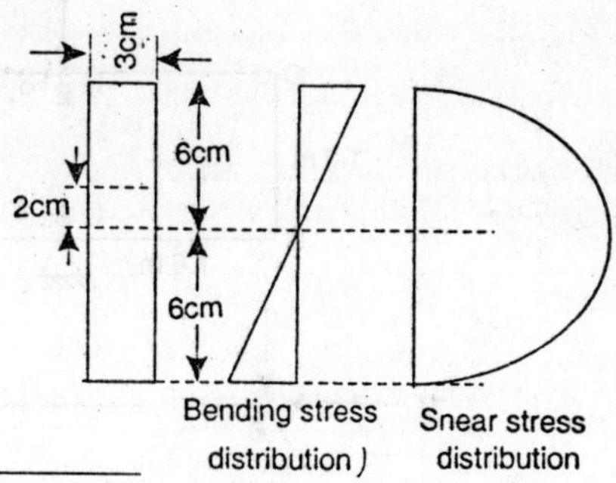

Fig. P. 13.44 (i)

Principal stress

$$\sigma_{1,2} = \frac{\sigma}{2} \pm \sqrt{\frac{\sigma^2}{4} + \tau^2}$$

$$= \frac{833.33}{2} \pm \sqrt{\frac{(833.33)^2}{4} + (222.22)^2}$$

$$= 416.665 \pm 472.22$$

$\therefore$ $\sigma_l = 416.665 + 472.22 = 888.88$ kg/cm² (compressive)

and $\sigma_2 = 416.665 - 472.22 = -55.55$ kg/cm² (tensile)

Direction of principal planes are given by

$$\tan 2\theta = \frac{2\sigma_s}{\sigma_b} = \frac{2 \times 222.22}{833.33}$$

or $2\theta = 28.05°$, or $\theta = 14.02°$

Hence $\theta_1 = 14.02°$ and $= 104.02°$

The principal stresses are 88.88 kg/cm² compressive and 55.555 kg/cm² (tensile) acting at an angle of 14.02° and 104.02° respectively.

Problem 13.45: Assuming all members of the trues shown in Fig. P. 13.45 to be pin-jointed, calculated forces in them.
(b) Analyse the rigid jointed frames shown in Fig. 13.45 (i), (ii) and (iii) and draw bending morent diagrams. [IES 1997]

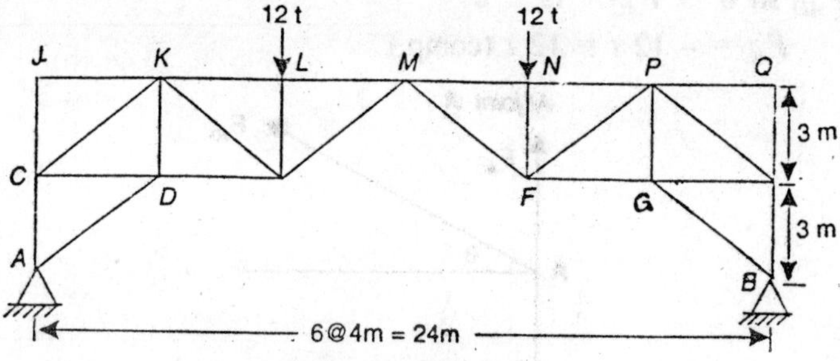

Fig. P. 13.45 (i)

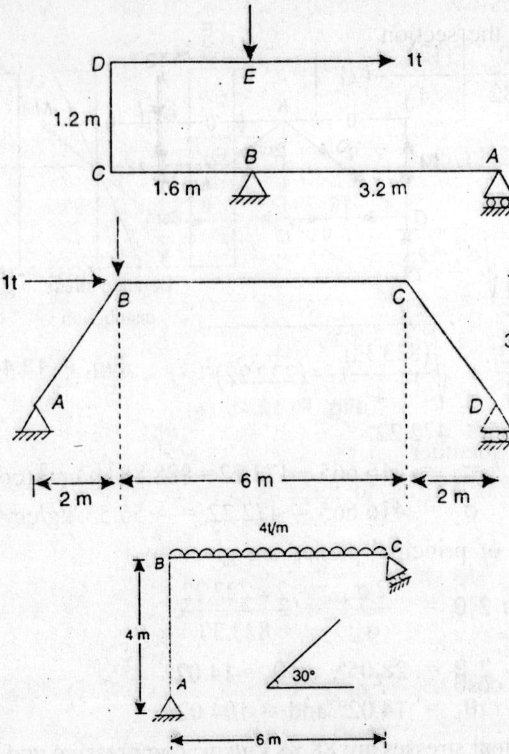

Fig. P. 13.45 (ii) & (iii)

Solution :

At joint A

$$\Sigma H = 0$$

$$F_{AD} \cos \theta = 0$$

$$\Rightarrow \quad F_{AD} = 0$$

and F_{AD} sn θ $+ F_{AC} + 12 = 0$

$$\Rightarrow \quad F_{AC} = -\ 12\ t = 12\ t \text{ (comp.)}$$

$$\tan \theta = 0.75$$

$$\theta = \tan^{-1} 0.75$$

$$= 36.869$$

$$\sin \theta = 0.6$$

$$\cos \theta = 0.8$$

At joint A

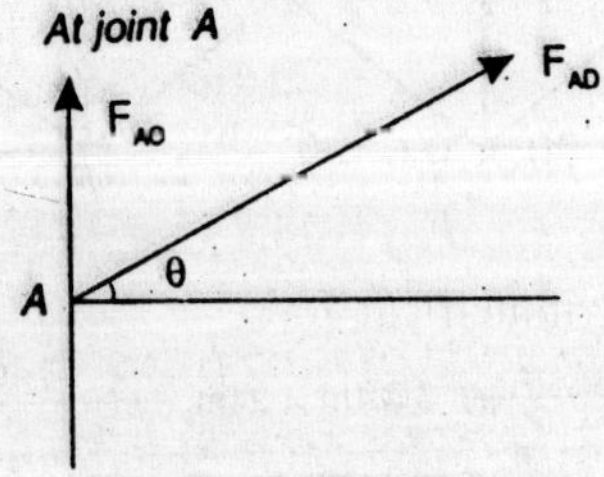

Fig. P. 13.45 (a)

Consider the section

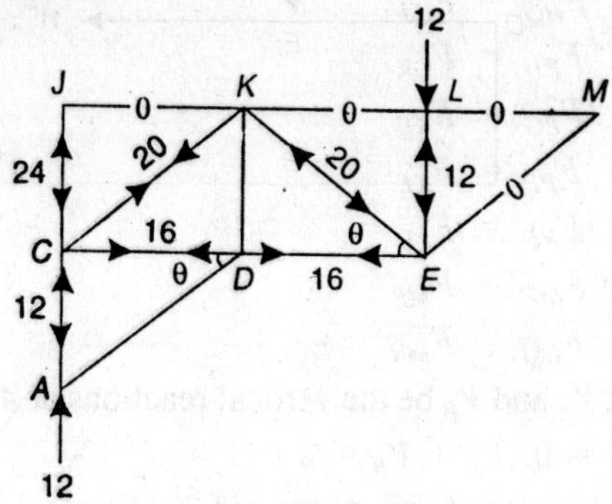

Fig. P. 13.45 (b)

Forces in member

ML and *EM* = 0 as no force is external acting.

At joint *L*

$$F_{KL} = 0$$

$$F_{LE} = -12 \text{ t} = 12 \text{ t (comp.)}$$

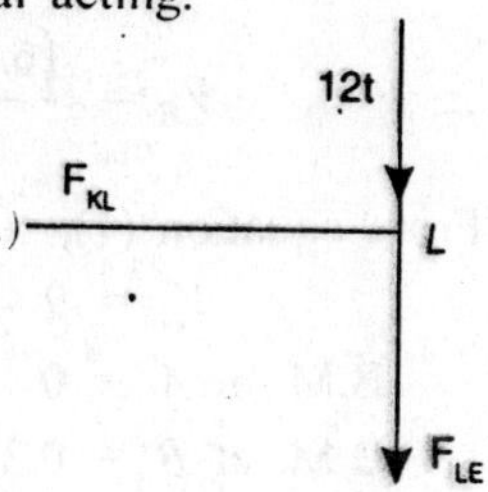

Fig. P. 13.45 (c)

At joint *E*

$$F_{KE} \cos\theta = -F_{DE} - 1$$

and $$F_{KE} = \frac{12}{\sin\theta} = 20 \text{ t (tensile)}$$

and $$F_{DE} = -20 \times 0.8 = 16 \text{ t (comp.)}$$

At joint *D*

$$F_{DK} = 0$$

and $$F_{DC} + 16\,t = 0$$

$$F_{DC} = -16\,t = 16 \text{ t (comp)}$$

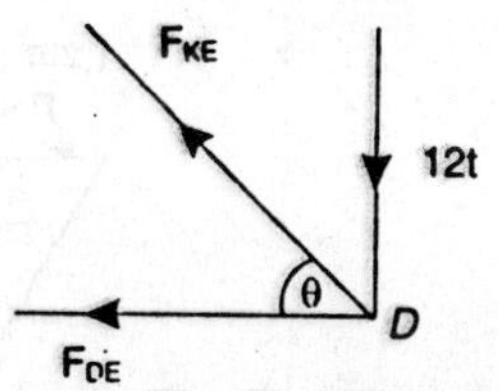

Fig. P. 13.45 (d)

At joint *C*

$$F_{CK} \cos\theta - 16 = 0$$

$$F_{CK}\ 0.8 = 16$$

$$F_{CK} = 20 \text{ (tensile)}$$

$$F_{CJ} + 12 + 20 \times 0.6 = 0$$

$$\Rightarrow \qquad F_{CJ} = 24 \text{ t (comp.)}$$

As the structure is symmetrical force in member

$$F_{BH} = F_{AG}$$

$$F_{BG} = F_{Ad}$$

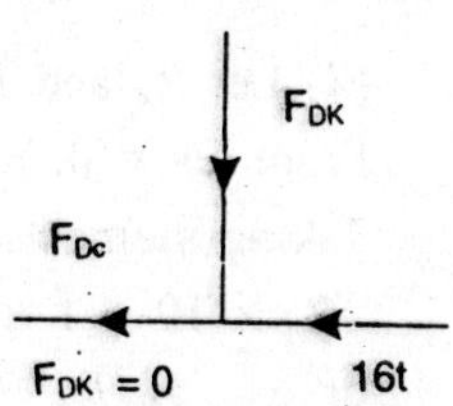

Fig. P. 13.45 (e)

$F_{GH} = F_{Cd}$
$F_{HQ} = F_{CJ}$
$F_{PH} = F_{CK}$
$F_{PG} = F_{KD}$
$F_{PF} = F_{KE}$
$F_{NF} = F_{LE}$
$F_{MF} = F_{ME}$
$F_{LM} = F_{MN}$

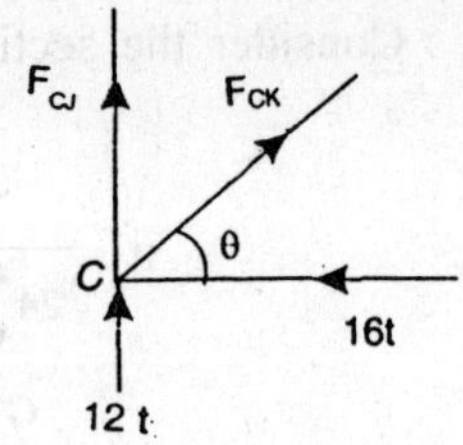

Fig. P. 13.45 (f)

(b) (i) Let V_A and V_B be the vertical reactions at A and B respectively.

From $\Sigma V = 0$, $V_A + V_B = 2$ t ...(1)

Taking moments about A, we get

$V_B \times 3.2 + 1 \times 1.2 = 2 \times 3.2$,

or $3.2\ V_B + 1.2 = 6.4$

$$\therefore \quad V_B = \frac{(6.4 - 1.2)}{3.2} = 1.625 \text{ t}$$

From equation (1),

$V_A = 2 - 1.625 = 0.375$ t

B.M. at $A = 0$

B.M. at $B = 0.375 \times 3.2 = 1.2$ t.m

B.M. at $C = 0.375 \times 4.8 + 1.625 \times 1.6 - 1 \times 1.2 - 2 \times 1.6 = 0$

B.M. at $D = 0.375 \times 4.88 + 1.625 \times 1.6 = 1.2$ t.m

B.M. at $E = 0$

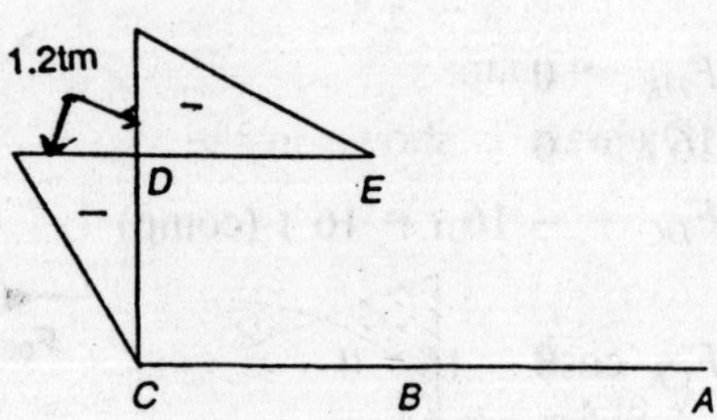

Fig. P. 13.45 (g)

(ii) Let R_A and R_D be the reactions at A and D respectively.

From $\Sigma V = 0$, $R_N + R_D = 6$ t ...(2)

Taking moments about D, we get

$R_A \times 10 + 1 \times 3 = 6 \times 8$

or $10\ R_A + 3 = 48$

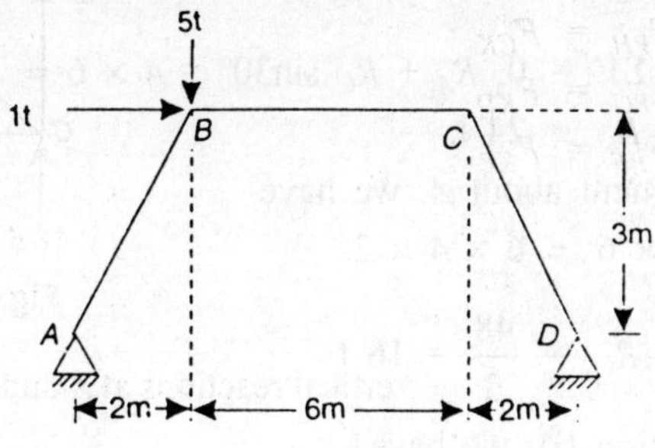

Fig. P. 13.45 (h)

or $\quad R_N = 4.5$ t

From equation (2),

$$R_D = 6 - 4.5 = 1.5 \text{ t}$$

Again, $\Sigma H = 0$

$$H_A = 1\ t$$

Now B.M. at A and $D = 0$

Taking moment about B, we get

B.M. at B $= R_A \times 2 + 1 \times 3 + 1.5 \times 8$

$= 4.5 \times 2 + 1 \times 3 + 1.5 \times 8$

$= 24$ t.m

B.M at C $= 1.5 \times 2 - 4.5 \times 8 - 1 \times 3$

$= 42$ t.m

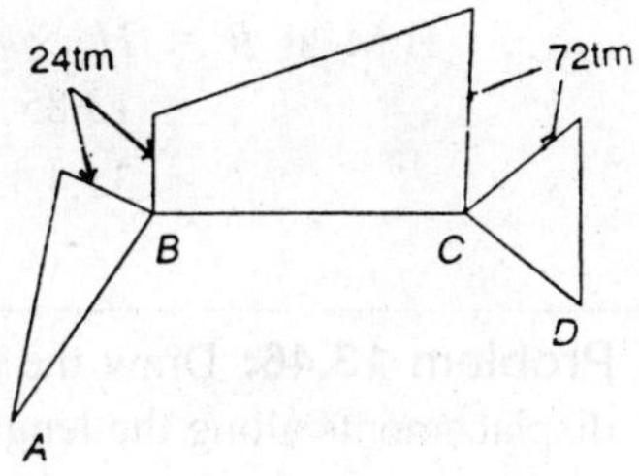

Fig. P. 13.45 (i)

The B.M. Diagram is shown in the Fig.

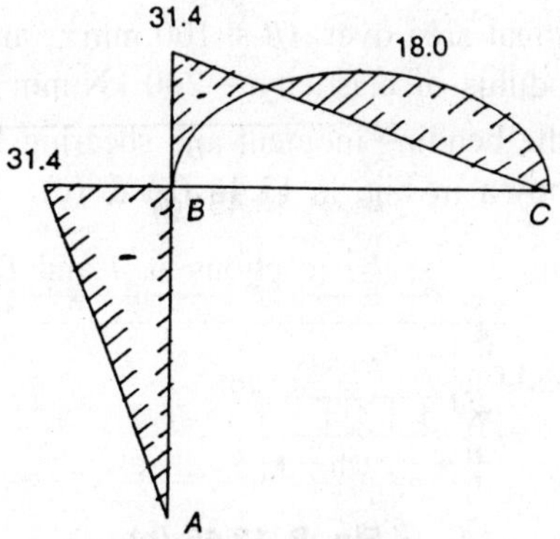

Fig. P. 13.45 (j)

(iii) Let R_A and R_C be the reactions at A and C respectively.

From

$$\Sigma V = 0, \; R_A + R_C \sin 30^\circ = 4 \times 6 = 24 \text{ t}$$

$$R_A + 0.5 \, R_C = 24 \text{ t} \qquad \ldots(i)$$

Taking moment about A, we have

$$R_C \sin 30^\circ \times 6 = 6 \times 4 \times 2$$

or
$$R_C = \frac{48}{3} = 16 \text{ t}$$

From equation (i), we have

$$\therefore \qquad R_A = 24 - 0.5 \times 16 = 16 \text{ t}$$

From $\Sigma V = 0$, $H_A = R_C \cos 30^\circ = 16 \times \cos 30^\circ = 13.85$ t

B.M. at A and $C = 0$

$$\therefore \quad \text{B.M at } B = H_A \times 4 + R_C \sin 30^\circ \times 6 - 4 \times 6 \times 3$$

$$= 13.85 \times 4 + 16 \times 0.5 \times 6 - 4 \times 6 \times 3$$

$$= 31.4 \text{ t.m}$$

Problem 13.46: Draw the diagrams of normal forces, stresses and displacements along the length of the stepped bar ABC shown in Fig. 13.46 (a).

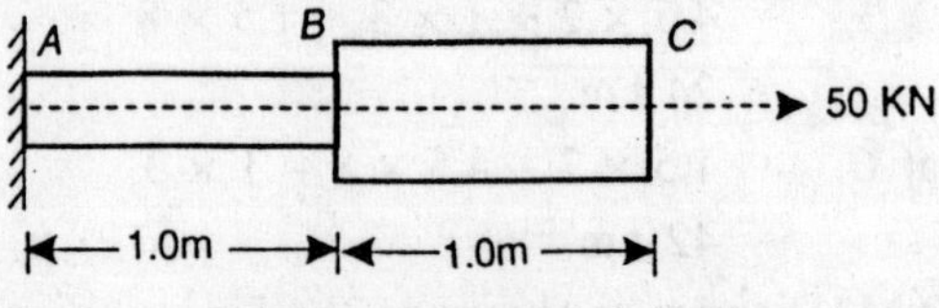

Fig. P. 13.46 (a)

Cross-sectional area over AB = 100 mm^2; and area over BC = 200 mm^2; modulus of elasticity = 200 kN/mm^2.

(b) Draw the bending moment and shearing force diagrams for the members shown in Fig. P. 13.46 (b) & (c)

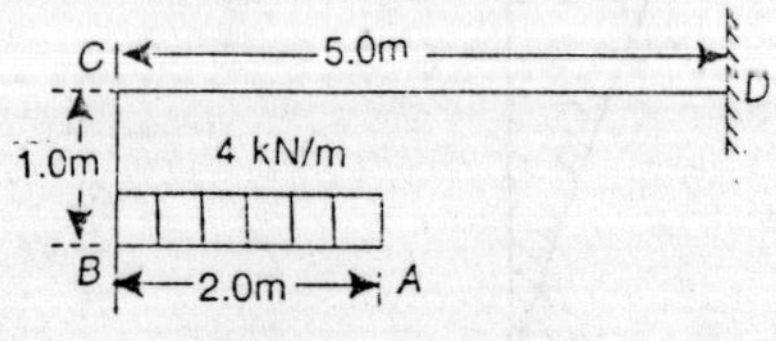

Fig. P. 13.46 (b)

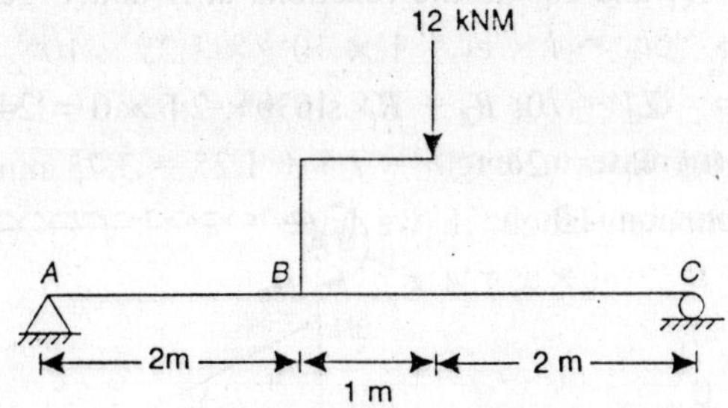

Fig. P. 13.46 (c)

(c) What is Mohr's of stress ?

At a point in an elastic material, a direct tensile stress of 60 N/mm^2 and a direct compressive stress of 40 N/mm^2 are applied on planes at right angles to each other. If the maximum principal stress is limited to 75 N/mm^2 (tensile), find the shear stress that may be allowed on the planes. Also determine the maximum principal stress and the minimum shear stress. [IES 1998]

Solution :

(a) Normal force diagram :

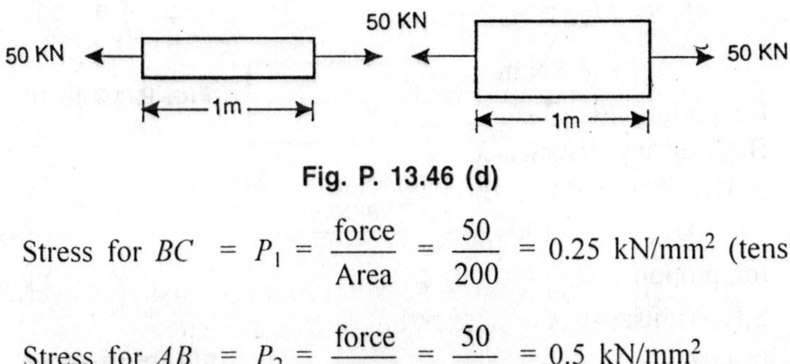

Fig. P. 13.46 (d)

$$\text{Stress for } BC = P_1 = \frac{\text{force}}{\text{Area}} = \frac{50}{200} = 0.25 \text{ kN/mm}^2 \text{ (tensile)}$$

$$\text{Stress for } AB = P_2 = \frac{\text{force}}{\text{Area}} = \frac{50}{100} = 0.5 \text{ kN/mm}^2$$

Strain in portion

$$BC \rightarrow \quad e_1 = \frac{P1}{E} = \frac{0.25 \text{kN/mm}^2}{200 \text{kN/mm}^2} = 1.25 \times 10^{-3}$$

$$\text{AB} \rightarrow \quad e_2 \doteq \frac{P_2}{E} = \frac{0.5 \text{kN/mm}^2}{200 \text{kN/mm}^2} = 2.5 \times 10^{-3}$$

Increase in length in portion

1. $BC \Rightarrow \quad \delta l_1 = l \times e_1 = 1 \times 10^{-3} \times 1.25 \times 10^{-3} = 1.25$ mm

2. AB $\Rightarrow \quad \delta l_2 = l \times e_2 = 1 \times 10^{-3} \times 2.5 \times 10^{-3} = 2.5$ mm

$\therefore$ Total increase in length = 2.5 + 1.25 = 3.75 mm.

(b) For portion *AB*

$\Rightarrow F_x = W_x = 4 \times x = 4x$

at $X = 0$

$F_X = 0$

at $X = 2$ m

$F_X = 8$ kN

for B.M.

$$- M_{XX} - \frac{WX^2}{2} = 0$$

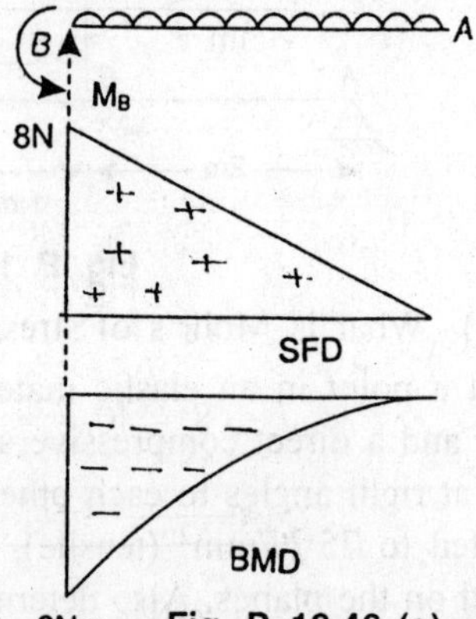

Fig. P. 13.46 (e)

or $M_{XX} = -\dfrac{Wx^2}{2} = -\dfrac{4x^2}{2}$

at $X = 0, M_{XX} = 0$

at $X = 2$

$$M_B = M_{XX} = -\frac{4 \times 4}{2}$$

$$= -8 \text{ kN.m}$$

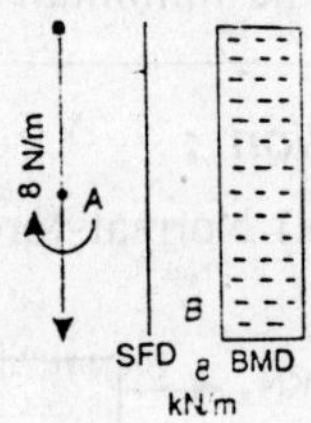

Fig. P. 13.46 (f)

for portion *BC*

B.M. at any distance *X*

$- M_{XX} - B = 0$

$M_{XX} = -8$ kN.m

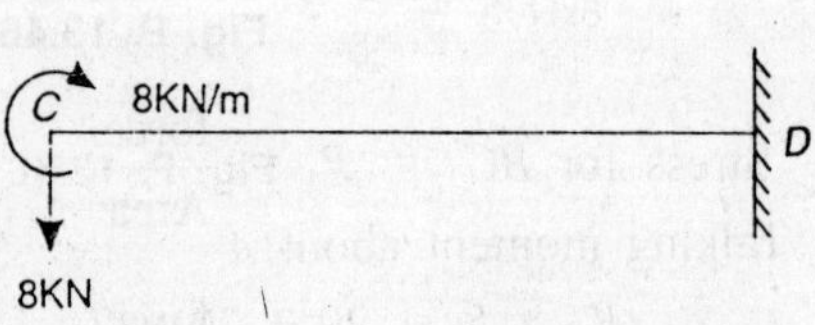

Fig. P. 13.46 (g)

for protion *CD*

S.F. at ditsance *X*

from *C*

$F_x + 8 = 0$

$F_x = -8$ kN

For B.M. at distance *X*

from *C*

$- M_{XX} + 8 - 8x = 0$

$M_{XX} = 8 - 8x$

for $X = 0 \quad M_{XX} = 8$ kN.m

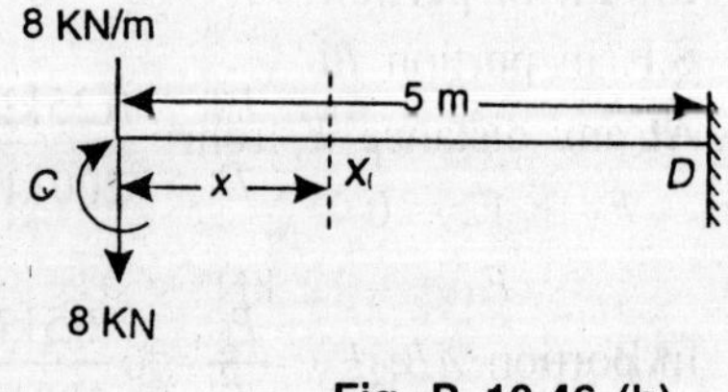

Fig. P. 13.46 (h)

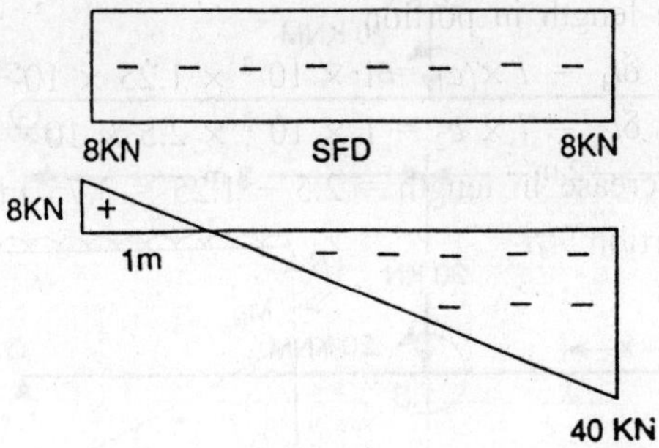

Fig. P. 13.46 (i)

for $X = 5$

$$M_{XX} = 8 - 40 = -32 \text{ kN.m}$$

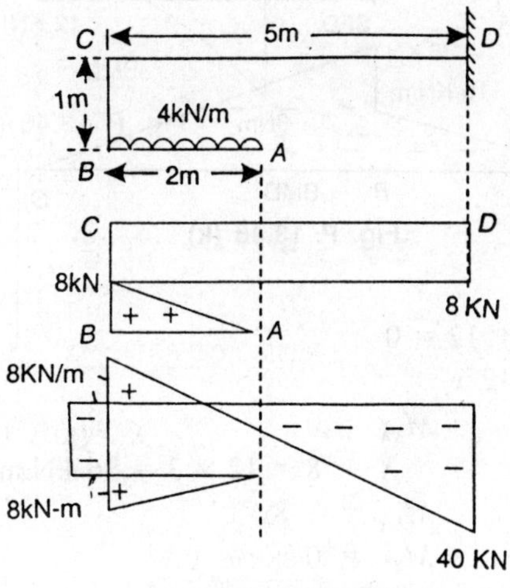

Fig. P. 13.46 (j)

Talking moment about A

$$R_C \times 5 = 20 + 20 \times 2$$

$$R_C = 60/5 = 12 \text{ kN}$$

$$\Rightarrow \quad R_A = 20 - 12 = 8 \text{ kN}$$

S.F. in portion BC

At any distance X from

$$F_X + 12 = 0$$

$$F_X = -12 \text{ kN}$$

In portion AB, $F_X - 8 = 0$

$$F_X = +8$$

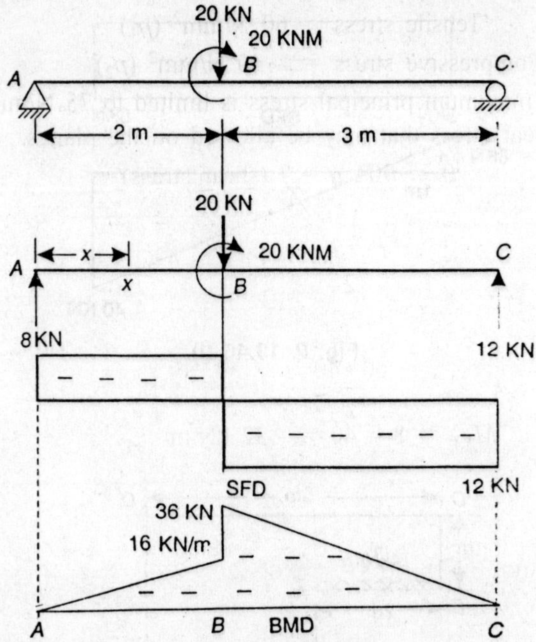

Fig. P. 13.46 (k)

For B.M.D

In BC, $-M_{XX} + 12 = 0$

$M_{XX} = 12\,x$

At $x = 0$ $\quad M_{XX} = 0$

$X = 8 = 12 \times 3 = 36$ kN.m

In AB, $\quad M_{XX} = +\,8x$

At $x = 0$ $\quad M_{XX} = 0$

At $x = 2$ $\quad M_{XX} = 16$ kN.m

(c) Mohr's circle of stress

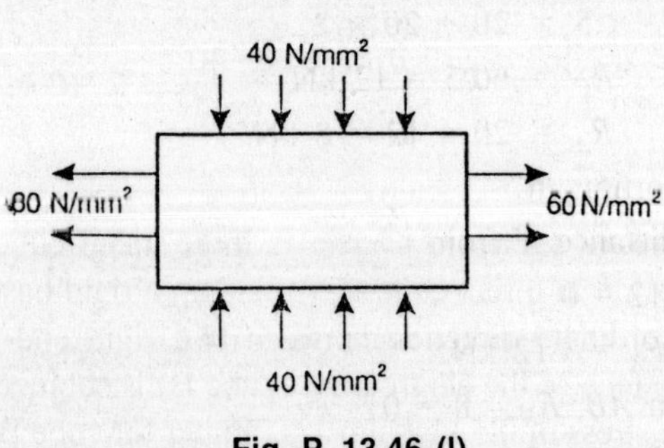

Fig. P. 13.46 (l)

Tensile stress = 60 N/mm² (p_1)

Compressive stress = – 40 N/mm² (p_2)

If the maximum principal stress is limited to 75 N/mm² (tensile), find the shear stress that may be allowed on the planes.

θ = 90°, q = ? (shear stress)

$$\text{Principal steess} = \frac{p_1+p_2}{2} + \sqrt{\left(\frac{p_1-p_2}{2}\right)^2 + q^2}$$

$$= \frac{60-40}{2} - \sqrt{\left(\frac{60+40}{2}\right)^2 + q^2}$$

$$= 75$$

$$= 10 + \sqrt{2500 + q^2} = 75$$

$$\Rightarrow 2500 + q^2 = (65)^2 = 4225$$

$$q^2 = 1725 \Rightarrow q = 41.53 \text{ N/mm}^2$$

Thus a shear stress of 41.53 N/mm² may be allowed on the planes.

Minimum principal stress

$$= \left(\frac{p_1+p_2}{2}\right) - \sqrt{\left(\frac{p_1-p_2}{2}\right)^2 + q^2}$$

$$= \left(\frac{60-40}{2}\right) - \sqrt{\left(\frac{60+40}{2}\right)^2 + (41.53)^2}$$

$$= 10 - \sqrt{2500+1725} = -55 \text{ N/mm}^2.$$

Maximum shear stress

$$= \frac{p_1+p_2}{2} = \frac{60-40}{2} = -10 \text{ N/mm}^2$$

Problem 13.47 : Fig. 13.47 shows three metal cubes *A*, *B* and *C*, of side 100 mm in direct contact, resting an a rigid base and confined in the x-coordinate direction between two rigid end-plates.

If the upper face of the centre cube (cube B) is subjuected to uniform compresive stress of 0.5 kN/ mm², computer for cube B, the following

(i) The direct stress in the x- direction (σ_x)
(ii) The direct strains in the three corrdinate directions x, y and z.
(iii) The volumetric strain.
The elastic properties for the threee cubes A, B and C are given in Fig. P. 13.47.
(b) State all assumptions made. [IES 1999]

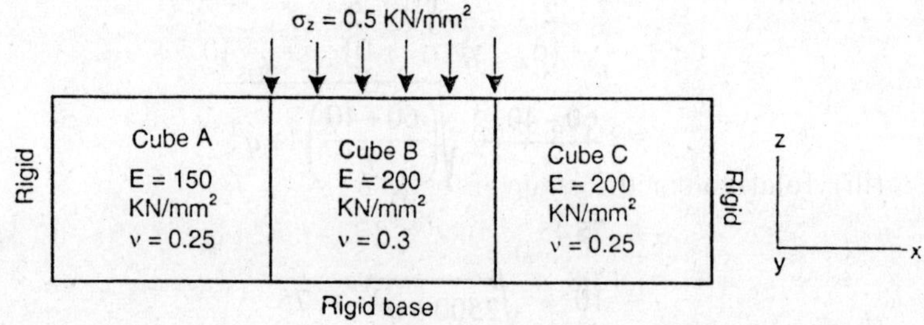

Fig. P. 13.47

Solution :

(i) cube B

$$E_{XB} = \frac{1}{E_B}[\sigma_x - \nu_B(\sigma_z + \sigma_y)] = \frac{1}{E_B}[\sigma_x - \nu_B\,\sigma_z]$$

Cubes C and A

$$\varepsilon_{XA} = \varepsilon_{XC} = \frac{1}{E_C}[\sigma_x - \nu_c(\sigma_z + \sigma_y)] = \frac{1}{E_C}\sigma_x$$

Now total charge in length in X – direction is zero.

$$\therefore \quad 100 \times [\varepsilon_{XB} + \varepsilon_{XC} + \varepsilon_{XA}] = 0$$

$$\Rightarrow \quad \varepsilon_{XB} + 2\,\varepsilon_{XC} = 0$$

$$\Rightarrow \quad \varepsilon_{XB} = -\,2\,\varepsilon_{XC}$$

$$\frac{1}{E_B}[\sigma_X - \nu_B\,\sigma_z] = -\,2\,\frac{1}{E_C}\sigma_x$$

$$\frac{1}{200}[\sigma_x - 0.3 \times 0.5] = -\frac{2}{150}\sigma_x$$

$$\Rightarrow \quad \sigma_x = 0.041 \text{ kN/mm}^2.$$

(ii) Strain in cube B

$$\varepsilon_x = \frac{1}{E_B}[\sigma_x - \nu_B\,\sigma_z] = \frac{1}{200}[0.41 - 0.3 \times 0.5]$$

$$= -5.45 \times 10^{-4}$$

$$\varepsilon_y = \frac{1}{E_B}[0 - \nu_B(\sigma_z + \sigma_x)] = \frac{1}{200}[-0.3\,(0.5 + 0.041]$$

$$= -8.115 \times 10^{-4}$$

$$\varepsilon_z = \frac{1}{E_B}[\sigma_z - \gamma_B(\sigma_x + 0] = \frac{1}{200}[0.5 - 0.3 \times 0.041]$$

$$= 2.438 \times 10^{-3}$$

(iii) Total volumetric strain in cube *B*

$$= -5.45 \times 10^{-4} - 8.115 \times 10^{-4} + 2.438 \times 10^{-3}$$

$$= 1.08 \times 10^{-3}$$

(b) Assumptions

(i) Meterial is isotropic and homogeneous.

(ii) Material obeys Hook's law

(iii) There is no friction between the cube surface and rigid end plates.

Problem 13.48 : Show a plane pin-jointed truess *BCDEFG* of the given dimensions and carying the loading shown. All numbers have the same axial stiffness.

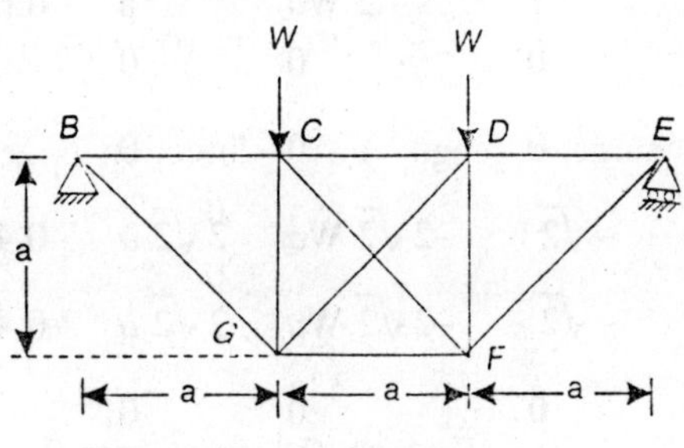

Fig. P. 13.48 (a)

(a) Calculate the axial force in member *FG*.

(b) Evaluate the force in the same member if it is fadricated 0.1 % too short, and the structure carries no external load. [IES 1999]

Solution :

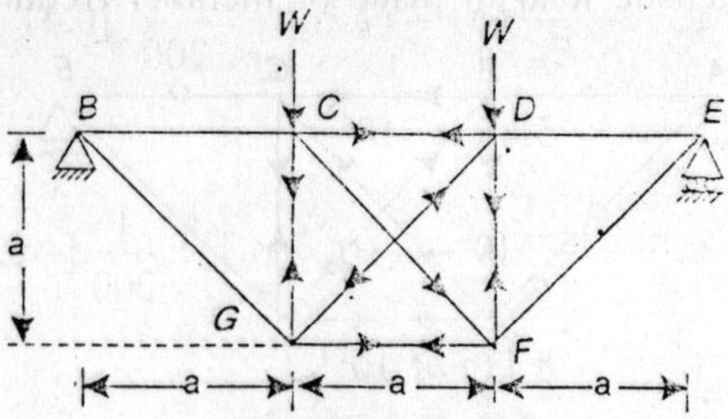

Fig. P. 13.48 (b)

Assume *FG* to be redundant.

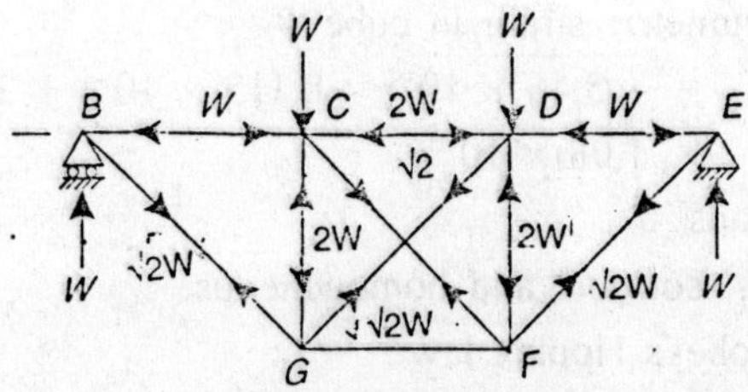

Fig. P. 13.48 (c)

Member	*P*	k	P_{kl}	k^2l	*Final force* (*P* + kl)	*Length l*
BC	–W	0	0	0	0	*a*
CD	–2W	1	– 2 W*a*	a	– 0.654W	*a*
DE	–W	0	0	0	0	*a*
BG	$\sqrt{2}$ W	0	0	0	0	$\sqrt{2}\,a$
CF	$\sqrt{2}$ W	$-\sqrt{2}$	$-2\sqrt{2}$ W*a*	$2\sqrt{2}\,a$	–0.489W	$\sqrt{2}\,a$
DG	$\sqrt{2}$ W	$-\sqrt{2}$	$-2\sqrt{2}$ W*a*	$2\sqrt{2}\,a$	–0.489W	$\sqrt{2}\,a$
FE	$\sqrt{2}$ W	0	0	0	0	$\sqrt{2}\,a$
CG	–2W	1	– 2 W*a*	*a*	– 0.654W	*a*
DF	–2W	1	– 2 W*a*	*a*	– 0.654W	*a*
			Σ = –11.657 W*a*	7.657*a*		

Apply unit tensile load in place of member (redundant) FG.

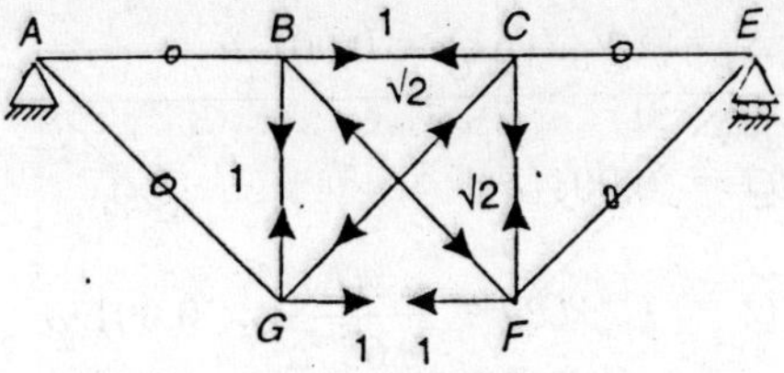

Fig. P. 13.48 (d)

Using the relation,

$$T = \frac{\Sigma \frac{Pkl}{AE}}{\Sigma\left(\frac{k^2 l}{AE}\right) + \frac{l_o}{A_o E}}$$

$$= -\frac{\Sigma Pkl}{\Sigma k^2 l + l_o}$$

$$= -\frac{(-11.657\,Wa)}{7.657\,a + a}$$

$$= \frac{11.657Wa}{8.657\,a} = 1.346 \text{ W}.$$

(b) Evaluate the force in same member if it is fabricated 0.1% too short, and the structure carries no external load.

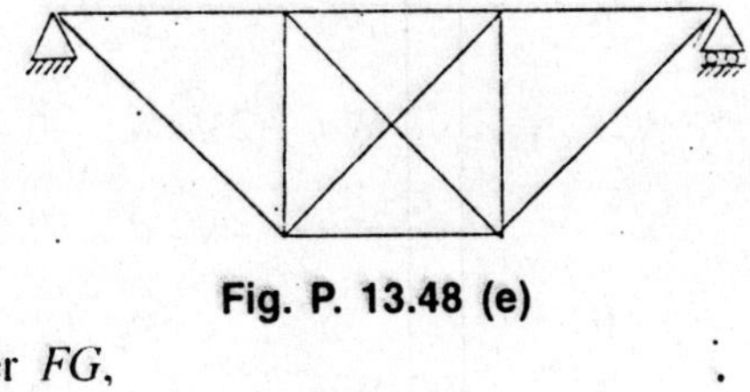

Fig. P. 13.48 (e)

For member FG,

$$\Delta L = \frac{0.1}{100} \times a = 0.001a.$$

When finding the decrease in distance between joint G and F due to shortening on member FG is to be found

$$N_{U_1} = 1$$

$$\Rightarrow \quad \Delta = \Sigma N_U.\ (\Delta L) = 0.001\ a$$

$$DQL = 0$$

$$DQ = 0.001\ a$$

$$Q = F^{-1}\ DQ = \frac{AE}{8.657a} \times 0.001\ a$$

$$= 1.16 \times 10^{-4}\ AE.$$

Problem 13.49 : The combined angle and channel section shown in Fig. 13.49 (a) forms part of a runway beam. calculate:

(a) Coordinates of the centroid

(b) Second moment of area about *X-X*

(c) Second moment of area about *Y-Y*

(d) Product of inertia about O. [IES 1999]

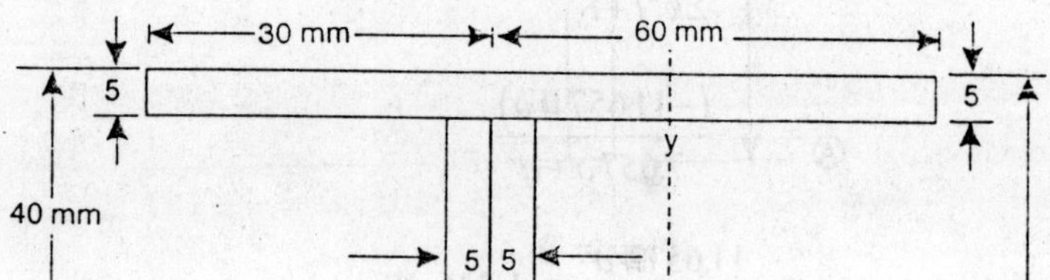

Fig. P. 13.49 (a)

Solution :

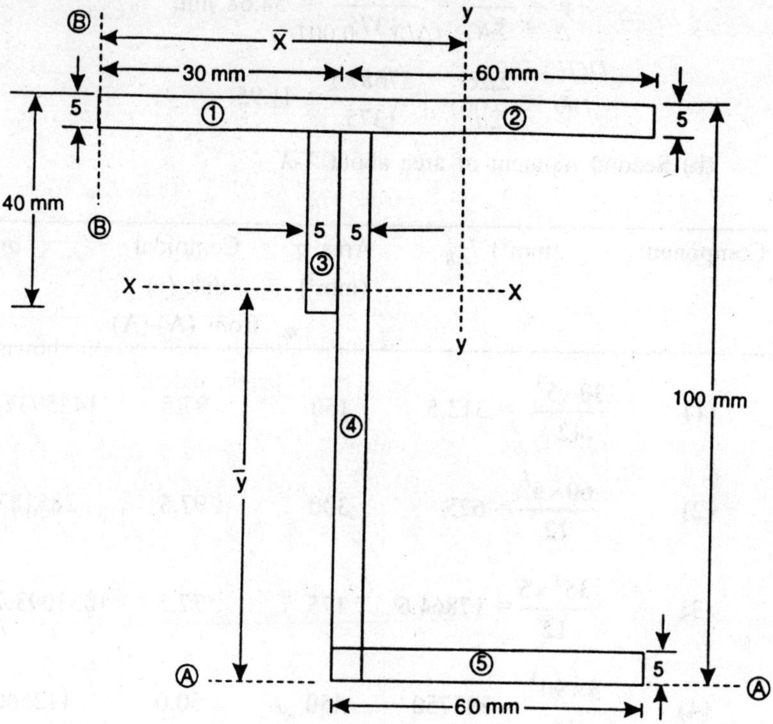

Fig. P. 13.49 (b)

Determine :

(a) Coordinates of the centroid

The whole area is split into five rectangles.

Component	Area, a (mm^2)	Centroidal dist (y) from (A)-(A) mm	Centoidal dist (x) from (B)-(B) mm	ay (mm^3)	ax (mm^3)
(1) 30 × 5	150	97.5	15.0	14625	2250
(2) 60 × 5	300	97.5	60.0	29250	18000
(3) 35 × 5	175	77.5	27.5	13562.5	4812.5
(4) 90 × 5	450	50.0	32.5	22500	14625
(5) 60 × 5	300	2.5	60.0	750	18000
Total	1375			80687.5	57687.5

$$\bar{y} = \frac{\Sigma ay}{\Sigma a} = \frac{80687.5}{1375} = 58.68 \text{ mm}$$

$$\bar{x} = \frac{\Sigma ax}{\Sigma a} = \frac{57687.5}{1375} = 41.95.$$

(b) Second moment of area about *X-X*

Component	(mm^4) I_{self}	Area a (mm^2)	Centoidal dist (y) from (A)-(A)	ay^2
(1)	$\frac{30 \times 5^3}{12} = 312.5$	150	97.5	1425937.5
(2)	$\frac{60 \times 5^3}{12} = 625$	300	97.5	2851875
(3)	$\frac{35^3 \times 5}{12} = 17864.6$	175	77.5	1051093.75
(4)	$\frac{5 \times 90^3}{12} = 303750$	450	50.0	1125000
(5)	$\frac{5^3 \times 60}{12} = 625$	300	2.5	1875
Total	323177.1	1375		6455781.25

Moment of inertia about axis (*A*) - (*A*)

$$= I_{A.A} = \Sigma I_{self} + \Sigma ay^2$$

$$= 323177.1 + 6455781.25$$

$$= 6778958.35$$

But $I_{A-A} = I_{XX} + A\bar{y}^2$

$\Rightarrow \quad 6778958.35 = I_{XX} + 1375 \times 58.68^2$

$\Rightarrow \quad I_{XX} = 2044362.55 \text{ mm}^4$

$= 2.044 \times 10^6 \text{ mm}^4.$

(c) Second moment of inertia about *Y-Y*

Component	(mm⁴) I_{self}	Area a	Centoidal dist (x) from (B-B)	ax^2
(1)	$\frac{30^3 \times 5}{12} = 11250$	150	15.0	33750
(2)	$\frac{60^3 \times 5}{12} = 90000$	300	60.0	1080000
(3)	$\frac{35 \times 5^3}{12} = 364.58$	175	27.5	132343.75
(4)	$\frac{5^3 \times 90}{12} = 937.5$	450	32.5	475312.5
(5)	$\frac{60^3 \times 5}{12} = 90000$	300	60.0	1080000
	192552.088	1375		2801406.25

Moment of inertia about axis (*B-B*)

$$= I_{B.B} = \Sigma I_{self} + \Sigma ax^2$$

$$= 192552.08 + 2801406.25 = 29933958.33 \text{ mm}^4$$

But $\quad I_{B-B} = I_{YY} + A\bar{x}^2$

$\Rightarrow \quad I_{YY} = I_{BB} - A\bar{x}^2$

$$= 2993958.33 - 1375 \times 41.95^2$$

$$= 574229.8925 \text{ mm}^4$$

$\Rightarrow \quad I_{YY} = 5.74 \times 10^5 \text{ mm}^4.$

Problem 13.50 :In the structure shown in the Fig. 13.50 (i) calculate the distance by which the points a move away under the action of forces *P*.

(b) When fabricating the pin jointed plane frame shown in the Fig. P. 13.50 (ii) it was found that the member AC is fabricated D too long. Determine the forces in the members after assembly. All members have same axial stiffness (AE).

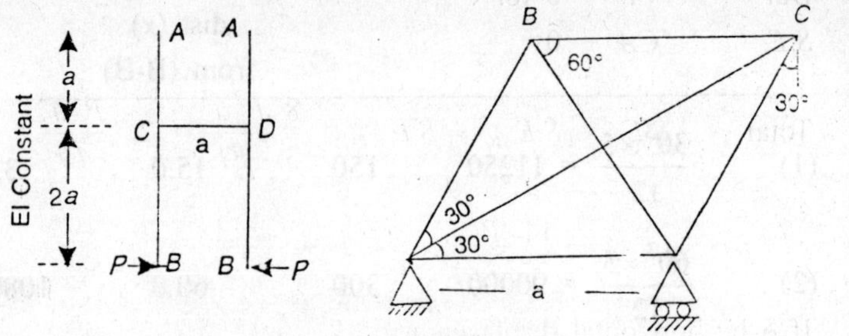

Fig. P. 13.50 (i) and (ii)

Solution :

S.E. stored by the colume BC

$$= \int \frac{M^2}{2\,EI} \cdot dy = \int_0^{2a} \frac{p^2 Y^2}{2\,EI} \cdot dy$$

$$S.E._{BC} = 2 \int_0^{2a} \frac{P^2 Y^2}{2\,EI} \cdot dy = \frac{P^2}{EI} \cdot (2\alpha)^3/3$$

$$= \frac{8}{3} \cdot \frac{P^2 a^3}{EI} \qquad \ldots\text{(i)}$$

$$S.E._{CD} = \int \frac{M^2}{2\,EI} \cdot dx \qquad \left[M = Pl = P(2a)\right]$$

$$= \int_0^{a} \frac{P^2 . l^2}{2\,EI} . dx = \int_0^{\alpha} \frac{4\,P^2 . a^2}{2\,EI} \cdot dx$$

$$S.E._{CD} = \frac{2\,P^2}{EI} a^3$$

$$S.E._{CA} = \int \frac{M^2}{2\,EI} \cdot dy \qquad [\because M = P\ (2a + Y)]$$

$$S.E._{CA} = \int_0^{a} \frac{P^2 (2a+Y)^2}{2\,EI} \cdot dy$$

But $\quad P = 0$ for CA

S.E. $\quad CA = 0$

Total $\quad S.E. = S.E._{BC} + S.E._{CD} = \frac{8}{3}\frac{P^2a^3}{EI} + 2.\ \frac{P^2a^3}{EI}$

$$= \frac{14}{3}\frac{P^2a^3}{EI} \qquad \text{...(iii)}$$

If δ be horizontal displacemnt of B

$$\frac{1}{2}P\delta = \frac{14}{3}\cdot\frac{P^2a^3}{EI}$$

$$\Rightarrow \qquad \delta = \frac{7}{3}\frac{Pa^3}{EI}$$

so horizontal movement of A – A′ is $\frac{1}{2}$ of δ.

$$\Delta A - A' = \frac{1}{2}\cdot\delta = \frac{7}{6}\cdot\frac{Pa^3}{EI}.$$

(b) $\quad AE$ = same

$R_A + R_D = 1$

$R_a = 1.5\ a$

$R_D = 1.5$

$R_A = -\ 0.5$

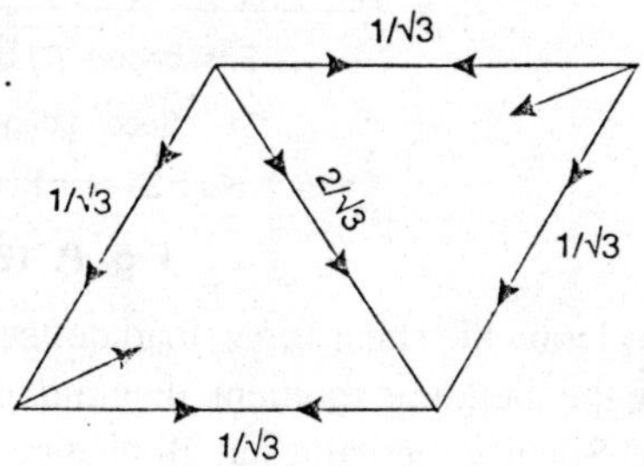

Fig. P. 13.50 (a)

Member	l	A	K	K^2l/A	Forces
1	a		$1/\sqrt{3}$	a/3	0.3/a AΔ
2	a		$1/\sqrt{3}$	a/3	0.3 AΔ
3	a		$1/\sqrt{3}$	a/3	0.3 AΔ
4	a		$1/\sqrt{3}$	a/3	0.3AΔ
5	$4a/\sqrt{3}$		$2/\sqrt{3}$	2.58a	0.3 AΔ
				3.97a	

$$x = \frac{-\delta}{\Sigma \frac{K^2 l}{AE} + \frac{l_o}{AE}} = \frac{\Delta}{\frac{2.67a}{AE} + \frac{2a}{AE}}$$

$$= \frac{AE\Delta}{5.975a} = \frac{0.17}{a} AE\Delta.$$

Problem 13.51 : (*a*) A simply supported beam of rectangular cross-section of of size 200 × 300 mm (deep) supports a uniformly distrributed load of 6.0 kN/m over an effective span of 4.0 m. Calculate the magnitude and direction of principal stresses at a point located 0.50 m from the left supports and 50 mm above the natural axis.

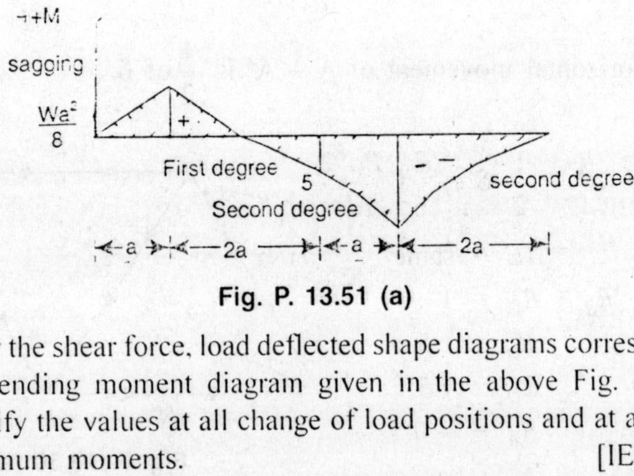

Fig. P. 13.51 (a)

(*b*) Draw the shear force, load deflected shape diagrams corresponding to the bending moment diagram given in the above Fig. P. 13.51 (a) Specify the values at all change of load positions and at all points of maximum moments. [IES 2000]

Solution :

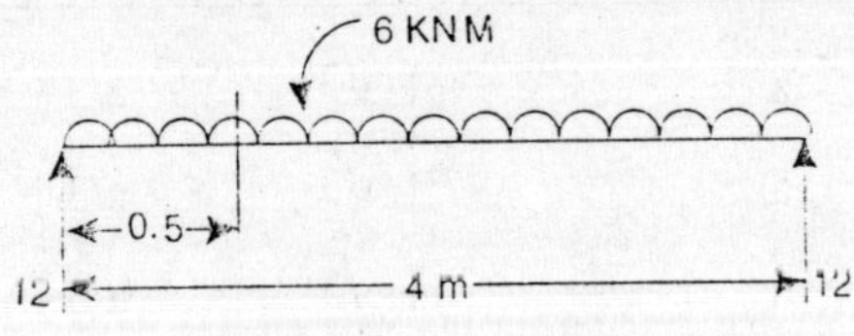

Fig. P. 13.51 (b)

$$M = 12 \times 0.5 - 6 \times 0.5 \times \frac{0.5}{2} = 5.25 \text{ kN.m}$$

$$SF = 12 - 6 \times 0.5 = 9 \text{ kN}$$

$$I = \frac{200 \times 300^3}{12} = 450 \times 10^6 \text{ mm}^4$$

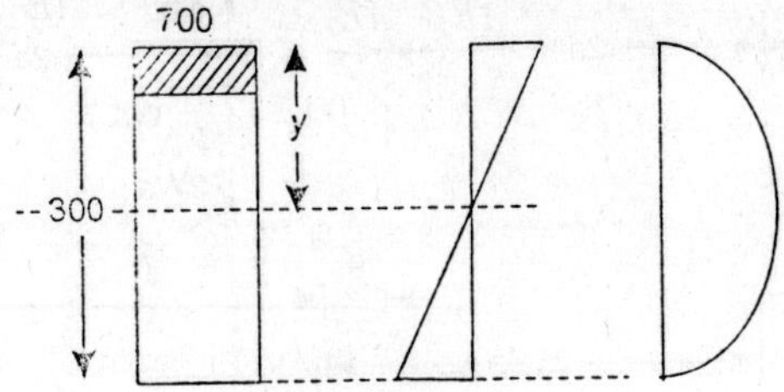

Fig. P. 13.51 (c)

$$P = \frac{M}{I} \times y = \frac{5.25 \times 10^6}{450 \times 10^6} \times 50 = 0.583 \text{ N/mm}^2$$

$$q = \frac{9 \times 10^3 \times 200 \times 100 \times 100}{450 \times 10^6 \times 200} = 0.2 \text{ N/mm}^2$$

$$\text{Principal stress} = \frac{P}{2} \pm \sqrt{\frac{P^2}{4} + q^2} = \frac{0.583}{2} \pm \sqrt{\frac{0.583^2}{4} + 0.2^2}$$

$$P_1 = 0.645 \text{ N/mm}^2. \; P_2 = -0.062 \text{ N/mm}^2.$$

(b) AB → Linear variation of BM. Hence at A →Simple support

$$Mb = R_A \cdot a = 9 \cdot \frac{wa^2}{8} \Rightarrow Ra = \frac{9}{8} wa$$

From B, BM changes, Hence concentrated load at B.

$$Me = -\frac{5}{8} wa^2 = \frac{9}{8} wa \,.\, 3a - R'2a$$

$$2a \cdot R' = \frac{27}{8} wa^2 + \frac{5}{8} wa^2 = 4wa^2 = R_B' = 2wa$$

DE → parabolic variation

$$-16\frac{wa^2}{8} = \frac{w' 4a}{2} \quad w' = w$$

BM changed at D, at D → simple supported

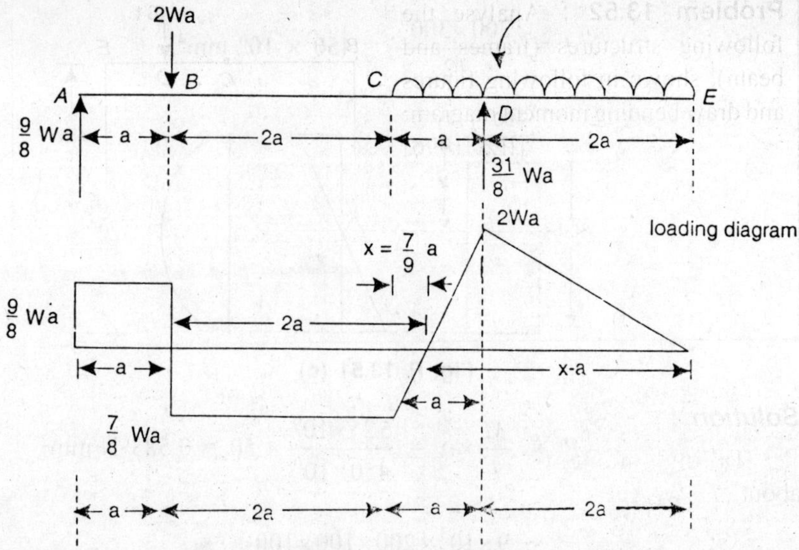

Fig. P. 13.51 (d)

$$Me = -\frac{5wa^2}{8} = w \cdot 3a.\frac{3a}{2} - R_D \cdot a$$

$$R_D = -\frac{5wa}{8} + \frac{9}{2}\ wa = \frac{3}{8} \cdot wa$$

$$\frac{9}{8}\ wa.x = 2\ wa\ (x - a)$$

$$\frac{9}{8}\ wa.x = 2\ wax - 2wa^2$$

$$2wa^2 = 2\ wax - \frac{9}{8}\ wax = \frac{7wax}{8}$$

$$x = \frac{16}{7}\ a = 2.3a \Rightarrow \text{point of contraflexure}$$

point of change of shear force,

$$\frac{7}{9}\ \frac{wa}{x} = \frac{2wa}{(x-a)}$$

$$x = \frac{7}{9}\ a.$$

Problem 13.52 : Analyse the following structures (frames and beam), shown in following figures and draw bending moment diagram: [IES 1996]

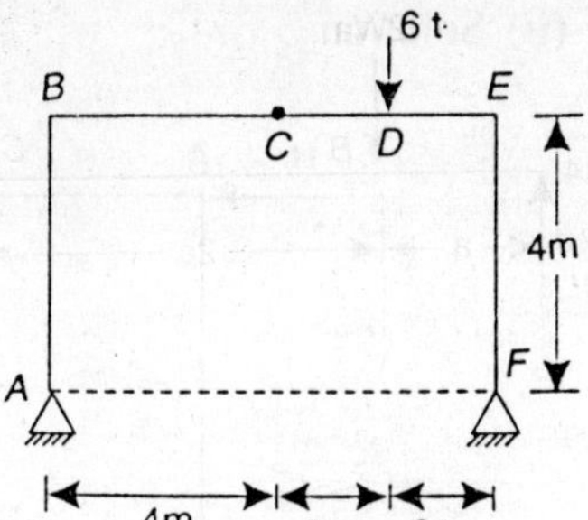

Fig. P. 13.52

Solution :

Taking moments about 'F' we get

$$8\ V_A = 16 \times 2$$

$$\Rightarrow \quad V_A = 4\ \text{t}$$

$$\Sigma V = 0$$

$$\Rightarrow \quad V_F = 16 - 4 = 12\ \text{t}$$

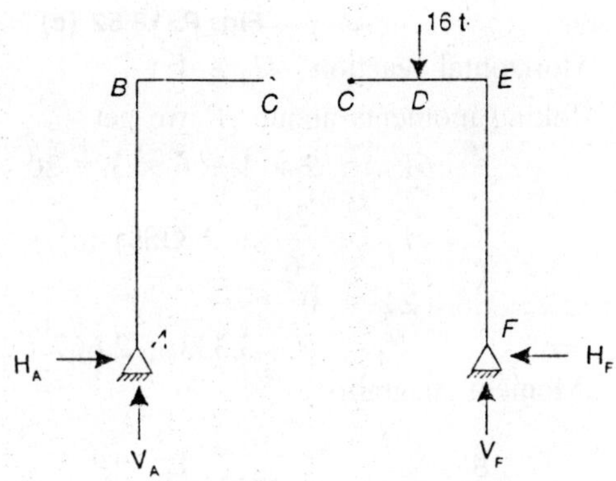

Fig. P. 13.52 (a)

Taking moments about 'C' of right portions

$$16 \times 2 + 4\ H_F = 12 \times 4$$

$$\Rightarrow \quad H_F = 4\ t$$

$$\Sigma H = 0$$

$$\Rightarrow \quad H_A = 4\ t$$

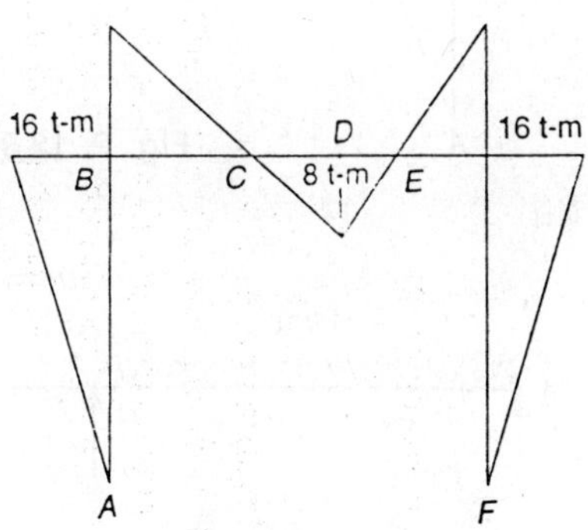

Fig. P. 13.52 (b)

(ii) Solution.

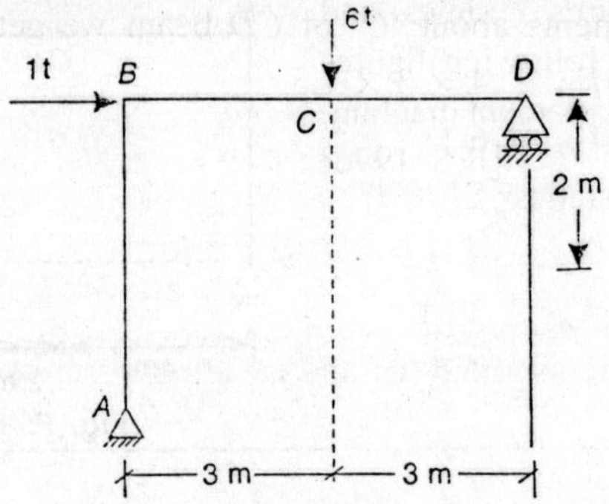

Fig. P. 13.52 (c)

Horizontal reaction $H_A = 1$ t

Taking moments about 'A' we get

$$6V_D = 2 \times 1 + 6 \times 3 = 20$$

$$V_D = \frac{20}{6} = 3.333 \text{ t}$$

$$\Sigma V = 0$$

$$\Rightarrow \quad V_A = 6 - 3.333 = 2.667 \text{ t}$$

Moment diagram

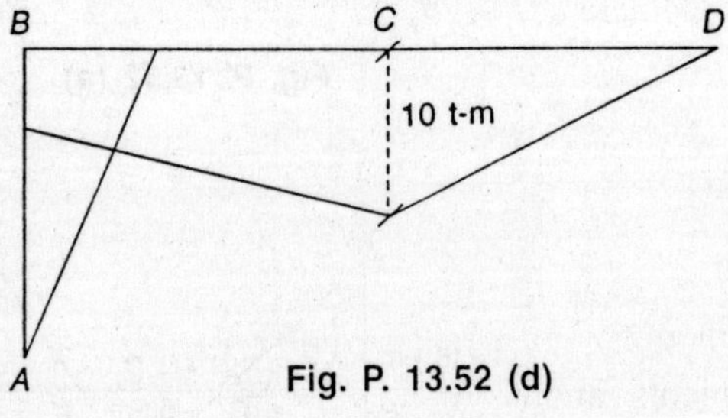

Fig. P. 13.52 (d)

(iii)

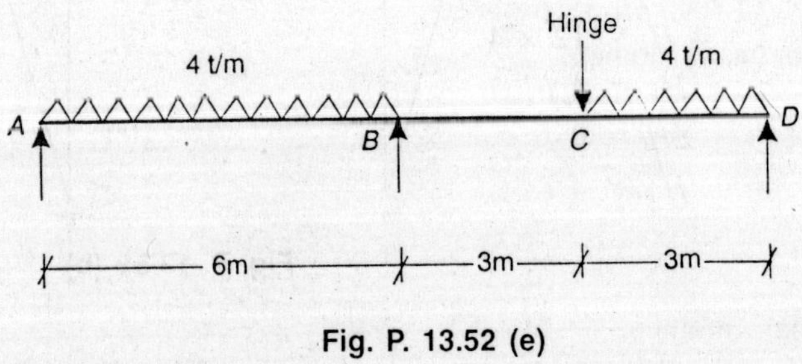

Fig. P. 13.52 (e)

Solution:

Taking moments about 'C' of CD beam we get

$$3V_D = 4 \times 3 \times 1.5$$

$$\Rightarrow \quad V_D = 6 \text{ t}$$

By simple statics $\Sigma V = 0$

$$\Rightarrow \quad V_A + V_B = 4 \times 6 + 4 \times 3 - 6$$

$$V_A + V_B = 30 \text{ t} \quad \text{...(1)}$$

Taking moments about 'C' and equating to zero since 'C' is a hinge we get

$$9\,V_A + 3\,V_B = 4 \times 6 \times 6 = 144$$

$$\Rightarrow 3\,V_A + V_B = 48 \quad \text{...(2)}$$

From equation (1) and (2) we get

$$V_A = 9 \text{ t}$$

$$V_B = 21 \text{ t}$$

Moment along AB at x distance from 'A' is given by

$$M_x = 9x - \frac{4\,xx}{2} = 9x - 2x^2$$

Moment at 'B'at 6 m from

$$A = 9 \times 6 - 2 \times 6^2 = -18 \text{ t.m}$$

For max. moment

$$\frac{dM_x}{dx} = 9 - 4x = 0$$

$$\Rightarrow \quad x = \frac{9}{4} = 2.25 \text{ m}$$

Max. moment

$$= 9 \times 2.25 - 2 \times 2.25^2 = 10.125 \text{ t.m}$$

Moment along CD at x distance from D is given by

$$M_x = 6x - 2x^2$$

For max. moment

$$\frac{dM_x}{dx} = 6 - 4\,x = 0$$

$$\Rightarrow \quad x = \frac{6}{4} = 1.5 \text{ m}$$

Max. moment $= 6 \times 1.5 - 2 \times 1.52 = 4.5$ t.m

Plotting moment diagram

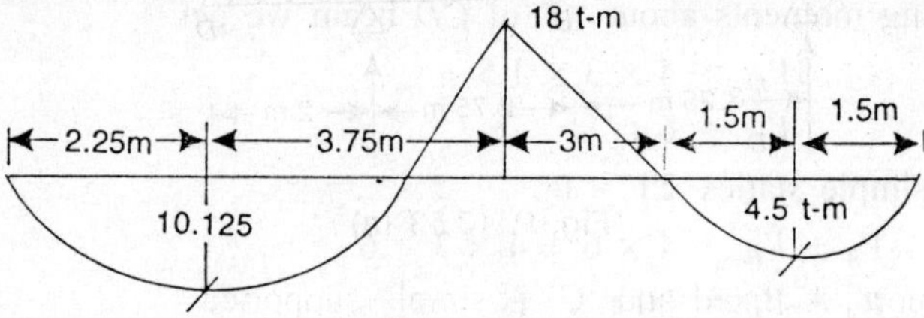

Fig. P. 13.52 (f)

Problem 13.53 : Shear force diagram of beam is shown in the figure. Identify the location and nature of supports. Draw loading and bending moment showing values of bending moment every 2 m and also showing position and magnitude of maximum bending moment. [IES 1996]

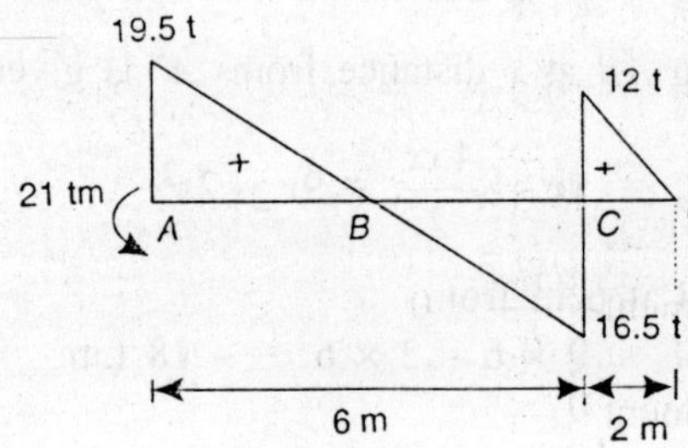

Fig. P. 13.53

Solution :

At 'A' and 'C' there is change in shear force i.e. there will be support with reaction of 19.5 t at 'A' and 18.5 t at 'C'.

Along AC the shear force in linearly varying, hence it will be loaded by *u.d.l* of value

$$\frac{(19.5+16.5)}{6} = 6 \text{ t/m}$$

Along CD the shear force is linerly varying and hence it will be loaded by *u.d.l* of value $\frac{12}{2}$ = 6 t/m.

Loading diagram

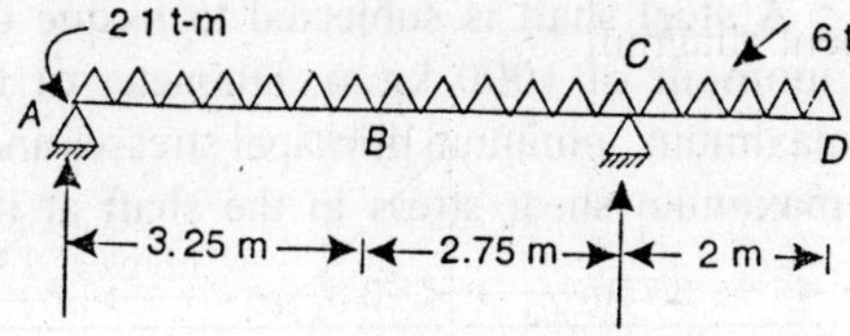

Fig. P. 13.53 (a)

Support 'A' fixed and 'C' is simply supported

Bending moment at 2 m from A $= 19.5 \times 2 - 6 \times 4 \times 2 - 21$

$= 6$ t.m

Bending moment at 4 m from A $= 19.5 \times 4 - 6 \times 4 \times 2 - 21$

$= 9$ t.m

Bending moment at 6 m from A $= 19.5 \times 6 - 6 \times 6 \times 3 - 21$

$= - 12$ t.m

Moment along 'AC' at a distance of x from 'A' is

$$M_x = 19.5\,x - 6 \times x \times \frac{x}{2} - 21$$

$$= 19.5\,x - 3\,x^2 - 21$$

For max. moment $\frac{dM_x}{dx} = 0$

$\Rightarrow +19.5 - 6x = 0$

$$x = \frac{19.5}{6} = 3.25 \text{ m}$$

$$\text{Max. moment} = 19.5 \times 3.25 - 3.25 \times 6 \times \frac{3.25}{2} - 21$$

$$= 10.69 \text{ t.m}$$

Moment diagram

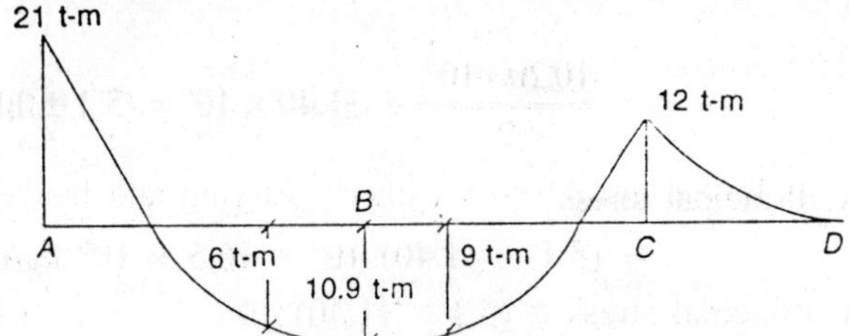

Fig. P. 13.53 (b)

Problem 13.54 : A steel shaft is subjected to torque of 2000 kg-m and a bending moment of 1000 kg.m. Diameter or the shaft = 10 cm. Calculate maximum, minimum principal stresses and maximum shear stress and maximum shear stress in the shaft at its surface. [IES 1996]

Solution :

$$T = 2000 \text{ kg.m}$$

$$Dia = 10 \text{ cm} = 0.1 \text{ m}$$

$$I_p = \frac{\pi D^4}{32} = \frac{\pi \times 0.1^4}{32} = 9.81 \times 10^6$$

$$I = \frac{\pi D^4}{64} = 4.9 \times 10^{-6}$$

$$\tau_{xy} = \text{Shear stress} = \frac{T \cdot r}{I_P}$$

$$= \frac{2000}{9.81 \times 10^{-6}} \cdot \frac{5}{100} = 10.20 \times 10^6$$

$$\sigma_b = \text{Bending stress} = \frac{M}{I} \cdot y = \frac{1000}{4.9 \times 10^{-6}} \times \frac{5}{100}$$

$$= 10.20 \times 10^6$$

$$\text{Principal stress} = \frac{\sigma_b}{2} \pm \sqrt{\left(\frac{\sigma_b}{2}\right)^2 + \tau_{xy}^2}$$

$$= \frac{10.20 \times 10^6}{2} \pm \sqrt{\left(5.1 \times 10^6\right)^2 + \left(10.2 \times 10^6\right)^2}$$

$$= \frac{10.20 \times 10^6}{2} \pm 11.40 \times 10^6 = (5.1 \pm 11.40) \times 10^6$$

Max. Principal stress

$$= (5.1 + 11.40)\ 10^6 = 16.5 \times 10^6 \text{ kg/m}^2$$

Min. Principal stress $= (5.1 - 11.40)\ 10^6$

$$= -6.3 \times 10^6 \text{ kg/m}^2$$

Max. shear stress

$$= \left(\frac{16.5 + 6.3}{2}\right) \times 10^6 = 11.4 \times 10^6 \text{ kg/m}^2$$

Problem 13.55 : A vertical tapered rod of length 'L' has its diameter varying linearly from 'd' at lower end to 'D' at the upper end which has fixed support. Young modulus of the material is 'E'. Show that the elongation of the rod at its lower end when subjected to a longitudinal force 'F' is given by

$$\delta = \frac{4FL}{\pi dDE}$$ [IES 1996]

Solution :

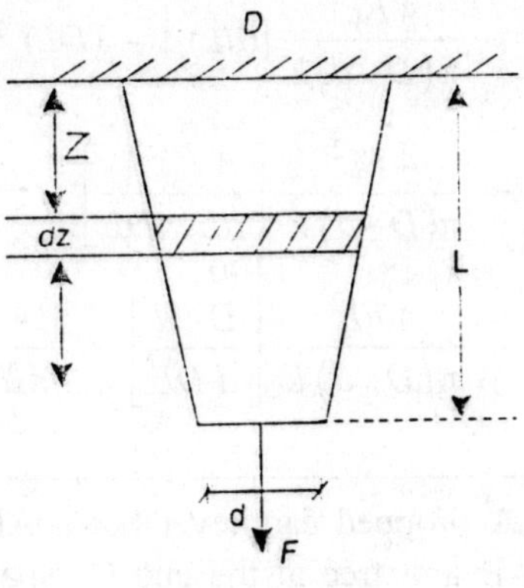

Fig. P. 13.55

Consider an element of thickness dz at a distance z from top. Let the total length be L,

$$\text{Stress} = \frac{\text{force}}{\text{area}}$$

$$= \frac{F}{A_Z} = \frac{4F}{\pi \cdot \left(\frac{D(L-z)+z.d.}{L}\right)^2}$$

$$\text{Strain} = \frac{\text{Stress}}{E}$$

Elongationn of element = Strain × dz

$$= \frac{4FL^2}{\pi[D(L-z)+z.d.]^2 E} \cdot dz$$

$$\text{Total elongation} = \int_0^L \frac{4FL^2}{\pi\left[D(L-z)+z.d.\right]^2 E}$$

$$= \int_0^L \frac{4FL^2}{\pi E}[D(L-z)+z.d.]^{-2}\, dz$$

$$= \frac{4FL^2}{\pi E}\left[\frac{\left[D(L-z)+zd\right]^{-1}}{(D-d)}\right]_0^L$$

$$= \frac{4FL^2}{\pi(D-d)E}[(dL)^{-1} - (DL)^{-1}]$$

$$= \frac{4FL^2}{\pi(D-d)E}\left[\frac{1}{dL}-\frac{1}{DL}\right]$$

$$= \frac{4FL^2}{\pi(D-d)E}\left[\frac{D-d}{d\,DL}\right] = \frac{4FL}{\pi d\,DE}$$

Problem 13.56 : A propped cantilever beam ABC is fixed at the end A, on roller at B and free at the end C. Span AB = 6 m, and BC = 2 m. A concentrated load of 24 kN is acting at mid-span of AB and another 4 kN at free end C. The moment of inertia of span AB is 2 I and that of BC is I. Analyse the beam using the principla of consistent deformation and determine the deflection at the free end C. [IES 1993]

Solution :

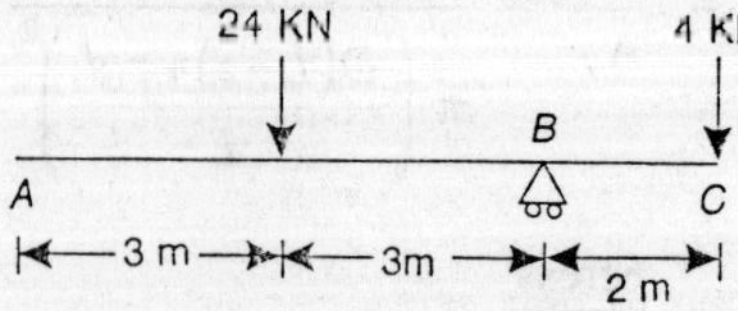

Fig. P. 13.56 (a)

The above problem is statically indeterminate to single degree. Releasing moment at 'A', then the relased structure will become.

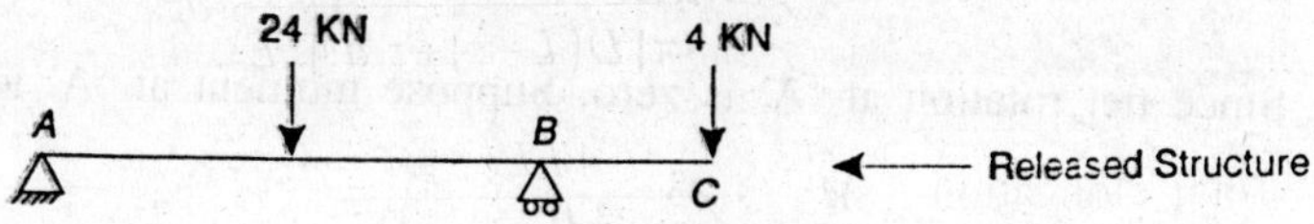

Fig. P. 13.56 (b)

Now, finding rotation at 'A' in the released structure by unit load method, i.e. by applying unit moment at *A*.

Bending Moment diagram of released structure

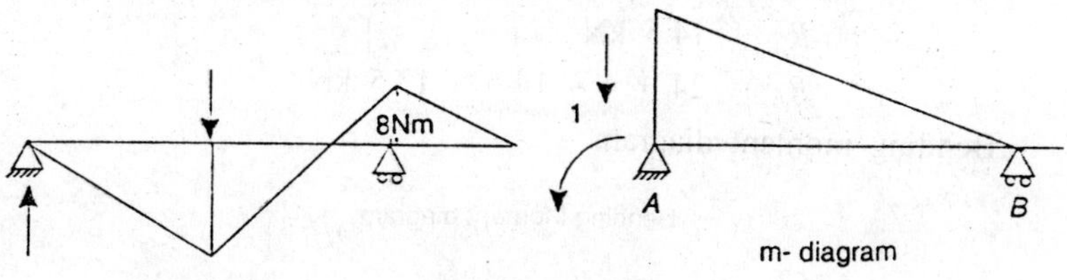

Fig. P. 13.56 (c)

Rotation at $A = \theta_A$ in released structure $= \int \cdot \frac{Mm\,dx}{EI}$

$$= -\int_0^3 \frac{10.667x \times \left(1-\frac{x}{6}\right)dx}{2\,EI} + \int_3^6 \frac{[10.667x - 24(x-3)]\left[1-\frac{x}{6}\right](-dx)}{2\,EI}$$

$$= -\frac{1}{2\,EI}\left[\frac{10.667x^2}{2} - \frac{1.778x^3}{3}\right]_0^3$$

$$-\frac{1}{2\,EI}\left[\frac{-13.333x^2}{2} + \frac{2.222x^3}{3} + 72x - \frac{12x^2}{2}\right]_3^6$$

$$= -\frac{1}{2\,EI} \times 32 - \frac{1}{2\,EI}\,[-180 + 140 + 216 - 162]$$

$$= -\frac{23}{EI}$$

Rotation at 'A' due to unit at 'A'

$$= \int_0^6 \frac{x \cdot x\,dx}{36\times 2\,EI} = \frac{x^3}{216\,EI}\Bigg|_0^6 = \frac{1}{EI}$$

Since net rotation at 'A' is zero. Suppose moment at 'A' is M.

$$= \frac{M}{EI} - \frac{23}{EI} = 0$$

$\Rightarrow$ $M = 23$ kN.m

Taking moment about B

$\Rightarrow$ $6\,R_A - 23 - 24\times 3 + 4\times 2 = 0$

$R_A = 14.5$ kN

$R_B = 24 + 4 - 14.5 = 13.5$ kN

Bending moment diagram

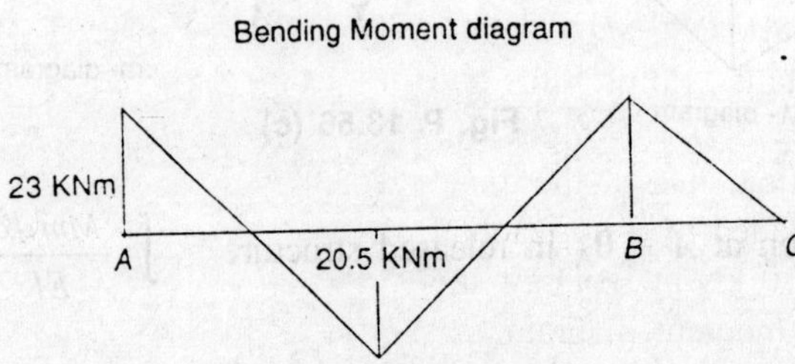

Fig. P. 13.56 (d)

Now to determine deflection at 'C', applying a unit load at 'C'

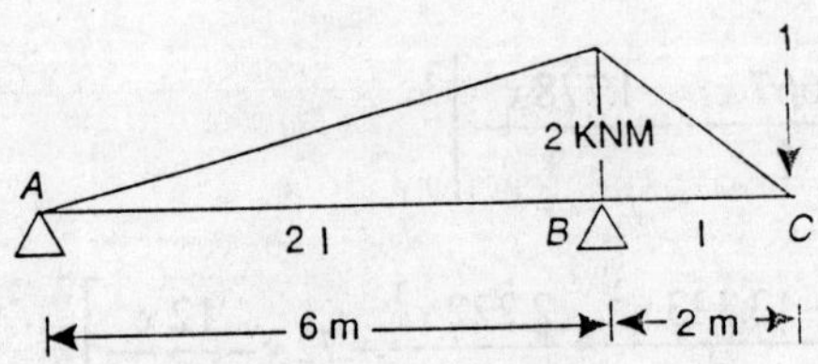

Fig. P. 13.56 (e)

Deflection at

$$C = \int \frac{Mm\,dx}{EI}$$

$$= \int_0^3 (23 - 14.5\,x)\left(\frac{x}{3}\right)\frac{dx}{2\,EI} + \int_0^3 (8 - 9.5\,x)\left(2 - \frac{x}{3}\right)x$$

$$\frac{dx}{2\,EI} + \int_0^2 \frac{x \cdot 4\,x\,dx}{EI}$$

$$= \frac{1}{2\,EI}[34.5 - 43.5] + \frac{1}{2\,EI}[48 - 12 - 85.5 + 28.5] + \frac{10.67}{EI}$$

$$= \frac{-4.5}{EI} - \frac{10.5}{EI} + \frac{10.67}{EI} = \frac{-4.33}{EI}$$

Problem 13.57 : A rigid portal *ABCD* has ends *A* & *D* hinged. *AB* and *CD* are vertical and *BC* horizongal *AB* = *CD* = 4 m, *I* = Constant, Employing the principle of consistent deformation, analyse the frame if it is loaded by a horizaontal load of 10 kN at B. Draw the bending moment diagram.

[IES 1999]

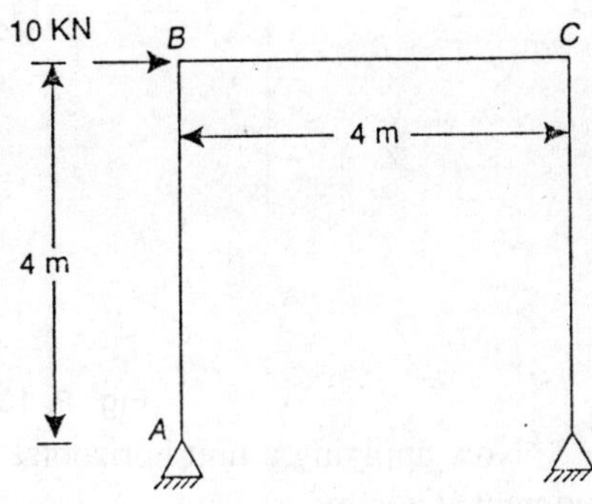

Fig. P. 13.57

Solution :

The above frame is statically indeterminate to single degree. Releasing the horizontal reaction at D.

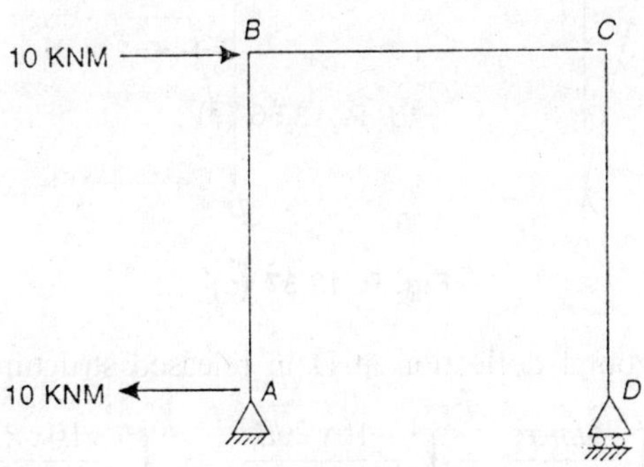

Fig. P. 13.57 (a)

Taking moment about D,

$$R_A = \frac{10 \times 4}{4}$$

$$= 10 \text{ kN (downwared)}$$

$$R_D = 10 \text{ kN (upwards)}$$

Plotting Bending moment diagram of released structure

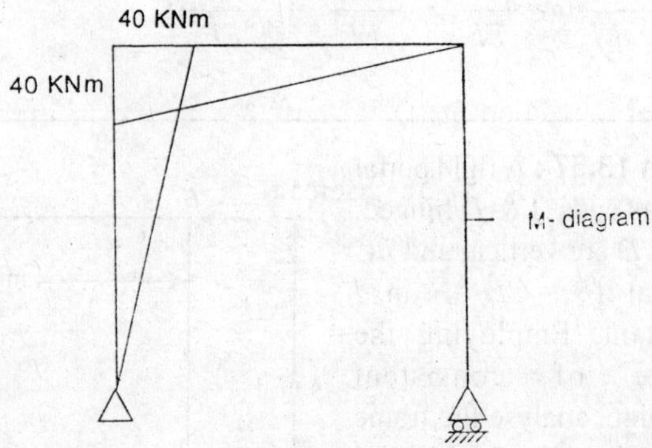

Fig. P. 13.57 (b)

Now applying a unit horizontal load at D and plotting the bending moment diagram

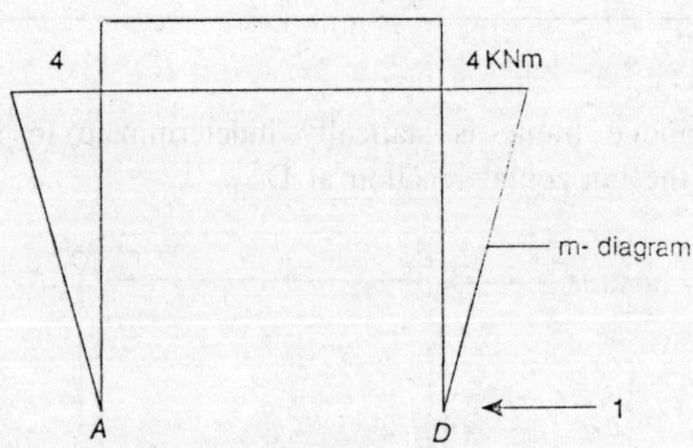

Fig. P. 13.57 (c)

Now horizontal deflection at D in released structure

$$= \int \frac{Mm\,dx}{EI} = \int_0^4 \frac{-10x \cdot x\,dx}{EI} + \int_0^4 \frac{-10x \times 4\,dx}{EI}$$

$$= \frac{-10}{EI} \cdot \frac{64}{3} - \frac{40}{EI} \cdot \frac{16}{2} = \frac{-533.33}{EI}$$

Horizontal deflection at 'D' due to unit load at D

$$= 2\int_0^4 \frac{x^2\,dx}{EI} + \int_0^4 \frac{4 \times 4\,dx}{EI} = \frac{2}{EI} \cdot \frac{4^3}{3} + \frac{16 \times 4}{EI}$$

$$= \frac{106.67}{EI}$$

Since, net deflection at 'D' is zero. so

$$\frac{-533.33}{EI} + H_D \times \frac{106.67}{EI} = 0$$

$$\Rightarrow \quad H_D = 5 \text{ kN}$$

$$H_A = 5 \text{ kN}$$

Taking moment about 'D'

$$V_A = \frac{10 \times 4}{4} = 10 \text{ kN (dowmwards)}$$

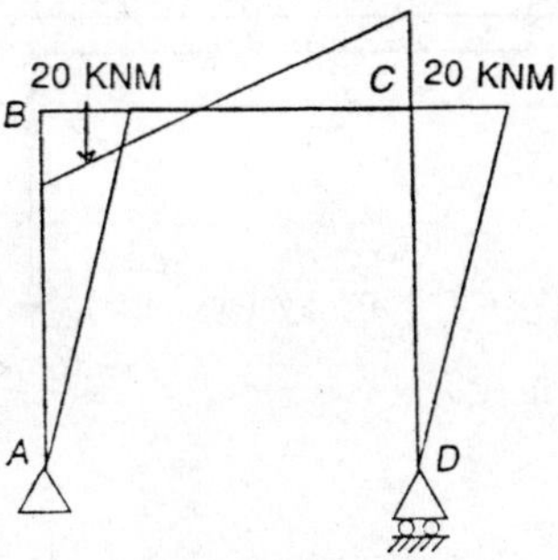

Fig. P. 13.57 (d)

Problem 13.58 : Employing the method of consistent deformation obtain the force in the members a, b and c of the structure loaded as in figure. The section areas and the second movement of area are indicated in the figure $E = 2 \times 10^5$ N/m^2. [CSE 1996]

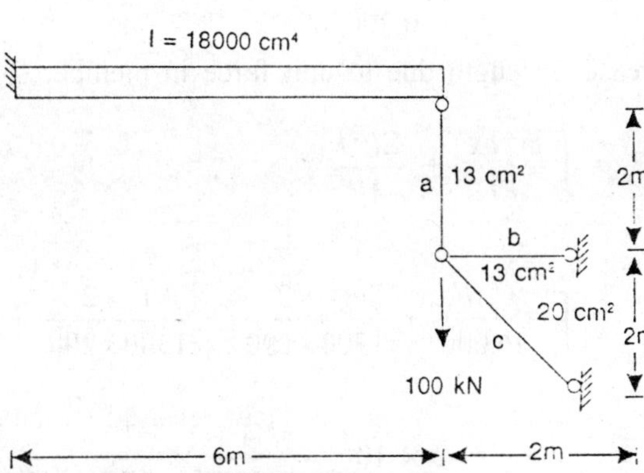

Fig. P. 13.58

Solution :

Let member 'b' be made redundant. Then decreases in length of member 'b' due to given loading be δ_1.

$$E = 2 \times 10^5 \text{ N/mm}^2 = 2 \times 10^8 \text{ kN/m}^2$$
$$I = 18000 \text{ cm}^4 = 18,000 \times 10^{-8} \text{ m}^4$$
$$EI = 36,000 \text{ kN.m}^2$$

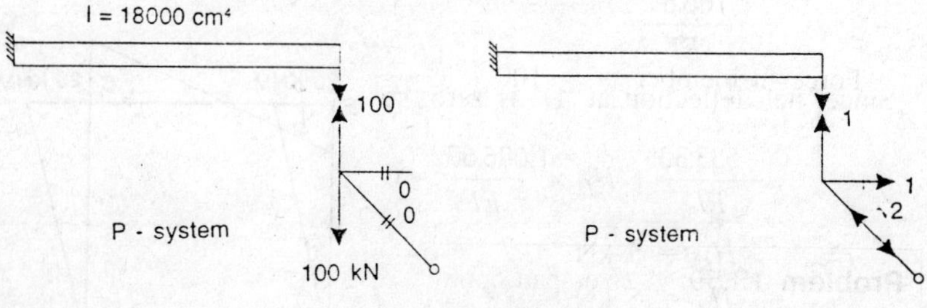

Fig. P. 13.58 (a)

$$\delta_1 = \int \frac{Mm\,dx}{EI} + \Sigma \frac{Ppl}{AE}$$

$$= \int_0^6 \frac{100x.(-x)\,dx}{EI} + \frac{100 \times (-1) \times 2}{1300 \times 200}$$

$$= \frac{-100 \times 72}{2 \times 10^8 \times 18000 \times 10^{-8}} - \frac{200}{1300 \times 200}$$

$$= -0.200769$$

Decrease in length due to unit force in member 'b' be δ_2.

$$\delta_2 = \int \frac{m^2\,dx}{EI} + \frac{\Sigma P^2 l}{AE}$$

$$= \int_0^6 \frac{x^2\,dx}{36,000} + \frac{1^2 \times 2}{1300 \times 200} + \frac{1^2 \times 2}{1300 \times 200} + \frac{(\sqrt{2})^2 \times 2\sqrt{2}}{2000 \times 200}$$

$$= 2 \times 10^{-3} + \frac{4}{1300} \times \frac{1}{200} + \frac{4\sqrt{2}}{2000 \times 200}$$

$$= 2.02952 \times 10^{-3}$$

Net decrease in length of member 'b' is zero. Let's force in member be 'R'

$\Rightarrow\ d_1 + R\delta_2 = 0$

$\Rightarrow\ -0.200769 + 2.02952 \times 10^{-3}\ R = 0$

$$R = 98.925 \text{ kN}$$

Force in member 'b' = 98.925 kN (tensile)

Force in member 'c' = $98.925 \times \sqrt{2}$ kN (compressive)

= 139.88 kN (compressive)

Force in member 'a' = $100 - \dfrac{139.88}{\sqrt{2}}$

= 1.075 kN (tensile)

Problem 13.59: A two-span continuous beam ABC is fixed at the end A, hinged at C and on roller at B. Span AB = 6 m and span BC is 4 m. A concentrated load of 36 kN is acting at 2 m from A in span AB and UDL of 12 kN/m over the entire span BC. EI is constant for both spans. Analyse the beam using slope deflection method and draw the bending moment and shear force diagram. [CSE 1993]

Solution :

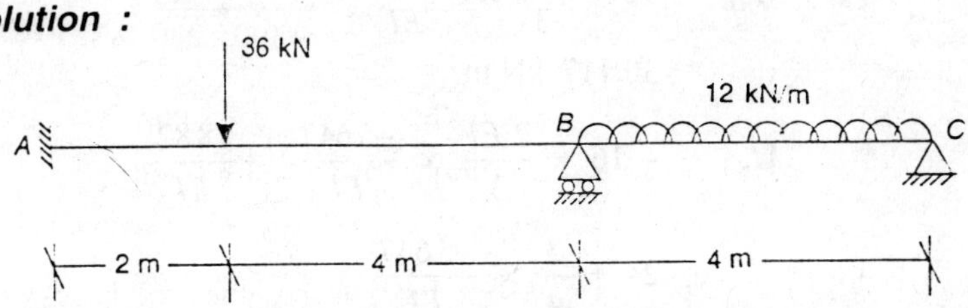

Fig. P. 13.59 (a)

There are tow degrees of freedom, i..e rotations at B and C.

$$FEM_{AB} = \frac{36 \times 2 \times 4^2}{36} = 32 \text{ kN.m (anticlockwise as +ve)}$$

$$FEM_{BA} = -\frac{36 \times 2^2 \times 4}{12} = -16 \text{ kN.m}$$

$$FEM_{BC} = \frac{12 \times 4 \times 4}{12} = 16 \text{ kN.m}$$

$$FEM_{CB} = -16 \text{ kN.m}$$

Now developing moment equations at different ends in terms of unknowns θ_B and θ_C.

$$M_{AB} = 32 + \frac{2\,EI}{6}\,\theta_B$$

$$M_{BA} = -16 + \frac{4\,EI}{6}\,\theta_B$$

$$M_{BC} = 16 + \frac{4\,EI}{4}\,\theta_B + \frac{2\,EI}{4}\,\theta_C$$

$$M_{CB} = -16 + \frac{2\,EI}{4}\,\theta_B + \frac{4\,EI}{4}\,\theta_C$$

By equilibrium at joint B

$$M_{BA} + M_{BC} = 0$$

$$\Rightarrow -16 + 24 + \frac{2}{3}\,EI\,\theta_B + \frac{3\,EI}{4}\,\theta_B = 0$$

$$\Rightarrow \quad \theta_B = \frac{-5.647}{EI}$$

$$M_{AB} = 32 + \frac{EI}{3} \times \frac{-5.647}{EI}$$

$$= 30.117 \text{ kN.m}$$

$$M_{BA} = -16 + \frac{2\,EI}{3} \times \frac{-5.647}{EI} = \frac{18.82}{EI}$$

$$= 24 + \frac{3}{4} \times \frac{-5.647}{EI}$$

$$= 19.764 \text{ kN.m}$$

Now drawing free body diagram

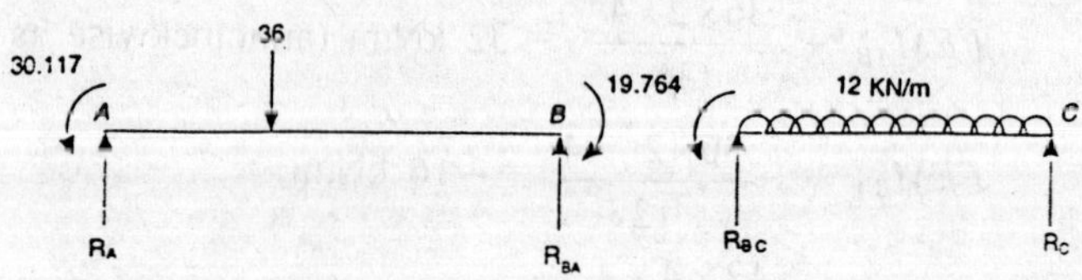

Fig. P. 13.59 (b)

Taking moments about B in left free body,

$$6\ R_A = 36 \times 4 + 30.117 - 19.764$$

$$R_A = 25.73\ \text{kN}$$

$$R_{BA} = 36 - 35.73 = 10.27\ \text{kN}$$

Taking moment about C in right free body

$$4\ R_{BC} = 19.764 + 12 \times 4 \times 2$$

$$R_{BC} = 28.94\ \text{kN}$$

$$R_C = 12 \times 4 - 28.94 = 19.06\ \text{kN}$$

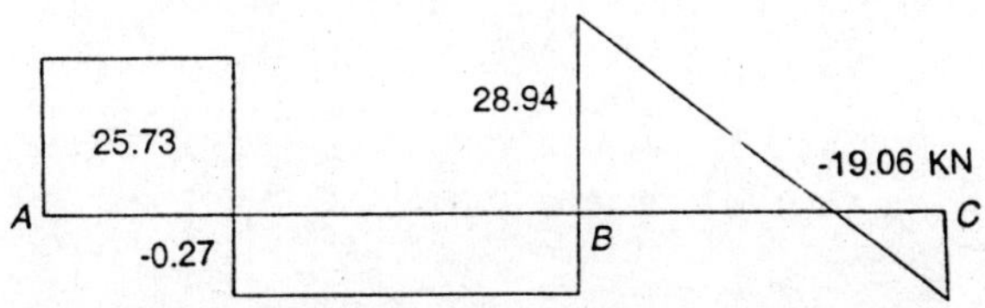

Fig. P. 13.59 (c)

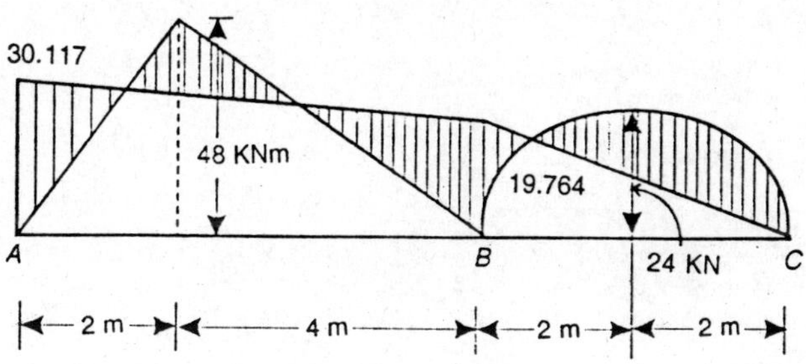

Fig. P. 13.59 (d)